Classic Texts Series

PROBLEMS IN ELEMENTARY PHYSICS

B. Bukhovtsev, V. Krivchenkov,
G. Myakishev, V. Shalnov

arihant
ARIHANT PRAKASHAN (Series), MEERUT

arihant

ARIHANT PRAKASHAN (Series), MEERUT

卐 **Administrative & Production Offices**

Regd. Office

'Ramchhaya' 4577/15, Agarwal Road, Darya Ganj, New Delhi -110002
Tele: 011- 47630600, 43518550

卐 **Head Office**

Kalindi, TP Nagar, Meerut (UP) - 250002
Tel: 0121-7156203, 7156204

卐 **Sales & Support Offices**

Agra, Ahmedabad, Bengaluru, Bareilly, Chennai, Delhi, Guwahati, Hyderabad, Jaipur, Jhansi, Kolkata, Lucknow, Nagpur & Pune.

卐 **ISBN** 978-93-24191-77-9

Published by Arihant Publications (India) Ltd.

For further information about the books published by Arihant, log on to www.arihantbooks.com or e-mail at info@arihantbooks.com

Follow us on

Preface

First Published 1971

Second Printing 1978

This collection of 816 problems is based on the textbook "Elementary Physics" edited by Academician G. S. Landsberg. For this reason the content and nature of the problems and their arrangement mainly conform with this textbook. There is no section devoted to "Atomic Physics", however since the exercises in Landsberg's book illustrate the relevant material in sufficient detail. Some problems on this subject have been included in other chapters.

The problems, most of which are unique, require a fundamental knowledge of the basic laws of Physics, and the ability to apply them in the most diverse conditions. A number of problems in the book have been revised from those used at the annual contests organized by the Physics faculty of the Moscow University.

The solutions of all the difficult problems are given in great detail. Solutions are also given for some of the simpler ones.

The book is recommended for self-education of senior pupils of general and special secondary and technical schools. Many problems will be useful for first and second year students of higher schools.

Contents

Problems

1

MECHANICS

1-1. Kinematics of Uniform Rectilinear Motion

1. A motor-boat travelling upstream met rafts floating downstream. One hour after this the engine of the boat stalled. It took 30 minutes to repair it, and during this time the boat freely floated downstream. When the engine was repaired, the boat travelled downstream with the same speed relative to the current as before and overtook the rafts at a distance of $s = 7.5$ km from the point where they had met the first time. Determine the velocity of the river current, considering it constant.

2. A man walking with a speed v constant in magnitude and direction passes under a lantern hanging at a height H above the ground. Find the velocity which the edge of the shadow of the man's head moves over the ground with if his height is h.

3. The distance between a town and a mill is 30 km. A man started to walk from the mill to the town at 6:30 a. m., while a cyclist left the town for the mill at 6:40 a.m., riding at a speed of 18 km/h. The man met the cyclist after walking 6 km. Determine at what time they met and the man's speed.

   Also find the place where the man was when he met the twelfth bus coming from the town and the number of buses which overtook the cyclist if bus traffic begins at 6 a. m. The buses leave their terminals every 15 minutes and their speed is 45 km/h.

4. Two trains left Moscow for Pushkino at an interval of $t = 10$ minutes and with a speed of $v = 30$ km/h. What was the speed u of a train bound for Moscow if it met these two trains at an interval of $\tau = 4$ minutes?

5. An engineer works at a plant out-of-town. A car is sent for him from the plant every day that arrives at the railway station at the same time as the train he takes.

One day the engineer arrived at the station one hour before his usual time and, without waiting for the car, started walking to work. On his way he met the car and reached his plant 10 minutes before the usual time. How long did the engineer walk before he met the car? Give a graphical solution.

6. Two landing-stages M and K are served by launches that all travel at the same speed relative to the water. The distance between the landing-stages is 20 km. It is covered by each launch from M to K in one hour and from K to M in two hours. The launches leave the two landing-stages at the same time at an interval of 20 minutes and stop at each of them also for 20 minutes.

Determine: (1) the number of launches in service, (2) the number of launches met by a launch travelling from M to K, (3) the number of launches met by a launch travelling from K to M.

7. Two tourists who are at a distance of 40 km from their camp must reach it together in the shortest possible time. They have one bicycle which they decide to use in turn. One of them started walking at a speed of $v_1 = 5$ km/h and the other rode off on the bicycle at a speed of $v_2 = 15$ km/h. The tourists agreed to leave the bicycle at intermediate points between which one walks and the other rides. What will the mean speed of the tourists be? How long will the bicycle remain unused?

8. Two candles of equal height h at the initial moment are at a distance of a from each other. The distance between each candle and the nearest wall is also a (Fig. 1). With what speed will the shadows of the candles move along the walls if one candle burns down during the time t_1 and the other during the time t_2?

9. A bus is running along a highway at a speed of $v_1 = 16$ m/s. A man is at a distance of $a = 60$ metres from the highway and $b = 400$ metres from the bus. In what direction should the man run to reach any point of the highway at the same time as the bus or before it? The man can run at a speed of $v_2 = 4$ m/s.

10. At what minimum speed should the man run (see Problem 9) to be able to meet the bus, and in what direction?

11. A man is on the shore of a lake at point A, and has to get to point B on the lake in the shortest possible time (Fig. 2). The distance from point B to the shore $BC = d$ and the distance $AC = s$. The man can swim in the water with a speed of v_1 and run along the shore with a speed of v_2, greater than v_1.

Which way should he use—swim from point A straight to B or run a certain distance along the shore and then swim to point B?

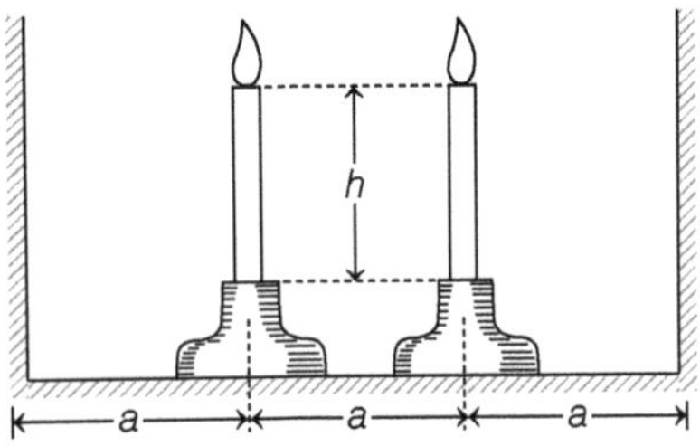

Fig. 1

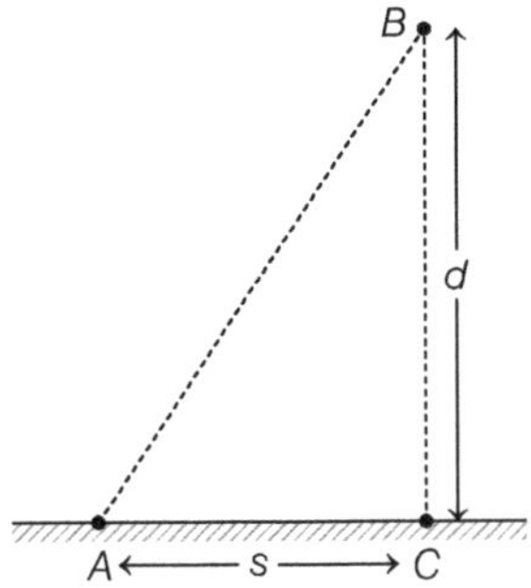

Fig. 2

12. A motor-ship travelling upstream meets rafts floating downstream. Forty-five minutes (t_1) later the ship stops at a landing-stage for $t_2 = 1$ hour. Then the ship travels downstream and overtakes the rafts in $t_3 = 1$ hour. The speed of the ship relative to the water is constant and equal to $v_1 = 10$ km/h. Determine the river current velocity v_2, considering it constant. Consider both graphical and analytical methods of solution.

13. Mail is carried between landing-stages M and K by two launches. At the appointed time the launches leave their landing-stages, meet each other, exchange the mail and return. If the launches leave their respective landing-stages at the same time, the launch departing from M spends 3 hours both ways and that from K – 1.5 hours. The speeds of both launches are the same relative to the water. Determine graphically how much later should the launch depart from M after the other one leaves K for the two to travel the same time.

14. Use the conditions of the previous problem to determine the speed of the launches with respect to the water, the velocity of the river current and the point where the launches will meet if they leave their respective landing-stages simultaneously. The distance between the stages is 30 km.

15. A rowboat travels from landing-stage C to T at a speed of $v_1 = 3$ km/h with respect to the water. At the same time a launch leaves T for C at a speed of $v_2 = 10$ km/h. While the boat moves between the landing-stages the launch covers this distance four times and reaches T at the same time as the boat. Find the direction of the current.

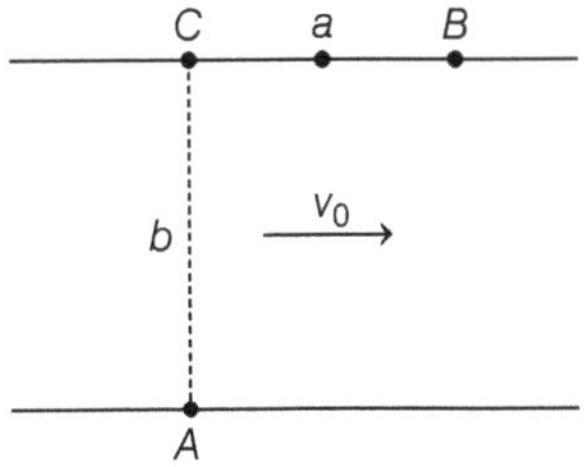

Fig. 3

16. A man in a rowboat must get from point A to point B on the opposite bank of the river (Fig. 3). The distance $BC = a$. The width of the river $AC = b$. At what minimum speed u relative to the water should the boat travel to reach point B? The current velocity is v_0.

17. A man must get from point A on one bank of a river to point B on the other bank moving along straight line AB (Fig. 4). The width of the river $AC = 1$ km, the distance $BC = 2$ km, the maximum speed of the boat relative to the water $u = 5$ km/h and the river current velocity $v = 2$ km/h. Can the distance AB be covered in 30 minutes?

18. A launch travels across a river from point A to point B on the opposite bank along a straight line AB forming an angle α with the bank (Fig. 5). The wind blows with a velocity of u at right angles to the bank. The flag on the mast of the launch forms an angle β with the direction of its motion. Determine the speed of the launch with respect to the bank. Can the data in this problem be used to find the river current velocity?

19. To find the speed of an airplane it is necessary to determine how long it takes it to fly around a closed loop of a known length. How long will it take a plane to fly around a square with a side a, with the wind blowing at a velocity u, in two cases: (1) the direction of the wind coincides with one of the sides of the square, (2) the direction of the wind coincides with the diagonal of the square? With no wind the speed of the plane is v, greater than u.

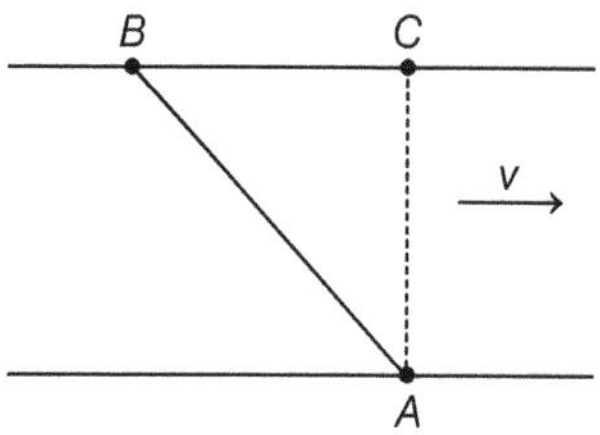

Fig. 4

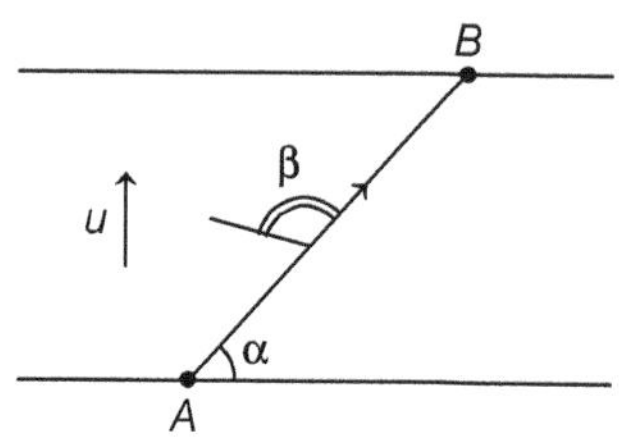

Fig. 5

20. Two motor vehicles run at constant speeds v_1 and v_2 along highways intersecting at an angle α. Find the magnitude and direction of the speed of one vehicle with respect to the other. In what time after they meet at the intersection will the distance between the vehicles be s?

21. Two intersecting straight lines move translationally in opposite directions with velocities v_1 and v_2 perpendicular to the corresponding lines. The angle between the lines is α. Find the speed of the point of intersection of these lines.

1-2. Kinematics of Non-Uniform and Uniformly Variable Rectilinear Motion

22. A motor vehicle travelled the first third of a distance s at a speed of $v_1 = 10$ km/h, the second third at a speed of $v_2 = 20$ km/h and the last third at a speed of $v_3 = 60$ km/h. Determine the mean speed of the vehicle over the entire distance s.

23. Determine the mean velocity and the mean acceleration of a point in 5 and 10 seconds if it moves as shown in Fig. 6.

24. A man stands on a steep shore of a lake and pulls a boat in the water by a rope which he takes up at a constant speed of v. What will the speed of the boat be at the moment when the angle between the rope and the surface of the water is α?

25. A point source of light S is at a distance l from a vertical screen AB. An opaque object with a height h moves translationally at a constant speed v from the source to the screen along straight line SA. Determine the instantaneous velocity with which the upper edge of the shadow of the object moves along the screen (Fig. 7).

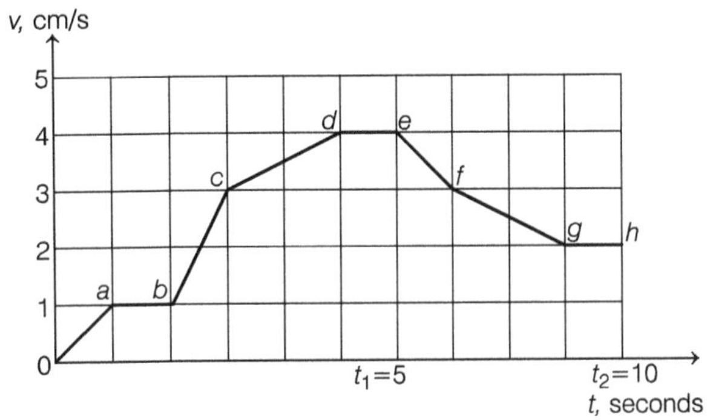

Fig. 6

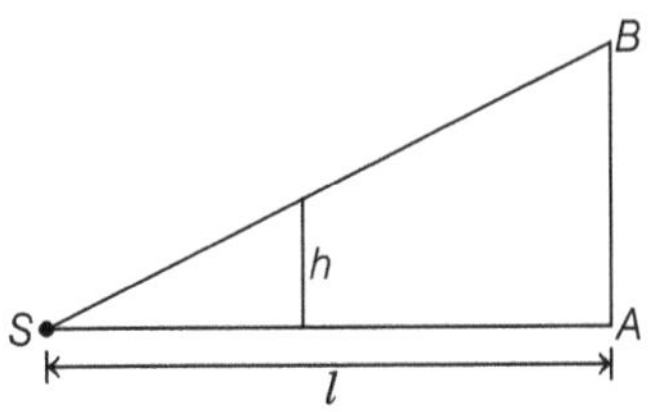

Fig. 7

26. The coordinate of a point moving rectilinearly along the x-axis changes in time according to the law $x = 11 + 35t + 41t^2$ (x is in centimetres and t is in seconds). Find the velocity and acceleration of the point.

27. Figures 8 and 9 show the velocity of a body and the change in its coordinate (parabola) with time. The origin of time reading coincides on both charts. Are the motions shown in the charts the same?

28. Two motor vehicles left point A simultaneously and reached point B in $t_0 = 2$ hours. The first vehicle travelled half of the distance at a speed of $v_1 = 30$ km/h and the other half at a speed of $v_2 = 45$ km/h. The second vehicle covered the entire distance with a constant acceleration. At what moment of time were the speeds of both the vehicles the same? Will one of them overtake the other en route?

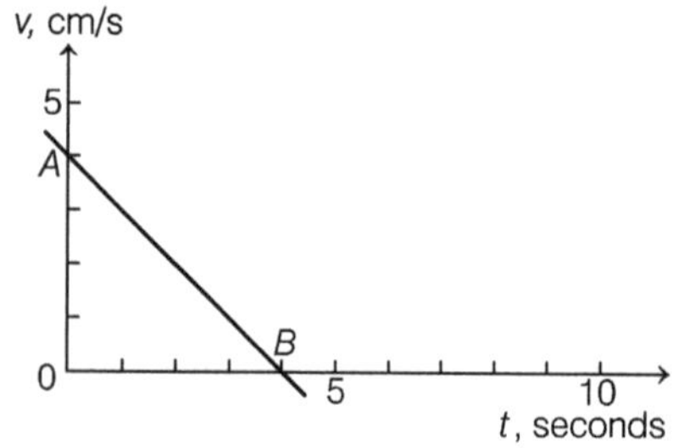

Fig. 8

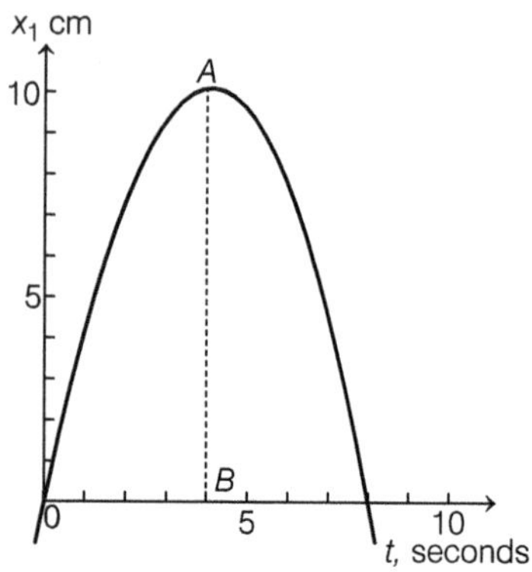

Fig. 9

29. A ball freely drops from a height H onto an elastic horizontal support. Plot charts showing the change in the coordinate and velocity of the ball versus the time neglecting the duration of the collision. The impact is absolutely elastic.

30. Two steel balls freely drop onto an elastic plate, the first one from a height of $h_1 = 44$ cm and the second in $\tau > 0$ seconds after the first from a height of $h_2 = 11$ cm. After a certain time the velocities of the balls coincide in magnitude and direction. Find the time τ and the interval during which the velocities of the two balls remain the same. The balls do not collide.

31. How long will a body freely falling without any initial velocity pass the n-th centimetre of its path?

32. Two bodies are thrown one after the other with the same velocities v_0 from a high tower. The first body is thrown vertically upward, and the second one vertically downward after the time τ. Determine the velocities of the bodies with respect to each other and the distance between them at the moment of time $t > \tau$.

33. At the initial moment three points A, B and C are on a horizontal straight line at equal distances from one another. Point A begins to move vertically upward with a constant velocity v and point C vertically downward without any initial velocity at a constant acceleration a. How should point B move vertically for all the three points to constantly be on one straight line? The points begin to move simultaneously.

34. Two trucks tow a third one by means of a pulley tied to it (Fig. 10). The accelerations of the trucks are a_1 and a_2. Find the acceleration a_3 of the truck being towed.

35. A lift moves with an acceleration a. A passenger in the lift drops a book. What is the acceleration of the book with respect to the lift floor if: (1) the lift is going up, (2) the lift is going down?

36. A railway carriage moves over a straight level track with an acceleration a. A passenger in the carriage drops a stone. What is the acceleration of the stone with respect to the carriage and the Earth?

37. A man in a lift moving with an acceleration a drops a ball from a height H above the floor. In t seconds the acceleration of the lift is reversed, and in $2t$ seconds becomes equal to zero. Directly after this the ball touches the floor. What height from the floor of the lift will the ball jump to after the impact? Consider the impact to be absolutely elastic.

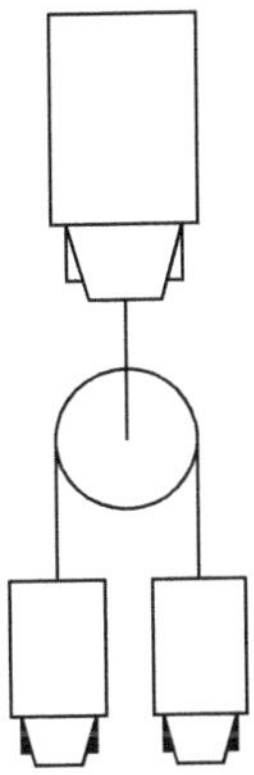

Fig. 10

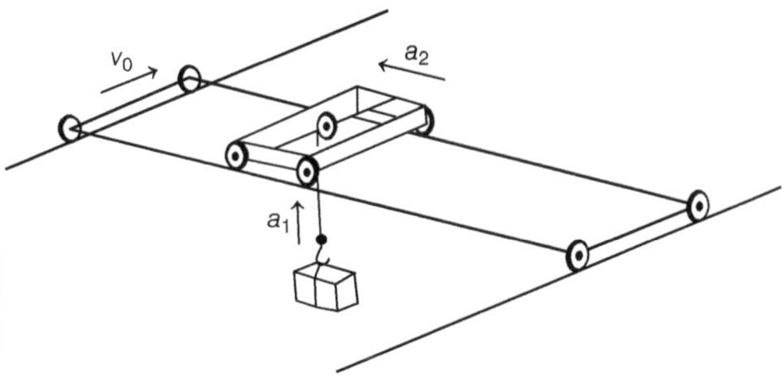

Fig. 11

38. An overhead travelling crane lifts a load from the ground with an upward acceleration of a_1. At the same time the hook of the crane carrying the load moves in a horizontal direction with an acceleration a_2 relative to the crane. Besides, the crane runs on its rails with a constant speed v_0 (Fig. 11). The initial speed of the hook relative to the crane is zero. Find the speed of the load with respect to the ground when it reaches the height h.

39. Body A is placed on a wedge forming an angle α with the horizon (Fig. 12). What acceleration should be imparted to the wedge in a horizontal direction for body A to freely fall vertically?

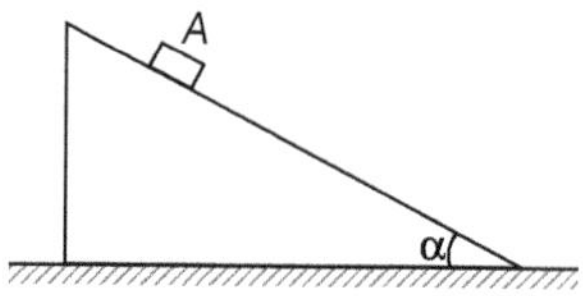

Fig. 12

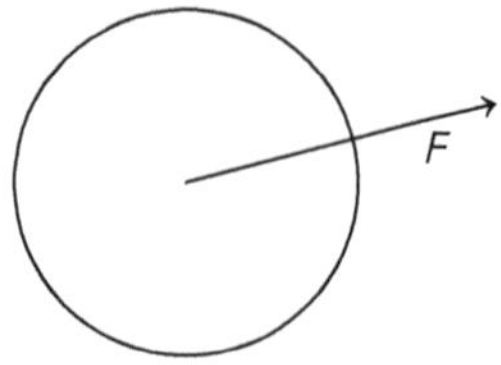

Fig. 13

1-3. Dynamics of Rectilinear Motion

40. A force F is applied to the centre of a homogeneous sphere (Fig. 13). In what direction will the sphere move?

41. Six forces lying in one plane and forming angles of 60° relative to one another are applied to the centre of a homogeneous sphere with a mass of $m = 4$ kg. These forces are consecutively 1, 2, 3, 4, 5 and 6 kgf (Fig. 14). In what direction and with what acceleration will the sphere move?

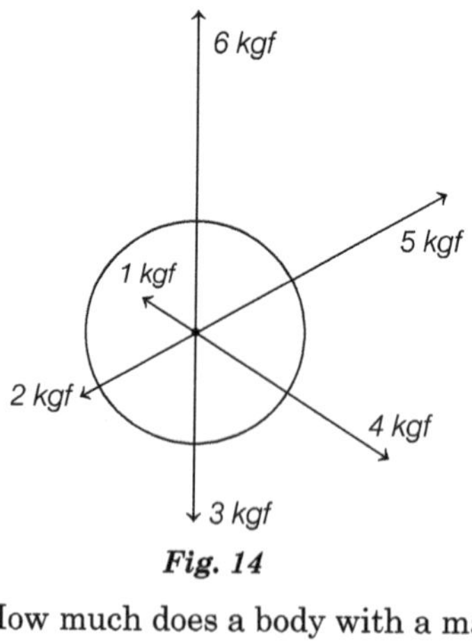

Fig. 14

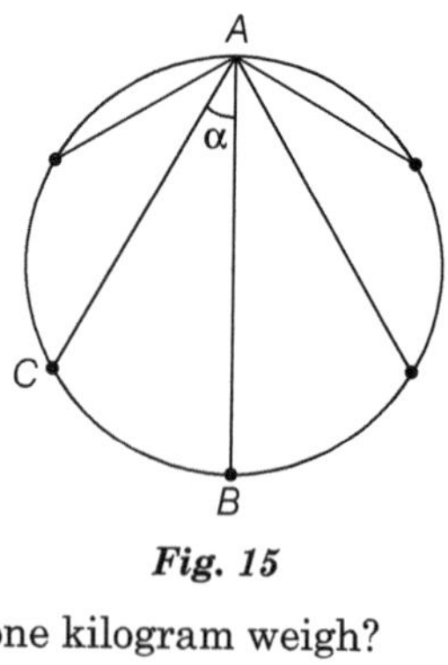

Fig. 15

42. How much does a body with a mass of one kilogram weigh?

43. The resistance of the air causes a body thrown at a certain angle to the horizon to fly along a ballistic curve. At what angle to the horizon is the acceleration of the body directed at the highest point of the trajectory A, if the mass of the body is m and the resistance of the air at this point is F?

44. A disk arranged in a vertical plane has several grooves directed along chords drawn from point A (Fig. 15). Several bodies begin to slide down the respective grooves from point A simultaneously. In what time will each body reach the edge of the disk? Disregard friction and the resistance of the air (Galileo's problem).

45. What is the minimum force of air resistance acting on a parachutist and his parachute when the latter is completely opened? The two weigh 75 kgf.

46. What is the pressure force N exerted by a load weighing G kgf on the floor of a lift if the acceleration of the lift is a? What is this force equal to upon free falling of the lift?

47. A puck with an initial velocity of 5 m/s travels over a distance of 10 metres and strikes the boards. After the impact which may be considered as absolutely elastic, the puck travelled another 2.5 metres and stopped. Find the coefficient of friction between the puck and the ice. Consider the force of sliding friction in this and subsequent problems to be equal to the maximum force of friction of rest.

48. The path travelled by a motor vehicle from the moment the brakes are applied until it stops is known as the braking distance. For the types of tyres in common use in the USSR and a normal air pressure in the tubes, the dependence of the braking distance on the speed of the vehicle at the beginning of braking and on the state and type of the pavement can be tabulated (see Table 1).

Find the coefficient of friction for the various kinds of pavement surface to an accuracy of the first digit after the decimal point, using the data in this table.

49. Determine the difference in the pressure of petrol on the opposite walls of a fuel tank when a motor vehicle travels over a horizontal road if its speed increases uniformly from $v_0 = 0$ to v during t seconds. The distance between the tank walls is l. The tank has the form of a parallelepiped and its side walls are vertical. It is completely filled with petrol. The density of petrol is ρ.

50. A homogeneous rod with a length L is acted upon by two forces F_1 and F_2 applied to its ends and directed oppositely (Fig. 16). With what force F will the rod be stretched in the cross section at a distance l from one of the ends?

Table 1. Braking Distance in Metres

Pavement surface	Speed, km/h									
	10	20	30	40	50	60	70	80	90	100
Ice	3.9	15.6	35.3	62.9	98.1	141.4	192.6	251.6	318.2	393.0
Dry snow	1.9	7.8	17.6	31.4	49.0	70.7	96.3	125.8	159.1	196.6
Wet wood block	1.3	5.2	11.7	20.9	32.7	47.1	64.2	83.8	106.0	131.0
Dry wood block	0.78	3.1	7.0	12.5	19.6	28.2	38.5	50.3	63.6	78.6
Wet asphalt	0.97	3.9	8.8	15.7	24.5	35.3	48.1	62.9	79.5	98.2
Dry asphalt	0.65	2.6	5.8	10.4	16.3	23.5	32.1	41.9	53.0	65.5
Dry concrete	0.56	2.2	5.0	9.0	14.0	20.2	27.5	35.9	45.4	56.1

51. A light ball is dropped in air and photographed after it covers a distance of 20 metres. A camera with a focal length of 10 cm is placed 15 metres away from the plane which the ball is falling in. A disk with eight equally spaced holes arranged along its circumference rotates with a speed of 3 rev/s in front of the open lens of the camera. As a result a number of images of the ball spaced 3 mm apart are produced on the film. Describe the motion of the ball. What is the final velocity of another ball of the same radius, but with a mass four times greater than that of the first one? Determine the coefficient of friction if the weight of the first ball is 4.5 gf. At high velocities of falling, the resistance of the air is proportional to the square of the velocity.

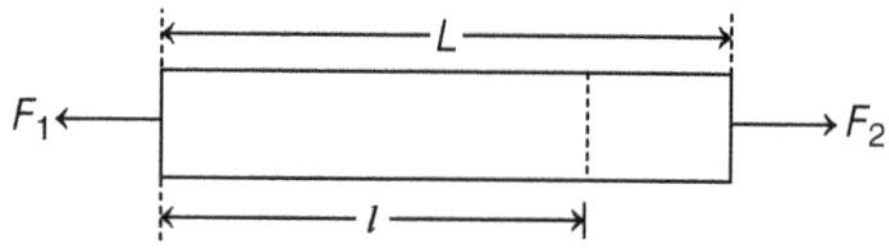

Fig. 16

52. Two weights with masses $m_1 = 100$ g and $m_2 = 200$ g are suspended from the ends of a string passed over a stationary pulley at a height of $H = 2$ metres from the floor (Fig. 17). At the initial moment the weights are at rest. Find the tension of the string when the weights move and the time during which the weight m_2 reaches the floor. Disregard the mass of the pulley and the string.

53. A weight G is attached to the axis of a moving pulley (Fig. 18). What force F should be applied to the end of the rope passed around the second pulley for the weight G to move upwards with an acceleration of a? For the weight G to be at rest? Disregard the mass of the pulleys and the rope.

54. Two weights are suspended from a string thrown over a stationary pulley. The mass of one weight is 200 g. The string will not break if a very heavy weight is attached to its other end. What tension is the string designed for? Disregard the mass of the pulley and the string.

55. Two pans with weights each equal to $G = 3$ kgf are suspended from the ends of a string passed over two stationary pulleys. The string between the pulleys is cut and the free ends are connected to a dynamometer (Fig. 19). What will the dynamometer show? What weight G_1 should be added to one of the pans for the reading of the dynamometer not to change after a weight $G_2 = 1$ kgf is taken off the other pan? Disregard the masses of the pans, pulleys, thread and dynamometer.

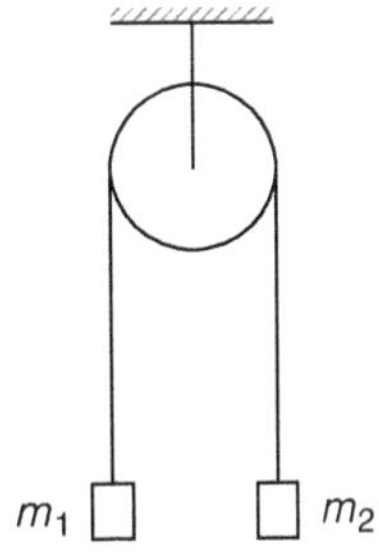

Fig. 17

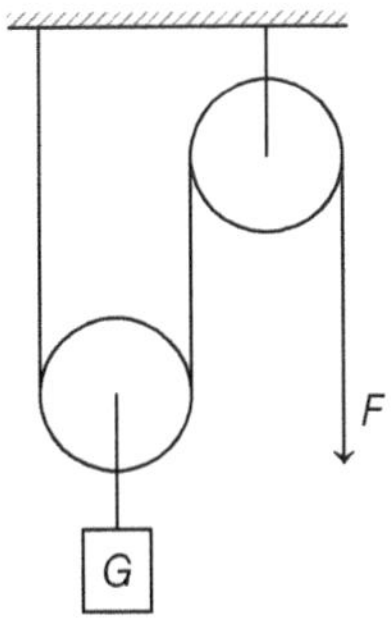

Fig. 18

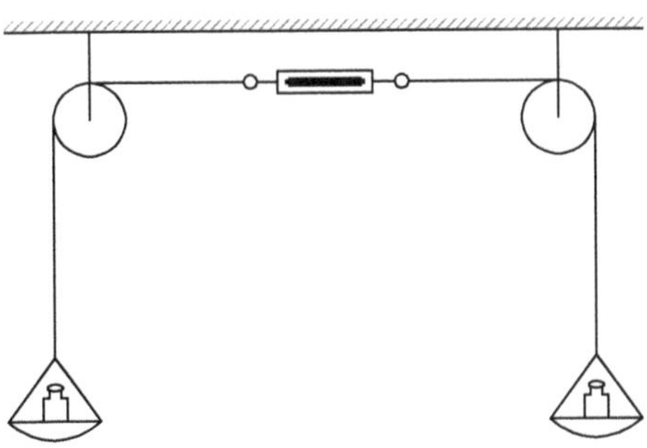

Fig.19

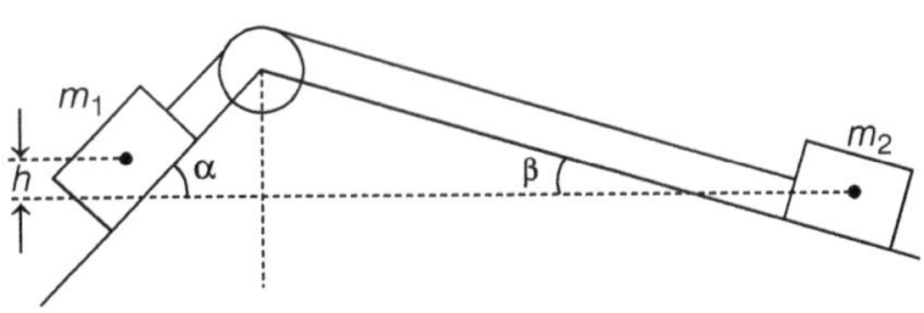

Fig. 20

56. A heavy sphere with a mass m is suspended on a thin rope. Another rope as strong as the first one is attached to the bottom of the sphere. When the lower rope is sharply pulled it breaks. What acceleration will be imparted to the sphere?

57. Two weights with masses m_1 and m_2 are connected by a string passing over a pulley. The surfaces on which they rest form angles α and β with the horizontal (Fig. 20). The right-hand weigh, is h metres below the left-hand one. Both weights will be at the same height in τ seconds after motion begins. The coefficients of friction between the weights and the surfaces are k. Determine the relation between the masses of the weights.

58. A slide forms an angle of $\alpha = 30°$ with the horizon. A stone is thrown upward along it and covers a distance of $l = 16$ metres in $t_1 = 2$ seconds, after which it slides down. What time t_2 is required for the return motion? What is the coefficient of friction between the slide and the stone?

59. A cart with a mass of $M = 500$ g is connected by a string to a weight having a mass $m = 200$ g. At the initial moment the cart moves to the left along a horizontal plane at a speed of $v_0 = 7$ m/s (Fig. 21). Find the magnitude and direction of the speed of the cart, the place it will be at and the distance it covers in $t = 5$ seconds.

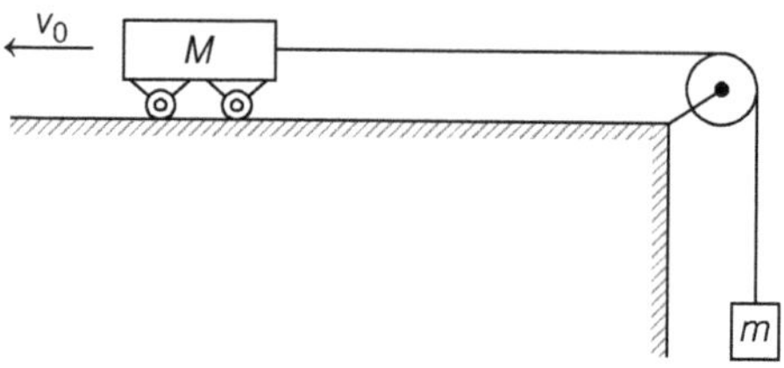

Fig. 21

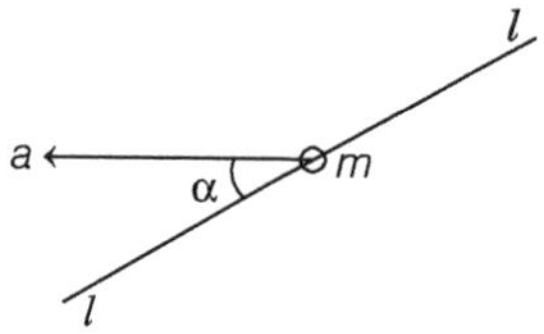

Fig. 22

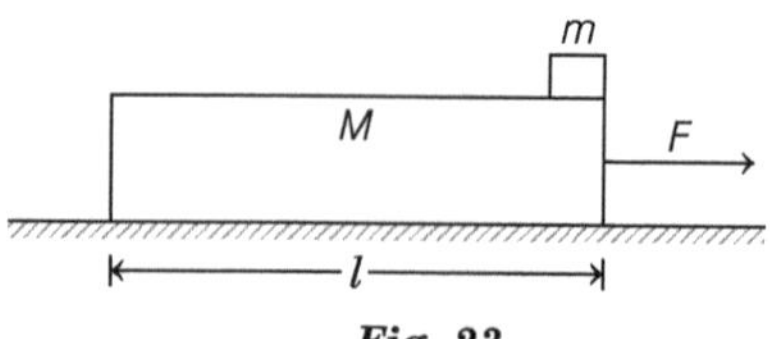

Fig. 23

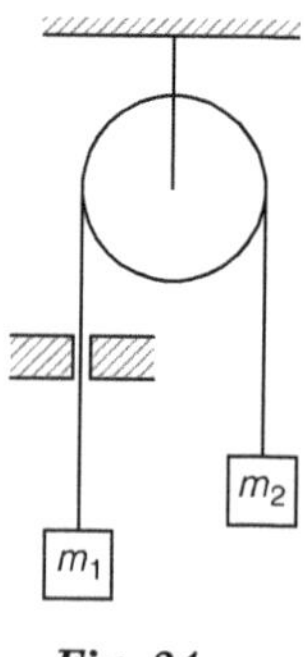

Fig. 24

60. Can an ice-boat travel over a level surface faster than the wind which it is propelled by?

61. The fuel supply of a rocket is $m = 8$ tons and its mass (including the fuel) is $M = 15$ tons. The fuel burns in 40 seconds. The consumption of fuel and the thrust $F = 20{,}000$ kgf are constant.

(1) The rocket is placed horizontally on a trolley. Find its acceleration at the moment of launching. Find how the acceleration of the rocket depends on the duration of its motion and show the relation graphically. Use the graph to find the velocity acquired by the rocket in 20 seconds after it begins to move. Disregard friction.

(2) The rocket is launched vertically upward. Measurements show that in 20 seconds the acceleration of the rocket is 0.8 g. Calculate the force of air resistance which acts on the rocket at this moment. Consider the acceleration g to be constant.

(3) The acceleration of the rocket is measured by an instrument having the form of a spring secured in a vertical tube. When at rest, the spring is stretched a distance of $l_0 = 1$ cm by a weight secured to its end. Determine the relation between the stretching of the spring and the acceleration of the rocket. Draw the scale of the instrument.

62. A bead of mass m is fitted onto a rod with a length of $2l$, and can move on it without friction. At the initial moment the bead is in the middle of the rod. The rod moves translationally in a horizontal plane with an acceleration a in a direction forming an angle α with the rod (Fig. 22). Find the acceleration of the bead relative to the rod, the reaction force exerted by the rod on the bead, and the time when the bead will leave the rod.

63. Solve the previous problem, assuming that the moving bead is acted upon by a friction force. The coefficient of friction between the bead and the rod is k. Disregard the force of gravity.

64. A block with the mass M rests on a smooth horizontal surface over which it can move without friction. A body with the mass m lies on the block (Fig. 23). The coefficient of friction between the body and the block is k. At what force F applied to the block in a horizontal direction will the body begin to slide over the block? In what time will the body fall from the block if the length of the latter is l?

65. A wagon with the mass M moves without friction over horizontal rails at a speed of v_0. A body with the mass m is placed on the front edge of the wagon. The initial speed of the body is zero. At what length of the wagon will the body not slip off it? Disregard the dimensions of the body as compared with the length l of the wagon. The coefficient of friction between the body and the wagon is k.

66. A weightless string thrown over a stationary pulley is passed through a slit (Fig. 24). As the string moves it is acted upon by a constant friction force F on the side of the slit. The ends of the string carry two weights with masses m_1 and m_2, respectively. Find the acceleration a of the weights.

67. A stationary pulley is secured to the end of a light bar. The bar is placed onto a balance pan and secured in a vertical direction. Different weights are attached to the ends of a string passed over the pulley. One of the weights slides over the bar with friction and therefore both weights move uniformly (Fig. 25). Determine the force which the pulley acts on the bar with and the readings of the balance when the weights move. Disregard the masses of the pulley, bar, string and the friction in the pulley axis. Consider two cases: (1) $m_1 = 1$ kg, $m_2 = 3$ kg, and (2) $m_1 = 3$ kg, $m_2 = 1$ kg.

68. A system consists of two stationary and one movable pulleys (Fig. 26). A string thrown over them carries at its ends weights with masses m_1 and m_3, while a weight with a mass m_2 is attached to the axis of the movable pulley. The parts of the string that are not on the pulleys are vertical. Find the acceleration of each weight, neglecting the masses of the pulleys and the string, and also friction.

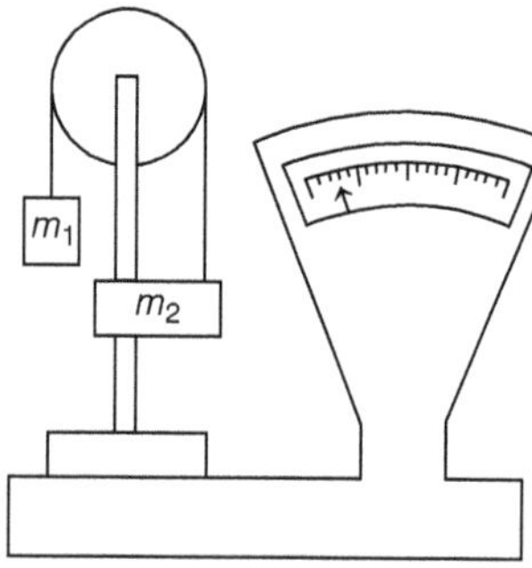

Fig. 25

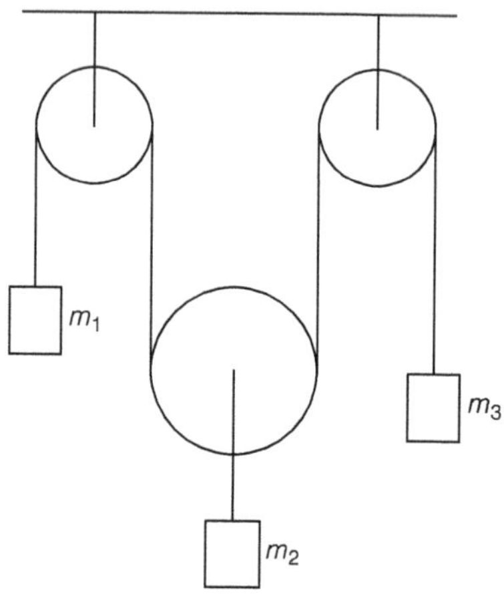

Fig. 26

69. Two monkeys of the same weight are hanging at the ends of a rope thrown over a stationary pulley. One monkey begins to climb the rope and the other stays where it is. Where will the second monkey be when the first one reaches the pulley? At the initial moment both monkeys were at the same height from the floor. Disregard the mass of the pulley and rope, and also friction.

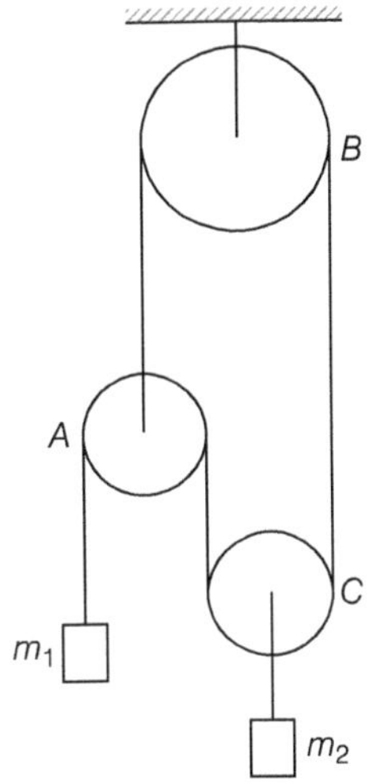

Fig. 27

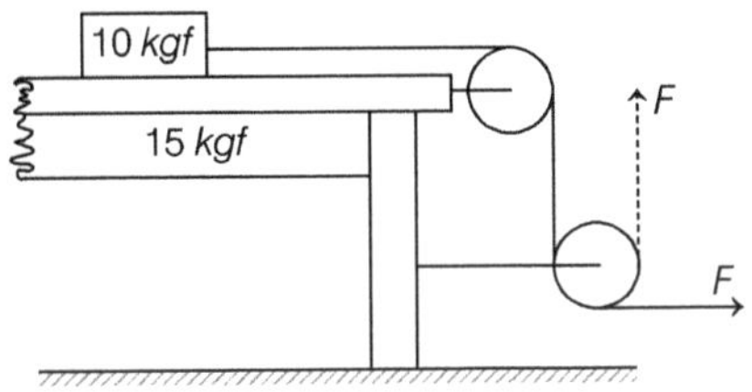

Fig. 28

70. Determine the accelerations of the weights in the pulley system depicted in Fig. 27. Disregard the masses of the pulleys and string, and also friction. In what direction will the pulleys rotate when the weights move?

71. A table with a weight of $G_1 = 15$ kgf can move without friction over a level floor. A weight of $G_2 = 10$ kgf is placed on the table, and a rope passed over two pulleys fastened to the table is attached to it (Fig. 28). The coefficient of friction between the weight and the table $k = 0.6$. What acceleration will the table move with if a constant force of 8 kgf is applied to the free end of the rope? Consider two cases: (1) the force is directed horizontally, (2) the force is directed vertically upward.

72. An old cannon without a counter-recoil device rests on a horizontal platform. A ball with a mass m and an initial velocity v_0 is fired at an angle of α to the horizon. What velocity v_1 will be imparted to the cannon directly after the shot if the mass of the cannon is M and the acceleration of the ball in the barrel is much greater than that of free fall? The coefficient of friction between the cannon and the platform is k.

1-4. The Law of Conservation of Momentum

73. A meteorite burns in the atmosphere before it reaches the Earth's surface. What happens to its momentum?

74. Does a homogeneous disk revolving about its axis have any momentum? The axis of the disk is stationary.

75. The horizontal propeller of a helicopter can be driven by an engine mounted inside its fuselage or by the reactive forces of the gases ejected from special nozzles at the ends of the propeller blades. Why does a propeller-engine helicopter need a tail rotor while a jet helicopter does not need it?

76. A hunter discharges his gun from a light inflated boat. What velocity will be imparted to the boat when the gun is fired if the mass of the hunter and the boat is $M = 70$ kg, the mass of the shot $m = 35$ g and the mean initial velocity of the shot $v_0 = 320$ m/s? The barrel of the gun is directed at an angle of $\alpha = 60°$ to the horizon.

77. A rocket launched vertically upward explodes at the highest point it reaches. The explosion produces three fragments. Prove that the vectors of the initial velocities of all three fragments are in one plane.

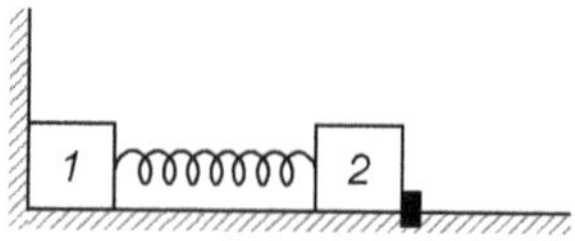

Fig. 29

78. A man in a boat facing the bank with its stern walks to the bow. How will the distance between the man and the bank change?

79. A boat on a lake is perpendicular to the shore and faces it with its bow. The distance between the bow and the shore is 0.75 metre. At the initial moment the boat was stationary. A man in the boat steps from its bow to its stern. Will the boat reach the shore if it is 2 metres long? The mass of the boat $M = 140$ kg and that of the man $m = 60$ kg.

80. Two identical weights are connected by a spring. At the initial moment the spring is so compressed that the first weight is tightly pressed against a wall (Fig. 29) and the second weight is retained by a stop. How will the weights move if the stop is removed?

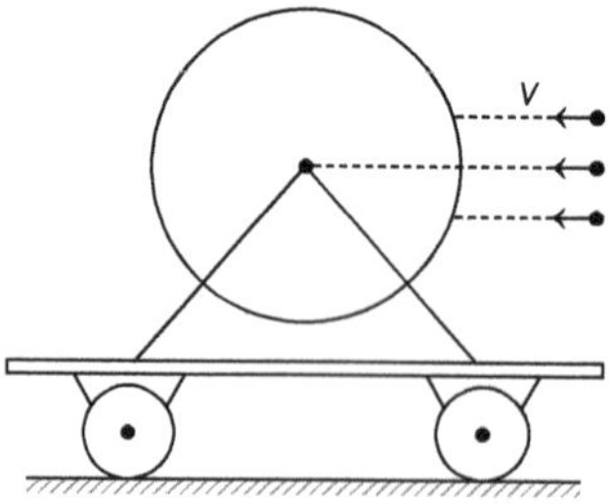

Fig. 30

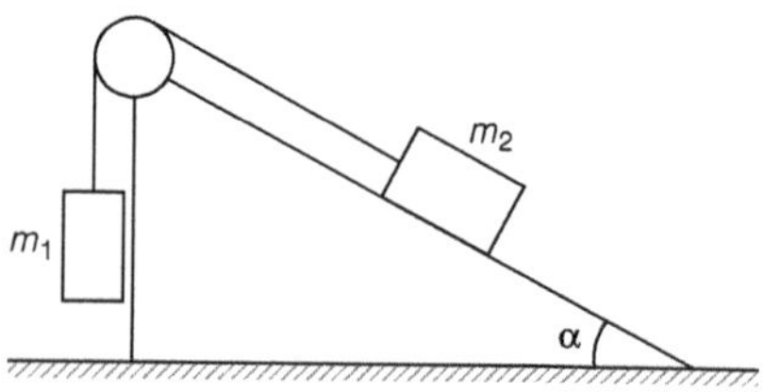

Fig. 31

81. A massive homogeneous cylinder that can revolve without friction around a horizontal axis is secured on a cart standing on a smooth level surface (Fig. 30). A bullet flying horizontally with a velocity v strikes the cylinder and drops onto the cart. Does the speed of the cart that it acquires after the impact depend on the point where the bullet strikes the cylinder?

82. At the initial moment a rocket with a mass M had a velocity v_0. At the end of each second the rocket ejects a portion of gas with a mass m. The velocity of a portion of gas differs from that of the rocket before the given portion of gas burns by a constant value u i. e., the velocity of gas outflow is constant. Determine the velocity of the rocket in n seconds, disregarding the force of gravity.

83. Will the velocity of a rocket increase if the outflow velocity of the gases with respect to the rocket is smaller than the velocity of the rocket itself, so that the gases ejected from the nozzle follow the rocket?

84. Two boats move towards each other along parallel paths with the same velocities. When they meet, a sack is thrown from one boat onto the other and then an identical sack is thrown from the second onto the first. The next time this is done simultaneously. When will the velocity of the boats be greater after the sacks are thrown?

85. A hoop is placed on an absolutely smooth level surface. A beetle alights on the hoop. What trajectory will be described by the beetle and the centre of the hoop if the beetle begins to move along the hoop? The radius of the hoop is R, its mass is M and the mass of the beetle m.

86. A wedge with an angle α at the base can move without friction over a smooth level surface (Fig. 31). At what ratio between the masses m_1 and m_2 of the weights, that are connected by a string passed over a pulley, will the wedge remain stationary, and at what ratio will it begin to move to the right or left? The coefficient of friction between the weight of mass m_2 and the wedge is k.

1-5. Statics

87. A homogeneous chain with a length l lies on a table. What is the maximum length l_1 of the part of the chain hanging over the table if the coefficient of friction between the chain and the table is k?

88. Two identical weights are suspended from the ends of a string thrown over two pulleys (Fig. 32). Over what distance will a third weight of the same mass lower if it is attached to the middle of the string? The distance between the axes of the pulleys is $2l$. The friction in the axes of the pulleys is negligible.

89. An isosceles wedge with an acute angle α is driven into a slit. At what angle α will the wedge not be forced out of the slit if the coefficient of friction between the wedge and the slit is k?

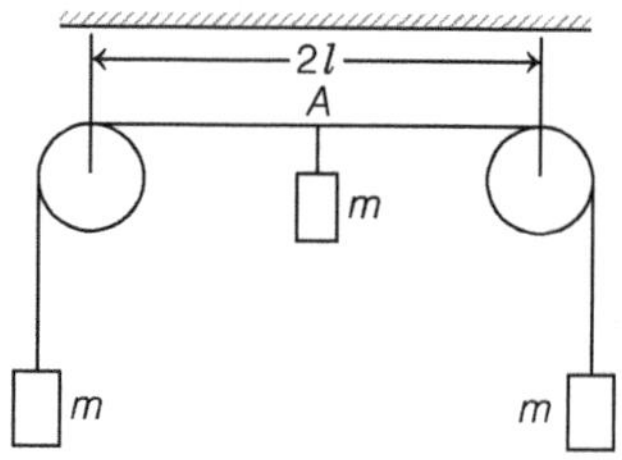

Fig. 32

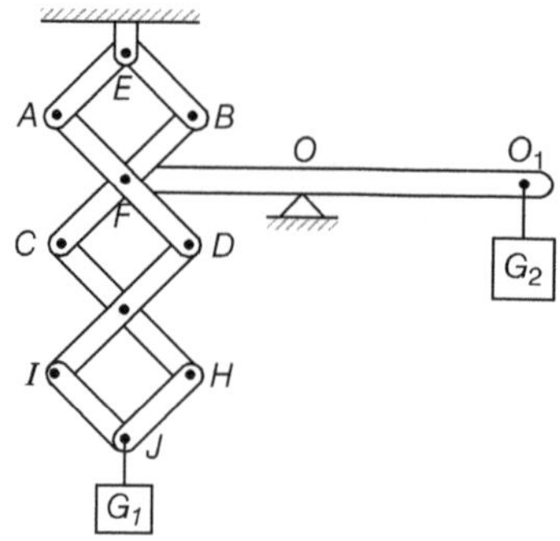

Fig. 33

90. What is the ratio between the weights G_1 and G_2 if the system shown in Fig. 33 is in equilibrium. Bars AD, BC, CH, DI and arm OO_1 of the lever are twice as long as bars AE, EB, IJ, JH and arm FO, respectively. Disregard the weight of the bars and the arm.

91. A horizontal force F is applied perpendicularly to an upper edge of a rectangular box with a length l and a height h to move it. What should the coefficient of friction k between the box and the floor be so that the box moves without overturning?

92. A homogeneous beam whose weight is G lies on a floor. The coefficient of friction between the beam and the floor is k. What is easier for two men to do—turn the beam about its centre or move it translationally?

93. An overhead travelling crane (see Fig. 11) weighing $G = 2$ tonf has a span of $L = 26$ metres. The wire rope carrying a load is at a distance of $l = 10$ metres from one of the rails. Determine the force of pressure of the crane on the rails if it lifts a load of $G_0 = 1$ tonf with an acceleration of $a = 9.8\ \text{m/s}^2$.

94. A lever is so bent that its sides AB, BC and CD are equal and form right angles with one another (Fig. 34). The axis of the lever is at point B. A force of $G = 1$ kgf is applied at point A at right angles to arm AB. Find the minimum force that should be applied at D for the lever to be in equilibrium. Disregard the weight of the lever.

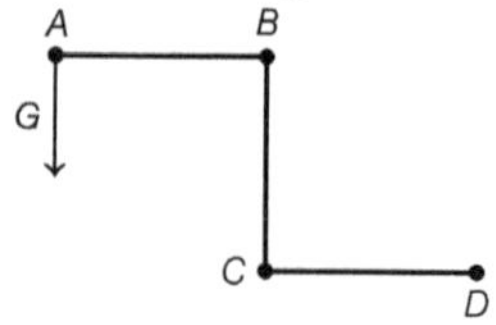

Fig. 34

95. A rod not reaching the floor is inserted between two identical boxes (Fig. 35). A horizontal force F is applied to the upper end of the rod. Which of the boxes will move first?

98. A heavy homogeneous sphere is suspended from a string whose end is attached to a vertical wall. The point at which the string is fastened to the sphere lies on the same vertical as the centre of the sphere. What should the coefficient of friction between the sphere and the wall be for the sphere to remain in equilibrium?

97. A homogeneous rectangular brick lies on an inclined plane (Fig. 36). What half of the brick (left or right) exerts a greater pressure on it?

98. A horizontally directed force equal to the weight of a heavy cylindrical roller with a radius R is applied to its axis to lift it onto a rectangular step. Find the maximum height of the step.

99. A sphere weighing $G = 3$ kgf lies on two inclined planes forming angles $\alpha_1 = 30°$ and $\alpha_2 = 60°$ with the horizon. Determine the pressure exerted by the sphere on each plane if there is no friction between the sphere and one of the planes.

100. The front wall of a drawer in a cabinet is provided with two symmetrical handles. The distance between the handles is l and the length of the drawer a. The coefficient of friction between the drawer and the cabinet is k. Can the drawer always be pulled out by applying a force perpendicular to the wall of the drawer only to one handle?

101. A homogeneous board is balanced on a rough horizontal log (Fig. 37). After a weight has been added to one of the ends of the board, equilibrium can be obtained when the board forms an angle α with the horizon. What is the coefficient of friction between the board and the log?

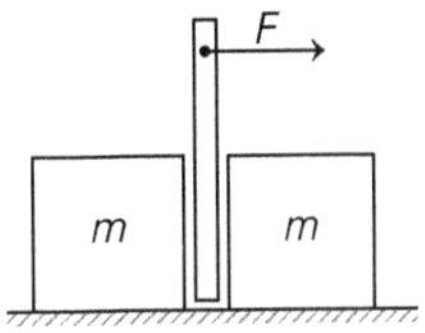

Fig. 35

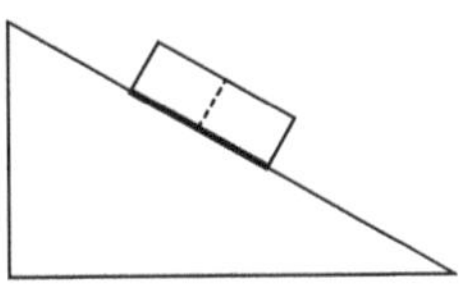

Fig. 36

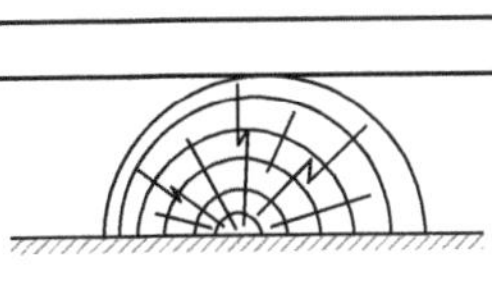

Fig. 37

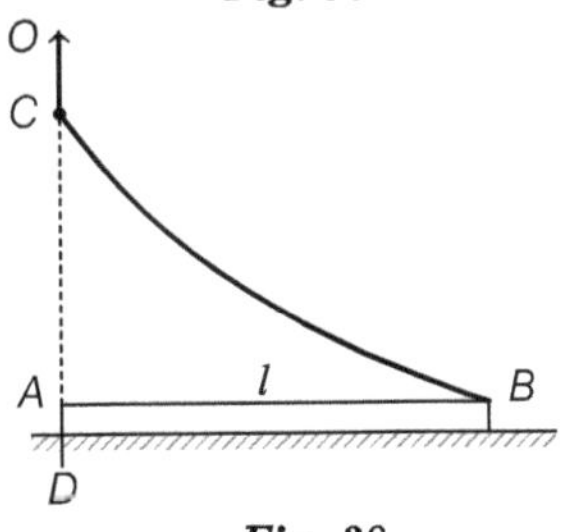

Fig. 38

102. The upper end of a ladder rests against a smooth vertical wall and its bottom end stands on a rough floor. The coefficient of friction between the ladder and the floor is k. Determine the angle α between the ladder and the wall at which the former will be in equilibrium.

103. Solve the previous problem, assuming that the wall is rough and the coefficient of friction between the ladder and the wall is also equal to k.

104. A homogeneous thin rod AB with a length l is placed onto the horizontal surface of a table. A string with a length of $2l$ is attached to end B of the rod (Fig. 38). How will the rod move if the other end C of the string is slowly lifted up a stationary vertical straight line DO passing through end A of the rod. Disregard the weight of the string.

105. At what coefficient of friction will a man not slip when he runs along a straight hard path? The maximum angle between a vertical line and the line connecting the man's centre of gravity with the point of support is α.

106. A ladder leans against a smooth vertical wall of a house. The angle between the ladder and the horizontal surface of the Earth is $\alpha = 60°$. The length of the ladder is l, and its centre of gravity is at its middle. How is the force acting on the ladder from the Earth directed?

107. A ladder with its centre of gravity at the middle stands on an absolutely smooth floor and leans against a smooth wall (Fig. 39). What should the tension of a rope tied to the middle of the ladder be to prevent its falling down?

108. A man climbs up a ladder leaning against a smooth vertical wall. The ladder begins to slip only when the man reaches a certain height. Why?

109. A picture is attached to a vertical wall by means of string AC with a length l forming an angle α with the wall. The height of the picture $BC = d$ (Fig. 40). The bottom of the picture is not fastened. At what coefficient of friction between the picture and the wall will the picture be in equilibrium?

110. Four homogeneous rods are pin-connected to one another at points B, C and D (Fig. 41). The two extreme rods AB and DE can freely revolve with respect to stationary points A and E on a horizontal straight line. The lengths of the rod $AB = ED$ and $BC = CD$. The masses of the rods are the same. Show that the angles α and β are related by the ratio $\tan\alpha = 3\tan\beta$ when in equilibrium.

111. What is the coefficient of friction between a floor and a box weighing one ton-force if a minimum force of 600 kgf is required to start the box moving?

112. A weightless unstretchable string is wound around a cylinder with a mass m (Fig. 42). With what minimum force F_{min} and at what angle

α_1 to the horizon should the string be pulled for the rotating cylinder to remain in place? The coefficient of friction between the cylinder and the floor is k.

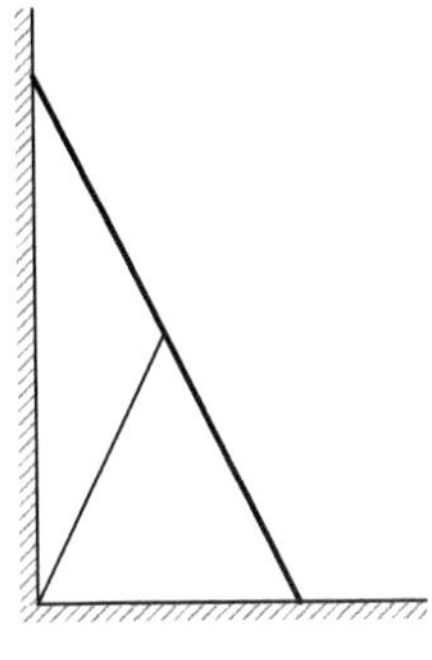

Fig. 39

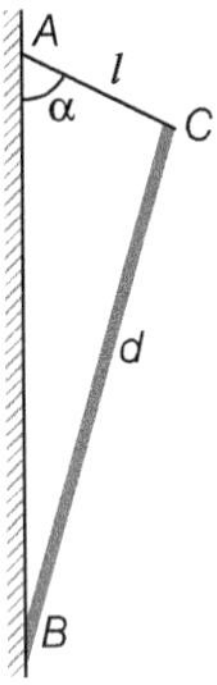

Fig. 40

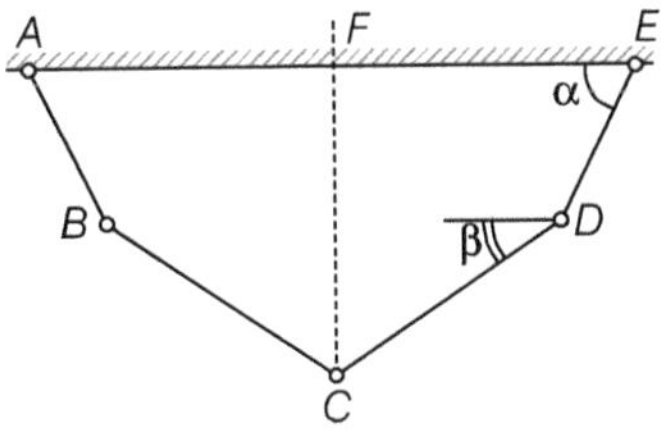

Fig. 41

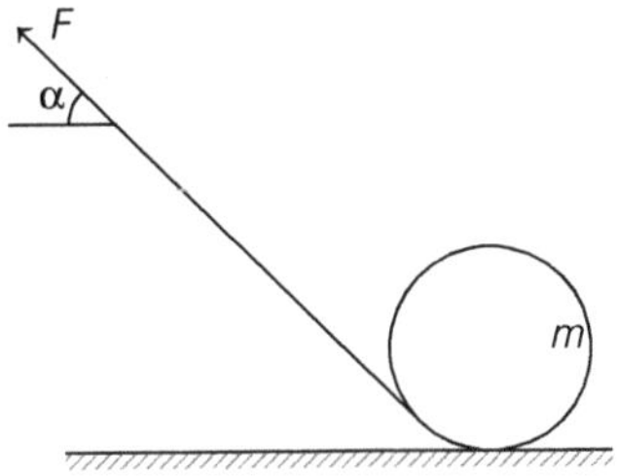

Fig. 42

113. Figure 43 shows a simplified diagram of the steam engine and crank gear of a steam locomotive. Fig. 43*a* and *b* correspond to the moments when steam has been admitted into the left and right-hand parts of the cylinder, respectively. Calculate the tractive effort for these cases when point C is on one vertical line with the axis of the driving wheel. The pressure of the steam in the cylinder is p, the area of the piston is A, the radius of the driving wheel is R and the distance $OC = r$. Disregard the masses of the crank gear, the piston and the driving wheel.

114. Bricks are so laid without a binder that a part of each following brick extends over the one below (Fig. 44). Over what maximum distance can the right-hand edge of the upper brick extend over the right-hand edge of the lowermost brick that serves as the base for the entire brickwork? The length of each brick is l.

115. Find the centre of gravity of a thin homogeneous wire bent to a semicircle with a radius r.

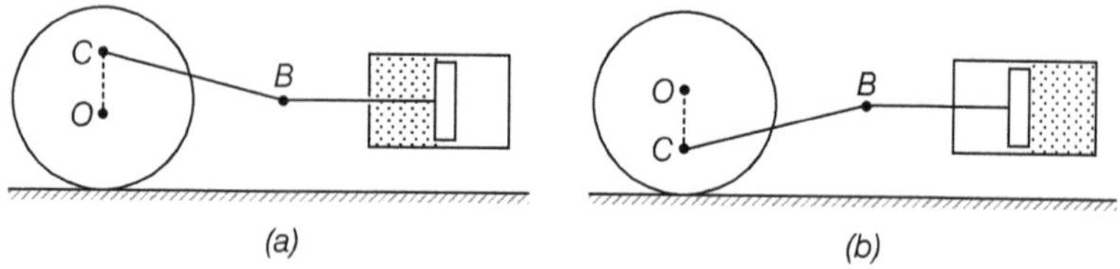

Fig.43

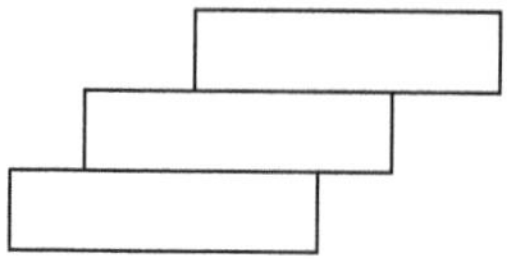

Fig. 44

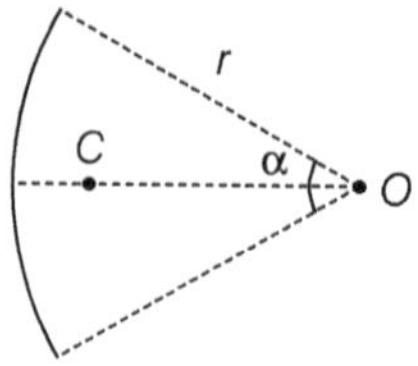

Fig. 45

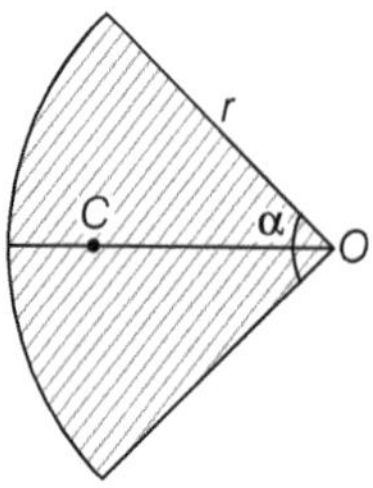

Fig. 46

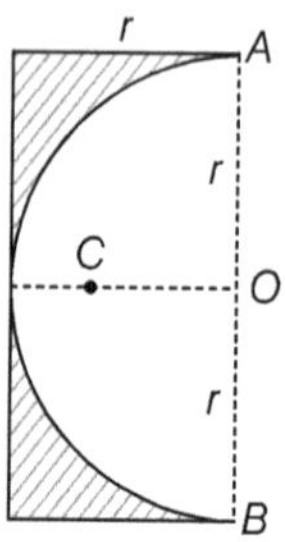

Fig. 47

116. Determine the centre of gravity of a homogeneous thin semicircle with a radius r.

117. Determine the centre of gravity of a thin homogeneous wire bent over an arc with a radius r (Fig. 45).

118. Determine the centre of gravity of a thin homogeneous plate cut in the form of a sector with a radius r and a central angle α (Fig. 46).

119. Determine the centre of gravity of a thin homogeneous plate having the form of a rectangle with sides r and $2r$ which a semicircle with a radius r is cut out of (Fig. 47).

1-6. Work and Energy

120. What work will be performed if a force of 3 kgf is used to lift a load of 1 kgf to a height of 5 metres?

121. In the formula for work $W = kFs$, the coefficient $k = 1$ if all the quantities are given in the same system of units. What is the coefficient k if the work is measured in joules, the force in kgf and the distance in centimetres?

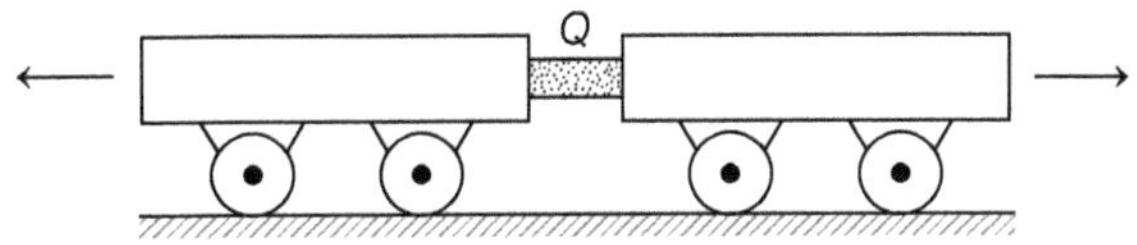

Fig. 48

122. In Guericke's experiment two copper hemispheres were tightly fitted to each other to form a hollow sphere from which the air was pumped out. The hemispheres were held so tightly together by the atmospheric pressure that they could be pulled apart only with the aid of horses. How many horses are required to do this if each horse pulled with a force F? The radius of each hemisphere was R and the atmospheric pressure p.

123. How do you explain the fact that when a stone is dropped onto the ground, the change in the momentum of the Earth is equal to that of the stone, while the change in the kinetic energy of the Earth is neglected?

124. A pile with a weight of 100 kgf is forced into the ground by a pile driver weighing 400 kgf. The latter freely drops from a height of 5 metres and drives the pile to a depth of 5 cm upon each impact. Determine the resistance of the soil, assuming it to be constant.

125. A box with sand having the mass M is suspended from a rope with a length L. The length of the rope is much greater than the linear dimensions of the box. A bullet with a mass m flies in a horizontal direction, strikes the box and gets stuck in it. After this the rope is deflected by angle α from the vertical. Determine the velocity of the bullet.

126. Two carts are pushed apart by an explosion of a powder charge Q placed between them (Fig. 48). The cart weighing 100 gf travels a distance of 18 metres and stops. What distance will be covered by the other cart weighing 300 gf? The coefficients of friction k between the ground and the carts are the same.

127. Solve Problem 65 using the law of conservation of momentum and taking into account the change in the kinetic energy of the wagon and the body.

128. A rocket launched vertically upward ejects hot gases consecutively in two equal portions. The outflow velocity of the gases relative to the rocket is constant and equal to u. What should the time interval between combustion of the portions be

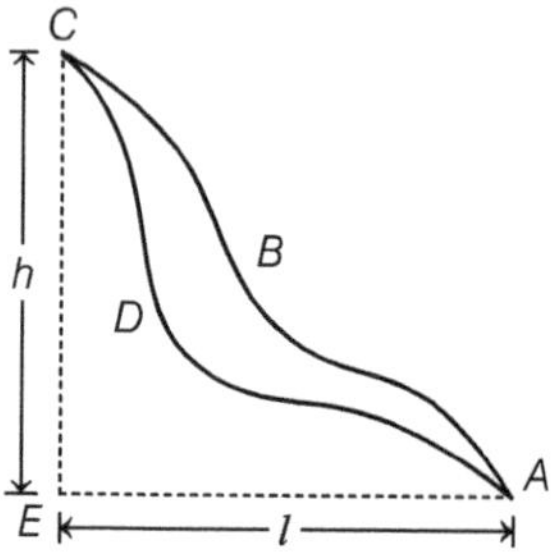

Fig. 49

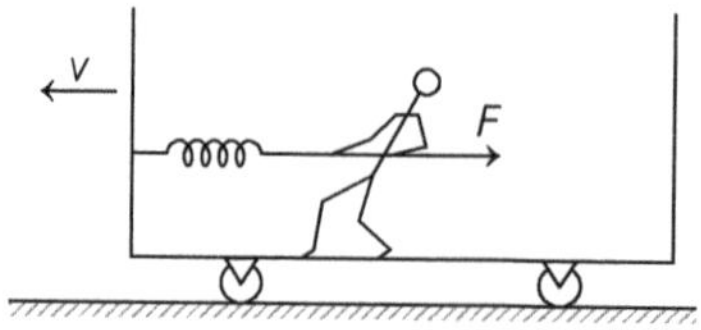

Fig. 50

for the rocket to reach a maximum altitude? The fuel burns instantaneously. Disregard the resistance of the air.

129. The fuel in a rocket burns in equal portions with a mass m. Combustion takes place instantaneously. Will the outflow velocity of the gases be constant with respect to the rocket if the mechanical energy of the system changes the same amount upon the combustion of each portion?

130. A body is first lifted to the top of a mountain over the path ADC, and again over the path ABC (Fig. 49). Prove that the work done will be the same provided the body is lifted slowly if the coefficient of friction is the same on both slopes.

131. What force should be applied to the handle of a screw jack to hold in equilibrium the load G lifted by the jack? The pitch of the screw is h and the length of the handle is R. There is no friction.

132. Find the maximum efficiency of a screw jack having no special device to prevent back travel.

133. A rope ladder with a length l carrying a man with a mass m at its end is attached to the basket of a balloon with a mass M. The entire system is in equilibrium in the air. Find the work the man should do to climb into the basket. What is the velocity of the balloon if the man climbs the ladder with a velocity v with respect to it?

134. How should the power of a pump motor change for the pump to deliver twice as much water in a unit of time through a narrow orifice? Disregard friction.

135. A rectangular pit with an area at its base A and a depth H is half filled with water. A pump forces the water up onto the surface through a cylindrical pipe with a radius R.

(1) What work is done by the pump if it pumps out all the water during the time τ?

(2) What work is done by the pump during the same time if a rectangular stone slab with an area at its base A_1 and a height h lies at the bottom of the pit? The depth of the water in the pit is $H/2$ as before.

136. What work should a man do to walk up a subway escalator moving down? The height of the escalator is h, its speed is constant and equal to v, and it is inclined at an angle α to the horizon.

137. Calculate the potential energy of a deformed spring if its elastic force $F = kx$, where k is the coefficient of spring elasticity, and x is the deformation.

138. A man acting with a force F upon a stretched spring stands in a railway carriage of a uniformly moving train (Fig. 50). The train covered a distance L. What work will be done by the man in a coordinate system related to the Earth?

139. A man stretches a spring attached to the front wall of a railway carriage over a distance l in a uniformly moving train. During this time the train covered the distance L. What work will be performed by the man in a coordinate system related to the Earth? What will this work be in a system related to the train? When the man stretches the spring he moves in a direction opposite to that of the train.

140. Two absolutely elastic spheres with masses m_1 and m_2 collide. Their initial velocities are v_1 and v_2. Find the velocities of the spheres after collision.

Consider the collision as central, the velocities of the spheres being directed along the line connecting their centres. Analyse two cases: (1) the velocity of the second sphere before collision is zero, (2) the masses of the spheres are the same.

141. Two elastic blocks of equal mass m and connected by a spring with a length l rest on an absolutely smooth horizontal surface (Fig. 51). The coefficient of elasticity of the spring is k.

A third block of the same mass m strikes the left-hand block with a velocity v. Prove that the blocks connected by the spring will always move in the same direction. Determine the velocities of the blocks when the spring is stretched as much as possible.

142. Two plates whose masses are m_1 and m_2, respectively, are connected by a spring (Fig. 52). What force should be applied to the upper plate for it to raise the lower one after the pressure is removed? Disregard the mass of the spring.

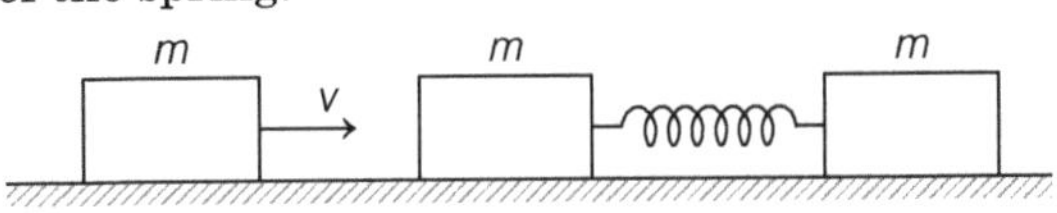

Fig. 51

143. A ball moving with a velocity v strikes a wall moving toward the ball with a velocity u (Fig. 53). An elastic impact occurs. Determine the velocity of the ball after the impact. What is the cause of the change in the kinetic energy of the ball? Consider the mass of the wall to be infinitely great.

144. Two stones identical in mass and connected by a rope with a length of $l = 39.2$ metres are dropped from a height of $h = 73.5$ metres with an initial velocity of zero. The first stone begins to fall $\tau = 2$ seconds before the second one. In what time will the stones reach the ground?
(1) Consider the rope to be absolutely elastic.
(2) Consider the rope to be absolutely inelastic.

145. Several identical elastic balls are so suspended in a row on strings of equal length (Fig. 54) that the distances between adjacent balls are very small. How will the balls behave if an extreme ball is moved aside and then released; two, three, etc., balls are moved aside and released at the same time?

146. A row of balls of identical size are placed at small intervals on a level surface (Fig. 55). The middle ball is made of steel and the others of ivory (the mass of the steel ball is greater). An ivory ball (of the same mass) strikes the balls from the right along the line of centres. How will the balls move after the impact?

147. Equal weights of mass m are suspended from the ends of a very long string thrown over two small stationary pulleys at a distance of $2l$ from each other (Fig. 56). Find the velocity of the weights after a sufficiently long time if a weight with a mass $2m$ is attached to the middle of the string.

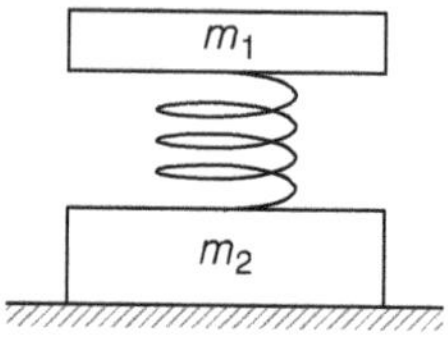

Fig. 52

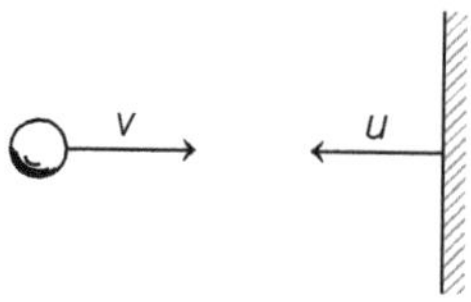

Fig. 53

148. A weight of mass $m_1 = 536$ g first kept near the ceiling between points A and B begins to lower (Fig. 57). At what angle ANB will its velocity be equal in absolute value to the velocity of a weight with a mass $m_2 = 1{,}000$ g? How will the weights move afterward?

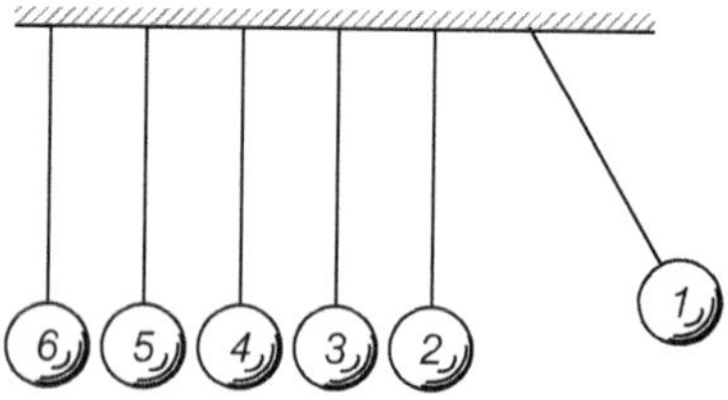

Fig. 54

149. A heavy board forming an angle α with the horizon rests on two rollers of different radii. Determine how the board will move if it does not slip. Disregard the mass of the rollers.

150. A homogeneous chain with a length $2l$ and mass M lies on an absolutely smooth table. A small part of the chain hangs from the table. At the initial moment of time the part of the chain lying on the table is held and then released, after which the chain begins to slide off the table under the weight which the chain begins to slide off the table under the weight of the hanging end. Find the velocity of the chain when the length of the hanging part is equal to x $(x < l)$.

Determine (for the same moment of time) the acceleration of the chain and the force with which it acts on the edge of the table.

151. A wagon with a mass M can move without friction along horizontal rails. A mathematical pendulum (a sphere with a mass m suspended from a long string l) is fastened on the wagon (Fig. 58). At the initial moment the wagon and the pendulum were at rest and the string was deflected through an angle α from the vertical. What will the velocity of the wagon be when the pendulum string forms an angle β $(\beta < \alpha)$ with the vertical?

152. A wedge with a mass M rests on an absolutely smooth horizontal surface. A block with a mass m is placed on the wedge. The block can slide along the wedge without friction under the force of gravity. Assuming that the system was at rest at the initial moment of time, find the velocity of the wedge when the block lowers vertically through a height h.

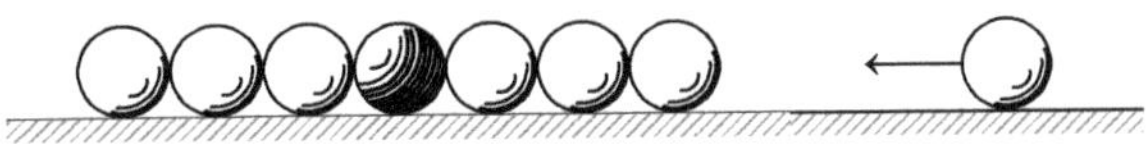

Fig. 55

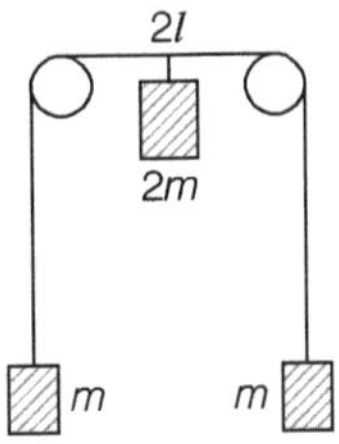

Fig. 56

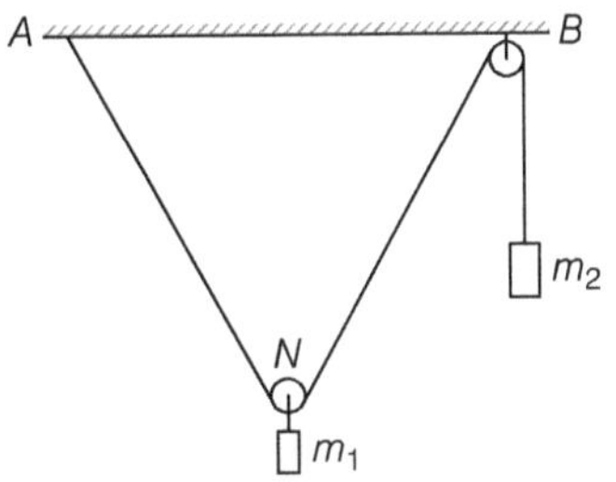

Fig. 57

153. A rod secured between two couplings can travel freely in a vertical direction (Fig. 59). The lower tip of the rod bears against a smooth wedge lying on a horizontal surface. The mass of the rod is m and that of the wedge is M. There is no friction. At the initial moment the rod and the wedge were at rest.

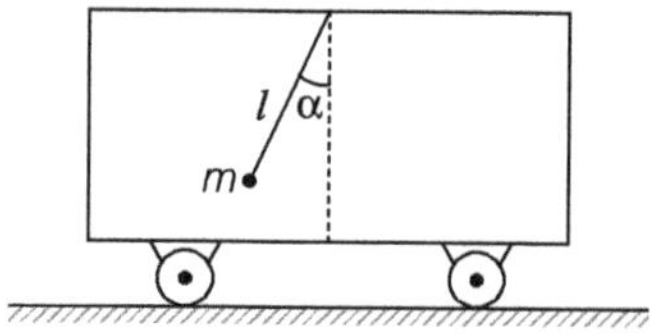

Fig. 58

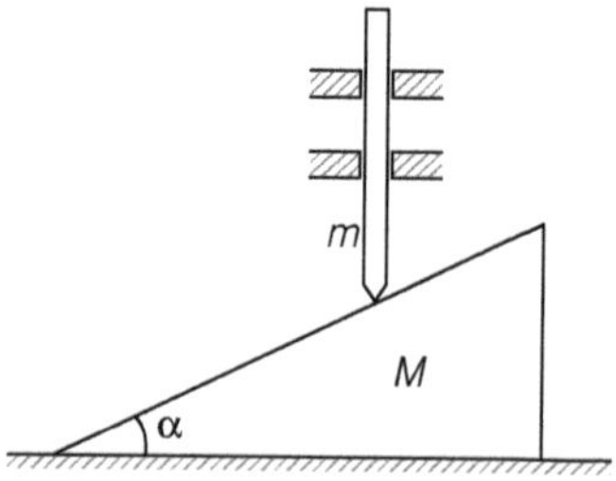

Fig. 59

Find: the velocity v of the wedge at the moment the rod lowers through the height h, the velocity u_{rel} of the rod relative to the moving wedge and the acceleration a of the rod.

1-7. Kinematics of Curvilinear Motion

154. Two shafts A and B are joined by an endless belt that transmits rotation from A to B. The speed of the driving shaft is $n_1 = 3000$ rpm. A pulley with a diameter of $D_2 = 500$ mm is fitted onto the driven shaft which should rotate at $n_2 = 600$ rpm. What is the diameter of the pulley to be fitted onto the driving shaft?

155. The crawler of a tractor consists of n links, each with a length of a. The radii of the wheels which the crawler is placed on are R. The tractor moves at a speed v. It is assumed that the crawler does not sag.

(1) What number of links move at this moment translationally, how many of them are at rest (relative to the Earth) and how many rotate on the wheels?

(2) The tractor covered a distance $s >> na$. How much time did each link of the crawler move translationally, remain at rest and participate in rotational motion?

156. The following device is used to determine the velocities of molecules. A silver-coated wire heated by current is arranged on a common axis of two cylinders fastened to each other and rotating with an angular velocity ω (Fig. 60). The internal cylinder is provided with a slit which the molecules evaporating from the wire fly into. The entire device is placed in a vacuum. If the cylinders are at rest, the trace of the deposited silver molecules appears at point A. If the cylinders rotate, the trace appears at point B at a distance l from A. Find the velocity of the molecules. The radii of the cylinders are r and R.

157. To turn a tractor moving at a speed of $v_0 = 18$ km/h the driver so brakes one of the crawlers that the axle of its driving wheel begins to move forward at a speed of $v_1 = 14$ km/h. The distance between the crawlers $d = 1.5$ metres. An arc of what radius will be described by the centre of the tractor?

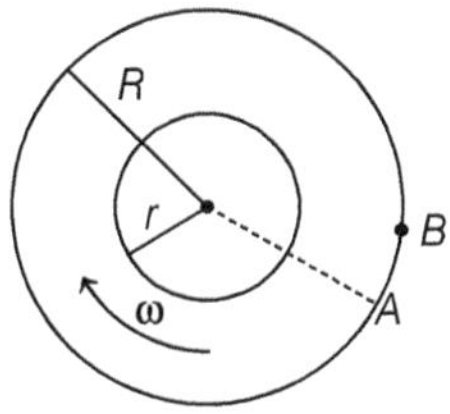

Fig. 60

158. In mountains the following phenomenon can be observed: a star being watched quickly disappears behind a remote summit. (The same phenomenon can also be observed on a plain if there is a sufficiently remote tall structure.) At what speed should the observer run to constantly see the star at the same angular distance from the mountain? The distance between the observer and the summit is 10 km. Assume that observations are being made at the pole.

159. The current velocity of a river grows in proportion to the distance from its bank and reaches its maximum v_0 in the middle. Near the banks the velocity is zero. A boat is so moving on the river that its velocity u relative to the water is constant and perpendicular to the current. Find the distance through which the boat crossing the river will be carried away by the current if the width of the river is c. Also determine the trajectory of the boat.

160. Four tortoises are at the four corners of a square with a side a. They begin to move at the same time with a constant speed v, the first one always moving in a direction toward the second, the second toward the third, the third toward the fourth and the fourth toward the first. Will the tortoises meet, and if they do, in what time?

161. Two ships A and B originally at a distance of $a = 3$ km from each other depart at the same time from a straight coastline. Ship A moves along a straight line perpendicular to the shore while ship B constantly heads for ship A, having at each moment the same speed as the latter. After a sufficiently great interval of time the second ship will obviously follow the first one at a certain distance. Find this distance.

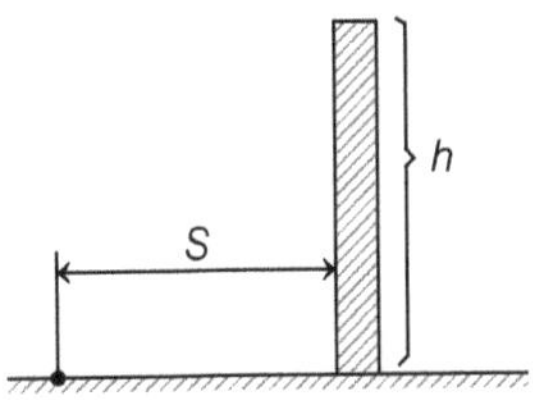

Fig. 61

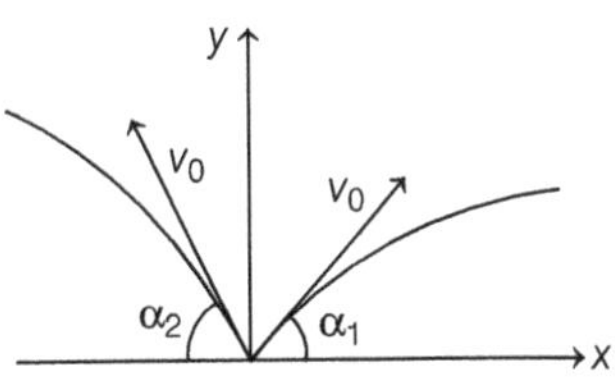

Fig. 62

162. A body is thrown with an initial velocity v_0 at an angle α to the horizon. How long will the body fly? At what distance from the spot where it was thrown will the body fall to the ground? At what angle α will the distance of the flight be maximum? At what height will the body be in the time τ after the motion begins? What will the velocity of the body be in magnitude and direction at the given moment of time?
Consider τ to be greater than the time during which the body reaches the maximum height. Disregard the resistance of the air.

163. Find the trajectory of the body thrown at an angle to the horizon (see Problem 162).

164. A rubber ball is to be thrown from the ground over a vertical wall with a height h from a distance s (Fig. 61). At what minimum initial velocity is this possible? At what angle α to the horizon should the velocity be directed in this case?

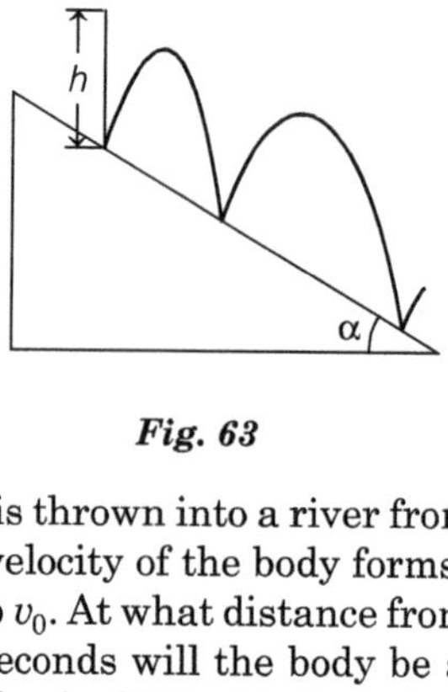

Fig. 63

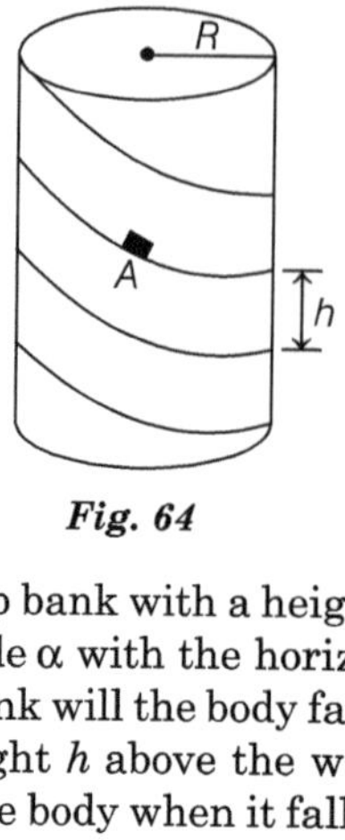

Fig. 64

165. A body is thrown into a river from a steep bank with a height H. The initial velocity of the body forms an angle α with the horizon and is equal to v_0. At what distance from the bank will the body fall? In how many seconds will the body be at a height h above the water after motion begins? What is the velocity of the body when it falls into the water?

166. At what angle to the horizon should a stone be thrown from a steep bank for it to fall into the water as far as possible from the bank? The height of the bank $h_0 = 20$ metres and the initial velocity of the stone $v_0 = 14$ m/s.

167. Two bodies are thrown at the same time and with an equal initial velocity v_0 from point $x = y = 0$ at various angles α_1 and α_2 to the horizon (Fig. 62). What is the velocity with which the bodies move relative to each other? What will the distance between the bodies be after the time τ elapses?

168. A dive-bomber drops a bomb from a height h at a distance l from the target. The speed of the bomber is v. At what angle to the horizon should it dive?

169. A passenger car is driving over a level highway behind a truck. A stone got stuck between double tyres of the rear wheels of the truck. At what distance should the car follow the truck so that the stone will not strike it if it flies out from between the tyres? Both vehicles have a speed of 50 km/h.

170. A ball freely falls from a height h onto an inclined plane forming an angle α with the horizon (Fig. 63). Find the ratio of the distances between the points at which the jumping ball strikes the inclined plane. Consider the impacts between the ball and the plane to be absolutely elastic.

171. Find the acceleration of body A which slides without initial velocity down a helical groove with a pitch h and a radius R at the end of the n-th turn (Fig. 64). Disregard friction.

172. Two steel slabs M and N with a height h are placed on sand (Fig. 65). The distance between the slabs $l = 20$ cm. A ball whose velocity has not been determined exactly moves over slab M. It is only known that this velocity ranges between 200 cm/s and 267 cm/s.

(1) At what height h is it impossible to predict the horizontal direction of the velocity of the ball at the moment when it falls onto the sand? (Before it touches the sand, the ball strikes slab N at least once.)

(2) At what minimum height of the slabs is it impossible to predict the spot on section l which the ball will strike? Disregard the duration of the impact between the ball and the slab, and consider the impact to be absolutely elastic.

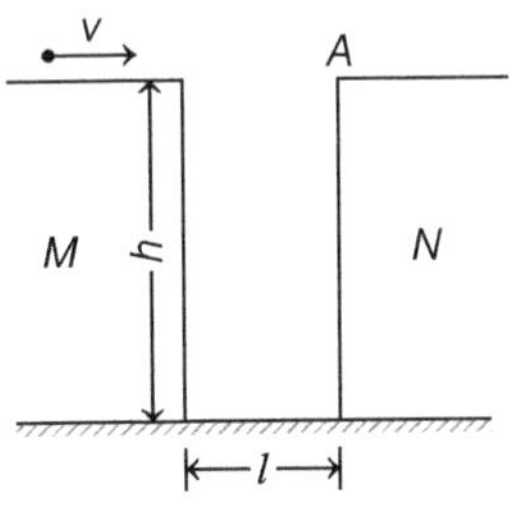

Fig. 65

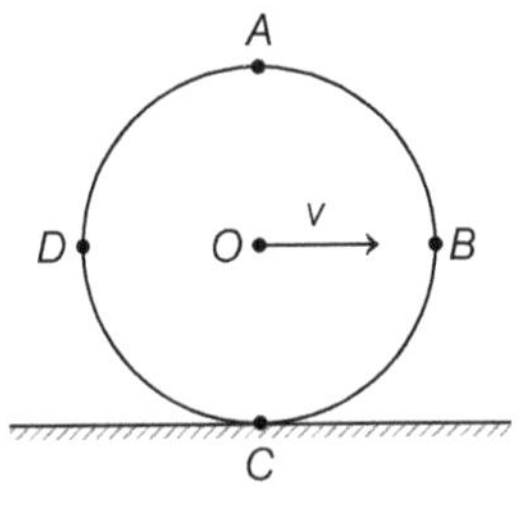

Fig. 66

173. A solid homogeneous disk rolls without slipping over a horizontal path with a constant velocity v (Fig. 66).

(1) Prove that the linear velocity of rotation of any point of the disk on its rim with respect to the centre O is equal to the translational velocity of the disk.

(2) Find the magnitude and direction of the velocities of points A, B, C and D on the rim of the disk with respect to a standing observer.

(3) What points of the disk have the same absolute velocity as the centre of the disk relative to a standing observer?

174. A moving cart is shown on a cinema screen. The radius of the front wheels of the cart $r = 0.35$ metre and that of the rear ones $R = 1.5\,r$. The front wheels have $N_1 = 6$ spokes. The film in the cinema camera moves with a speed of 24 frames per second.

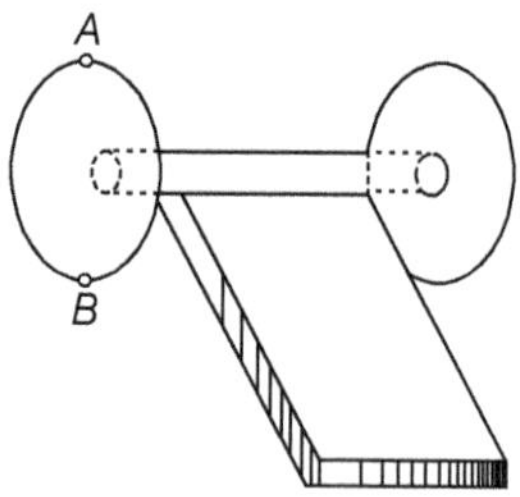

Fig. 67

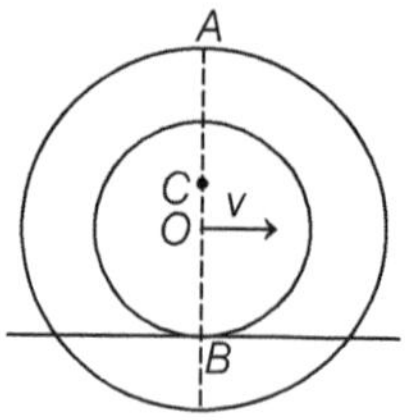

Fig. 68

Assuming that the wheels of the cart do not slip, find the minimum speed with which the cart should move for its front wheels to seem stationary on the screen. What minimum number of spokes N_2 should the rear wheels have for them also to seem stationary?

175. At what speeds of the cart moving from the left to the right (see Problem 174) will it seem to the audience that:

(1) the spokes of the wheels rotate counterclockwise?

(2) the spokes of the front and rear wheels rotate in opposite directions? There are six spokes on each front and rear wheel.

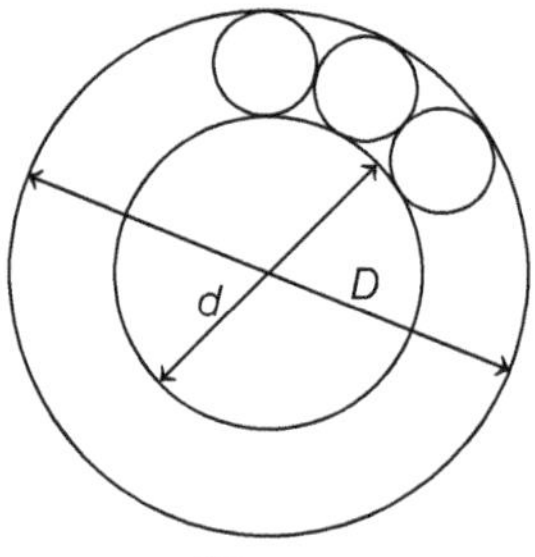

Fig. 69

176. A spool consists of a cylindrical core and two identical solid heads. The core rolls without slipping along a rough horizontal block with a constant velocity v (Fig. 67). The radius of the core is r and that of the heads is R.

What instantaneous velocity will points A and B on the rim of one of the heads have? What points on the heads have an instantaneous velocity equal in magnitude to the velocity of the spool core?

177. Draw the trajectories of points A, B and C of a spool (Fig. 68) rolling with its cylindrical core without slipping along a block (see Problem 176).

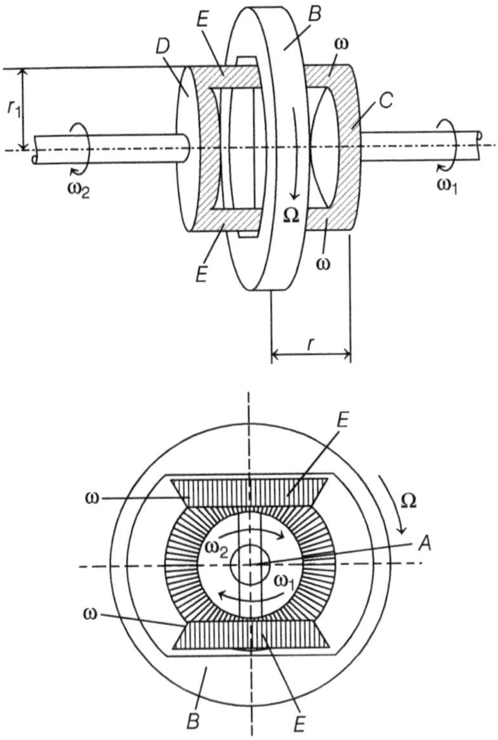

Fig. 70

178. A ball bearing supports the end of a shaft rotating with an angular velocity ω. The diameter of the shaft axis is d and that of the race is D (Fig. 69). Find the linear velocity of motion of the centre of one of the balls if the race is stationary, or rotates with an angular velocity of Ω. Assume that in both cases the balls do not slip when they roll over the shaft.

179. A cone rolls without slipping along a plane. The axis of the cone rotates with a velocity ω around a vertical line passing through its apex. The height of the cone is h and the angle between the axis and the generatrix of the cone is α. What is the angular velocity with which the cone rotates around its axis? Also determine the linear velocity of an arbitrary point on the diameter of the cone base lying in a vertical plane.

180. Figure 70 shows schematically a differential transmission of a motor vehicle that does not allow the driving wheels of the vehicle to slip when negotiating a curve. (In this case the wheels should revolve at different speeds.) The engine rotates wheel B rigidly mounted on axle A around which a pair of bevel gears E can freely rotate. These gears mesh with another pair of bevel gears along which they run when axle A rotates. The axle of the driving wheels (usually the rear ones) is divided into two halves that carry on their ends gears C and D. These halves can rotate with various angular velocities, remaining connected by the differential transmission.

Find the relationship between the angular velocities Ω, ω, ω_1 and ω_2 of the differential transmission if the radii of gears E are equal to r and the radii of gears D and C are r_1.

1-8. Dynamics of Curvilinear Motion

181. Determine the tension of the rope of a ballistic pendulum (see Problem 125) at the moment when it is struck by a bullet.

182. Four identical weights are secured on a flexible un stretchable string whose weight may be neglected (Fig. 71). The entire system rotates with an angular velocity ω around a vertical axis passing through point O. The weights move over a smooth level surface. Determine the tension of the string in various sections.

183. Masses m_1 and m_2 are fastened to the ends of a weightless rod with a length l. The velocities of these masses are in one plane and are equal to v_1 and v_2, respectively (Fig. 72). Find the velocity which the centre of gravity of the system moves with and the angular velocity which the rod rotates with relative to an axis passing through the centre of gravity.

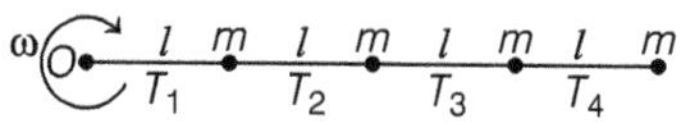

Fig. 71

184. A cannon is at the centre of a platform freely rotating around a vertical axis. The axis of rotation passes through the breech of the cannon. A shell is fired in a horizontal direction along the radius of the platform. Will the velocity of rotation change in this case?

185. A small body begins to slide without initial velocity down an inclined plane with a height H (Fig. 73). Assuming that there is no friction and the impact of the body against horizontal plane AB is absolutely elastic, determine the nature of motion of the body after it leaves the inclined plane. Answer the same question if the impact is absolutely inelastic.

186. What is the minimum radius of an arc that can be negotiated by a motor-cyclist if his speed is $v = 21$ m/s and the coefficient of friction between the tyres and the ground $k = 0.3$? To what angle α to the horizon should the motor-cycle be inclined in this case?

187. A massive sphere is fitted onto a light rod (Fig. 74). When will the rod fall faster: if it is placed vertically on end A or on end B? The end of the rod on the ground does not slip.

188. A massive sphere is secured on the end of a light rod placed vertically on a floor. The rod begins to fall without any initial velocity. At what angle α between the rod and a vertical will the end of the rod no longer press on the floor?

At what coefficient of friction will the end of the rod not slip up to this moment?

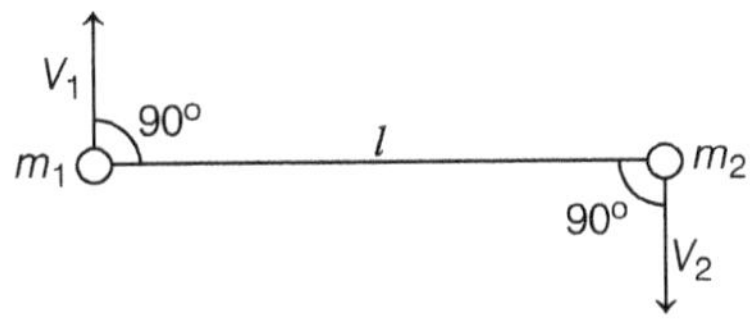

Fig. 72

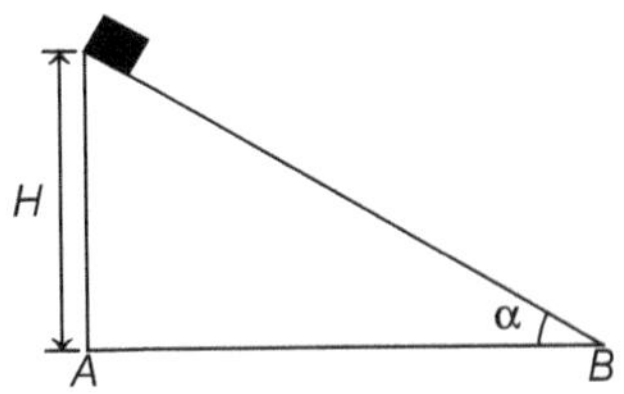

Fig. 73

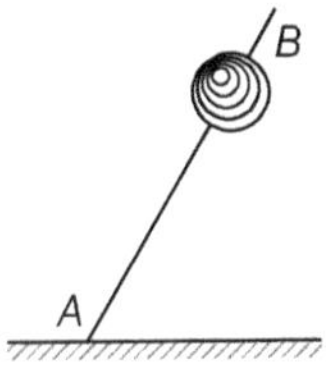

Fig. 74

189. At what distance from the bottom of the rod will the sphere (see Problem 188) fall if the coefficient of friction $k > \frac{\sqrt{5}}{2}$.

190. A wire is bent along an arc with a radius R (Fig. 75). A bead is placed on the wire that can move along it without friction. At the initial moment the bead was at point O. What horizontal velocity should be imparted to the bead for it to get onto the wire again at point B after flying a certain distance (AB) through the air?

191. A small body slides down an inclined surface passing into a loop from the minimum height ensuring that the body does not leave the surface of the loop (Fig. 76). What symmetrical segment with an angle $\alpha < 90°$ can be cut out of the loop for the body to reach point B after travelling a certain distance in the air?

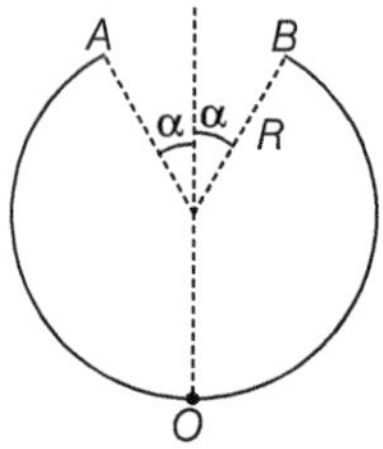

Fig. 75

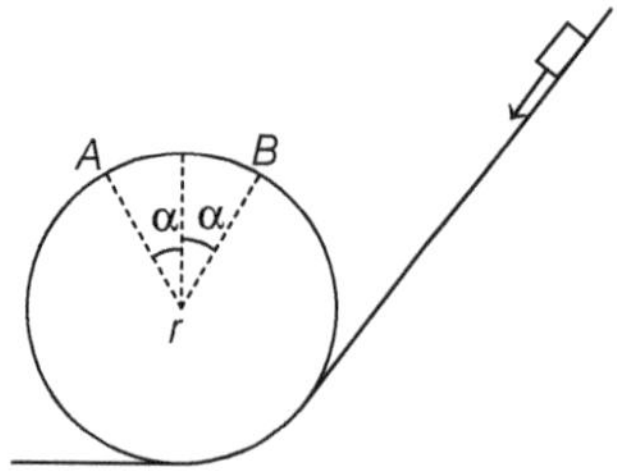

Fig. 76

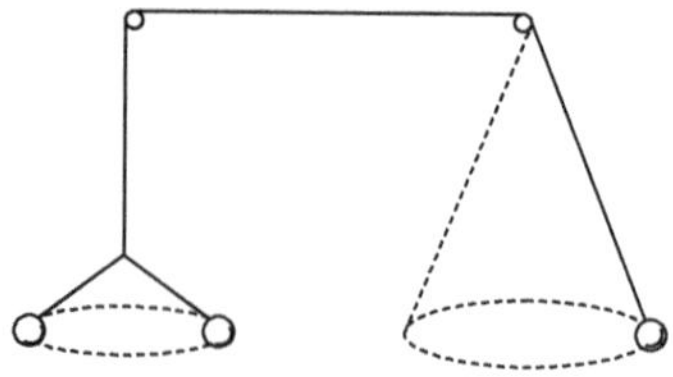

Fig. 77

How will the body move if the angle α is greater or smaller than the found one? Disregard friction and the resistance of the air.

192. Three weights moving circularly are attached to the ends of a string passing over two nails (Fig. 77). Two weights with a mass m each hang on the left and one weight with a mass $2m$ on the right. Will this system be in equilibrium?

193. A ball is suspended from a very thin string. The latter is brought into a horizontal position and is then released. At what points of the trajectory is the acceleration of the ball directed vertically downward, vertically upward and horizontally? The string is not tensioned at the initial moment.

194. A weightless rod can rotate in a vertical plane with respect to point O. The rod carries masses m_1 and m_2 at distances r_1 and r_2 from O (Fig. 78). The rod is released without any initial velocity from a position forming an angle α with the vertical. Determine the linear velocities of masses m_1 and m_2 when the rod reaches a vertical position.

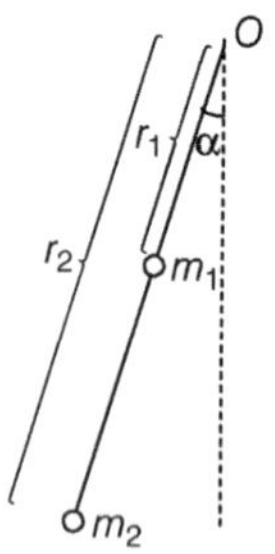

Fig. 78

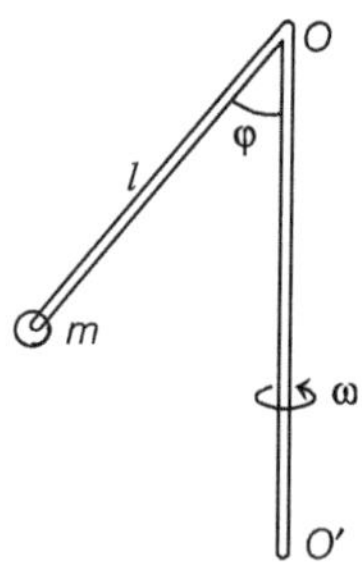

Fig. 79

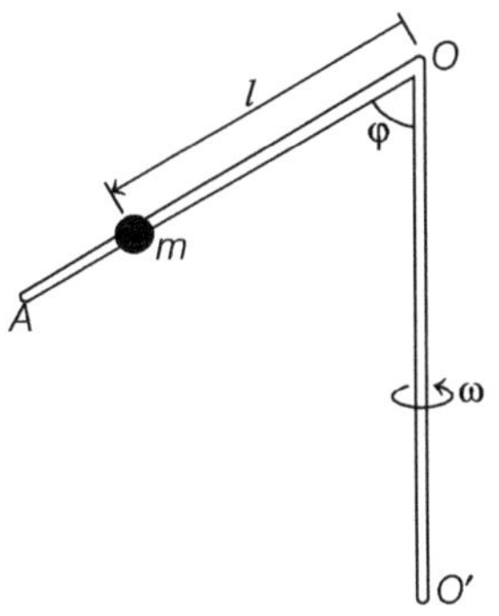

Fig. 80

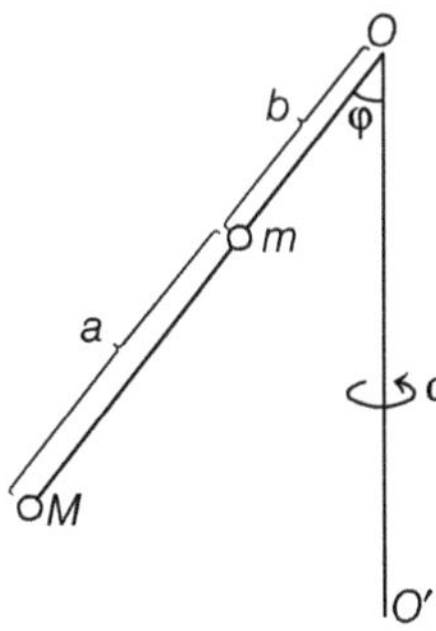

Fig. 81

195. A plumb to which a small ball is attached by means of a string with a length of $l = 12.5$ cm is secured on the axis of a centrifugal machine. Find the angle α through which the string deflects from the vertical if the machine makes one revolution per second, two revolutions per second.

196. A rigid rod bent as shown in Fig. 79 rotates with an angular velocity ω around axis OO'. A weight with a mass m is attached to the end of the rod. Find the force which the rod acts with on the mass m.

197. A rigid rod AOO' bent as shown in Fig. 80 rotates with an angular velocity ω around axis OO'. A bead is fitted onto the rod. Find the distance l from point O at which the bead will be in equilibrium if the coefficient of friction between the bead and the rod is k.

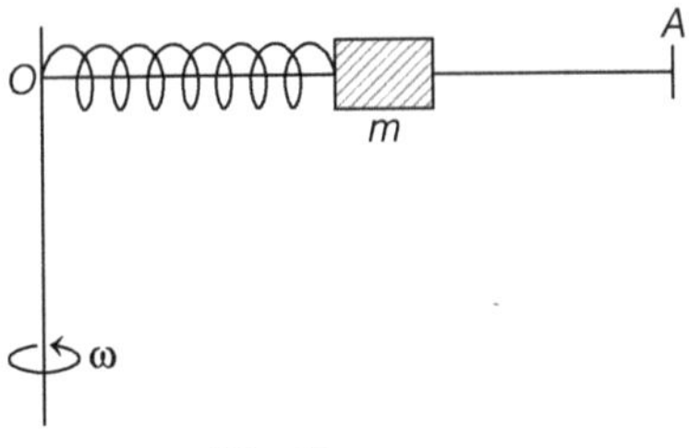

Fig 82

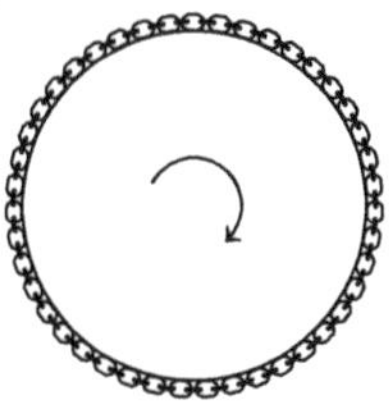

Fig 83

198. A weight with a mass m is attached to the end of a string with a length l fastened to a vertical rod rotating with an angular velocity ω. Another string of the same length as the first and carrying on its end another weight with a mass m is secured to the first weight. Prove that during rotation the angle between the first string and the vertical will be smaller than the angle between the vertical and the second string. Disregard the weight of the string.

199. A weightless rod carries two weights of mass m and M. The rod is hinge-jointed to vertical axis OO' (Fig. 81), which rotates with an angular velocity ω. Determine the angle φ formed by the rod and the vertical.

200. A horizontal straight bar rotates with a constant angular velocity around a vertical axis. A body can move without friction over the bar. Initially, the body is retained in equilibrium by a spring (Fig. 82). What will happen to the body if an initial velocity is imparted to it along the bar? The length of the spring in an unstretched state can be neglected.

201. A metallic chain with a length of $l = 62.8$ cm and whose ends are joined together is fitted onto a wooden disk (Fig. 83). The disk rotates with a speed of $n = 60$ rps. Find the tension of the chain T if its mass is $m = 40$ g.

202. Water flows with a velocity v along a rubber tube having the form of a ring (Fig. 84). The radius of the ring is R and the diameter of the tube $d \ll R$. What force is the rubber tube stretched with?

203. A homogeneous rod with a length l and a mass m rotates with an angular velocity ω in a horizontal plane around an axis passing through its end. Find the tension of the rod at a distance x from its axis of rotation.

204. A ball with the mass m secured on a weightless rod rotates with a constant velocity v in a horizontal plane (Fig. 85). Its kinetic energy in a coordinate system that is stationary with respect to the axis of rotation is constant and equal to $mv^2/2$. The kinetic energy changes with time from zero to $4mv^2/2$ with respect to a reading system that moves rectilinearly in a horizontal plane with a velocity v relative to the axis. What is the reason for this change in the energy?

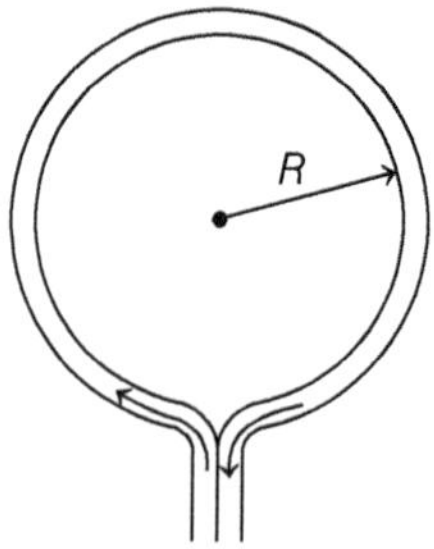

Fig. 84

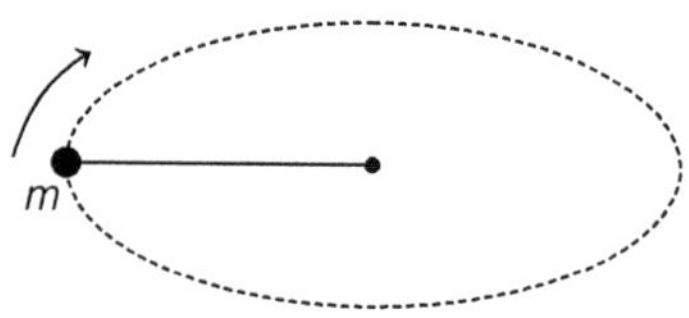

Fig. 85

205. A thin homogeneous hoop rolls over a horizontal surface with a constant velocity v. How and under the action of what forces does the total energy of a small section AB that is at the highest point of the loop at the given moment change?

206. A heavy spool with a thread wound on it lies on a rough horizontal surface over which it can roll without slipping. If the thread is pulled to the left in a horizontal direction, the spool will also move to the left. If the direction of the thread is changed (Fig. 86), the spool will begin to roll to the right at a certain angle α between the direction of the thread and a vertical line.

Determine this angle. What will happen to the spool at the given value of this angle? The radius of the spool heads is R and that of its core is r.

207. Find the kinetic energy of a hoop with a mass M and a radius R if it moves uniformly with a velocity v and rotates with an angular velocity ω around an axis passing through its centre.

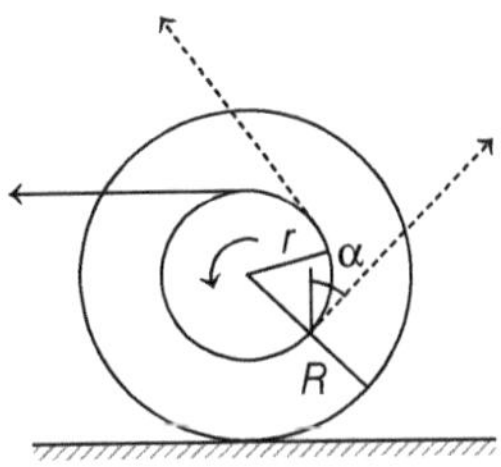

Fig. 86

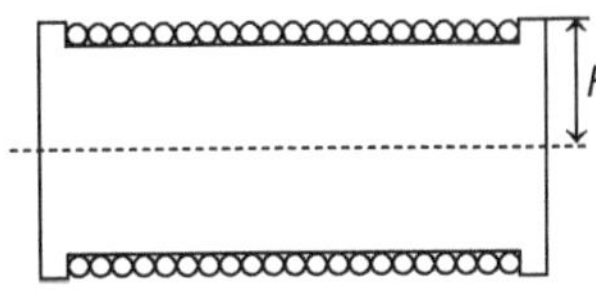

Fig. 87

208. Find the kinetic energy of the crawler of a tractor moving with a speed v. The distance between the centres of the wheels which the crawler is placed onto is l. The radius of the wheels is r. A unit of crawler length weighs G.

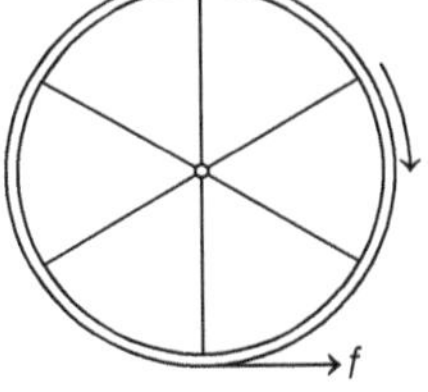

Fig. 88

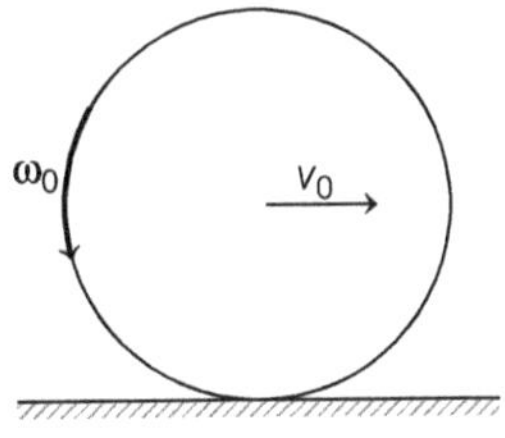

Fig. 89

209. Two cylinders of equal mass and size are made of unknown materials of different density. How is it possible to tell which of the two is hollow?

210. A flexible cable is wound in one row around a drum with a radius R (Fig. 87). The weight of a unit of cable length is ρ. The entire cable weighs G. The drum moves by inertia without slipping along a horizontal surface, and the cable is wound off onto it. At the initial moment, when the cable was completely wound on the drum, the velocity of the drum centre was v.

Appraise the velocity of the drum centre at the moment of time when a part of the cable with a length x lies on the surface, neglecting the radius of the cable cross section (in comparison with R) and the mass of the drum. What force changes the momentum of the cable?

211. A friction force f is applied to a pulley with a radius r rotating around a stationary axis (Fig. 88). Determine the change in the angular velocity of the pulley with time if this velocity is ω_0 at the initial moment. The mass of the pulley is m and the mass of the spokes can be neglected.

212. A hoop of radius r rotating with an angular velocity ω_0 is placed on a rough horizontal surface. Determine the velocity v of the centre of the hoop when the hoop ceases to slip. At the initial moment the velocity of the centre of the hoop is zero.

213. A translational velocity v_0 is imparted in a horizontal direction to a hoop with a radius r placed on a rough horizontal surface. Find the angular velocity ω of rotation of the hoop after it stops slipping.

214. A hoop with a radius r rotating with an angular velocity ω_0 is placed on a rough horizontal surface. A translational velocity v_0 is imparted to the hoop (Fig. 89). Determine the nature of the motion of the hoop, assuming that the force of sliding friction is f.

215. A cylindrical tube with a radius r is connected by means of spokes to two hoops with a radius R. The mass of both the hoops is M. The mass of the tube and the spokes in comparison with the mass M can be neglected. A string passed over a weightless pulley is wound around the tube. A weight with a mass m is attached to the end of the string (Fig. 90).

Find the acceleration of the weight, the tension of the string and the force of friction acting between the hoops and the surface. (Assume that the hoops do not slip.) Also determine the coefficient of friction at which the hoops will slip.

216. A spool lies on an inclined plane. A thread is wound on the spool and its free end carrying a weight having a mass m is thrown over a weightless pulley (Fig. 91). It is assumed that the mass of the spool M is uniformly distributed over a circle with a radius R and there is no friction. Determine the angle of inclination α at which the centre of gravity of the spool will be at rest.

217. A board with a mass M is placed on two identical cylindrical rolls with a radius R. The rolls rest on a horizontal surface. At the initial moment the system is at rest. Then, a force Q is applied to the board in a horizontal direction. Find the acceleration of the board and the forces of friction acting between the rolls and the board and also between the rolls and the horizontal surface. The rolls have the form of thin-walled cylinders with a mass m each and do not slip.

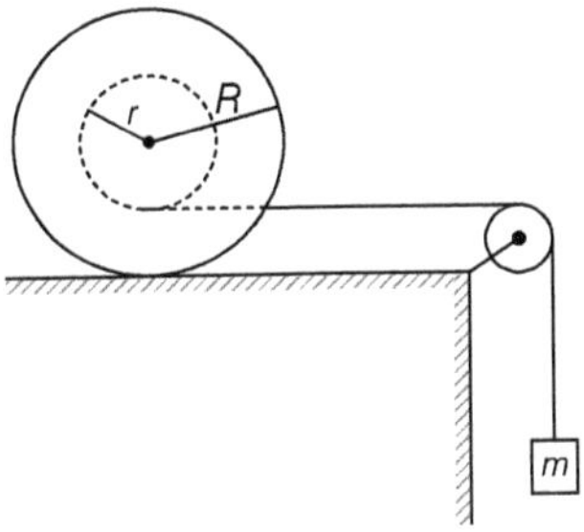

Fig. 90

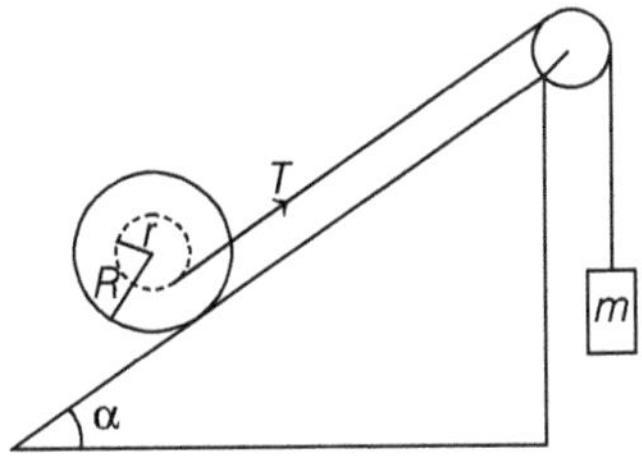

Fig. 91

218. A double-step pulley consists of two rigidly connected thin hoops with radii R and r and masses M_1 and M_2, respectively. Strings carrying weights m_1 and m_2 at their ends are passed around each step (Fig. 92). Find the acceleration of the weights m_1 and m_2, the tension of the strings and the force with which the system acts on the axis of the pulley.

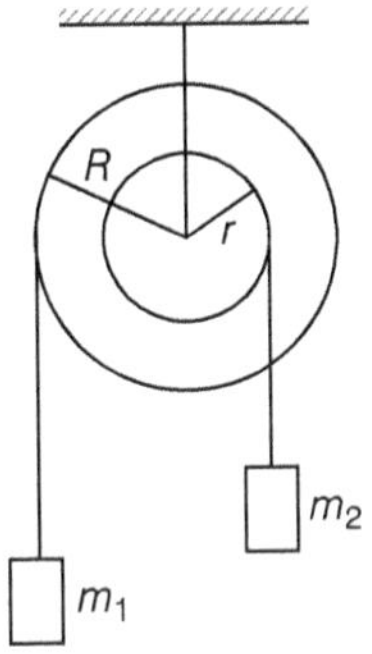

Fig. 92

219. A homogeneous thin-walled cylinder with a radius R and a mass M rolls without slipping under the force of gravity down an inclined plane forming an angle α with the horizon.

Using the law of conservation of energy, determine the velocity of the centre of gravity and the angular velocity of cylinder rotation when the time t elapses after motion begins. (It is assumed that the cylinder is at rest at the initial moment.) Also find the acceleration of the centre of gravity of the cylinder.

1-9. The Law of Gravitation

220. Why does the Earth impart to all bodies the same acceleration irrespective of their mass?

221. Find the magnitude and dimension of the gravitational constant in the CGS system, bearing in mind that the mean radius of the Earth is $R = 6.4 \times 10^8$ cm and its mass $M = 6 \times 10^{27}$ g.

222. In what conditions will bodies inside a spaceship be weightless, i. e., cease to exert any pressure on the walls of the cabin?

223. A light pendulum consisting of a rod and a disk (Fig. 93) is attached to a wooden block which can freely fall along guiding wires.

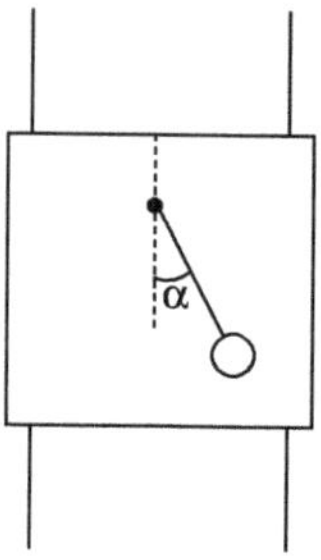

Fig. 93

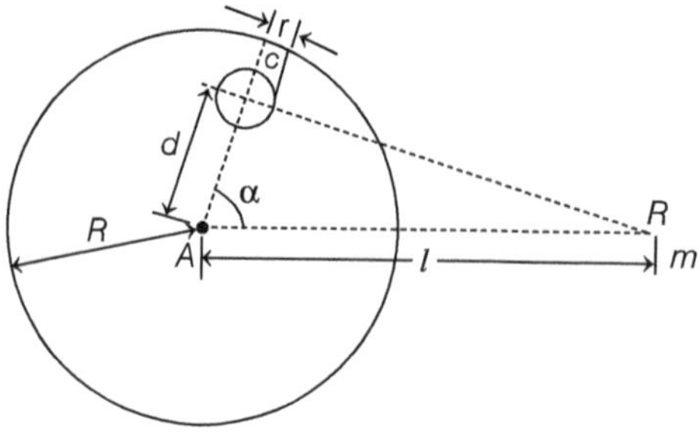

Fig. 94

The pendulum is brought out of equilibrium through an angle α and released. At the moment when the pendulum passed through its lowermost position, the block was released and began to fall freely. How will the pendulum move with respect to the block? Disregard friction and the resistance of the air.

224. A planet moves along an ellipse having the Sun in its focus. Taking into account the work of the force of gravity, indicate the points on the trajectory at which the velocity of the planet will be maximum and minimum.

225. An artificial satellite of the Earth moves at an altitude of $h = 670$ km along a circular orbit. Find the velocity of the satellite.

226. How will the velocity of an artificial satellite of the Earth change with time when it moves in the upper layers of the atmosphere?

227. Two satellites move along a circular orbit in the same direction at a small distance from each other. A container has to be thrown from the first satellite onto the second one. When will the container reach the second satellite faster: if it is thrown in the direction of motion of the first satellite or in the opposite direction? The velocity of the container with respect to the satellite u is much less than that of the satellite v.

228. Determine the mass of the Sun M if the mean radius of the Earth's orbit is $R = 149 \times 10^6$ km.

229. Determine the minimum distance h from the first Soviet artificial satellite launched on October 4, 1957, to the Earth's surface if the following data are known: the maximum distance between the satellite and the Earth $H = 900$ km, the period of revolution around the Earth $T = 96$ minutes, the major semiaxis of the Moon's orbit $R = 384{,}400$ km, the period of rotation of the Moon around the Earth $T = 27.3$ days and the Earth's radius $R_0 = 6{,}370$ km.

230. An air bubble with a radius r and an iron ball with the same radius are present in water. Will they be attracted or repulsed from each other? What is the force of interaction between them? The distance between the centre of the ball and the bubble is R.

231. Two air bubbles with a radius r are present in water. Are the bubbles attracted or repulsed? What is the force of interaction between them? The distance between the bubbles is R.

232. A lead ball with a radius of $R = 50$ cm has inside a spherical space with a radius $r = 5$ cm whose centre is at a distance of $d = 40$ cm from the centre of the ball (Fig. 94). With what force will a material particle with a mass of $m = 10$ g at a distance of $l = 80$ cm from the centre of the ball be attracted to it if the line connecting the centres of the ball and the space forms an angle $\alpha = 60°$ with the line connecting the centre of the ball and the material particle?

233. A body whose dimensions may be neglected is placed inside a thin homogeneous sphere. Prove that the force of attraction acting from the sphere on the body is zero irrespective of its position in the sphere.

234. What is the force with which a body with a mass m in a deep mine will be attracted to the centre of the Earth if the distance between the body and the Earth's centre is r? The density of the Earth is the same everywhere and equal to ρ.

1-10. Hydro- and Aerostatics

235. A block floats vertically in a glass filled with water. How will the level of the water in the glass change if the block assumes a horizontal position?

236. A vessel filled with water is placed on the edge of a board (Fig. 95). Will equilibrium be violated if a small board carrying a weight is placed on the surface of the water?

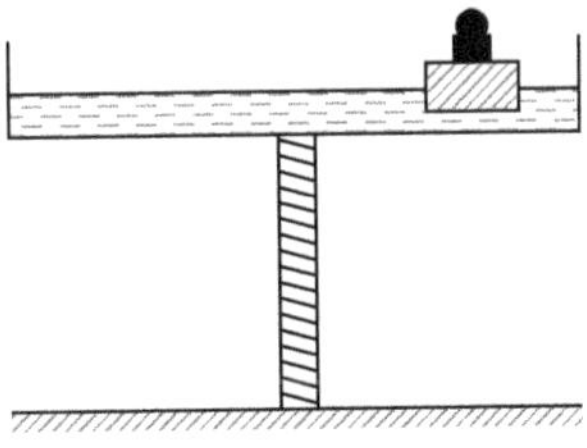

Fig. 95

237. A piece of ice floats in a glass filled with water. How will the level of the water in the glass change when the ice melts? Consider the following cases:

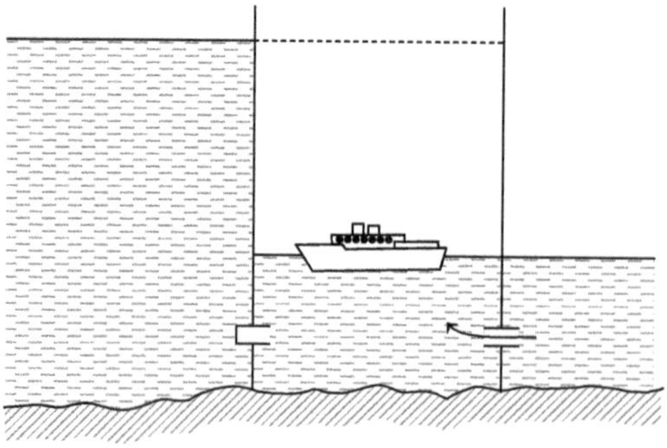

Fig. 96

(1) the ice is absolutely homogeneous;

(2) a stone is frozen in the ice;

(3) the ice contains an air bubble.

238. A solid homogeneous body submerged into a liquid with a specific weight of γ_1 weighs G_1, and in a liquid with a specific weight of γ_2, it weighs G_2. Determine the specific weight γ of the body.

239. A hole is cut in the ice in the middle of a large lake. The ice is 10 metres thick.What length of rope is required to scoop up a bucketful of water?

240. A match-box with a small stone on its bottom floats in a cup filled with water. Will the level of the water in the cup change if the stone is taken out of the box and dropped into the water?

241. A ship passing through a lock rises to a higher level in a chamber of the lock into which water is pumped from the side of the lower level (Fig. 96). When will the pumps perform more work: when a large ship is in the chamber, or a small boat?

242. A square with a side a and a rectangle with sides a and $2a$ are cut out from two plates of equal thickness with specific weights of $3.5\ \text{gf}/\text{cm}^3$ and $2\ \text{gf}/\text{cm}^3$, the square being cut out of the heavier material. The square and the rectangle are fastened together in the form of the letter L and placed upside down on the bottom of an empty vessel (Fig. 97). What will occur if the vessel is filled with water?

243. A tube floats vertically in water (Fig. 98). The portion of the tube protruding from the water is $h = 5$ cm. Oil with a specific weight of $\gamma = 0.9\ \text{gf}/\text{cm}^3$ is poured into the tube. What length can the tube have so that it can be completely filled with oil?

244. A vessel with water falls with an acceleration $a < g$. How does the pressure p in the vessel change with depth?

245. A vessel with a body floating in it falls with an acceleration $a < g$. Will the body rise to the surface?

246. A cart supports a cubic tank filled with water up to its top (Fig. 99). The cart moves with a constant acceleration a. Determine the pressure at point A which is at a depth h and a distance l from the front wall, if the tank is tightly closed with a lid. In uniform motion the lid does not exert any pressure on the water.

247. A rubber ball with a mass m and a radius R is submerged into water to a depth h and released. What height will the ball jump up to above the surface of the water? Disregard the resistance of the water and the air.

248. Mercury is poured into two communicating cylindrical vessels, and water is poured in above it. The level of the water is the same in both vessels. Will the level of the water and the mercury be the same if a piece of wood is dropped into one vessel and some water equal in weight to this piece is added to the other? Consider cases when the cross sections of the vessels are the same or different.

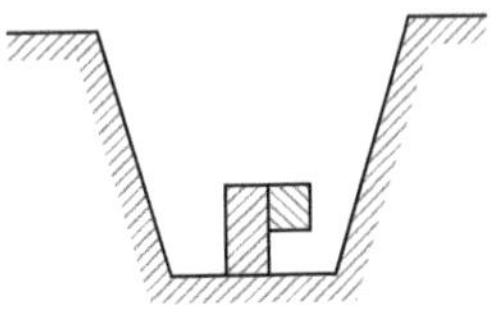

Fig. 97

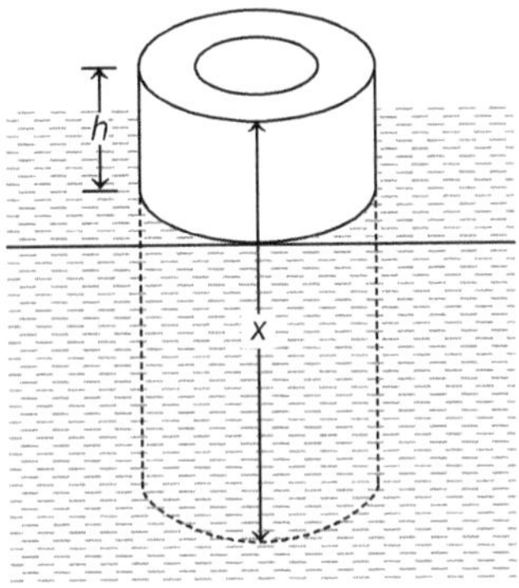

Fig. 98

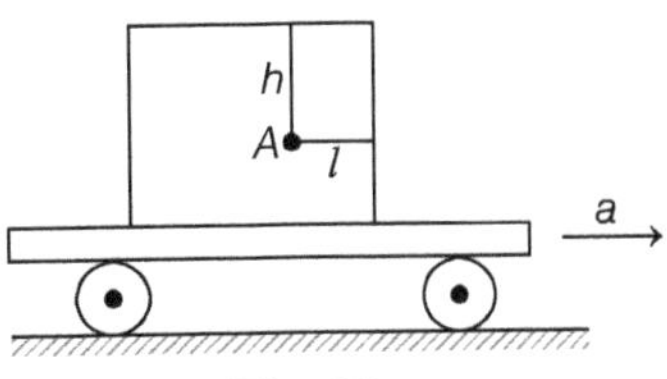

Fig. 99

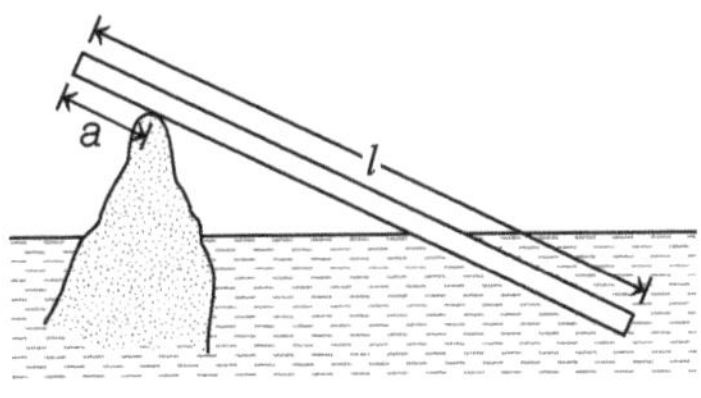

Fig. 100

249. Mercury is poured into cylindrical communicating vessels with different cross sections. An iron block with a volume V_0 is dropped into the broad vessel, and as a result the level of the mercury in it rises. Then water is poured into the vessel until the mercury reaches the previous level. Find the height of the water column h if the cross section of the narrow vessel is A_1.

250. One end of a board with a length l is placed on top of a stone protruding from water. Part a is above the point of support (Fig. 100). What part of the board is below the surface of the water if the specific weight of wood is γ?

251. A conical vessel without a bottom tightly stands on a table. A liquid is poured into the vessel and as soon as its level reaches the height h, the pressure of the liquid raises the vessel. The radius of the bottom greater base of the vessel is R, the angle between the cone generatrix and a vertical is α, and the weight of the vessel is G. What is the density of the liquid?

252. Can water be pumped over a wall 20 metres high with the aid of a syphon?

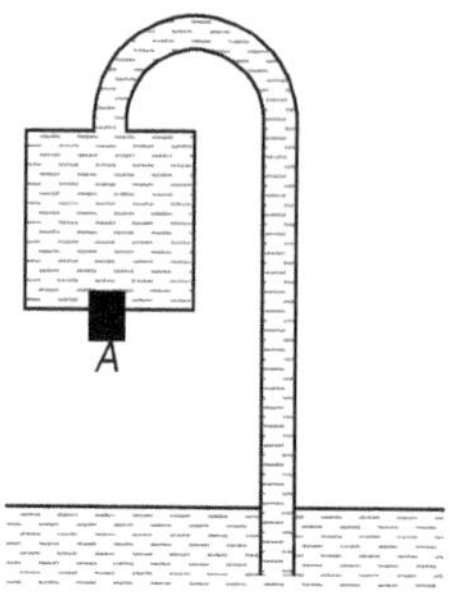

Fig. 101

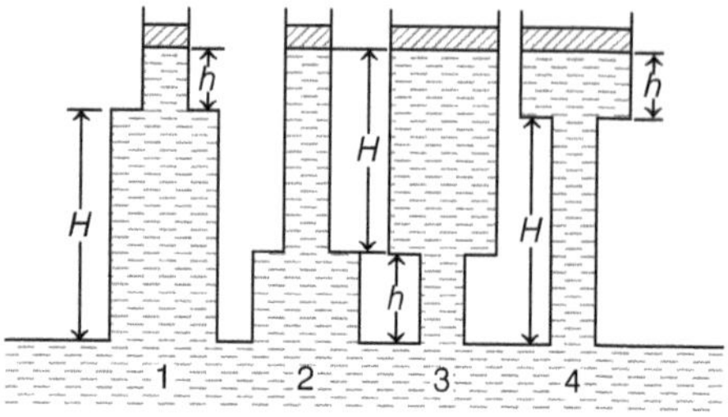

Fig. 102

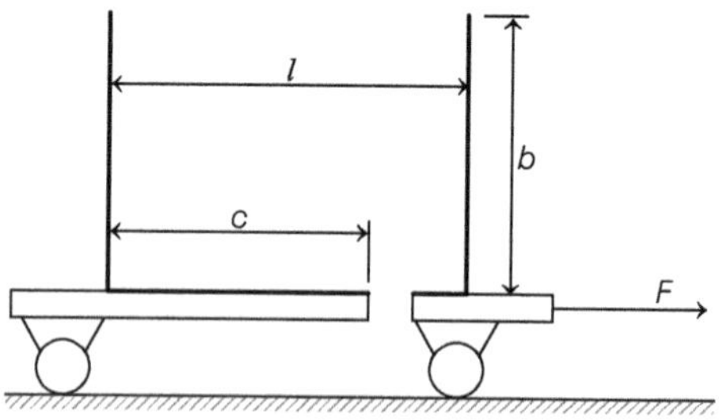

Fig. 103

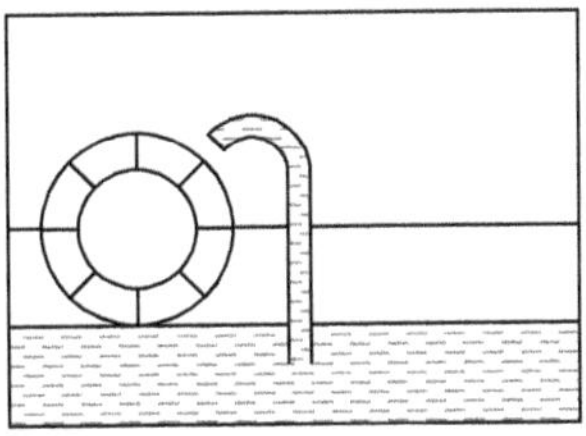

Fig. 104

253. The vessel shown in Fig. 101 is entirely filled with water. What will happen if plug A is removed? The radius of the hole is about 0.5 cm.

254. Four piston pumps are made of pipe sections with a large and a small diameters. The pumps lifted water to the same height $H + h$ (Fig. 102). Which of the pistons should be pulled with a greater force to keep it in equilibrium? Disregard the weight of the pistons.

255. A piston weighing $G = 3$ kgf has the form of a circular disk with a radius $R = 4$ cm. The disk has a hole into which a thin-walled pipe with a radius $r = 1$ cm is inserted. The piston can enter a cylinder tightly and without friction, and is initially at the bottom of the cylinder. What height H will the piston rise to if m = 700 g of water is poured into the pipe?

256. A vessel with a hole in its bottom is fastened on a cart. The mass of the vessel and the cart is M and the area of the vessel base is A. What force F should the cart be pulled with so that a maximum amount of water remains in the vessel? The dimensions of the vessel are shown in Fig. 103. There is no friction.

257. The following design of a perpetuum mobile was suggested (Fig. 104), A hermetic vessel is divided into two halves by an air-tight partition through which a tube and a water turbine of a special design are passed. The turbine is provided with chambers having covers which close and open automatically. The pressure p_1 in the lower part of the vessel is greater than the pressure p_2 in the upper part, and the water rises along the tube filling an open chamber of the turbine. After this the chamber closes and the disk turns. In the lower part of the vessel the chamber opens automatically and returns the water. After this the chamber closes hermetically, etc. Why will this machine not function perpetually?

258. The following alternative of the machine described in Problem 257 was suggested. Air-tight chambers (Fig. 105) are filled with water in the right-hand side of the disk and lower. At the bottom the chambers open and, in contrast to the machine in Problem 257, the walls of the chambers are automatically retracted into the disk. In the upper portion of the vessel the walls are automatically pushed out and the chambers are filled with water. Otherwise this perpetuum mobile is designed on the same lines as the previous one. Why will it also fail to work?

259. Three vessels with attached bottoms are submerged into water to the same depth (Fig. 106). Each bottom will fall off if the respective vessel is filled with 1 kgf of water. Will the bottom fall of if the vessels are filled with 1 kgf of oil? 1 kgf of mercury? or if a weight of 1 kgf is placed into each vessel?

260. A body is weighed on an accurate analytical balance placed under a glass hood. Will the reading of the balance change if the air is pumped out from the hood?

261. A man carries a tyre tube and decides to make it lighter by using the expulsive force of air (according to Archimedes' principle). For this purpose he inflates the tube, thus increasing its volume. Will his aim be achieved?

262. What error is made in weighing a body with a volume of $V = 1$ litre if when weighed in the air copper weights weighing $G_1 = 800$ gf are placed on the pan of the balance. The specific weight of copper $\gamma_1 = 8.8\ \text{gf/cm}^3$ and of air $\gamma_0 = 1.29$ gf/litre.

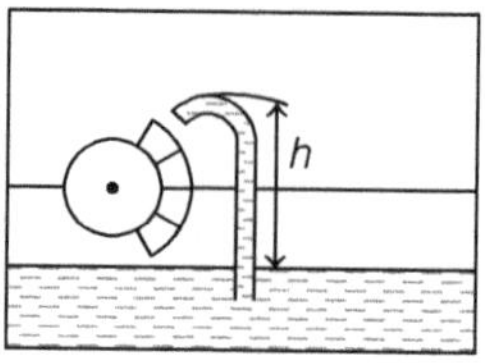

Fig. 105

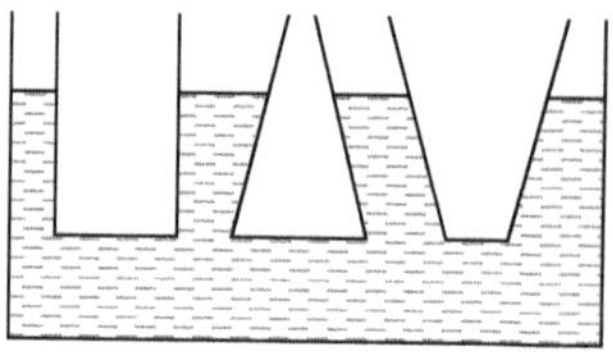

Fig. 106

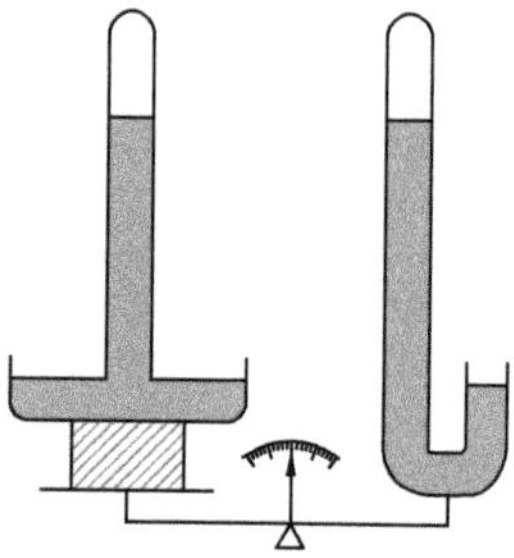

Fig. 107

263. A cup and a U-shaped mercury barometers are brought into equilibrium on a very sensitive balance (Fig. 107). The barometers are made of the same material, have tubes of the same diameter and contain the same amount of mercury. The distance between the soldered ends of the tubes and the upper levels of the mercury in them is also the same. How will the equilibrium of the balance change if the atmospheric pressure grows?

264. Appraise the weight of the Earth's atmosphere.

265. An air mattress is filled with air to a certain pressure exceeding the atmospheric pressure. When will the air pressure in the mattress be greater: when a man stands on it or when he lies on it?

266. A wheel of a motor vehicle is designed as follows. A rubber tube enclosed in a tyre casing is placed onto a metallic rim. The tube is then inflated with air. The air pressure in the top and the bottom of the tube is the same. Besides the air pressure, the rim is acted upon by the force of gravity (Fig. 108). Why does the rim not lower? What holds it in a state of equilibrium?

267. A steam boiler consists of a cylindrical portion and two hemispherical heads (Fig. 109), all of the same radius. The cylinder walls are 0.5 cm thick. All the parts of the boiler are made of the same material. How thick should the walls of the heads be for the strength of all the parts of the boiler to be identical?

268. What shape should a steam boiler be given to obtain the maximum strength with the given wall thickness?

269. Why is ballast always taken on a stratosphere balloon, although extra weight inevitably decreases its ceiling?

1-11. Hydro- and Aerodynamics

270. Two holes with an area of $A = 0.2 \text{ cm}^2$ each are drilled one above the other in the wall of a vessel filled with water. The distance between the holes $H = 50$ cm. Every second $Q = 140 \text{ cm}^3$ of water is poured into the vessel. Find the point where the streams flowing out of the holes intersect.

271. A broad vessel with water stands on a smooth surface. The level of the water in the vessel is h. The vessel together with the water weighs G. The side wall of the vessel has at the bottom a plugged hole (with rounded edges) with an area A. At what coefficient of friction between the bottom and the surface will the vessel begin to move if the plug is removed?

272. When a stream of liquid flows out of a vessel through a hole with an area A_0, the force that acts on the wall with the hole is $2pA_0$ smaller than the force acting on the opposite wall (see Problem 271).

If a tube is inserted into the hole as shown in Fig. 110, the difference between the forces acting on the opposite walls will approximately equal pA_0, since the tube will prevent the motion of the liquid near the walls. On the other hand, the change in the momentum of the liquid flowing out of the vessel in a unit of time is always $2pA$, where A is the cross-sectional area of the stream. How can these two facts be brought into agreement?

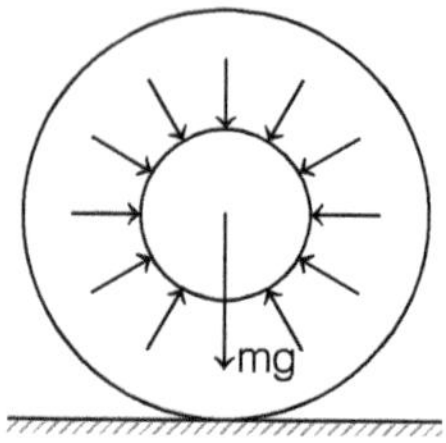

Fig. 108

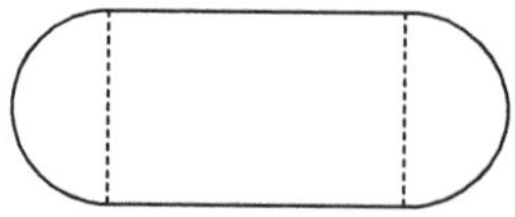

Fig. 109

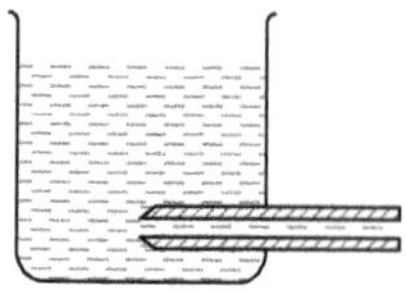

Fig. 110

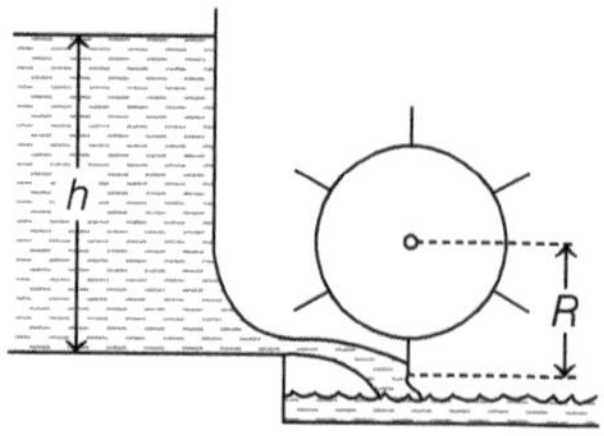

Fig. 111

273. A stream of water flowing out of a pipe with a diameter of $d = 1$ cm at a velocity of $v = 1$ m/s strikes a vertical wall. Determine the force acting on the wall, assuming that the pipe is perpendicular to it and neglecting splashing of the water.

274. A gas flows with a velocity v along a pipe with a cross section A bent through 90 degrees. The density of the gas is ρ. What force does the gas act on the pipe with? Disregard the compression of the gas and friction.

275. Find the force acting on the blade of an undershot wheel (Fig. 111) if the stream after impinging on the blade continues to move with the velocity of the blade. The height of the water head is h, the radius of the wheel is R, the angular velocity of the wheel is ω and the cross-sectional area of the stream is A.

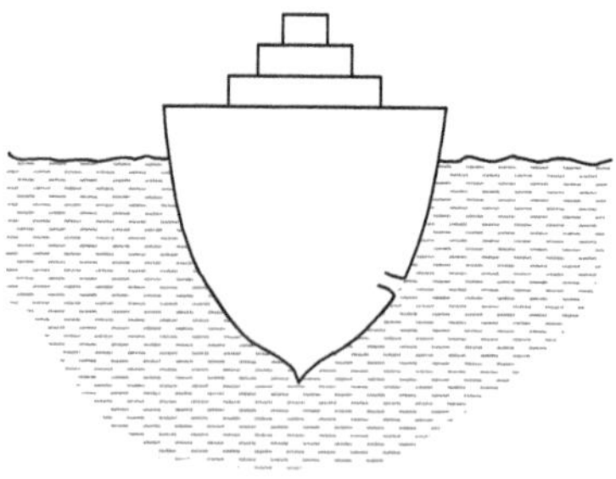

Fig. 112

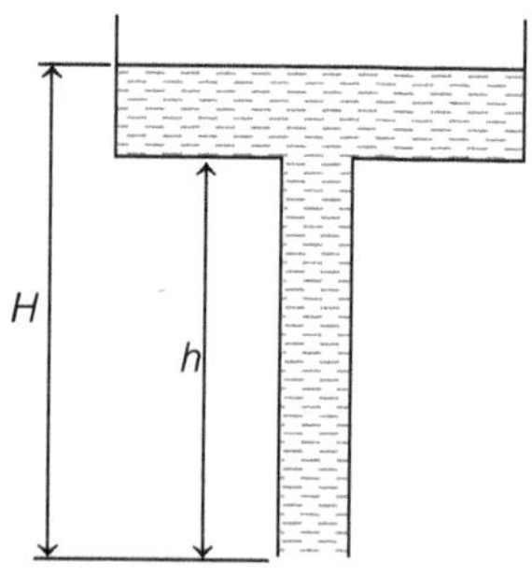

Fig. 113

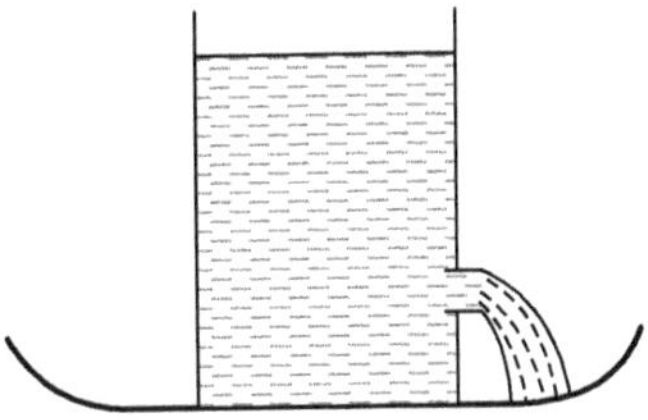

Fig. 114

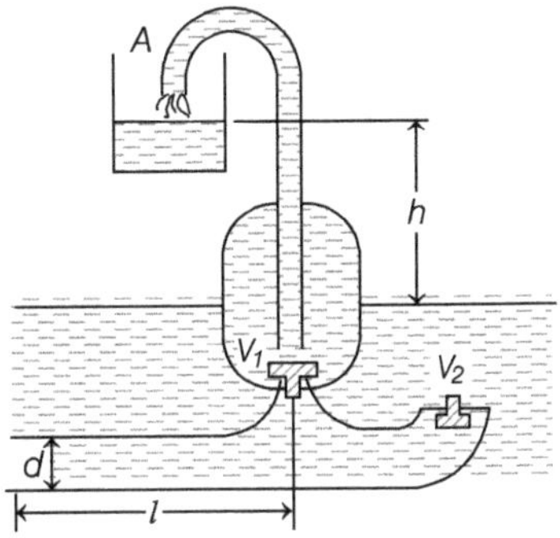

Fig. 115

276. A ship gets a large hole in its underwater portion (Fig. 112). In what direction will it begin to move as a result of this?

277. A liquid flows out of a broad vessel through a narrow pipe (Fig. 113). How are the pressure and velocity of the liquid in the vessel and in the pipe distributed along a vertical?

278. The vessel with water described in the previous problem is suspended from a spring balance. The lower end of the pipe is closed with a plug. How will the reading of the balance change at the first moment after the plug is removed and the liquid begins to flow out?

279. A vessel with water is placed on one pan of a balance in equilibrium (Fig. 114). Will the equilibrium change if the cock is opened? The outflowing water gets onto the same pan on which the vessel is placed.

280. Figure 115 shows a self-acting water lifting device called a hydraulic ram. The principle of its operation is based on the phenomenon of a hydraulic impact—a sharp increase in the pressure of a liquid flowing along a tube when its flow is suddenly stopped, for example, by the shutting of a valve that discharges the water from the tube.

A tube with a length of $l = 2$ metres and a diameter of $d = 20$ cm is lowered into a brook with a current velocity of $v = 400$ cm/s.

First let valve V_2 be open and valve V_1 be shut. A sharp increase in pressure will cause valve V_1 to open (valve V_2 will close at the same time) and the water will flow up into vessel *A*. The pressure drops, valve V_1 shuts and V_2 opens. The water in the tube assumes its course and the phenomenon is repeated in the previous sequence.

Find the amount of water raised by the ram in one hour to a height of $h = 30$ metres if each valve opens thirty times a minute.

281. During storms, when the velocity of the wind is very high, it tears off the roofs of buildings. Two cases can be distinguished: (1) if the roof is fastened more firmly at points *A* and *B* than at ridge *C*, the wind will break the roof along ridge *C* and raise both halves up (Fig. 116*a*);

(2) if the roof is secured more firmly at the ridge and less firmly at points A and B, the wind will first lift the roof up and then carry it aside (Fig. 116*b*). How do you explain this phenomenon?

282. Why will a light celluloid ball placed in a stream of a gas or water issuing at a high velocity from a tube with a narrow neck freely hover in this stream (Fig. 117)?

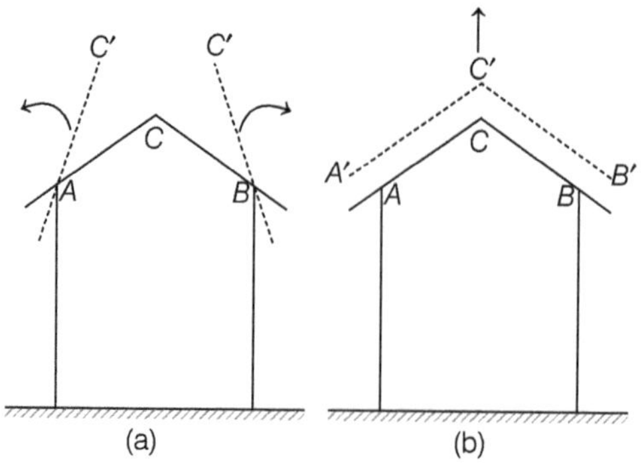

Fig. 116

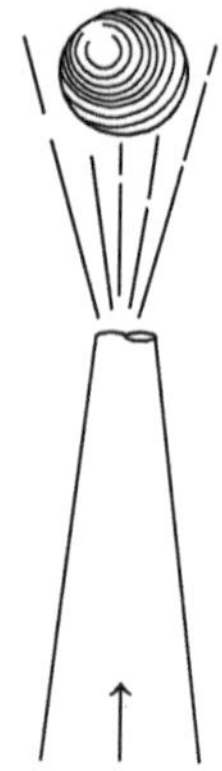

Fig. 117

283. The demonstration device shown in Fig. 118 consists of two disks A and B. The hole in the centre of disk A is connected by a pipe to a cylinder containing compressed air. Disk B hangs on three pins along which it can move freely up and down. If a stream of compressed air is passed through the pipe, the lower disk will begin to knock against the upper one. Explain the cause of this phenomenon.

284. The bottom of a broad vessel is provided with a narrow tube through which the water can flow out of the vessel (Fig. 119). A screen is placed between the vessel and the tube. If a light ball is submerged to the bottom of the vessel the water flows out of it, and the ball will not rise to the surface. If the outflow of the water is stopped, the ball will immediately rise to the surface. Why? (This experiment can easily be conducted in a sink with a ping pong ball).

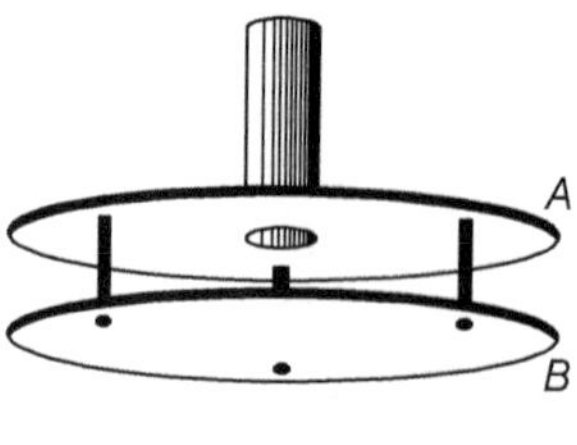

Fig. 118

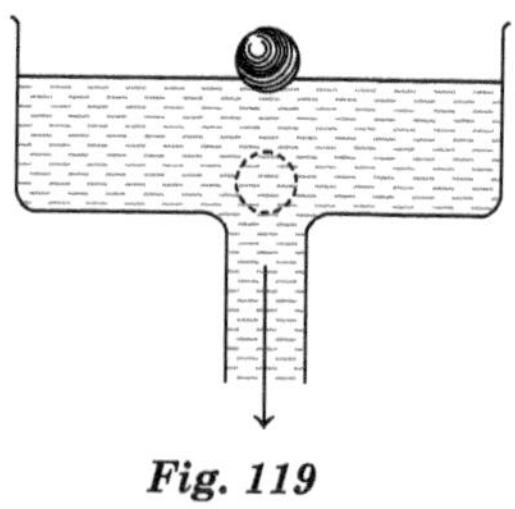

Fig. 119

285. A pump is designed as a horizontal cylinder with a piston having an area of A and an outlet orifice having an area of a arranged near the cylinder axis. Determine the velocity of outflow of a liquid from the pump if the piston moves with a constant velocity under the action of a force F. The density of the liquid is ρ.

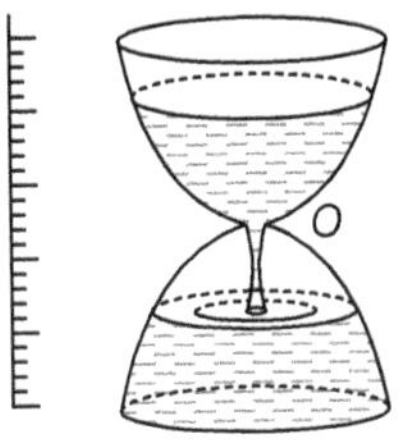

Fig. 120

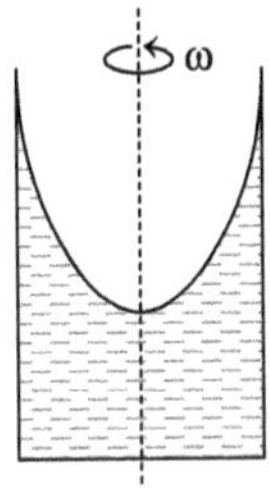

Fig. 121

286. In Problem 285, the velocity v may become infinitely high even with a small force F when $a \to A$.. Explain this paradoxical phenomenon.

287. A water clock (clepsydra) used in ancient Greece is designed as a vessel with a small orifice O (Fig. 120). The time is determined according to the level of the water in the vessel. What should the shape of the vessel be for the time scale to be uniform?

288. A cylindrical vessel with a liquid rotates with an angular velocity ω around a vertical axis (Fig. 121). Determine the change of pressure in the horizontal cross section of the vessel depending on the distance to the axis of rotation.

Note. Apply the method described in solving Problem 203.

289. Find the shape of the surface of a liquid in a cylindrical vessel rotating with an angular velocity ω around a vertical axis (i.e., the height of the liquid level depending on the distance to the axis of rotation).

290. Why do the tea leaves gather in the middle of a glass after stirring?

2

HEAT : MOLECULAR PHYSICS

2-1. Thermal Expansion of Solids and Liquids

291. An iron tyre is to be fitted onto a wooden wheel 100 cm in diameter. The diameter of the tyre is 5 mm smaller than that of the wheel. How much should the temperature of the tyre be increased for this purpose? The coefficient of linear expansion of iron $\alpha_1 = 12 \times 10^{-6}\ \text{deg}^{-1}$.

292. Why is only iron or steel used as reinforcement in concrete structures, while other metals, duralumin for example, are never employed?

293. Why is a greater time required to measure the temperature of a human body with a thermometer than to shake it down?

294. The height of a mercury column measured with a brass scale at a temperature of t_1 is H_1. What height H_0 will the mercury column have at $t_0 = 0°\text{C}$? The coefficients of linear expansion of brass α and of volume expansion of mercury γ are known.

295. How can the temperature of a human body be measured with a thermometer if the temperature of the ambient air is + 42°C?

296. Determine the lengths of an iron and a copper ruler l_0' and l_0'' at $t = 0°\text{C}$ if the difference in their lengths at $t_1 = 50°\text{C}$ and $t_2 = 450°\text{C}$ is the same in magnitude and equal to $l = 2$ cm. The coefficient of linear expansion of iron $\alpha_1 = 12 \times 10^{-6}\ \text{deg}^{-1}$ and of copper $\alpha_2 = 17 \times 10^{-6} \text{deg}^{-1}$.

297. The period of oscillations of a pendulum depends on the length, which changes with the temperature. How should the pendulum be suspended so that its length does not change with the temperature?

298. A glass cylinder can contain $m_0 = 100$ g of mercury at a temperature of $t_0 = 0°\text{C}$. When $t_1 = 20°\text{C}$, the cylinder can contain $m_1 = 99.7$ g of mercury. In both cases the temperature of the mercury is assumed to be equal to that of the cylinder. Use these data to find the coefficient of linear expansion of glass α, bearing in mind that the coefficient of volume expansion of mercury $\gamma_1 = 18 \times 10^{-5}\ \text{deg}^{-1}$.

299. A clock with a metallic pendulum is $\tau_1 = 5$ seconds fast each day at a temperature of $t_1 = +15°C$ and $\tau_2 = 10$ seconds slow at a temperature of $t_2 = +30°C$. Find the coefficient α of thermal expansion of the pendulum metal, remembering that the period of oscillations of a pendulum $T = 2\pi\sqrt{\frac{l}{g}}$, where l is the length of the pendulum and g is the gravity acceleration.

2-2. The Law of Conservation of Energy.

Thermal Conductivity

300. A brick with a mass m is lowered onto a cart with a mass M that moves rectilinearly with a constant velocity of v_0. Find the amount of heat liberated in this case.

301. An iron washer slides down a weightless rubber cord with a length l_0 (Fig. 122). The force of friction between the cord and the washer is constant and equal to f. The coefficient of elasticity of the cord k is known. Find the amount of heat Q evolved. What part of the work produced by friction on the cord will be converted into heat?

302. A refrigerator consuming P watts converts q litres of water into ice in τ minutes at a temperature t.

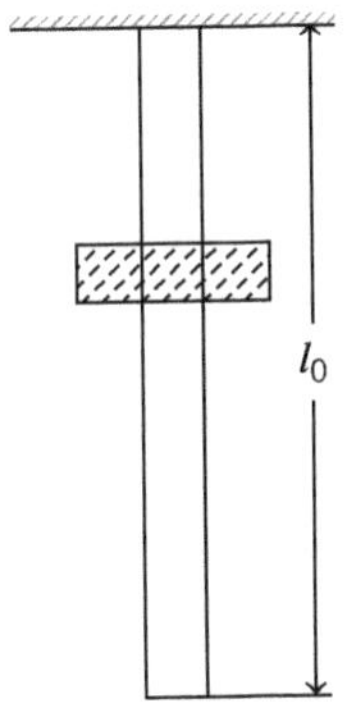

Fig. 122

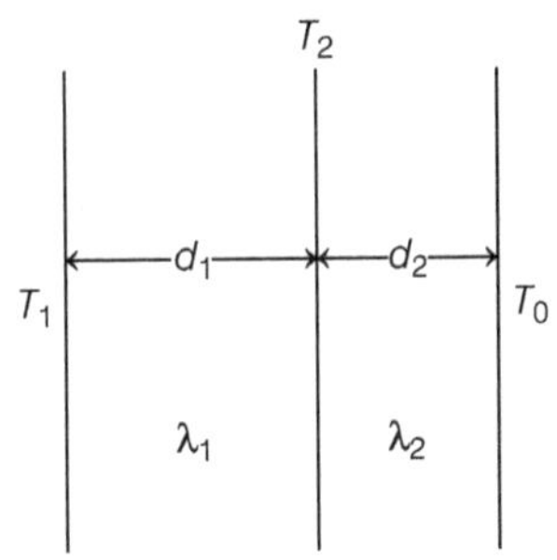

Fig. 123

What quantity of heat will be liberated by the refrigerator in a room during this time if the heat capacity of the refrigerator can be neglected?

303. Will the temperature in a room drop if the door of a working refrigerator is opened?

304. It is easiest to heat premises by means of electric appliances. Is this method the most advantageous from the energetic point of view?

305. Equal quantities of salt are dissolved in two identical vessels filled with water. In one case the salt is one large crystal and in the other—powder.

In which case will the temperature of the solution be higher after the salt is completely dissolved, if in both cases the salt and the water originally had the same temperatures?

306. It is known that if water is heated or cooled with certain care it can retain its liquid state at temperatures below 0°C and higher than + 100°C.

A calorimeter with a heat capacity of $q = 400$ cal/deg contains $m_1 = 1$ kg of water cooled to $t_1 = -10°$C. Next $m_2 = 100$ g of water overheated to $t_2 = +120°$C is added to it. What is the temperature in the calorimeter?

307. An incandescent lamp consuming $P = 54$ watts is immersed into a transparent calorimeter containing $V = 650$ cm^3 of water. During $\tau = 3$ minutes the water is heated by $t = 3.4°$C. What part Q of the energy consumed by the lamp passes out of the calorimeter in the form of radiant energy?

308. The area of a brick wall facing a street is $A = 12$ m^2 and its thickness $d = 1$ metre. The temperature of the outside air $T_0 = -15°$C and that of the air in the room $T = +15°$C. What amount of heat is lost from the room in 24 hours? The coefficient of thermal conductivity of brick is $\lambda = 0.003$ cal/cm $\times$ s$\cdot$deg.

309. A wall consists of two adjoining panels made of different materials. The coefficients of thermal conductivity and the thicknesses of the panels are λ_1, d_1 and λ_2, d_2, respectively (Fig. 123). The temperatures of the external surfaces of the wall are T_1 and T_0 (where $T_0 > T_1$) and are kept constant. Find the temperature T_2 on the plane between the panels.

310. Assuming in Problem 309 that the panels have the same thickness d, determine the coefficient of thermal conductivity of the wall.

311. A wall consists of alternating blocks with a length d and coefficients of thermal conductivity λ_1 and λ_2 (Fig. 124). The cross-sectional areas of the blocks are the same. Determine the coefficient of thermal conductivity of the wall.

312. Two walls *I* and *II* of the same thickness are made of heterogeneous metals, as shown in Figs. 125 and 126. In what case will the coefficient of thermal conductivity be greater?

313. During one second m grammes of water boils away in a pan. Assuming that heat is received by the water only through the bottom of the pan and neglecting the transfer of heat from the pan walls and the water surface to the ambient air, determine the temperature T of the pan bottom in contact with the heater.

The area of the pan bottom is A and its thickness d. The coefficient of thermal conductivity is λ.

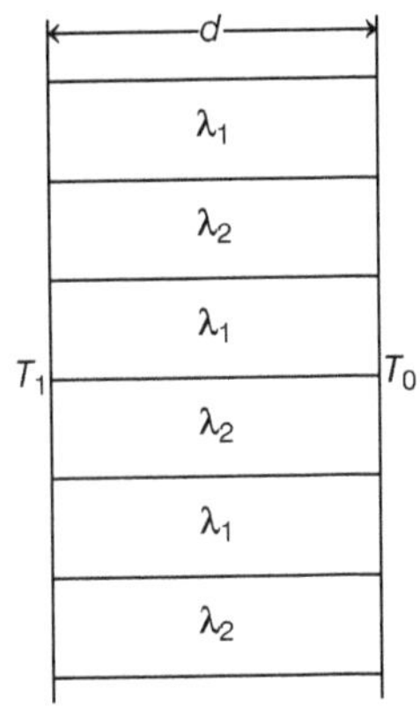

Fig. 124

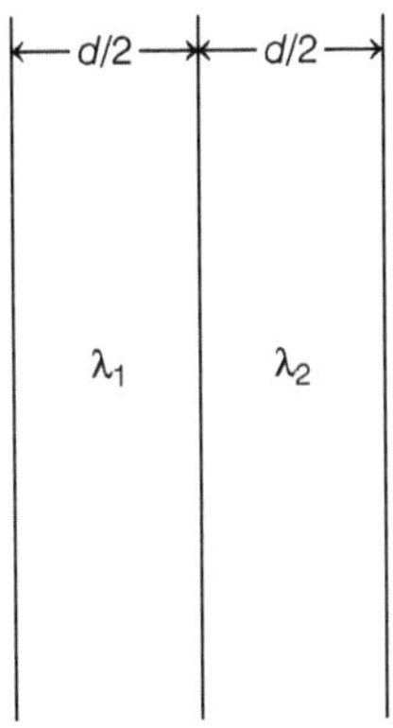

Fig. 125

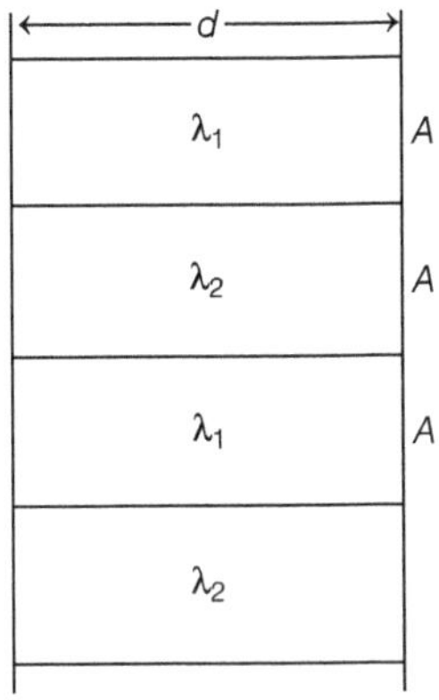

Fig. 126

2-3. Properties of Gases

314. The cap of a fountain-pen is usually provided with a small orifice. If it is clogged, the ink begins to leak out of the pen. What is the cause of this phenomenon?

315. A barometer gives wrong readings because some air is present above the mercury column. At a pressure of $p_{01} = 755$ mm Hg the barometer shows $p_1 = 748$ mm, and at $p_{02} = 740$ mm it shows $p_2 = 736$ mm. Find the length l of the barometer tube (Fig. 127).

316. A glass tube with a length of $l = 50$ cm and a cross section of $A = 0.5\ \text{cm}^2$ is soldered at one end and is submerged into water as shown in Fig. 128.

What force F should be applied to hold the tube under the water if the distance from the surface of the water to the soldered end is $h = 10$ cm and the atmospheric pressure $p_0 = 760$ mm Hg? The weight of the tube $G = 15$ gf.

317. A narrow tube open at both ends is passed through the cork of a vessel filled with water. The tube does not reach the bottom of the vessel (Mariotte's vessel shown in Fig. 129). Draw a diagram showing how the pressure p of the air in the vessel depends on the quantity of water Q that has flowed out.

318. Upon each double stroke a piston pump sucks in a volume v_0 of air. When this pump is used to evacuate the air from a vessel with a volume V it performs n double strokes.

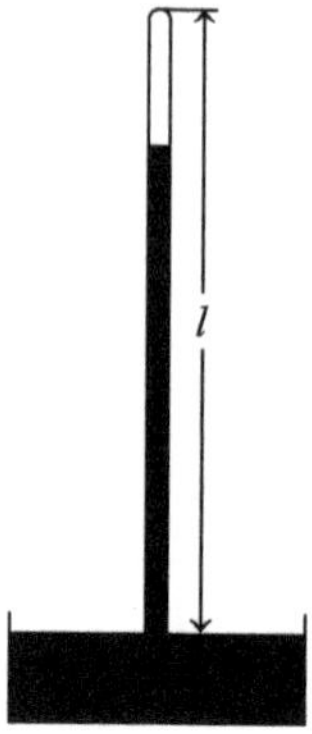

Fig.127

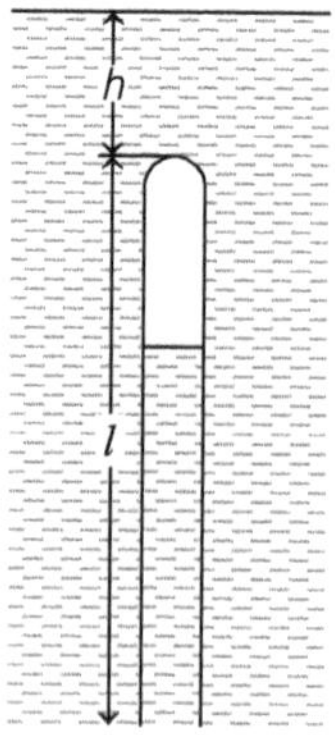

Fig. 128

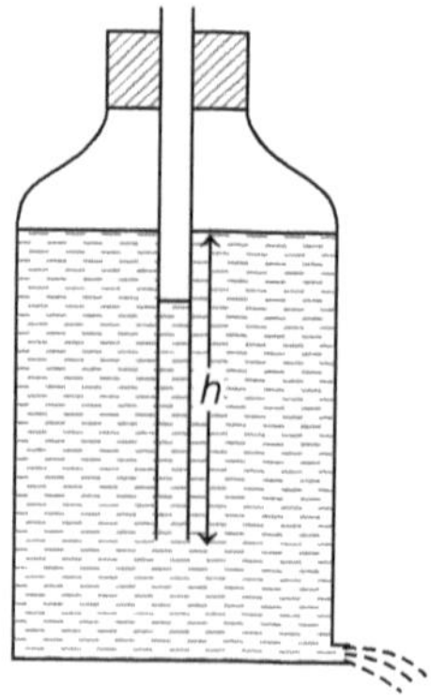

Fig. 129

The initial pressure inside the vessel p_0 is equal to atmospheric pressure.

After that, another pump with the same active volume v_0 begins to suck in the atmospheric air, also making n double strokes. What will the pressure in the vessel be?

319. A mercury column with a length l is in the middle of a horizontal tube with a length L closed at both ends. If the tube is placed vertically, the mercury column will shift through the distance Δl from its initial position.

At what distance will the centre of the column be from the middle of the tube if one end of the tube placed horizontally is opened? if the upper end of the tube placed vertically is opened? if the lower end of the tube placed vertically is opened?

The atmospheric pressure is H cm Hg. The temperature remains the same.

320. Bearing in mind that, according to Avogadro's law, the volume of one gramme-molecule of any gas under standard conditions (temperature 0°C and pressure 1 atm) is 22.4 litres, find the constant in the equation of state of an ideal gas (the Clapeyron-Mendeleyev equation) for a quantity of a gas equal to one mole and prove that this constant is the same for all gases.

321. Write the equation of state for an arbitrary mass of an ideal gas whose molecular weight μ is known.

322. How would the pressure inside a fluid change if the forces of attraction between the molecules suddenly disappeared?

323. A vessel contains one litre of water at a temperature of 27°C. What would the pressure in the vessel be if the forces of interaction between the water molecules suddenly disappeared?

324. Is the pressure the same inside a gas and at the wall of the vessel containing this gas?

325. Is the concentration of gas molecules inside a vessel and at its wall the same?

326. Find the temperature of a gas contained in a closed vessel if its pressure increases by 0.4 per cent of the initial pressure when it is heated by 1°C.

327. A thin-walled rubber sphere weighing $G = 50$ gf is filled with nitrogen and submerged into a lake to a depth of $h = 100$ metres. Find the mass of the nitrogen m if the sphere is in a position of equilibrium. Will equilibrium be stable? The atmospheric pressure $p_0 = 760$ mm Hg. The temperature in the lake $t = +4°$C. Disregard the tension of the rubber.

328. Two hollow glass balls are connected by a tube with a drop of mercury in the middle. Can the temperature of the ambient air be appraised from the position of the drop?

329. A cylinder closed at both ends is separated into two equal (42 cm each) parts by a piston impermeable to heat. Both the parts contain the same masses of gas at a temperature of 27°C and a pressure of 1 atm.

How much should the gas be heated in one part of the cylinder to shift the piston by 2 cm? Find the pressure p of the gas after shifting of the piston.

330. Dry atmospheric air consists of nitrogen (78.09 per cent by volume), oxygen (20.95 per cent), argon (0.93 per cent) and carbon dioxide (0.03 per cent). Disregarding the negligible admixtures of other gases (helium, neon, krypton, xenon), determine the composition of the air (in per cent) by weight.

331. Find the mean (effective) molecular weight of dry atmospheric air if the composition of the air in per cent is known (see Problem 330).

332. The density of the vapour of a compound of carbon and hydrogen is 3 g/lit at 43°C and 820 mm Hg. What is the molecular formula of this compound?

333. When will the change in the pressure of a gas be greater—if it is compressed to a certain extent in a heat-impermeable envelope or upon isothermal compression?

334. A gas that occupies a volume of $V_1 = 1$ lit at a pressure of $p_1 = 1$ atm expands isothermally to a volume of $V_2 = 2$ lit. Then, the pressure of the gas is halved at the same volume. Next the gas expands at a constant pressure to $V_4 = 4$ litres.

Draw a diagram of p versus V and use it to determine the process in which the gas performs the greatest work. How does the temperature change?

335. A cyclic process 1-2-3-1 depicted on a V-t diagram (Fig. 130) is performed with a certain amount of an ideal gas. Show the same process on a $p-V$ diagram and indicate the stages when the gas received and when it rejected heat.

336. A gas heated geyser consumes $V_0 = 1.8\ m^3$ of methane (CH_4) an hour. Find the temperature t_2, of the water heated by the geyser if the water flows out at a rate of $v = 50$ cm/s. The diameter of the stream $D = 1$ cm, the initial temperature of the water and the gas $t_1 = 11°C$ and the calorific value of methane $q_0 = 13{,}000$ cal/g. The gas in the tube is under a pressure of $p = 1.2$ atm. The efficiency of the heater $\eta = 60$ per cent.

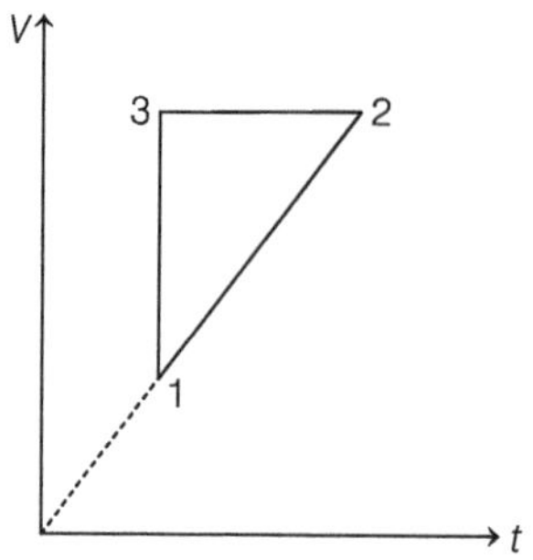

Fig. 130

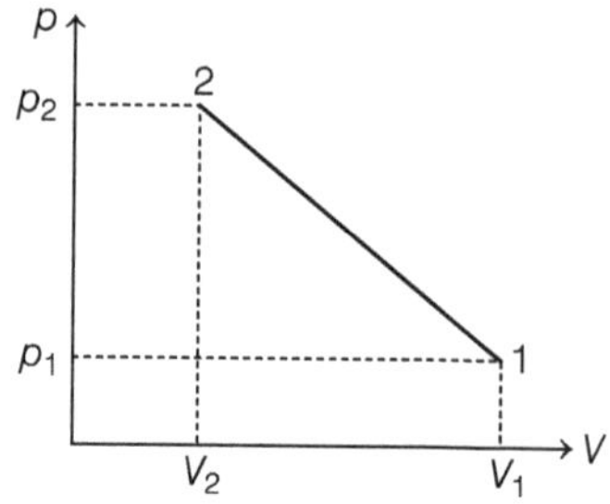

Fig 131

337. A closed vessel impermeable to heat contains ozone (O_3) at a temperature of $t_1 = 527°C$. After some time the ozone is completely converted into oxygen (O_2).

Find the increase of the pressure in the vessel if $q = 34{,}000$ cal have to be spent to form one gramme-molecule of ozone from oxygen. Assume that the heat capacity of one gramme-molecule of oxygen at a constant volume is equal to $C_V = 5$ cal/deg · mole.

338. Twenty grammes of helium in a cylinder under a piston are transferred infinitely slowly from a state with a volume of $V_1 = 32$ lit and a pressure of $p_1 = 4.1$ atm to a state with $V_2 = 9$ lit and $p_2 = 15.5$ atm. What maximum temperature will the gas reach in this process if it is depicted on the $p-V$ diagram as a straight line (Fig. 131)?

339. Will the energy of the air in a room increase if a stove is heated in it?

Note. Assume the energy of a unit of mass of the air u to be proportional to the absolute temperature, i.e., $u = cT$.

340. The temperature in a room with a volume of 30 m^3 rose from 15°C to 25°C. How much will the mass of the air in the room change if the atmosphere pressure $p = 1$ atm? The molecular (mean) weight of the air is $\mu = 28.9$ g/mole.

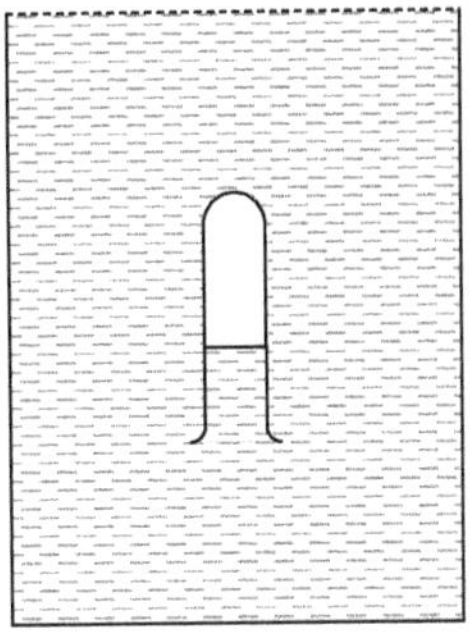

Fig. 132

341. A small tube filled with air and open at the bottom is placed into a water-filled open vessel with a screen on the top (Fig. 132). The tube cannot be turned over. Draw a diagram showing how the depth of submergence of the tube depends on the temperature of the water if the temperature first slowly increases and then gradually decreases.

342. A cylinder contains $m = 20$ g of carbon dioxide under a heavy piston. The gas is heated from the temperature $t_1 = 20°$C to $t_2 = 108°$C. What work is performed by the gas?

343. What quantity of heat should be imparted to carbon dioxide (see Problem 342) expanding at a constant pressure as a result of heating? The molar heat of the carbon dioxide (heat capacity of one gramme-molecule) at a constant volume is $C_v = 6.864$ cal/mole·deg.

2-4. Properties of Liquids

344. Is it more difficult to compress a litre of air to three atmospheres, or a litre of water?

345. How can a minimum and a maximum thermometers be made using the phenomena of wettability and unwettability?

346. The surface layer of a liquid is frequently likened to a stretched rubber film. In what respect does this analogy disagree with reality?

347. To remove a grease spot from a fabric it is good to apply some petrol to the edges of the spot, while the spot itself should never be wetted with petrol immediately. Why not?

348. Why is moisture retained longer in soil if it is harrowed?

349. Skiing boots are usually warmed up to ensure that they absorb grease better. Should the boots be warmed up outside or inside?

350. Why can an iron be used to remove greasy spots from clothing?

351. Why do drops of water appear at the end of a piece of firewood in the shade when it is being dried in the sun?

352. A vessel whose bottom has round holes with a diameter of $d = 0.1$ mm is filled with water. Find the maximum height of the water level h at which the water does not flow out. The water does not wet the bottom of the vessel.

353. A soapy film is stretched over a rectangular vertical wire frame (Fig. 133). What forces hold section *abcd* in equilibrium?

354. A cube with a mass $m = 20$ g wettable by water floats on the surface of water. Each face of the cube a is 3 cm long. What is the distance between the lower face of the cube and the surface of the water?

355. The end of a capillary tube with a radius r is immersed into water. What amount of heat will be evolved when the water rises in the tube?

356. A capillary tube is lowered into a vessel with a liquid whose vapour pressure may be neglected. The density of the liquid is ρ. The vessel and the tube are in a vacuum under the bell of an air pump (Fig. 134). Find the pressure inside the liquid in the capillary tube at a height h above the level of the liquid in the vessel.

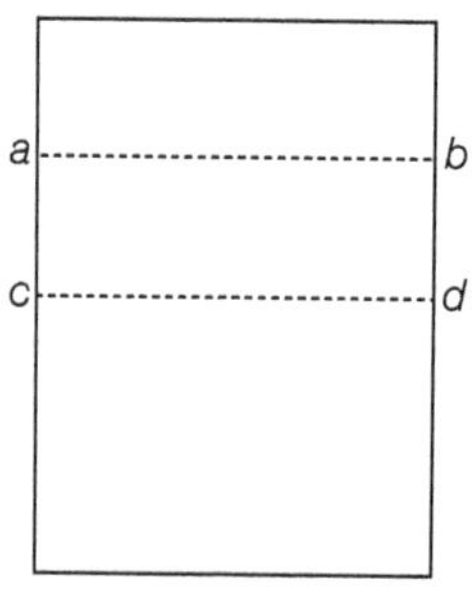

Fig. 133

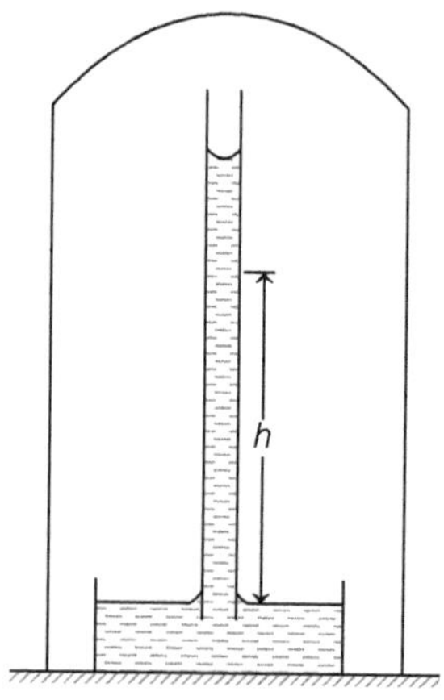

Fig. 134

357. The course of reasoning given below is usually followed to prove that the molecules of the surface layer of a liquid have surplus potential energy. A molecule inside the liquid is acted upon by the forces of attraction from the other molecules which compensate each other on the average. If a molecule is singled out on the surface, the resulting force of attraction from the other molecule is directed into the liquid. For this reason the molecule tends to move into the liquid, and definite work should be done to bring it to the surface. Therefore, each molecule of the surface layer has excess potential energy equal to this work.

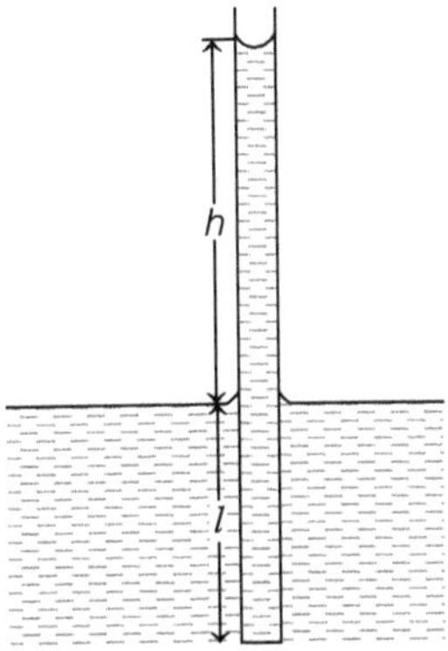

Fig. 135

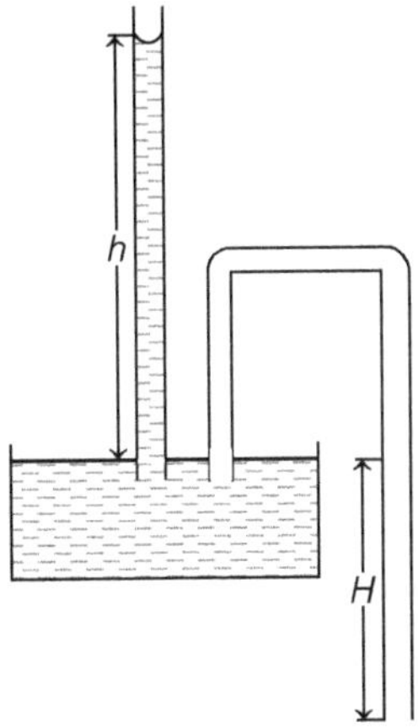

Fig. 136

The average force that acts on any molecule from the side of all the others, however, is always equal to zero if the liquid is in equilibrium. This is why the work done to move the liquid from a depth to the surface should also be zero. What is the origin, in this case, of the surface energy?

358. One end of a glass capillary tube with a radius $r = 0.05$ cm is immersed into water to a depth of $h = 2$ cm. What pressure is required to blow an air bubble out of the lower end of the tube?

359. A glass capillary tube with an internal diameter of 0.5 mm is immersed into water. The top end of the tube projects by 2 cm above the surface of the water. What is the shape of the meniscus?

360. Water rises to a height h in a capillary tube lowered vertically into water to a depth l (Fig. 135). The lower end of the tube is closed, the tube is then taken out of the water and opened again. Determine the length of the water column remaining in the tube.

361. Two capillary tubes of the same cross section are lowered into a vessel with water (Fig. 136). The water in the straight tube rises to a height h. What will the level of the water be in the bent tube and what form will the meniscus take? The lower end of the bent tube is at a depth H below the level of the water in the vessel.

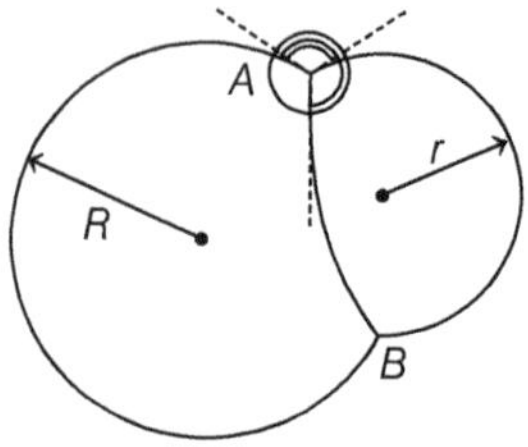

Fig. 137

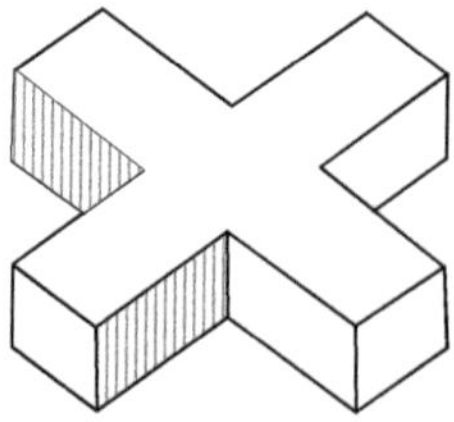

Fig. 138

Consider the following five cases:

(1) $H > h$

(2) $H = h$

(3) $0 < H < h$

(4) $H = 0$

(5) $H < 0$ (the end of the bent tube is above the level of the water in the vessel).

362. A soap-bubble with a radius r is placed on another bubble with a radius R (Fig. 137). What will the form of the soapy film separating the two bubbles be? What angles will be formed between the films at the points of contact?

363. A wooden cross floats in water. Each arm of the cross is coated on one side with varnish (Fig. 138). The water will rise to different heights on both sides of each arm owing to the different wettability of the wood and the varnish.

The wetting angle will be different, and therefore the horizontal component of the force of surface tension F will also be different on each side of an arm (Fig. 139). Will this cause the cross to rotate?

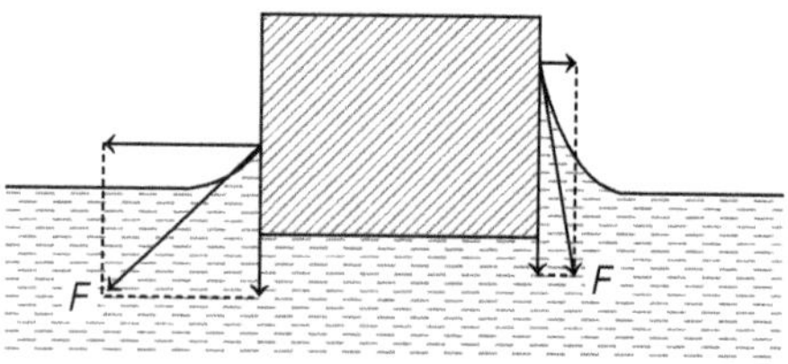

Fig. 139

364. Light bodies wetted by water (for example, two matches) and floating on its surface are mutually attracted. This also occurs if bodies are not wetted (matches coated with a thin layer of paraffine). If one body is wetted and the other not, they will be repulsed. How can these phenomena be explained?

2-5. Mutual Conversion of Liquids and Solids

365. Water in a glass freezes at 0°C. If this water is separated into fine drops, the water in them can be overcooled to – 40°C.

For example, water drops which clouds are composed of usually begin to freeze at a temperature below – 17°C. How can these facts be explained?

366. A vessel with 100g of water at a temperature of 0°C is suspended in the middle of a room. In 15 minutes the temperature of the water rises to 2°C. When ice equal in weight to the water is placed into the vessel, it melts during ten hours. May these data be used to appraise the specific heat of fusion of ice H?

367. Two identical pieces of ice fly toward each other with equal velocities and are converted into vapour upon impact. Find the minimum possible velocities of the pieces before the impact if their temperature is – 12°C.

368. A calorimeter contains ice. Determine the heat capacity of the calorimeter if $Q_1 = 500$ cal are required to heat it together with its contents from 270 to 272°K, and $Q_2 = 16{,}600$ cal from 272 to 274° K.

369. A calorimeter contains 400 g of water at a temperature of +5°C. Then, 200 g of water at a temperature of +10°C are added and 400 g of ice at a temperature of – 60°C are put in. What is the temperature in the calorimeter?

370. Ice with a mass of $m_2 = 600$ g and at a temperature of $t_2 = -10°$ C is placed into a copper vessel heated to $t_1 = 350°$ C. As a result, the vessel now contains $m_3 = 550$ g of ice mixed with water. Find the mass of the vessel. The specific heat of copper $c_1 = 0.1$ cal/deg · g.

371. When a small ice crystal is placed into overcooled water it begins to freeze instantaneously.

(1) What amount of ice is formed from $M = 1$ kg of water overcooled to $t = -8°$ C ?

(2) What should the temperature of the overcooled water be for all of it to be converted into ice? Disregard the relation between the heat capacity of the water and the temperature.

372. One hundred grammes of ice at a temperature of 0°C are placed into a heat-impermeable envelope and compressed to $p = 1{,}200$ atm.

Find the mass of the melted part of the ice if the melting point decreases in direct proportion to the pressure, and if it lowers by 1°C when the pressure is increased by 138 atm.

2-6. Elasticity and Strength

373. A copper ring with a radius of $r = 100$ cm and a cross-sectional area of $A = 4\ \text{mm}^2$ is fitted onto a steel rod with a radius $R = 100.125$ cm. With what force F will the ring be expanded if the modulus of elasticity of copper $E = 12{,}000\ \text{kgf/mm}^2$? Disregard the deformation of the rod.

374. What work can be performed by a steel rod with a length l and a cross-sectional area A when heated by t degrees?

375. A wire with a length of $2l$ is stretched between two posts. A lantern with a mass M is suspended exactly from the middle of the wire. The cross-sectional area of the wire is A and its modulus of elasticity E. Determine the angle α of sagging of the wire, considering it to be small (Fig. 140).

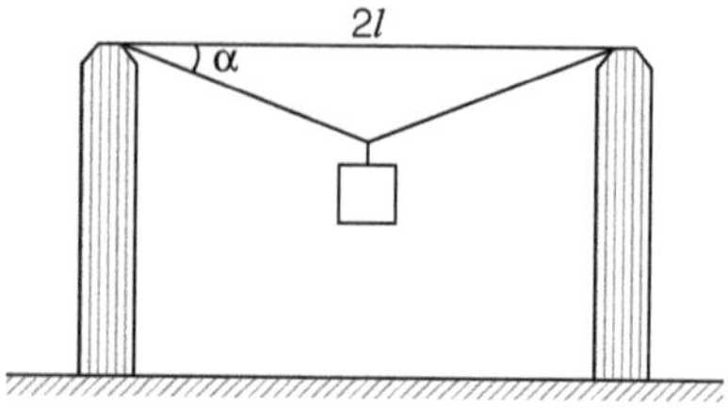

Fig. 140

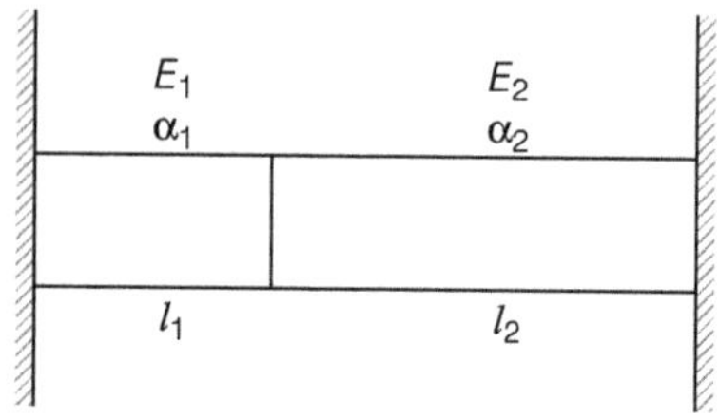

Fig. 141

376. A steel rod with a cross section $A = 1\ \text{cm}^2$ is tightly fitted between two stationary absolutely rigid walls. What force F will the rod act with on the walls if it is heated by $\Delta t = 5°\text{C}$? The coefficient of linear thermal expansion of steel $\alpha = 1.1 \times 10^{-5}\ \text{deg}^{-1}$ and its modulus of elasticity $E = 20{,}000\ \text{kgf/mm}^2$.

377. Two rods made of different materials are placed between massive walls (Fig. 141). The cross section of the rods is A and their respective lengths l_1 and l_2. The rods are heated by t degrees.

Find the force with which the rods act on each other if their coefficients of linear thermal expansion α_1 and α_2 and the moduli of elasticity of their materials E_1 and E_2 are known. Disregard the deformation of the walls.

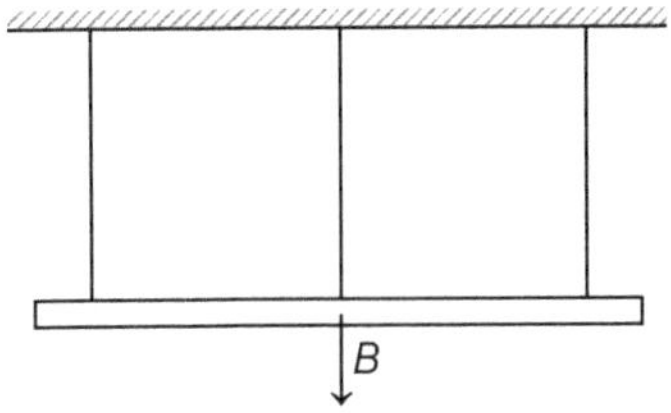

Fig 142

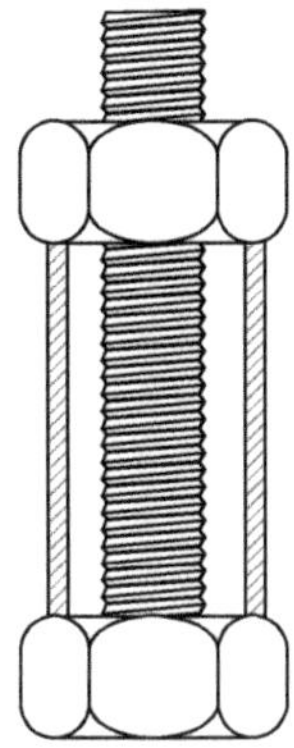

Fig 143

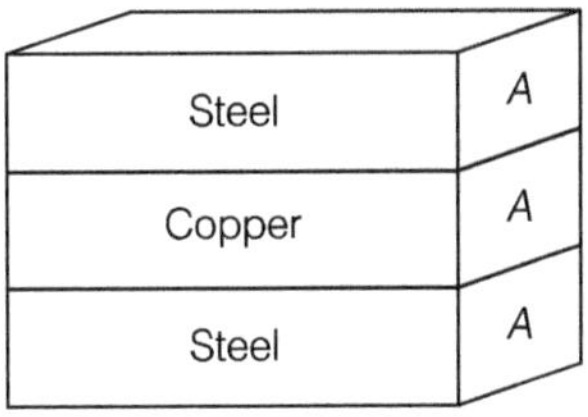

Fig. 144

378. A homogeneous block with a mass $m = 100$ kg hangs on three vertical wires of equal length arranged symmetrically (Fig. 142). Find the tension of the wires if the middle wire is of steel and the other two are of copper. All the wires have the same cross section. Consider the modulus of elasticity of steel to be double that of copper.

379. A reinforced-concrete column is subjected to compression by a certain load. Assuming that the modulus of elasticity of concrete E_c is one-tenth of that of iron E_i, and the cross-sectional area of the iron is one-twentieth of that of concrete, find the portion of the load acting on the concrete.

380. A steel bolt is inserted into a copper tube as shown in Fig. 143. Find the forces induced in the bolt and in the tube when the nut is turned through one revolution if the length of the tube is l, the pitch of the bolt thread is h and the cross-sectional areas of the bolt and the tube are A_b and A_t, respectively.

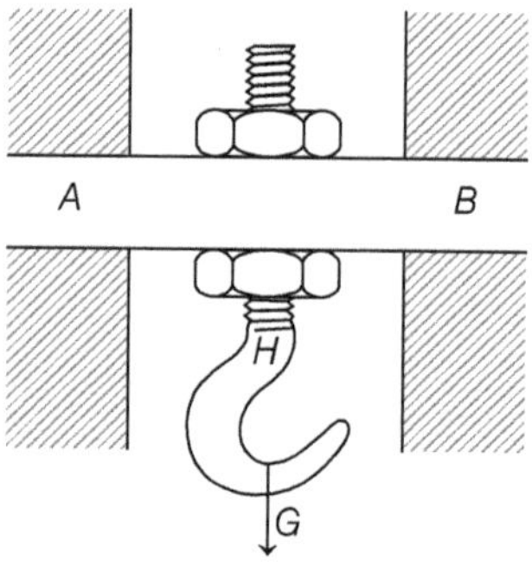

Fig. 145

381. A copper plate is soldered to two steel plates as shown in Fig. 144. What tensions will arise in the plates if the temperature is increased by t°C? All three plates have the same cross sections.

382. Find the maximum permissible linear velocity of a rotating thin lead ring if the ultimate strength of lead $\sigma_u = 200$ kgf/cm^2 and its density $\rho = 11.3$ g/cm^3.

383. An iron block AB has both ends fixed. Hook H is fastened with two nuts in a hole in the middle of the block (Fig. 145). The block is clamped by the nuts with a force F_0.

What forces will act on the upper and lower nuts from the side of the block if the hook carries a load whose weight can change from zero to $G = 2F_0$? Disregard sagging of the block and the weight of the hook.

2-7. Properties of Vapours

384. Water vapour amounting to 150 g at a temperature of +100°C is admitted into a calorimeter containing 100 g of ice at a temperature of – 20° C. What will the temperature inside the calorimeter be if its heat capacity is 75 cal/deg?

385. Why does a strong jet of steam burst out from a boiling kettle when the gas burner is switched off, while no steam was visible before?

386. Prove that the density of water vapour at a temperature near room temperature, expressed in g/m^3, is approximately equal to the pressure of water vapour expressed in millimetres of mercury.

387. The pressure of saturated water vapours in a hermetic vessel increases with the temperature as shown in Fig. 146. The pressure of an ideal gas at constant volume is directly proportional to the temperature. Using a table of the properties of saturating water vapours (Table 2), find whether the equation of state of an ideal gas can be used to calculate the density or the specific volume of saturated water vapours. Explain the result obtained.

388. Nine grammes of water vapour at a temperature of 30°C

Table 2. Properties of Saturating Water Vapours

t (°C)	Pressure (kgf/cm^2)	Specific volume of vapour (m^3/kg)	t (°C)	Pressure (kgf / cm^2)	Specific volume of vapour (m^3/kg)
17.2	0.02	68.3	151.1	5	0.3818
45.4	0.1	14.96	158.1	6	0.3214
59.7	0.2	7.80	164.2	7	0.2778
75.4	0.4	4.071	169.6	8	0.2448
85.45	0.6	2.785	174.5	9	0.2189
93.0	0.8	2.127	179.0	10	0.1980
96.2	0.9	1.905	187.1	12	0.1663
99.1	1	1.726	194.1	14	0.1434
100	1.0333	1.674	200.4	16	0.1261
116.3	1.8	0.996	206.2	18	0.1125
119.6	2	0.902	211.4	20	0.1015
132.9	3	0.617	232.8	30	0.0679
142.9	4	0.4708	249.2	40	0.0506

are compressed isothermally in a cylinder under a piston. At what volume will the vapour begin to condense?

Note. Use Table 2.

389. The relative humidity in a room is 10 per cent at a temperature of 15°C. How will the relative humidity change if the temperature in the room gradually increases by 10°C?

390. A cold autumn drizzle settled down for the day. Washing is hung up to dry in a room. Will it dry faster if the window is opened?

391. Two vessels connected by tubes with cocks are filled with water to different levels (Fig. 147). The air has been pumped out of the vessels. What will occur if the vessels are connected (1) by opening the cock in the lower tube, (2) by opening the cock in the upper tube?

392. What is the relative humidity of air at a temperature of $t_1 = 10°$ C if the moisture from this air, when the latter has been heated to $t_2 = 30°$ C, begins to condense after isothermal compression from 1 to 10 at.

Note. Use Table 2.

393. A porous body is placed for drying under the bell of a vacuum pump. The pressure under the bell is maintained at 6.5 mm Hg for one hour, after which it sharply drops. The capacity of the pump is 60 lit/min. The temperature under the pump bell is $t = 5°\,C$. What amount of water did the body contain?

394. Water with a mass of $m = 30$ g at a temperature of 0°C is in a heat-insulated cylinder under a weightless piston. The area of the piston $A = 512\ cm^2$ and the external pressure $p = 1$ atm. What height will the piston rise to if an electric heater contained in the cylinder supplies $Q = 5{,}760$ cal?

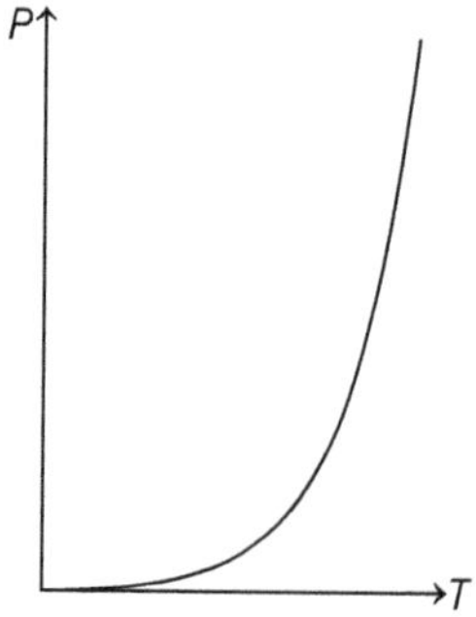

Fig. 146

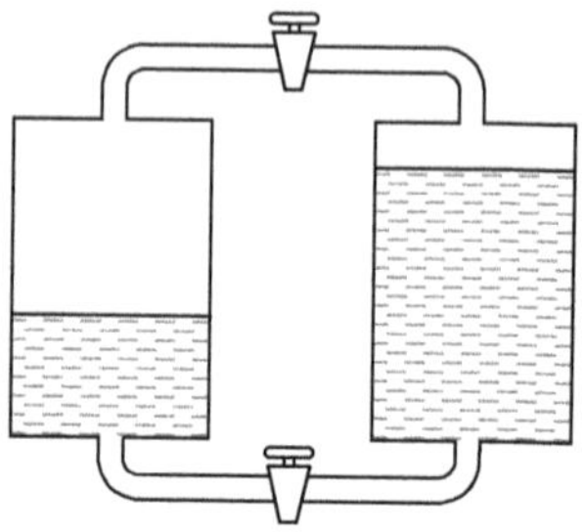

Fig. 147

3

ELECTRICITY AND MAGNETISM

3-1. Electrostatics

395. What is the force of interaction between point charges of one coulomb at a distance of 1 km from each other? Can a small (several centimetres) body have an electrostatic charge of one coulomb?

396. Three identical small balls, each weighing 0.1 gf, are suspended at one point on silk threads having a length of $l = 20$ cm. What charges should be imparted to the balls for each thread to form an angle of $\alpha = 30°$ with the vertical?

397. Two identical balls are suspended on threads at a distance from each other. The balls are given equal charges and immersed in kerosene. Determine the density of the material of the balls if the threads do not deflect from the vertical in a vacuum or in kerosene. The density of kerosene $\rho_0 = 0.8$ g/cm^3 and its permittivity (dielectric constant) $\varepsilon = 2$.

398. Two small balls with equal but opposite charges are secured in a horizontal plane at a distance a from each other. A third charged ball is suspended on a string. The point of suspension is first so moved that the third ball, when in a state of equilibrium, is precisely above the first ball at a distance a from it, and then it is so moved that the third ball is above the second one. Find the angles through which the string is deflected from the vertical if the angle of deflection above one of the balls is twice that above the other.

399. In classical experiments performed to measure the charge of an electron, a charged drop of oil is placed between the horizontal plates of a plane capacitor. Under the action of an electrostatic field, the drop moves uniformly upward, covering a certain distance during the time t_1, or downward, when the sign of the charge on the plates changes, covering the same distance during the time t_2.

Assuming the force of friction between the drop and the air to be proportional to the velocity of the drop, find the time t during which the drop travels the same distance after the field is switched off.

400. If only one charged body is available, can it be used to obtain a charge exceeding many times in absolute magnitude that which it itself has?

401. A charged body has some energy. What is the source of the energy in a body which receives a charge as a result of the process described in the solution to Problem 400?

402. Can two likely-charged balls be attracted to each other?

403. A thin wire ring with a radius R carries an electric charge q. The centre of the ring contains a charge Q of the same sign as q, and $Q \gg q$. Find the force which the ring is stretched with.

404. Two point charges Q_1 and Q_2 are at a distance d from each other. Find the intensity of the electric field at a point at a distance r_1 from the charge Q_1 and r_2 from the charge Q_2. Consider the cases of opposite and like charges.

405. Three identical positive charges Q are arranged at the vertices of an equilateral triangle. The side of the triangle is a. Find the intensity of the field at the vertex of a regular tetrahedron of which the triangle is the base.

406. A positive charge Q is uniformly distributed over a thin wire ring with a radius R. Find the intensity of the electric field at points on the axis of the ring at a distance r from its centre.

407. Find the points on the axis of the charged ring (see Problem 406) at which the intensity of the electric field is maximum. Determine the intensity of the field at these points.

408. Two parallel metal plates, each with an area A, carry the charges Q_1 and Q_2. The distance between the plates is much less than their linear dimensions. Determine the intensity of the electric field at points A, B, and C (Fig. 148).

409. Two large current-conducting plates are arranged parallel to each other. The distance between them is much less than their dimensions. One of the plates is given a charge $+Q$. What are the charges induced on the surfaces of the other plate?

410. A molecule is at a distance r from the axis of a charged infinitely long metallic cylinder. Find the force acting on the molecule if the intensity of the cylinder field is expressed by the formula $E = \frac{2q}{r}$ (q is the charge per unit of cylinder length) and the molecule has the form of a "dumb-bell" with a length l and with charges $+Q$ and $-Q$ at its ends.

411. Two molecules of equal mass are at a certain distance from the axis of a charged cylinder. One molecule has a constant electric moment $p = Ql$ (see Problem 410).

An "elastic" force acts between the charges of the other molecule, i.e. the distance l is determined from the expression $QE = kl$, where E is the mean intensity of the field acting on the molecule and k is a proportionality factor.

First, the electric moments of the molecules are the same and their velocities are zero. Which molecule will reach the surface of the cylinder quicker under the action of the force of attraction?

412. A charge $+Q$ is imparted to a rectangular metal plate with sides a and b. The thickness c of the plate is much smaller than a and b. Find the intensity of the field created by this charged plate at points in space close to the centre of the plate.

413. A point electric charge $+Q$ is at a distance d from a large current-conducting plate. Find the force with which the plate acts on the point charge.

414. A point charge $+Q_1$ is brought up to a distance d from the centre of the metal plate described in Problem 412. The distance d is much smaller than sides a and b of the plate. Determine the force with which the plate acts on the charge $+Q_1$. When will a positively charged plate attract a positive point charge?

415. What potential can an isolated metal sphere with a radius R in the air be charged to if the intensity of the electric field that causes a puncture in the air is $E_0 = 30{,}000$ V/cm?

416. A body with a charge $-q$ is introduced through a small orifice into a hollow current-conducting sphere with a radius R carrying a charge $+Q$. What is the potential of a point in space at a distance of $R > r$ from the centre of the sphere?

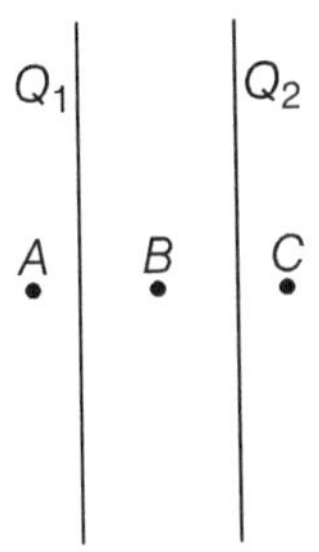

Fig. 148

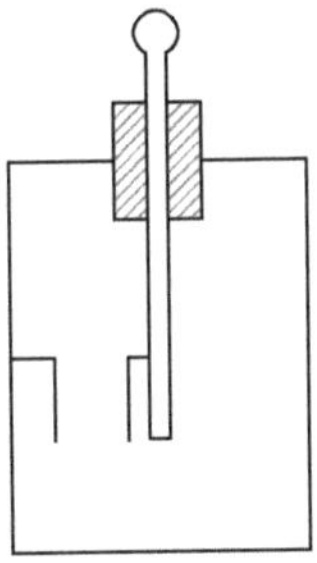

Fig. 149

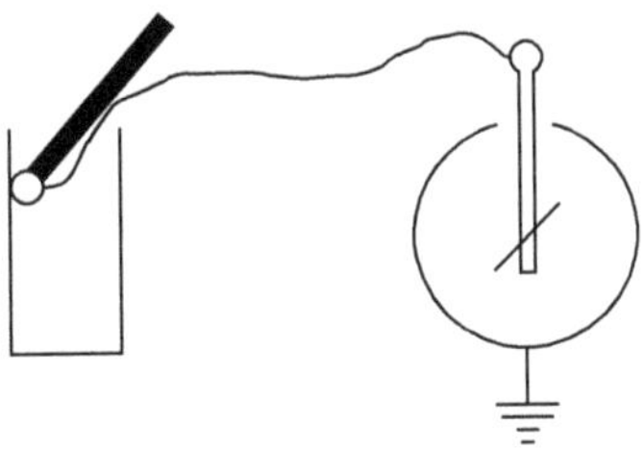

Fig. 150

417. A metal leaf is attached to the internal wall of an electrometer insulated from the earth (Fig. 149). The rod and the housing of the electrometer are connected by a conductor, and then a certain charge is imparted to the housing. Will the leaves of the electrometer deflect? What will happen to the leaves if the conductor is removed and the rod is then earthed?

418. The housing of the electrometer described in Problem 417 is given a charge (the conductor is absent). Will the leaves of the electrometer deflect in this case? Will the angle of deflection of the leaves change if the rod is earthed?

419. By touching different points on a metal bucket having a narrow bottom with a test ball connected by a wire to an earthed electrometer (Fig. 150), we can observe an identical deflection of the leaves of the electrometer at any position of the ball. If the wire is removed, the deflection of the leaves of the electrometer, whose rod the ball is made to contact, will depend on what point of the bucket surface (external or internal) we touched previously. Why?

420. Why does an electrometer connected by a wire to the metal body shown in Fig. 151 make it possible to measure the potential of the body? Why do the leaves deflect in proportion to the density of the charge on separate portions of the body when the charge is transferred from the body to the electrometer with the aid of an insulated current-conducting ball?

421. An uncharged current-conducting sphere with a radius R is at a distance d from a point charge Q. What is the potential of the sphere?

422. An isolated current-conducting sphere with a radius R carries a charge $+Q$. What is the energy of the sphere?

423. A metal sphere two metres in diameter is in the centre of a large room and charged to a potential of 100,000 V. What quantity of heat will be liberated if the sphere is connected to the earth with a conductor?

424. Two metal balls with radii of $r_1 = 1$ cm and $r_2 = 2$ cm at a distance of $R = 100$ cm from each other are connected to a battery with an electromotive force of $E = 3{,}000$ V. Find the force of interaction of the balls, disregarding the interaction of the connecting wires.

425. Two small balls carry charges different in magnitude and identical in sign. One of the balls is secured. If the second ball is released, it can perform the mechanical work W_1 as it moves away under the action of the electrostatic forces of repulsion.

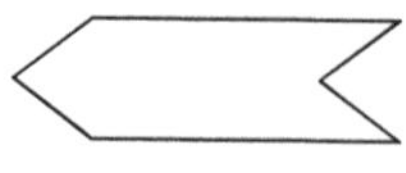

Fig. 151

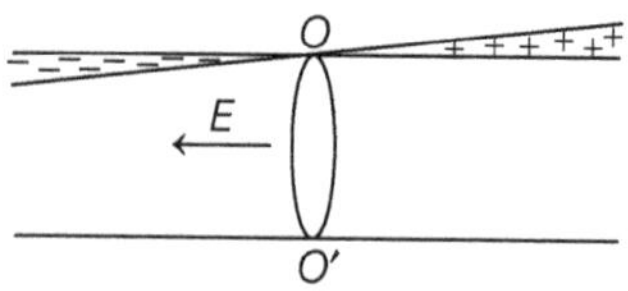

Fig. 152

If before the second ball is released, the balls are connected by a conductor for a certain time, the second ball upon moving away can perform the mechanical work W_2.

Calculate the quantity of heat liberated in the conductor when the balls are connected, and find the energy at the expense of which this heat is liberated and the mechanical work changed.

426. A spherical envelope with a radius R is charged uniformly with a charge Q. Find the expanding force per unit of envelope area.

427. There are three charges on a straight line: a positive one q and two negative ones Q. At what ratio between the charges can they be so arranged that the entire system is in equilibrium? Will this equilibrium be stable? Draw the relation between the potential energy for each charge and its position on the straight line, assuming that the other two charges are stationary.

428. An electric charge Q moves from infinity toward a metallic plate. Find the kinetic energy of the charge when it is at a distance d from the plate. The initial velocity of the charge is zero. The dimensions of the plate are infinitely great.

429. Charges are so arranged on the surface of an infinitely long cylinder that the right-hand half of the cylinder surface from section OO' is charged positively, and the left-hand half negatively (Fig. 152). The density of the charges increases in both directions directly proportional to the distance from section OO'.

Prove that the intensity of the electric field at all points inside the cylinder will be the same everywhere and directed along the axis of the cylinder as shown in the figure by the arrow.

430. An analogy is frequently drawn between capacitance and the capacity of a vessel. How should the vessel be shaped for this analogy to be true?

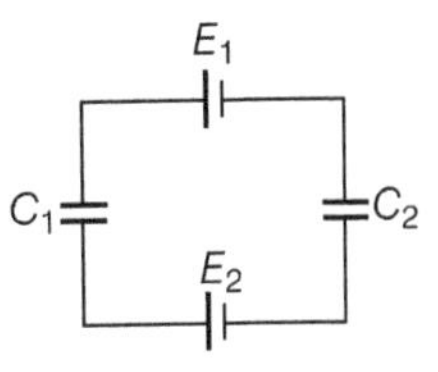

Fig. 153

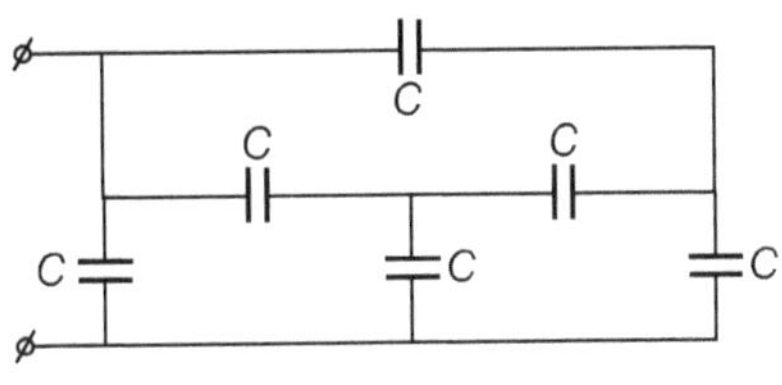

Fig. 154

431. Appraise the capacitance of a human body in the order of its magnitude.

432. Will the readings of an electrometer joined to a galvanic cell change if a capacitor is connected in parallel with it? Will the capacitance of the capacitor have any significance?

433. Four identical plane capacitors with an air dielectric are connected in series. The intensity of the field at which the air is punctured is $E_a = 3 \times 10^4$ V/cm. The distance between the plates $d = 0.7$ cm.

(1) What maximum voltage can be fed to this battery of capacitors?

(2) What will this maximum voltage be if one of the capacitors is replaced by a similar one in which glass is used as a dielectric?

The permittivity of glass $\varepsilon = 7$ and the puncturing field intensity for glass $E_g = 9 \times 10^4$ V/cm.

434. Determine the voltages U_1 and U_2 on the capacitors (Fig. 153) if $\varepsilon_1 = 12$ kV and $\varepsilon_2 = 13$ kV, $C_1 = 3\ \mu F$ and $C_2 = 7\ \mu F$.

Disregard the conductivity of the dielectrics.

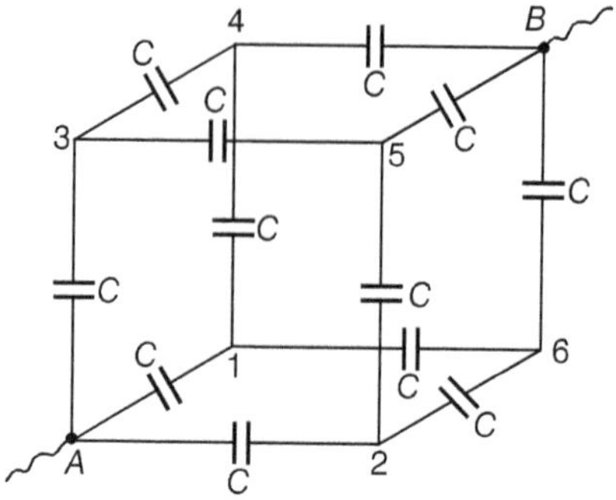

Fig. 155

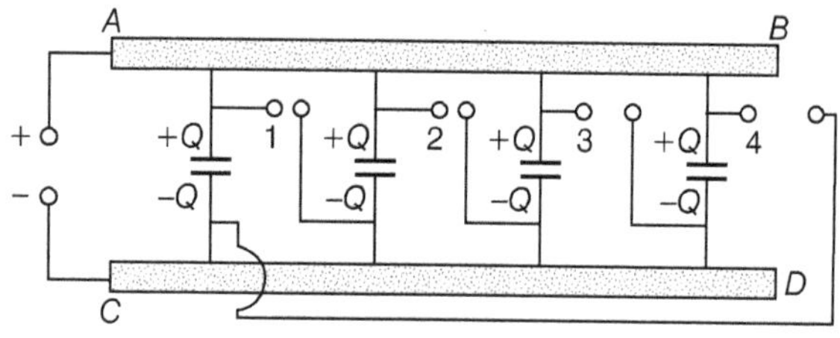

Fig. 156

435. Find the capacitance C_0 of the battery of identical capacitors shown in Fig. 154.

436. Each edge of a cube made of wire contains a capacitor C (Fig. 155). Find the capacitance of this battery if it is connected to the circuit by means of conductors joined to the opposite apices A and B of the cube.

437. The spark-capacitor transformer designed by Arkadyev can be used to obtain short-time high voltages. A diagram of the device is shown in Fig. 156.

A group of capacitors connected in parallel by conductors AB and CD having a very high resistance is joined to a high-voltage source.

The upper plate of each capacitor is connected through a spark gap to the lower plate of the following capacitor (gaps 1, 2, 3 and 4). Each following gap is greater than the preceding one.

A discharge occurs when the potential difference between the plates attains the puncturing voltage of the first gap. After this the second, third, etc., gaps will be punctured. What will the potential difference be when the last gap is punctured if there are n capacitors and the voltage applied is V_0?

438. The plates of a charged plane capacitor are alternately earthed. Will the capacitor be discharged?

439. A plane capacitor is charged to a potential difference U. Both plates are arranged symmetrically with respect to the earth so that their potentials relative to it are $+U/2$ and $-U/2$, respectively.

How will the potentials of the plates change relative to the earth if the first plate is earthed, then disconnected from the earth, after which the second plate is earthed?

440. One of the plates of a capacitor connected to a battery with an e.m.f. of E is earthed (Fig. 157). Will the potentials of the capacitor plates change with respect to the earth if the earthing wire is removed?

441. Two plane capacitors with capacitances C_1 and C_2 are charged to potential differences U_1 and $U_2 (U_1 \neq U_2)$.

Prove that if these capacitors are connected in parallel, their total electrostatic energy diminishes. Explain the drop in the energy.

442. A dielectric in the form of a sphere is introduced into a homogeneous electric field. How will the intensity of the field change at points A, B and C (Fig. 158).

443. One of the plates of a plane capacitor with a mica dielectric carries a positive charge of $Q = 1.4 \times 10^5\, CGS_Q$. The other plate isolated from earth is not charged. The area of each plate $A = 2{,}500\ \text{cm}^2$. The permittivity of mica $\varepsilon_r = 7$.

Find the intensity of the field in the space between the plates.

444. What is the force of interaction of the balls connected to the battery (Problem 424) if they are immersed in kerosene?. The permittivity of kerosene $\varepsilon_r = 2$.

445. The plates of a plane capacitor are connected to a storage battery whose e.m.f. is E. Calculate the mechanical work performed by the electric field when the plates are moved if the initial distance between the plates is d_1, the final distance is d_2, and $d_2 < d_1$. Disregard the evolution of heat in the battery and the feeding conductors.

446. Elongated pieces of a dielectric are usually placed along the force lines of an electric field. It would seem that the separate molecules of a non-polar dielectric only stretch along the field, and never rotate. In a dielectric consisting of dipole molecules the average number of molecules that rotate clockwise when the field is switched on is equal to the number of molecules rotating in the opposite direction. Why does the entire piece of dielectric rotate?

447. The space between the plates of a plane capacitor is filled with a dielectric as shown in Fig. 159. The area of each plate is A and the permittivity of the dielectric is E_r. Find the capacitance of the capacitor in both cases.

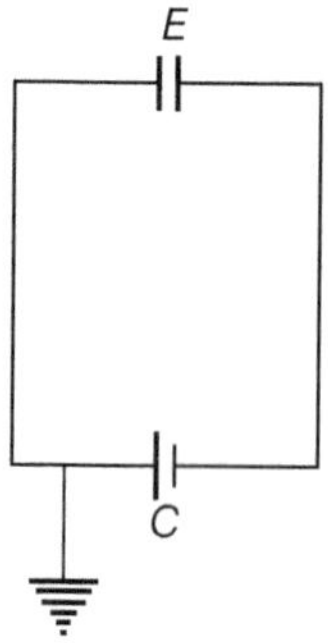

Fig. 157

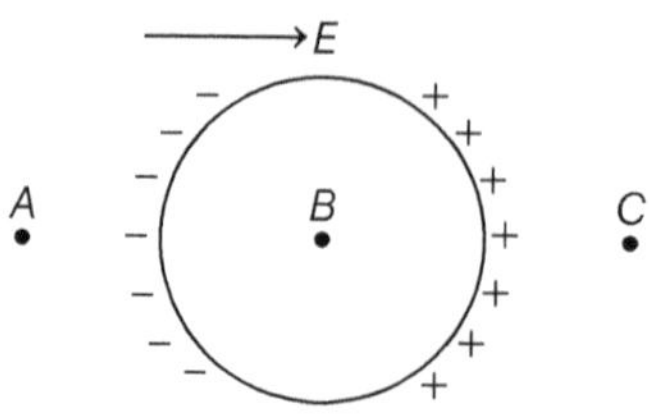

Fig. 158

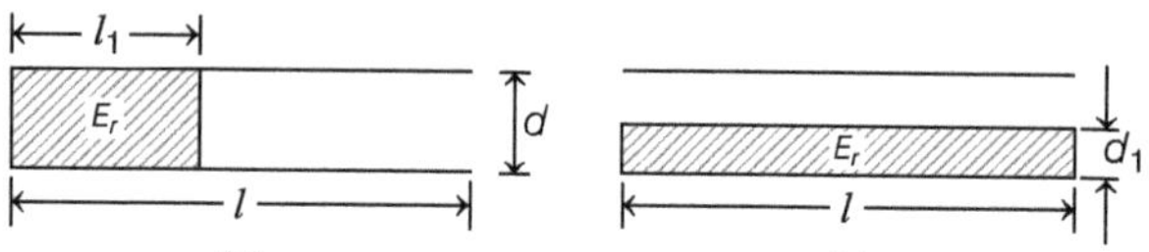

(a) (b)

Fig. 159

448. Determine the energy of a plane capacitor having the space between its plates filled with a dielectric.

449. A plane capacitor is filled with a dielectric whose permittivity is E_r. The intensity of the field in the dielectric is E. What is the intensity of the field in a space made inside the dielectric and having the form of a long thin cylinder directed along the field or the form of a parallelepiped one side of which is much smaller than the other two? The smaller side is directed along the field.

450. Two rectangular plates with a length l and an area A are arranged parallel to each other at a distance d. They are charged to a potential difference U (plane capacitor). A dielectric with a permittivity E_r whose thickness is d and whose width is equal to that of the plates is drawn into the space between the latter. The length of the dielectric is greater than l (Fig. 160). Find the resulting force F acting on the dielectric from the side of the field depending on the distance x.

451. Solve Problem 450 if the capacitor is connected to a battery whose e.m.f. is U. Disregard the resistance of the connecting wires.

452. The following design of a perpetuum mobile has been suggested. Kerosene is poured into communicating vessels (Fig. 161). One part of the vessel is placed into a strong electric field between the plates of a capacitor owing to which the level of the kerosene in this part is higher than in the other. A chain of balls is passed over two pulleys.

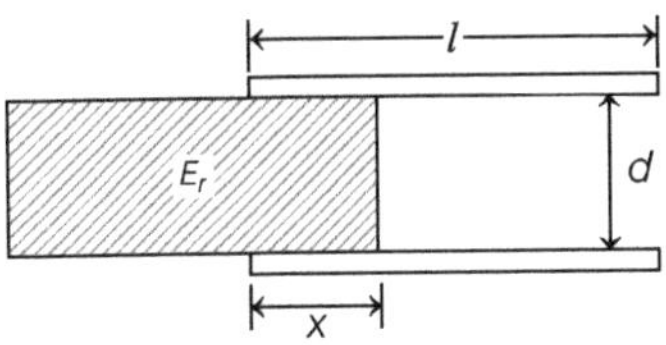

Fig. 160

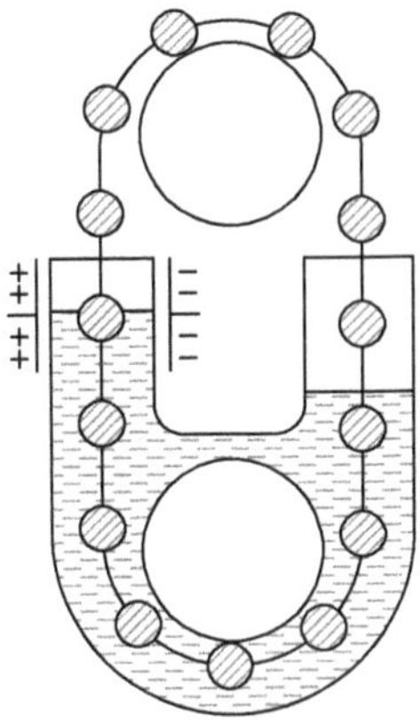

Fig. 161

The specific weight of the material of the balls is less than that of the kerosene. The lifting force acting on the balls will be greater in the left-hand part than in the right-hand one because more balls are immersed in the kerosene in the left-hand part. For this reason the inventor believes that the chain should start rotating clockwise. Why will there actually be no rotation?

453. The space between the plates of a plane capacitor is filled with a dielectric whose permittivity is E_r. One plate is given a charge $+Q$ and the other $-Q$. Determine the density of the bound electric charges that appear on the surface of the dielectric and the forces that are exerted by the field on the dielectric.

454. The space between the plates of a plane capacitor is filled with a dielectric. Each molecule of the dielectric is assumed to have the form of a "dumb-bell" with a length l whose ends carry charges $+Q$ and $-Q$. The number of molecules in a unit of volume (1 cm^3) is n.

Let us assume further that all the molecules have turned along the electric field under its action.

Find the intensity E of the field inside the capacitor filled with the dielectric if before filling it the intensity of the field was E_0.

455. A dielectric consists of molecules each of which can be represented as two charges $+Q$ and $-Q$ between which an "elastic force" acts. The latter term should be understood to mean that x (the distance between the charges $+Q$ and $-Q$) can be found from the equality $kx = QE$, where E is the intensity of the field acting on the charges, and k is a proportionality factor.

Assume that a unit of volume (1 cm^3) of the dielectric contains n molecules. Solve Problem 454, assuming that the space between the plates of the plane capacitor is filled with a dielectric of this type.

Determine the permittivity of the dielectric.

456. A capacitor is filled with the dielectric whose properties are described in Problem 455.

Find the energy stored in the dielectric owing to its polarization.

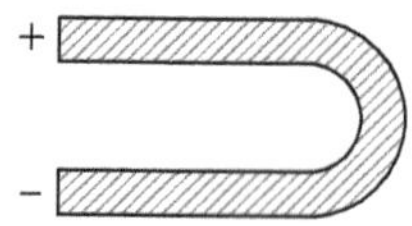

Fig. 162

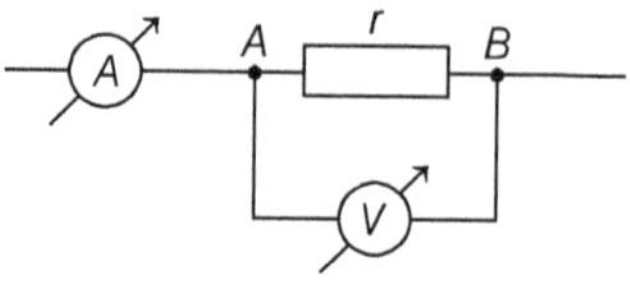

Fig. 163

3-2. Direct Current

457. Is there an electric field near the surface of a conductor carrying direct current?

458. Draw approximately the arrangement of the force lines of an electric field around a homogeneous conductor bent to form an arc (Fig. 162). The conductor carries direct current.

459. Determine the resistance r if an ammeter shows a current of $I = 5$ A and a voltmeter 100V (Fig. 163). The internal resistance of the voltmeter $R = 2{,}500\ \Omega$.

460. What resistance r should be used to shunt a galvanometer with an internal resistance of $R = 10{,}000$ ohm to reduce its sensitivity $n = 50$ times?

461. Determine the voltage across a resistance R using a voltmeter connected to its ends. What relative error will be made if the readings of the voltmeter are taken as the voltage applied before it was switched on? The current intensity in the circuit is constant.

462. An ammeter is connected to measure the current intensity in a circuit with a resistance R.

What relative error will be made if connection of the ammeter does not change the current intensity in the circuit? The voltage across the ends of the circuit is kept constant.

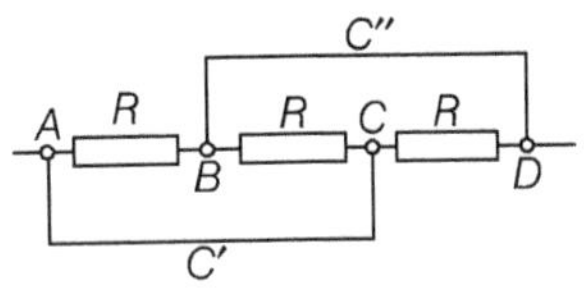

Fig. 164

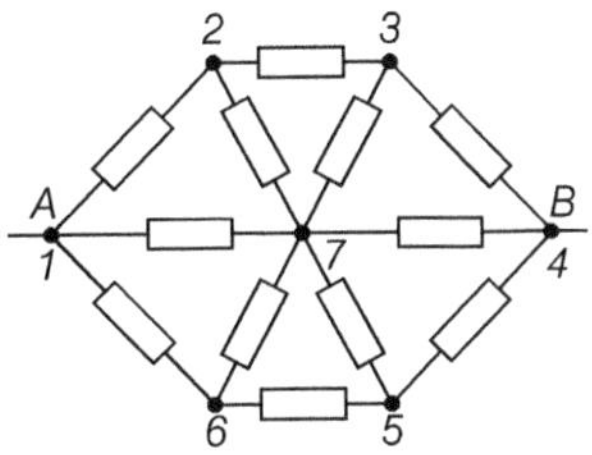

Fig. 165

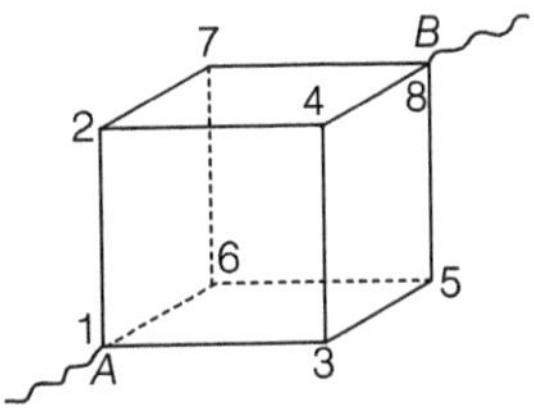

Fig. 166

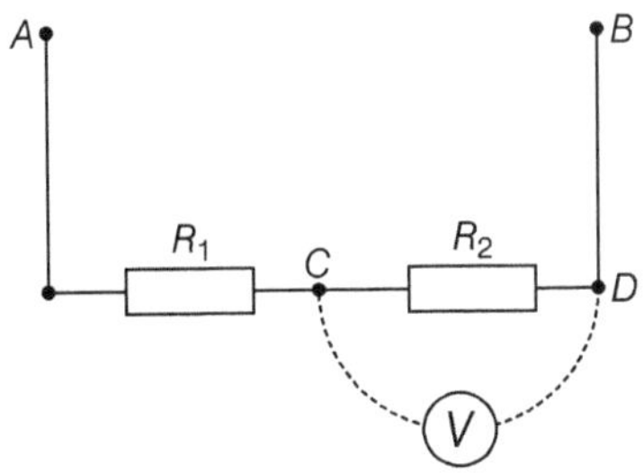

Fig. 167

463. Two conductors with temperature coefficients of resistance α_1 and α_2 have resistances R_{01} and R_{02} at 0°C. Find the temperature coefficient of a circuit consisting of these conductors if they are connected in series and in parallel.

464. Find the resistance of the circuit shown in Fig. 164. Disregard the resistance of the connecting wires $AC'C$ and $BC''D$.

465. Find the resistance of the hexagon shown in Fig. 165 if it is connected to a circuit between points A and B. The resistance of each conductor in the diagram is R.

466. Find the resistance of a wire cube when it is connected to a circuit between points A and B (Fig. 166). The resistance of each edge of the cube is R.

467. Resistances R_1 and R_2, each 60 Ω, are connected in series (Fig. 167). The potential difference between points A and B is $U = 120$ V. Find the reading U_1 of a voltmeter connected to points C and D if its internal resistance $r = 120$ Ω.

468. Wires identical in cross section A and resistivity ρ are soldered into a rectangle $ADBC$ (Fig. 168) with the diagonal AB of the same cross section and material. Find the resistance between points A and B and between C and D if $AD = BC = a$ and $AC = BD = b$.

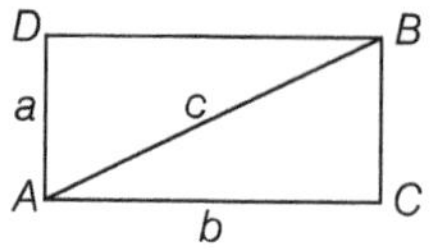

Fig. 168

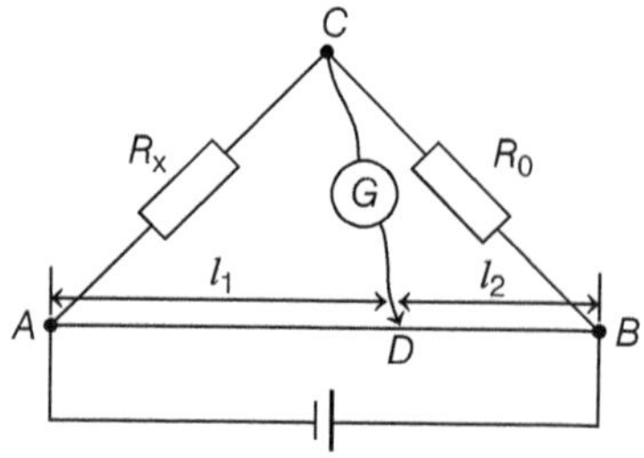

Fig. 169

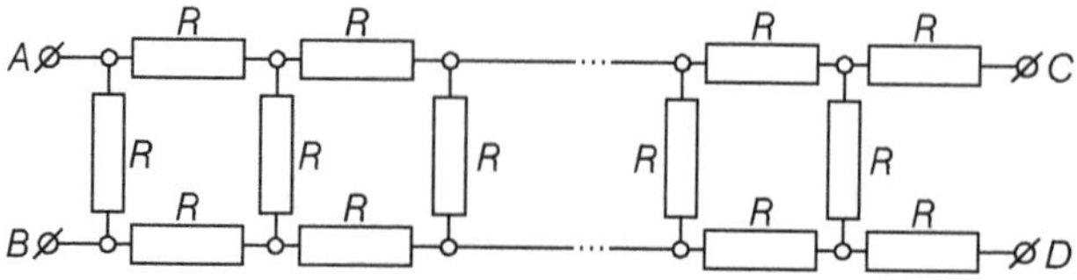

Fig. 170

469. Figure 169 shows the diagram of a Wheatstone bridge used to measure resistances. Here R_x is the unknown resistance, R_0 a standard resistance, G a galvanometer connected by a sliding contact D to a homogeneous conductor AB with a high resistance (a slide wire).

Prove that the equation $\frac{R_x}{R_0} = \frac{l_1}{l_2}$ is true when no current flows through the galvanometer. Disregard the resistance of the connecting wires.

470. What resistance should be connected between points C and D (Fig. 170) so that the resistance of the entire circuit (between points A and B) does not depend on the number of elementary cells?

471. The output voltage can be reduced in the output circuits of generators as desired by means of an attenuator designed as the voltage divider shown in Fig. 171.

A special selector switch makes it possible to connect the output terminal either to the point with a potential U_0 produced by the generator, or to any of the points $U_1, U_2, \dots U_n$, each having a potential k times smaller ($k > 1$) than the previous one. The second output terminal and the lower ends of the resistances are earthed.

Find the ratio between the resistances $R_1 : R_2 : R_3$ with any number of cells in the attenuator.

472. What devices are needed to verify Ohm's law experimentally, i.e., to show that the current intensity is directly proportional to the potential difference?

473. A charge Q is imparted to two identical plane capacitors connected in parallel.

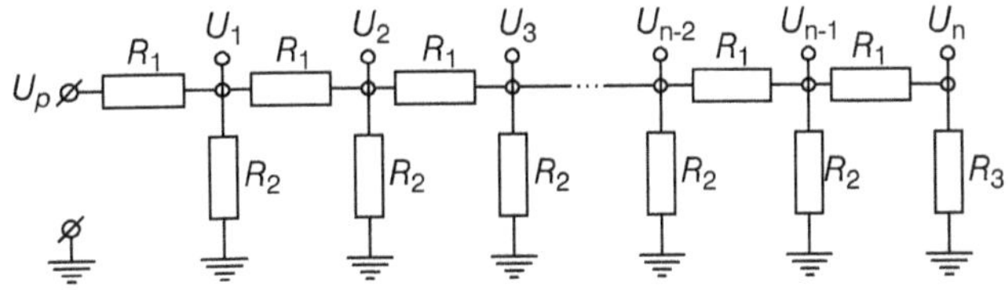

Fig. 171

At the moment of time $t = 0$ the distance between the plates of the first capacitor begins to increase uniformly according to the law $d_1 = d_0 + vt$, and the distance between the plates of the second capacitor to decrease uniformly according to the law $d_2 = d_0 - vt$. Neglecting the resistance of the feeding wires, find the intensity of the current in the circuit when the plates of the capacitors move.

474. Find the work performed by an electrostatic field (see Problem 473) when the distance between the plates of the first capacitor increases and that between the plates of the second capacitor simultaneously decreases by a.

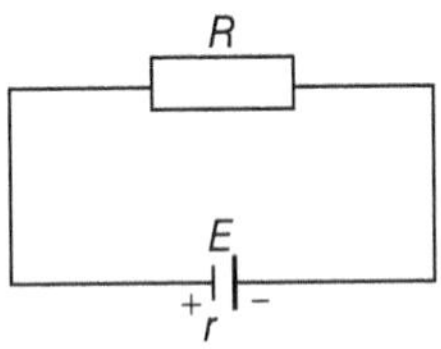

Fig. 172

475. A curious phenomenon was observed by an experimenter working with a very sensitive galvanometer while sitting on a chair at a table. (The galvanometer was secured on a wall and the ends of its winding connected to an open key on the table.) Upon rising from the chair and touching the table with his hand, the experimenter observed an appreciable deflection of the galvanometer pointer. If he touched the table while sitting on the chair there was no deflection. Also, the galvanometer showed no deflection when he touched the table without first sitting down. Explain this phenomenon.

476. The following effect was observed in a very sensitive galvanometer when the circuit was opened. If a charged body is brought up to one end of the winding of the galvanometer, its pointer deflects. If the body is brought up to the other end of the winding, the deflection is in the same direction. Explain this phenomenon.

477. How is the potential distributed in a Daniell cell when the external circuit is opened?

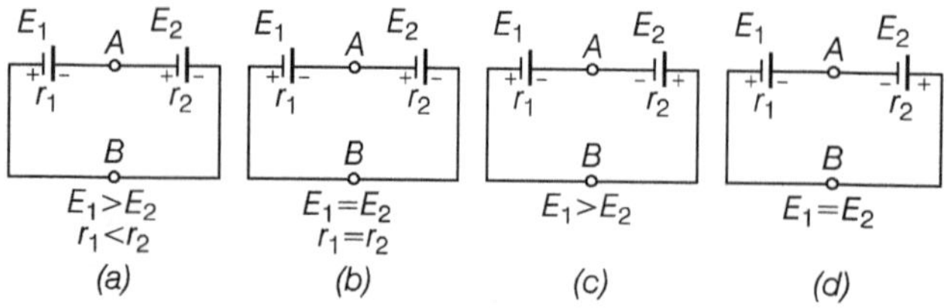

Fig. 173

478. Show graphically the distribution of a potential along the closed circuit illustrated in Fig. 172 and on this basis deduce Ohm's law for a closed circuit.

479. Show graphically the approximate distribution of a potential along the closed circuits depicted in Fig. 173.

Determine the current intensity for each circuit and the potential difference between points *A* and *B*. Disregard the resistance of the connecting wires.

480. Prove that an electromotive force in a circuit containing a galvanic cell is equal to the work of forces of non-electrostatic origin when a single positive charge moves along a closed circuit.

481. About 106,000 calories are evolved when one mole of zinc combines with sulphuric acid and about 56,000 calories are consumed when a mole of copper is liberated from blue vitriol. Use these data to find the e. m. f. of a Daniell cell.

482. Two Daniell cells with internal resistances of $r_1 = 0.8\ \Omega$ and $r_2 = 1.3\ \Omega$ and the same e. m. f. s are connected in parallel and across an external resistance *R*. Find the ratio between the quantities of the zinc dissolved in these elements during a definite interval of time.

483. A Daniell cell is made of absolutely pure materials. Find the consumption of zinc and crystals of blue vitriol $CuSO_4 \cdot 5H_2O$ if the cell produces a current of 0.1 A in eight hours.

484. In a Daniell cell, copper is replaced by wax coated with a layer of graphite. Describe the phenomena that will occur in such a cell if the zinc is connected to the graphite layer by a wire.

485. How will the e. m. f of the battery shown in Fig. 174 change if the partition between the vessels is removed? A solution of sulphuric acid is used as the electrolyte.

486. A homogeneous carbon rod lies on the bottom of a vessel filled with electrolyte. A voltmeter with a high resistance is connected to the ends of the rod. A zinc rod bears against the middle of the carbon rod

(Fig. 175). What will the voltmeter show if the zinc rod is placed vertically? How will the readings of the voltmeter change if the zinc rod is inclined to the right or the left?

487. A hollow current-conducting sphere with a radius of $R = 5$ cm is placed into an electrolytic bath filled with a solution of blue vitriol. The surface of the sphere has an orifice with a radius of $r = 0.5$ mm.

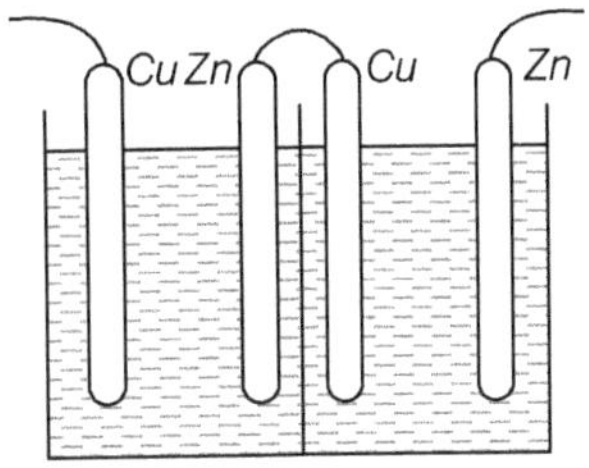

Fig. 174

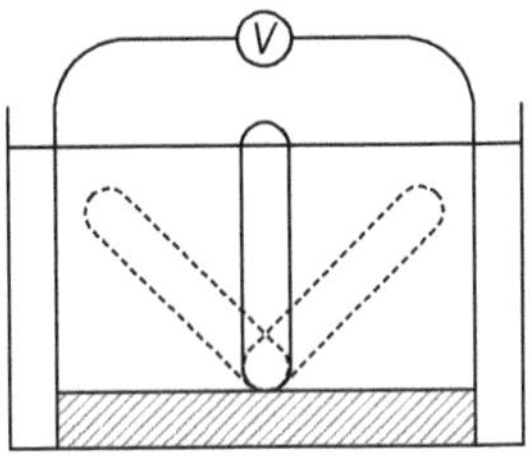

Fig. 175

How much will the weight of the sphere increase if the copper is deposited during $t = 30$ min with a current density in the electrolyte of $j = 0.1$ A/cm^2.

488. If a capacitor carrying a charge Q is discharged through an electrolytic bath with acidified water, m grammes of detonating gas will be liberated. According to Faraday's law, the quantity of substance evolved during electrolysis depends only on the amount of electricity passed through the electrolyte. This means that if the capacitor is discharged through N series-connected baths, then mN grammes of detonating gas will be liberated. N can be made as great as required to obtain any quantity of the gas. Combustion of this gas can produce any amount of energy, which is obviously inconsistent with the law of conservation of energy, since the initial energy of a charged capacitor is not infinitely great. Explain this fact.

489. When detonating gas explodes, 34,500 cal are liberated per gramme of reacted hydrogen. Use these data to find the minimum e. m. f. of a battery at which the electrolysis of water is possible.

490. In electrolysis, positive and negative ions are continuously neutralized on the respective electrodes. What maintains the concentration of the ions in the electrolytes at a constant level? In what sections of the electrolyte is the reduction in the number of the ions compensated?

491. The total density of the current in electrolytes is determined as the sum of the current of the positive ions and that of the negative ions:

$$j = e(n_+ v_+ + n_- v_-)$$

where e is the charge of an ion, and n and v are the concentrations and velocities of the positive and negative ions.

Why is the amount of substance evolved, for example, on the cathode considered to be proportional to the full current, and not to the current en_+v_+?

492. What minimum change in temperature can be determined with the aid of an iron-constantan thermocouple, if the measuring instrument (galvanometer) has a sensitivity of 10^{-9} A and a resistance of $R = 20\ \Omega$? The e.m.f. of the thermocouple is 50 microvolts (50×10^{-6} V) per degree and its resistance $r = 5\ \Omega$.

493. The temperature of the hot joints of a thermoelectric battery $t_1 = 127°$ C and of the cold ones $t_2 = 27°$ C. The e. m. f. of the battery $E = 4$ V. Two calories of heat are supplied to the heated joints in a unit of time to maintain a constant temperature.

An electrolytic bath with a solution of blue vitriol is connected to the battery. What maximum (theoretical) amount of copper can be deposited on the cathode in a unit of time?

494. A current with an intensity of I flows through a storage battery with an internal resistance of r and an e.m.f. of E. What is the potential difference across the terminals of the battery?

495. A voltmeter with a resistance of $R_1 = 100\ \Omega$ connected to the terminals of a cell shows a potential difference of 2 V. When this cell is connected across a resistance of $R = 15\ \Omega$, an ammeter connected to the circuit shows a current intensity of $I = 0.1$ A. Find the e. m. f. of the cell if the resistance of the ammeter $R_2 = 1\ \Omega$.

496. Why can a galvanic cell with an e. m. f. of several volts produce a large current, while an electrostatic machine having an e. m. f. rated at dozens of thousands of volts generates a negligible current?

497. In the circuit shown in Fig. 176, the capacitor C_2 is punctured and the resistance between its plates is finite. What is the potential difference between the plates of each of the capacitors if the key K is closed?

496. When will two series-connected galvanic cells connected to an external resistance give a lower current than one of these cells connected to the same resistance?

499. A storage battery with an e. m. f. of $E = 24$ V was connected to one end of a two-conductor telephone line with a length of $L = 5.6$ km to find an insulation breakdown between the wires. It was found that if the conductors at the other end of the line were opened, the current flowing through the battery was $I_1 = 1.5$ A, and if short-circuited the current was $I_2 = 2$ A.

The short-circuit current of the battery $I_3 = 96$ A and the resistance of each conductor of the line $r = 7\Omega$. Find the resistance of the insulation R at the point of breakdown.

500. Galvanic cells with e.m.f. s of $E_1 = 2$ V and $E_2 = 1.5$ V are connected as shown in Fig. 177*a*. A voltmeter with zero in the middle of the scale shows a voltage of $u_1 = 1$ V and its pointer deflects in the same direction as when key K is open. What will the voltmeter show if the cells are connected as shown in Fig. 177*b*? Disregard the current branched off into the voltmeter.

501. Solve Problem 500 if, with key K closed (Fig. 177*a*), the pointer of the voltmeter deflects in the direction opposite to the one in which it deflects with the key open.

502. Two cells with e. m. f.s of $E_1 = 2$ V and $E_2 = 1$ V are connected as shown in Fig. 178. The resistance $R = 0.5\ \Omega$. The internal resistances of the cells are the same and equal to $1\ \Omega$ each. Determine the currents flowing through the cells and the resistance R. Disregard the resistance of the feeding wires.

503. At what resistance R in the circuit of Problem 502 will the current not flow through a galvanic cell with an e. m. f. E_2? At what values of R will the current through this cell be directed against the e. m. f. of the cell?

504. Can a resistance of $R = 0.2\ \Omega$ and a current of $I = 21$ A be obtained in an external circuit with the aid of 24 storage battery cells, each with an e.m.f. of $E_0 = 2$ V and an internal resistance of $r = 0.3\ \Omega$ by connecting them into separate identical groups?

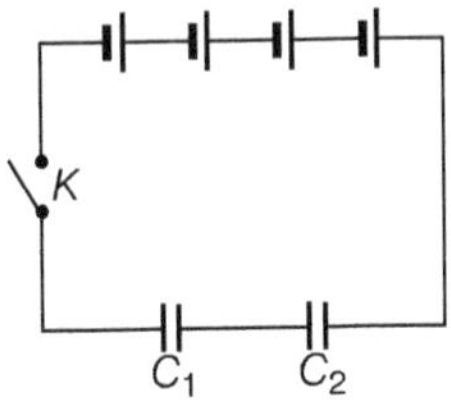

Fig. 176

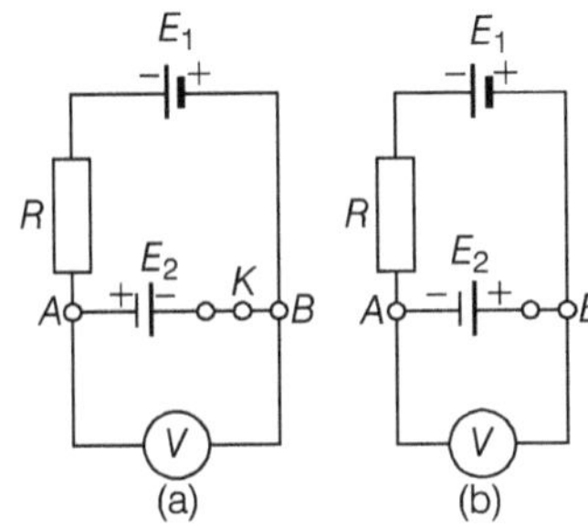

Fig. 177

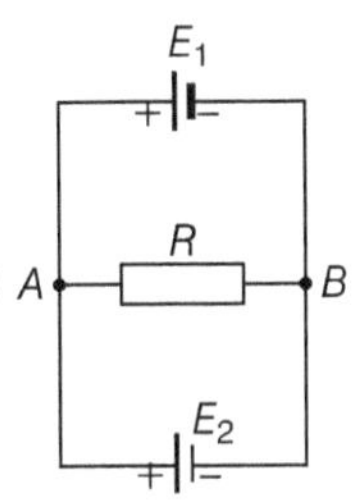

Fig. 178

505. An electric stove designed for 220 V is to be adapted for 110 V without changing or shortening the coil so that its power remains the same. What should be done for this?

506. A lamp the resistance of whose filament in a heated state is $R = 2.9\ \Omega$ is placed into a calorimeter containing a mixture of water and ice. In what time will the amount of water in the calorimeter increase by $m = 15$ g if the lamp is connected to mains with a voltage of $u = 220$ V ? The specific heat of fusion of ice $H = 80$ cal/g.

507. An electric lamp with a tungsten filament consumes 50 watts. When the lamp burns, the temperature of its filament is 2,500° C. What power will be consumed by the lamp at the first moment after it is switched on? The temperature coefficient of resistance of tungsten $\alpha = 4.5 \times 10^{-3}$ deg^{-1}.

508. Why does the incandescence of the lamps in a room noticeably drop as soon as a high-power device (for example, an electric iron) is switched on, and after a brief interval of time increase, reaching about the same value as before?

509. The wires leading from the mains into a building have a resistance of $R_0 = 0.5\ \Omega$. The voltage in the mains is constant and equal to $U_0 = 127$ V. What is the maximum permissible power of the electric energy consumed in the building if the voltage across the devices connected to the mains should not drop below $U = 120$ V ?

510. An electric tea-kettle has two windings. When one of them is switched on the kettle begins to boil in t_1 minutes, and when the other is switched on—in t_2 minutes. In what time will the kettle begin to boil if both windings are switched on simultaneously in series and in parallel?

511. An electric heater has three windings. If two windings are connected in parallel and the third is connected to them in series, then with various combinations of the windings the water in the tank will begin to boil in 20, 40 and 16 minutes, respectively.

In what time will the water begin to boil if all the windings are connected (1) in series? (2) in parallel?

512. When a direct current flows through a conductor, the amount of energy liberated is QU, where Q is the charge passing through the conductor and U the potential difference, while an energy of $QU/2$ is liberated when a capacitor is discharged. Why?

513. When electric energy is transmitted over great distances, a transformer is used to so increase the voltage as to reduce the current intensity at the same power. According to the Joule-Lenz law, the amount of heat evolved in the wires $Q = 0.24\, I^2Rt$, and therefore the losses due to heat evolution will be small with small currents. On the other hand, $Q = 0.24 \dfrac{U^2}{R} t$, i.e., the amount of heat evolved grows with an increase in the voltage. Explain why an increase in the voltage saves electric energy when it is transmitted over great distances.

514. When two identical lamps are connected in series to the circuit of a battery, the voltage drop across the internal resistance is k% of the e.m.f. The rated voltage of the lamp is U volts and the rated power p watts. Determine the internal resistance r of the battery.

515. A storage battery with an e.m.f. of $E = 10$ V and an internal resistance of $r = 1\ \Omega$ is connected across an external resistance R and liberates in it a power $P = 9$ watts. Find the potential difference U across the terminals of the battery. What is the cause of the ambiguity of the result?

516. What maximum useful power (evolved on an external resistance) can be produced by a storage battery with an e. m. f. of $E = 10$ V and an internal resistance of $r = 1\ \Omega$? What is the resistance of the external circuit?

517. Determine the efficiency η of the storage battery in Problems 515 and 516. How does the efficiency depend on the external resistance with a constant internal resistance? How does the useful power change in this case? Can η be equal to unity?

518. The efficiency of a source of current connected to an external resistance R is $\eta_1 = 60$ per cent. What will the efficiency η_2 be if the external resistance is increased six times?

519. A storage battery with an initial e. m. f. of E is charged from a station rated at U volts. The internal resistance of the battery is r. Find the useful power spent to charge the battery and the power used to liberate the heat in it.

520. Does the useful power spent to charge a storage battery exceed the heat evolved in it?

521. A conductor carries a current of $I = 10$ A. The cross-sectional area of the conductor $A = 5\ \text{cm}^2$ and the number of free electrons in $1\ \text{cm}^3$ of the conductor $n = 10^{23}$. Find the directed velocity of the electrons v assuming it to be the same for all the electrons.

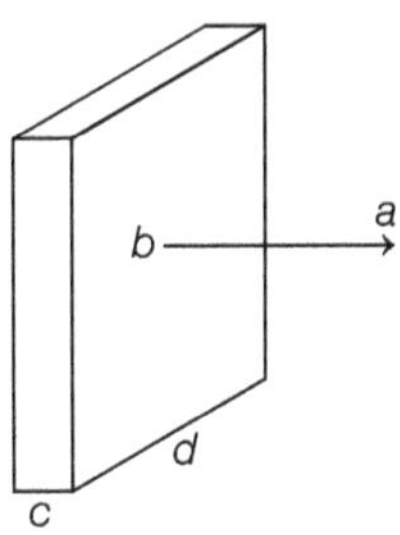

Fig. 179

522. A metal rectangular parallelepiped with sides d, b, c ($d \gg c, b \gg c$) moves with an acceleration a in the direction shown by the arrow in Fig. 179. Find the intensity of the electric field produced by the accelerated motion of the metal block and the density of the electric charges on its surfaces that are perpendicular to the direction of motion.

523. A solid metal cylinder whose radius is R rotates with a constant angular velocity ω. Find how the field intensity depends on the distance to the axis of the cylinder and determine the potential difference between the surface of the cylinder and the axis.

3-3. Electric Current in Gases and a Vacuum

524. Will a glow discharge occur if an anode is placed into a cathode dark space (the region of the cathode drop)?

525. Figure 180 shows a diagram of an X-ray tube with a cold cathode: C is the cathode, A the anode and A_c the anticathode. A high voltage is created between the anode and the cathode. The electrons accelerated near the cathode (region of the cathode drop) bombard the anticathode with a high velocity and initiate X-rays. Why is the tube provided with two electrodes, an anode and an anticathode, instead of one?

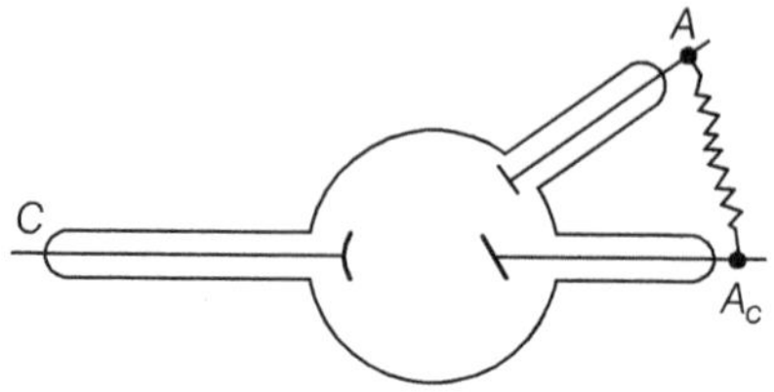

Fig. 180

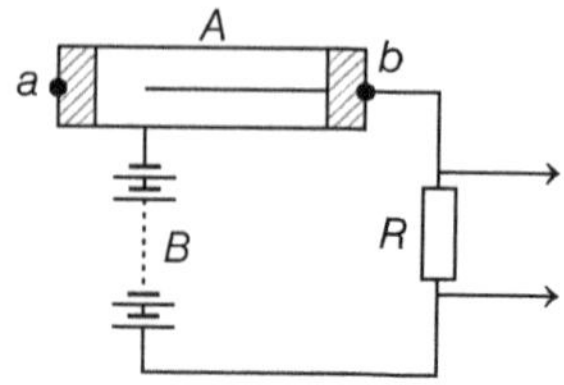

Fig. 181

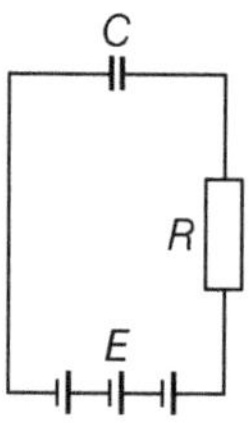

Fig. 182

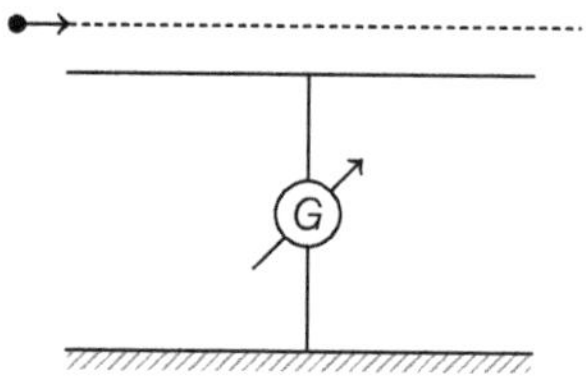

Fig. 183

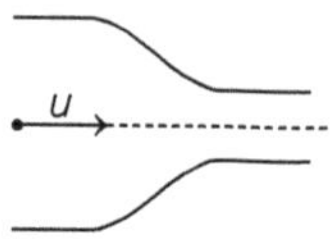

Fig. 184

526. Why are the anode and the anticathode of an X-ray tube connected by a wire (see Problem 525)? What will occur if the anticathode is isolated?

527. Figure 181 shows a Geiger-Muller counter of elementary particles. A high voltage is produced between the housing of tube A and a thin wire ab that is only slightly smaller than the critical voltage necessary to ignite the charge.

When a fast charged particle enters the counter, the molecules of the gas are ionized and a discharge begins.

The flow of current through the circuit is accompanied by a drop of voltage across the large resistance R. This voltage drop is recorded after amplification by corresponding instruments.

For the counter to answer its purpose, the discharge caused by the particle should be quickly extinguished. What extinguishes the discharge in the circuit in Fig. 181?

528. A capacitor with a capacitance $C = 8$ cm and a distance between the plates of $d = 3$ mm is connected to a high-voltage source through a resistance of $R = 10^3\ \Omega$ (Fig. 182).

The air in the space between the plates of the capacitor is ionized by X-rays so that $n = 10^4$ ionic pairs form in one cubic centimetre per second. The charge of each ion is equal to that of an electron.

Find the voltage drop across the resistance R assuming that all the ions reach the plates before they recombine.

529. What will occur to a burning electric arc if the negative carbon is intensively cooled? What will occur if the positive carbon is cooled?

530. Why is an electric iron with a thermocontroller used only with alternating current?

531. What energy in ergs will an electron acquire after passing through a potential difference of 1 V in a vacuum? (In atomic physics this energy is taken as the unit "electron-volt".)

532. Does the trajectory of a charged particle in an electrostatic field coincide with a force line?

533. A broad metal plate is connected to earth through a galvanometer. A charged ball flies along a straight line above the plate at a distance much less than the linear dimensions of the plate (Fig. 183).

Draw an approximate diagram showing how the current flowing through the galvanometer depends on time.

534. An electron moves along a metal tube with a variable cross section (Fig. 184). How will its velocity change when it approaches the neck of the tube?

535. A potential difference U is created between a filament emitting electrons and a current-conducting ring (Fig. 185). The electrons move with an acceleration along the axis of the ring.

Their kinetic energy increases while the battery producing the potential difference U performs no work, since no current flows through the circuit. (It is assumed that the electrons do not impinge upon the ring.) How can this be brought into agreement with the law of conservation of energy?

536. A charge $+Q$ is uniformly distributed over a thin ring with a radius R. Find the velocity of a negative point charge $-Q$ at the moment it passes through the centre O of the ring if the charge $-Q$ was initially at rest at point A sufficiently removed from the ring (Fig. 186). The mass of the charge $-Q$ is equal to m. The ring is stationary.

537. The plates of a plane capacitor with a capacitance C at a distance of l from each other carry charges $+Q$ and $-Q$. An electron flies into the middle of the capacitor with a velocity v_0 directed parallel to the plates. What is the velocity of the electron at a sufficiently great distance from the capacitor?

What is the nature of the change in the velocity of the electron (in absolute magnitude) when it moves inside and outside of the capacitor?

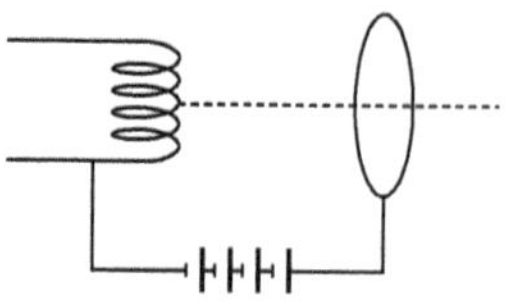

Fig. 185

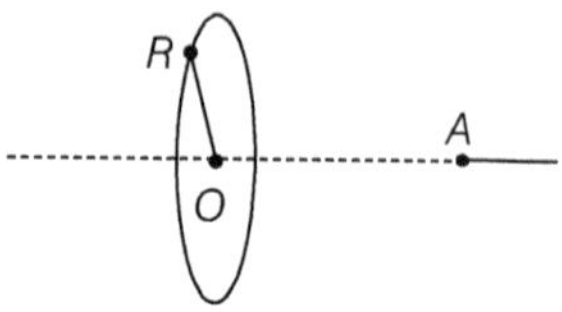

Fig. 186

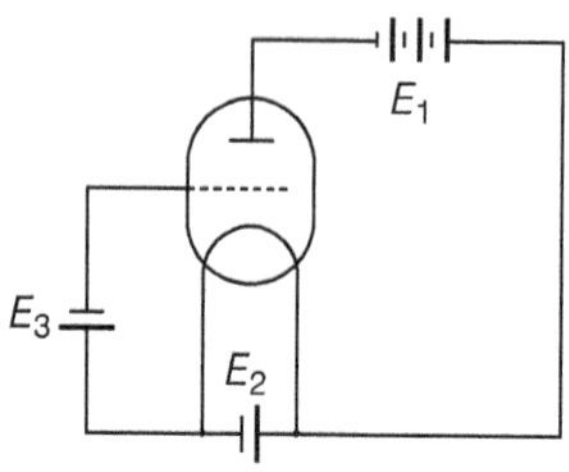

Fig. 187

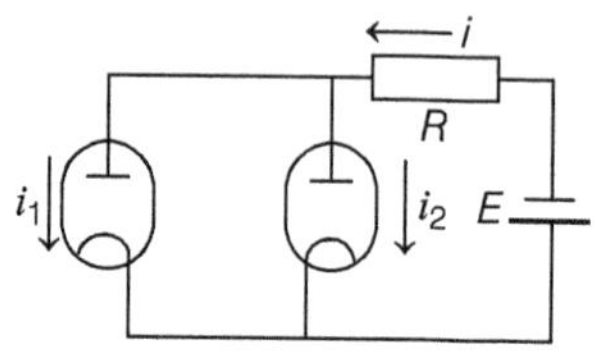

Fig. 188

Consider the following three cases:

(1) at the initial moment the electron is at the same distance from both plates of the capacitor;

(2) at the initial moment the electron is at a distance of $l/4$ from the positive plate;

(3) the electron is at a distance of $l/4$ from the negative plate.

538. A battery (directly heated) triode is connected to the circuit shown in Fig. 187. The e. m. f. of the B battery $E_1 = 80$ V, of the A battery $E_2 = 6$ V and of the C battery $E_3 = 2$ V.

With what energies will the electrons reach the anode? How will the energy of the electrons reaching the anode change if the e.m.f. E_3 changes in magnitude and even in sign?

Assume the anode current to be small as compared with the heating current.

539. The anode current of a diode can be related to the potential difference U_a between the electrodes within a certain voltage range by the equation $I_a = AU_a + BU_a^2$.

Find the anode current if the diode is connected in series with a resistance of $R_a = 20\ k\ \Omega$ to the circuit of a battery with an e. m. f, of $E = 120$ V. For the given diode $A = 0.15$ mA/V and $B = 0.005\ \text{mA}/\text{V}^2$. Disregard the internal resistance of the battery.

540. Two electronic valves are connected in parallel. They are connected to the circuit of a battery with an e. m. f. of $E = 300$ V in series with a resistance of $R = 4k\ \Omega$ (Fig. 188). The relation between the anode current i on the anode voltage U_a for each valve can be approximately expressed as $i = AU_a + BU_a^2$, where for one valve $A_1 = 0.07$ mA/V, $B_1 = 0.005\ \text{mA}/\text{V}^2$ and for the other $A_2 = 0.03$ mA/V, $B_2 = 0.01\ \text{mA}/\text{V}^2$. Determine the anode currents of the valves disregarding the internal resistance of the battery.

541. An electronic valve (one of the triodes 6H8C) is connected to the circuit of a battery with an e.m.f. of $E = 250$ V in series with a resistance of $R = 10^4 \Omega$ (Fig. 189).

The valve grid is connected to the negative pole of a small battery ($E_1 = 3$V) and the cathode to its positive pole. In this case the voltage drop across the resistance R reaches $U_1 = 95$ V.

If the grid circuit includes a battery with $E_2 = 6$ V, the potential difference across the resistance R will be $U_2 = 60$ V.

What will the potential difference between the anode and the cathode of the valve be if the grid and the cathode are short-circuited? Consider the grid characteristic of the valve as a straight line in the range of the grid potential change being considered.

542. Three identical diodes whose anode characteristics can be approximately represented by sections of straight lines:

$$I_a = 0 \text{ at } U_a \le 0$$
$$I_a = kU_a \text{ at } U_a > 0$$

where $k = 0.12$ mA/V, are connected to a circuit as shown in Fig. 190. Draw a diagram showing how the current I in the circuit depends on the voltage V if $E_1 = 2$ V, $E_2 = 5$V, $E_3 = 7$ V and V can change from -10 V to $+10$ V.

543. Calculate the sensitivity of a cathode-ray tube to voltage, i.e., the deflection of the light spot on the screen caused by a potential difference of 1V on the control grids. The length of the control grids is l, the distance between them is d, the distance from the end of the grids to the screen is L, and the accelerating potential difference is U_0.

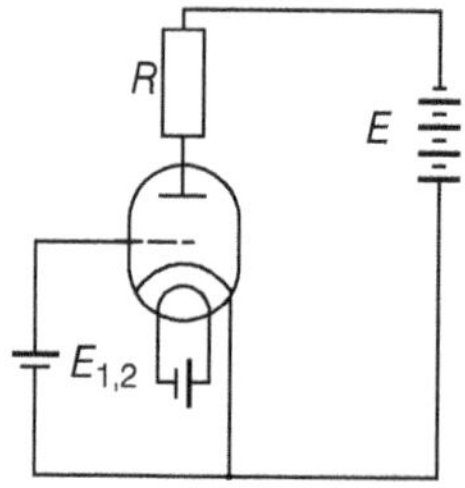

Fig. 189

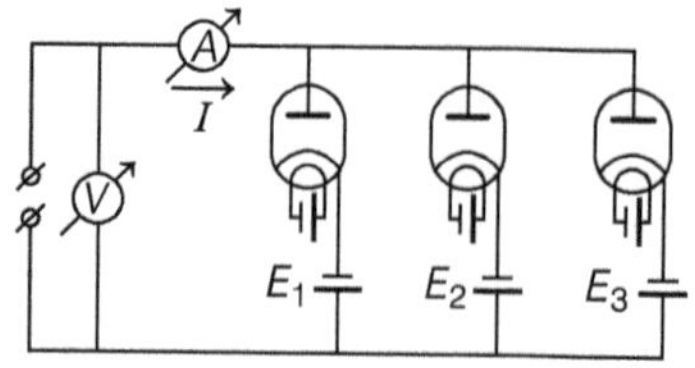

Fig. 190

3-4. Magnetic Field of a Current. Action of a Magnetic Field on a Current and Moving Charges

544. Determine the dimension and magnitude of the coefficient k in the expression for the intensity of the magnetic field of a solenoid $H = k\,4\pi I\frac{n}{l}$, if H is measured in oersteds and I in cgs electrostatic units.

The dimension of the oersted coincides with that of the electric field intensity in cgs units.

545. Two windings connected as shown in Fig. 191 are wound around a thin iron ring with a radius $R = 10$ cm. The first winding has 2,000 turns and the second 1,000 turns. Find the intensity of the magnetic field inside the ring if a current of $I = 10\ A$ flows through the windings.

546. A current I flows through an infinitely long conductor ABC bent to form a right angle (Fig. 192).

How many times will the intensity of the magnetic field change at point M if an infinitely long straight conductor BD is so connected to point B that the current I branches at point B into two equal parts and the current in the conductor AB remains the same?

Note. Take into account the fact that the intensity of a magnetic field induced at a certain point by a small element of current is perpendicular to the plane containing this element and a radius-vector drawn from this current element to the given point.

547. A current flows through a conductor arranged in one plane as shown in Fig. 193. Find the intensity of the magnetic field at an arbitrary point on line AB, which is the axis of symmetry of the conductor.

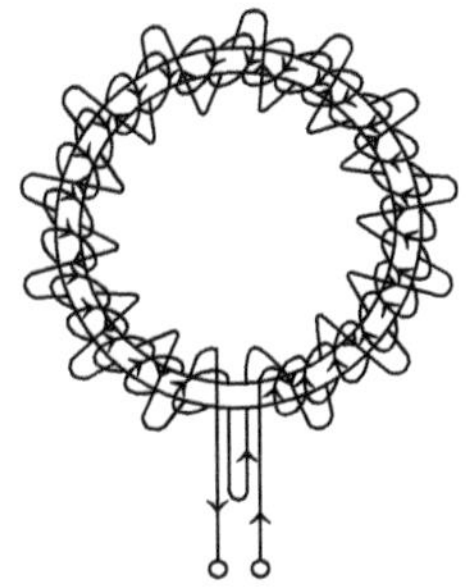

Fig. 191

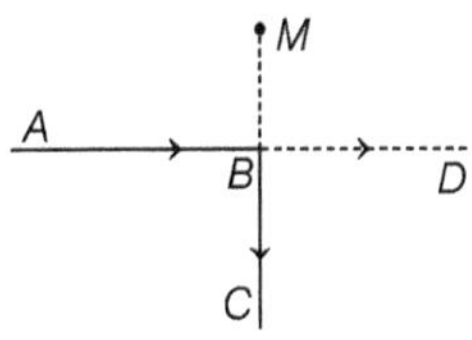

Fig. 192

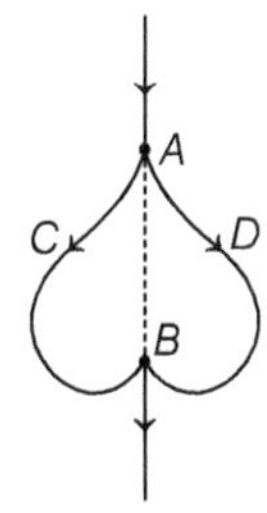

Fig. 193

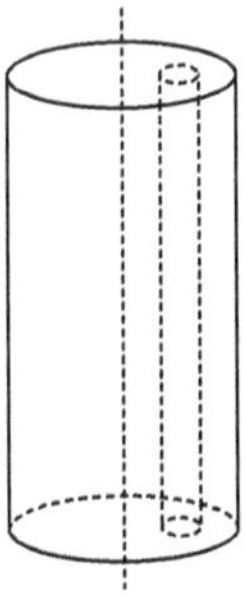

Fig. 194

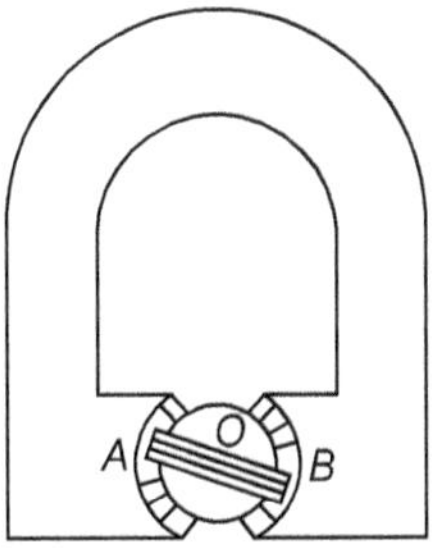

Fig. 195

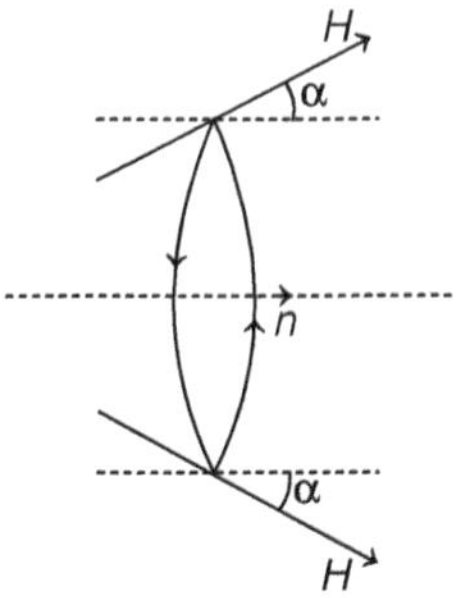

Fig. 196

548. How will a magnetic pointer be positioned if it is placed in the centre of a single-layer toroidal solenoid through which a direct current flows?

549. A current I flows along an infinite straight thin-walled pipe. Bearing in mind that the intensity of the magnetic field of an infinite straight conductor at a distance r from it is proportional to I/r, find the intensity of the magnetic field at an arbitrary point inside the pipe.

550. Remembering that the intensity of a magnetic field inside a long cylindrical conductor $H = k2\pi jr$, where j is the current density and r is the distance from the conductor axis, find the intensity of the field at an arbitrary point on a long cylindrical space inside the conductor (Fig. 194) through which a current with a density j flows. The axis of the space is parallel to the axis of the conductor and is at a distance d from it.

551. Draw the distribution of the force lines of a magnetic field in the space of the cylindrical conductor described in Problem 550.

552. Determine the dimension and the magnitude of the coefficient k in the expression for the force $F = kHIl \sin \psi$ acting from a magnetic field on a current if H is in oersteds and I in cgs electrostatic units.

553. Will the density of a direct current flowing in a cylindrical conductor be constant across the entire cross section of the conductor?

554. A lightning arrester is connected to earth by a circular copper pipe. After lightning strikes, it is discovered that the pipe became a circular rod. Explain the cause of this phenomenon.

555. A very great current is made to flow for a short time through a thick winding of a solenoid. Describe the deformation of the winding from the viewpoint of quality.

556. The magnetic system of a galvanometer consists of a magnet, pole shoes A and B, and a cylinder made of soft iron (Fig. 195). The magnetic force lines in the gap between the shoes and the cylinder are perpendicular to the surface of the cylinder. The intensity of the magnetic field is H. A rectangular coil with n turns is placed in the gap on axis O. The sides of the coil are parallel to the diameter and the generatrix of the cylinder. The area of each turn is A. One end of a spiral spring is so attached to the axis of the coil that when the latter rotates through an angle α, the deformation of the spring creates a rotational moment $k\alpha$ that tends to turn the coil to a position of equilibrium. Determine the angle through which the coil will turn if a current I passes through it.

557. A current of $I = 1$ A flows through a wire ring with a radius $R = 5$ cm suspended on two flexible conductors. The ring is placed in a homogeneous magnetic field with an intensity of $H = 10$ Oe whose force lines are horizontal. What force will the ring be tensioned with?

558. A wire ring with a radius $R = 4$ cm is placed into a heterogeneous magnetic field whose force lines at the points of intersection with the ring form an angle of $\alpha = 10°$ with a normal to the plane of the ring (Fig. 196). The intensity of the magnetic field acting on the ring $H = 100$ Oe. A current of $I = 5$ A flows through the ring. What force does the magnetic field act on the ring with?

559. A rectangular circuit $ABCD$ with sides a and b placed into a homogeneous magnetic field with an intensity H can revolve around axis OO' (Fig. 197). A direct current I constantly flows through the circuit.

Determine the work performed by the magnetic field when the circuit is turned through 180° if initially the plane of the circuit is perpendicular to the magnetic field and arranged as shown in Fig. 197.

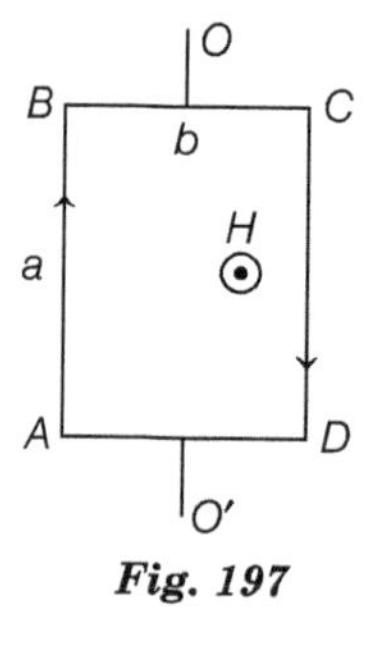

Fig. 197

560. A conductor is placed into a magnetic field whose intensity H forms an angle α with the conductor. A force $F = kHIl\sin\alpha$ acts on the section of the conductor with a length l when it carries a current I. If F is expressed in dynes, H in oersteds, I in amperes and l in centimetres, then $k = 0.1$.

This force is the resultant of all the forces that act on the moving electrons present at this moment in the volume of the section of the conductor with a length l.

Find the force which the magnetic force acts on one electron with.

561. Can a magnetic field independent of time change the velocity of a charged particle?

562. How will an electron move in a homogeneous magnetic field if the velocity of the electron at the initial moment is perpendicular to the force lines of the field?

563. How will an electron move in a homogeneous magnetic field if the velocity of the electron at the initial moment forms an angle α with the force lines of the field?

564. A current I flows along a metal band with a width $AB = a$ placed in a magnetic field with an intensity H perpendicular to the band (Fig. 198). Find the potential difference between points A and B of the band.

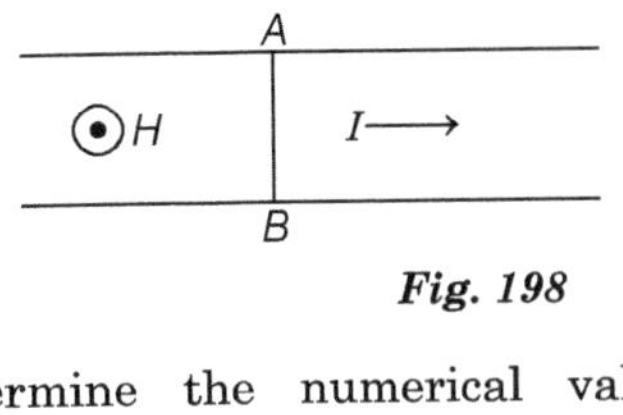

Fig. 198

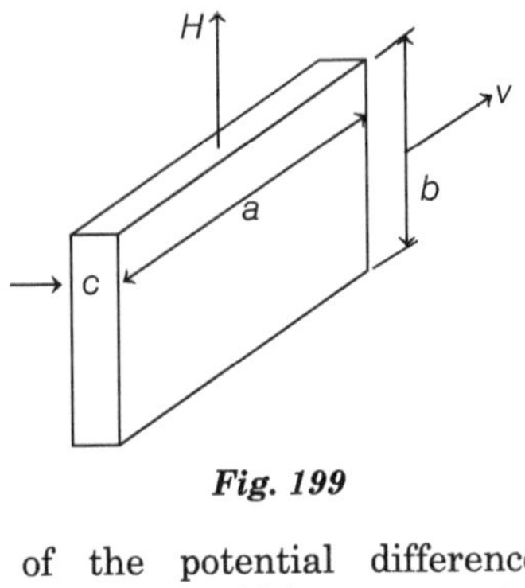

Fig. 199

565. Determine the numerical value of the potential difference (see Problem 564) if $H = 10{,}000$ Oe, the band width $a = 1$ cm, its thickness $d = 0.1$ mm, and the current $I = 10$ A. The number of electrons in a unit of volume is $n = 9 \times 10^{21}$ cm^{-3}.

566. An uncharged metal block has the form of a rectangular parallelepiped with sides a, b, and c ($a \gg c$, $b \gg c$). The block moves in a magnetic field in the direction of side a with a velocity v. The intensity of the magnetic field H is perpendicular to the base of the block with the sides a and c (Fig. 199).

Determine the intensity of the electric field in the block and the density of the electric charges on the surfaces of the parallelepiped formed by sides a and b.

567. An uncharged metal cylinder with a radius r revolves about its axis in a magnetic field with an angular velocity ω. The intensity of the magnetic field is directed along the axis of the cylinder.

What should the intensity of the magnetic field be for no electrostatic field to appear in the cylinder?

3-5. Electromagnetic Induction. Alternating Current

568. Determine the direction of the intensity of an electric field in a turn placed in a magnetic field (Fig. 200) directed away from us in a direction perpendicular to the plane of the turn. The intensity of the magnetic field grows with time.

569. A rectangular circuit $ABCD$ moves translationally in the magnetic field of a current flowing along straight long conductor OO' (Fig. 201). Find the direction of the current induced in the circuit if the turn moves away from the conductor.

570. A non-magnetized iron rod flies through a coil connected to a battery and an ammeter (Fig. 202). Draw an approximate diagram of the change of current in the coil with time as the rod flies through it.

571. A current in a coil grows directly with time. What is the nature of the relation between the current and time in another coil inductively connected to the first one?

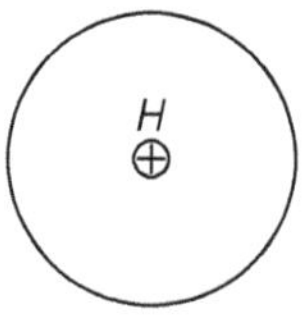

Fig. 200

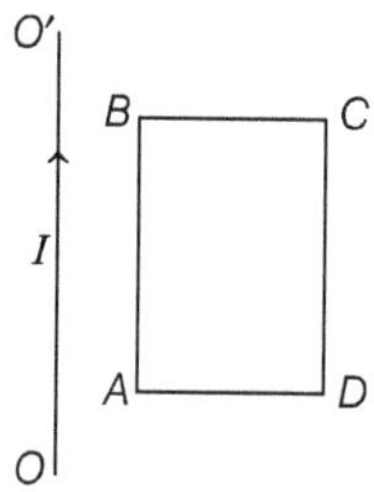

Fig. 201

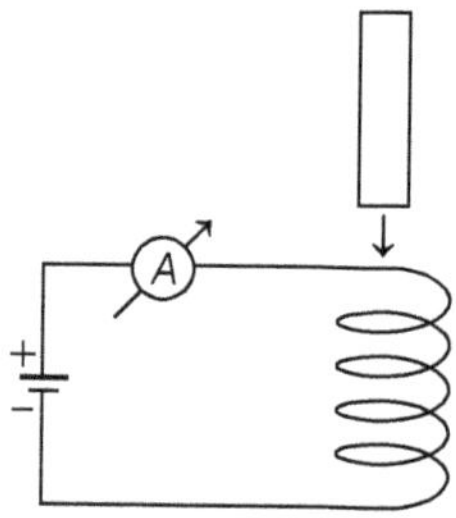

Fig. 202

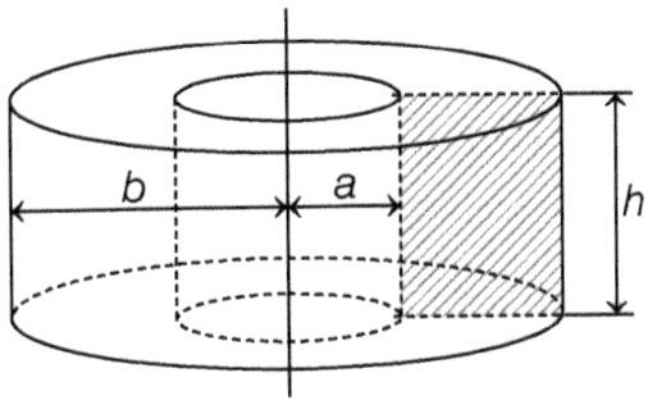

Fig. 203

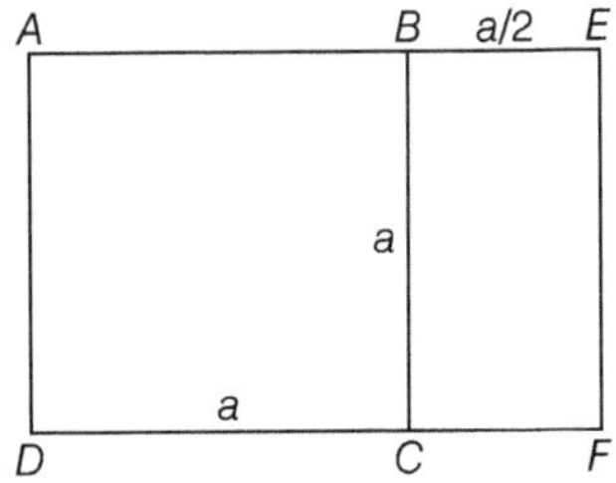

Fig. 204

572. Will the result of Problem 571 change if an iron core is inserted into the second coil?

573. A wire ring with a radius r is placed into a homogeneous magnetic field whose intensity is perpendicular to the plane of the ring and changes with time according to the law $H = kt$. Find the intensity of the electric field in the turn.

574. A ring of a rectangular cross section (Fig. 203) is made of a material whose resistivity is ρ. The ring is placed in a homogeneous magnetic field. The intensity of the magnetic field is directed along the axis of the ring and increases directly with time, $H = kt$. Find the intensity of the current induced in the ring.

575. A coil having n turns, each with an area of A, is connected to a ballistic galvanometer. (The latter measures thc quantity of electricity passing through it.) The resistance of the entire circuit is R. First the coil is between the poles of a magnet in a region where the magnetic field H is homogeneous and its intensity is perpendicular to the area of the turns. Then the coil is placed into a space with no. magnetic field. What quantity of the electricity passes through the galvanometer? (Express the answer in coulombs.)

576. Determine the current in the conductors of the circuit shown in Fig. 204 if the intensity of a homogeneous magnetic field is perpendicular to the plane of the drawing and changes in time according to the law $H = kt$. The resistance of a unit of length of the conductors is r.

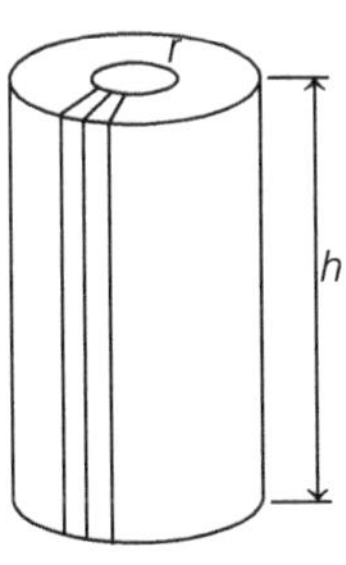

Fig. 205

577. The winding of a laboratory regulating autotransformer is wound around an iron core having the form of a rectangular toroid (Fig. 205). For protection against eddy (Foucault) currents the core is assembled of thin iron laminas insulated from one another by a layer of varnish. This can be done in various ways:

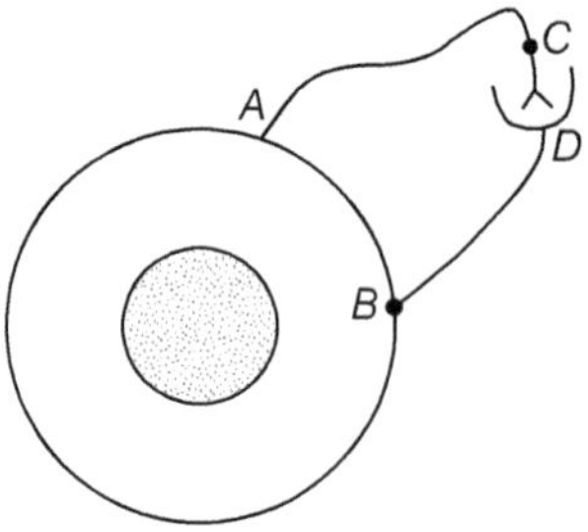

Fig. 206

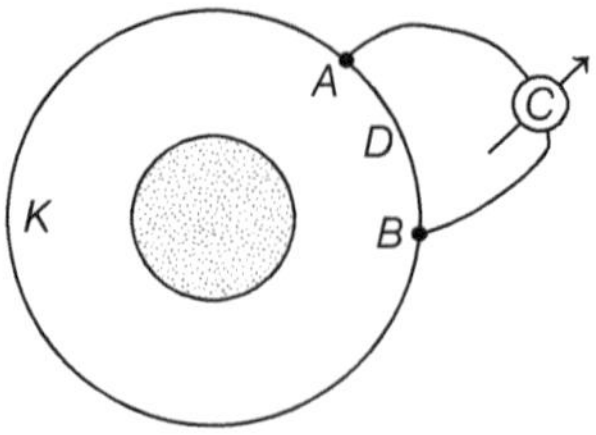

Fig. 207

(1) by assembling the core of thin rings piled on one another;

(2) by rolling up a long band with a width h;

(3) by assembling the core of rectangular laminas $l \times h$ in size, arranging them along the radii of the cylinder.Which is the best way?

578. A direct induced current I is generated in a homogeneous circular wire ring. The variable magnetic field producing this current is perpendicular to the plane of the ring, concentrated near its axis and has an axis of symmetry passing through the centre of the ring (Fig. 206). What is the potential difference between points A and B? What is the reading of an electrometer connected to these points?

579. A variable magnetic field creates a constant e.m.f. E in a circular conductor *ADBKA* (see Problem 578). The resistances of the conductors *ADB*, *AKB* and *ACB* (Fig. 207) are equal to R_1, R_2 and R_3, respectively. What current will be shown by ammeter C? The magnetic field is concentrated near the axis of the circular conductor.

580. The resistance of conductor *ACB* (see Problem 579) is $R_3 = 0$. Find the currents I_1, I_2 and I_3 and the potential difference $U_A - U_B$.

581. A medical instrument used to extract alien particles from an eye has the form of a strong permanent magnet or an electromagnet. When brought close to the eye (without touching it) it extracts iron and steel particles (filings, chips, etc.).

What current should flow through the electromagnet to extract, without touching the eye, metal objects made of non-ferromagnetic materials (aluminium, copper, etc.)?

582. A wire ring secured on the axis passing through its centre and perpendicular to the force lines is placed in a homogeneous magnetic field (Fig. 208). The intensity of the field begins to grow. Find the possible positions of equilibrium of the ring and show the position of stable equilibrium. What will happen if the intensity of the field decreases?

583. A conductor with a length l and mass m can slide without friction along two vertical racks AB and CD connected by a resistor R. The system is in a homogeneous magnetic field whose intensity H is perpendicular to the plane of the drawing (Fig. 209).

How will the movable conductor travel in the field of gravity if the resistance of the conductor itself and the racks is neglected?

584. A conductor with a mass m and length l can move without friction along two metallic parallel racks in a horizontal plane and connected across capacitor C. The entire system is in a homogeneous magnetic field whose intensity H is directed upward. A force F is applied to the middle of the conductor perpendicular to it and parallel to the racks (Fig. 210).

Determine the acceleration of the conductor if the resistance of the racks, feeding wires and conductor is zero. What kinds of energy will the work of the force F be converted into? Assume that the velocity of the conductor is zero at the initial moment.

585. Considering the motion of a straight magnet in a plane perpendicular to a wire and using the law of conservation of energy, prove that the field of a long forward current diminishes with the distance from the wire as $1/R$.

586. A cylinder made of a non-magnetic material has N turns of a wire (solenoid) wound around it. The radius of the cylinder is r and its length $l (r << l)$. The resistance of the wire is R. What should the voltage at the ends of the solenoid be for the current flowing in it to increase directly with time, i.e., $I = kt$?

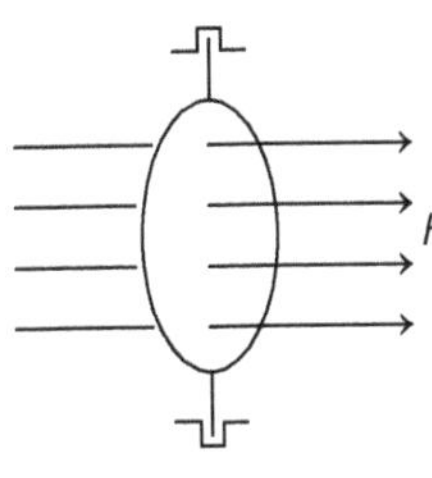

Fig. 208

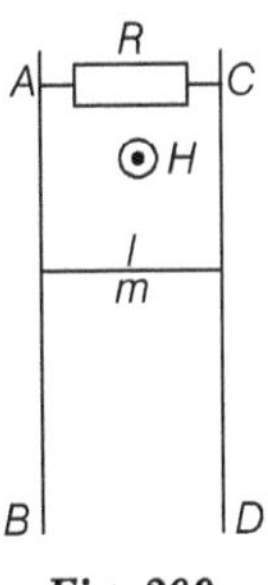

Fig. 209

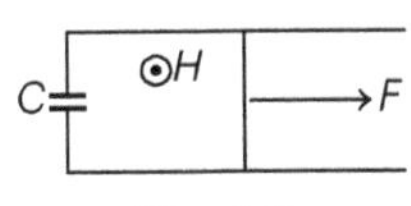

Fig. 210

587. A solenoid (see Problem 586) is connected to a battery whose e.m.f. is E. The key is closed at the moment $t = 0$. What is the intensity of the current flowing through the circuit of the solenoid if the resistance R of the solenoid, battery and feeding wires is neglected?

588. Calculate the work of the battery (see Problem 587) during the time τ. What kind of energy is this work converted into?

589. A ring made of a superconductor is placed into a homogeneous magnetic field whose intensity grows from zero to H_0. The plane of the ring is perpendicular to the force lines of the field. Find the intensity of the induction current appearing in the ring. The radius of the ring is r and its inductance L.

590. A superconductive ring with a radius r is in a homogeneous magnetic field with an intensity H. The force lines of the field are perpendicular to the plane of the ring. There is no current in the ring. Find the magnetic flux piercing the ring after the magnetic field is switched off.

591. Find the inductance of a coil wound onto the iron core shown in Fig. 211. The number of turns of the coil N, the cross-sectional area A, the perimeter of the core (medium line) l and the permeability of the core μ_r are known.

Note. Take into account the fact that the intensity of the magnetic field inside the core is practically constant and can be approximately expressed by the formula $H = 0.4\pi \frac{N}{l} I$.

592. Estimate approximately the coefficient of mutual inductance of the windings of a transformer. Consider the windings as coils of identical cross section. Disregard the dispersion of the force lines of the magnetic field.

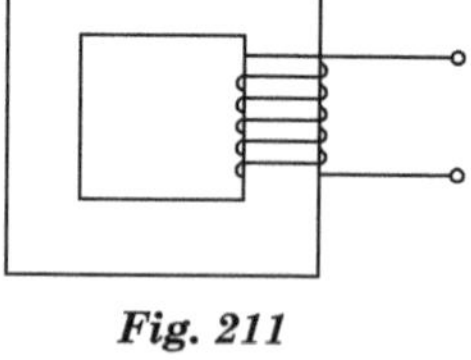

Fig. 211

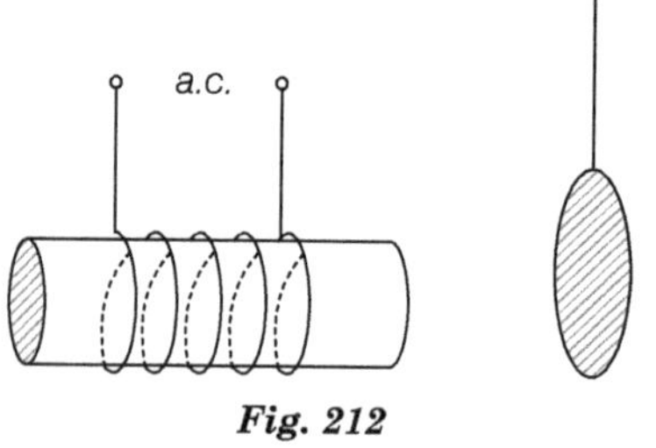

Fig. 212

Note. The coefficient of mutual inductance of two circuits is the ratio between the magnetic flux Φ induced by the magnetic field of the first circuit through the area limited by the second circuit and the magnitude of this current $M = \frac{\Phi}{I}$.

593. A light aluminium disk is suspended on a long string in front of the pole of an electromagnet (Fig. 212).
How will the disk behave if an alternating current is passed through the winding of the electromagnet? The resistance of the disk is small.

594. When a capacitor carrying a charge Q is connected in parallel with an uncharged capacitor having the same capacitance, the energy of the electric field of the system is halved (see Problem 441). Without resorting to the law of conservation of energy, prove by direct calculations that the amount of heat evolved in the wires is $W_e = \frac{W_{eo}}{2}$, where $W_{eo} = \frac{Q^2}{2c}$ is the initial energy of the system. Disregard the inductance of the connecting wires.

595. Find the effective magnitude of an alternating current that changes according to the law:

$$I = I_0 \text{ when } 0 < t < \frac{T}{8}$$
$$I = 0 \text{ when } \frac{T}{8} < t < \frac{T}{2}$$
$$I = -I_0 \text{ when } \frac{T}{2} < t < \frac{5}{8}T$$
$$I = 0 \text{ when } \frac{5}{8}T < t < T$$
$$I = I_0 \text{ when } T < t < \frac{9}{8}T$$

etc., (Fig. 213).

596. A d-c ammeter and an a-c thermal ammeter are connected to a circuit in series. When a direct current is passed through the circuit, the d-c ammeter shows $I_1 = 6$ A. When a sinusoidal alternating current flows through the circuit, the a-c ammeter shows $I_2 = 8$A. What will the reading of each ammeter be if a direct and an alternating currents flow simultaneously through the circuit?

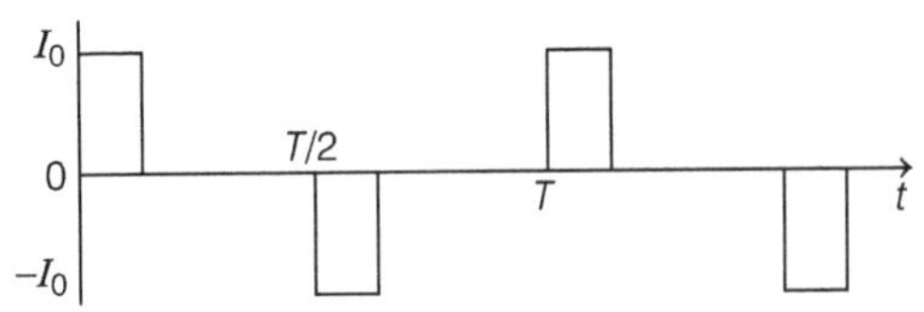

Fig. 213

597. An alternating sinusoidal current flows through a coil without any ohmic resistance. Draw a diagram showing the product of the current and the voltage (instantaneous power) changes depending on time. Explain the nature of the curve. What is the average power consumed by the coil per period?

598. Like an electric arc, daylight lamps have a dropping current-voltage characteristic, and for this reason a coil with a high inductance (choke) is connected in series with the lamp as a ballast resistance for stable burning. Why are ordinary less expensive resistors not used?

599. Why are capacitors connected in parallel with electric devices having a high inductance (chokes, for example), if there are many of them in a-c mains?

600. (a) A tap C is made from the middle of a coil with an iron core (the winding is a thick copper wire with many turns) (Fig. 214). A constant potential difference U_1 is created between points B and C. Find the voltage U_2 between points A and B.

(b) A variable potential difference with an amplitude U_1 (for example, from town mains) is applied between points B and C. Find the amplitude U_2 of the variable potential difference between points A and B.

601. Why does the presence of a very high voltage in the secondary winding of a step-up transformer (see Problem 513) not lead to great losses of energy due to the evolution of heat in the winding itself ?

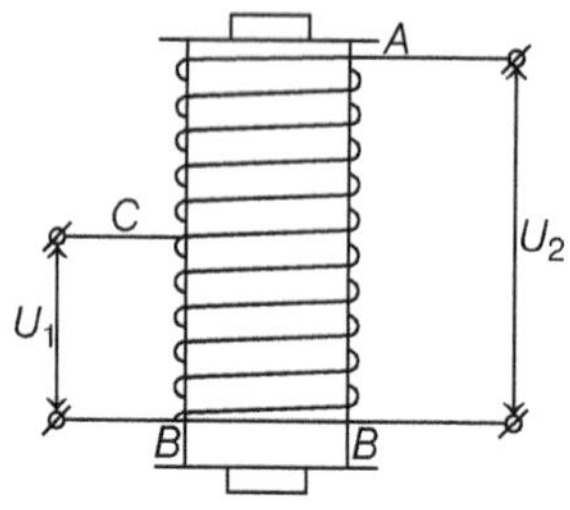

Fig. 214

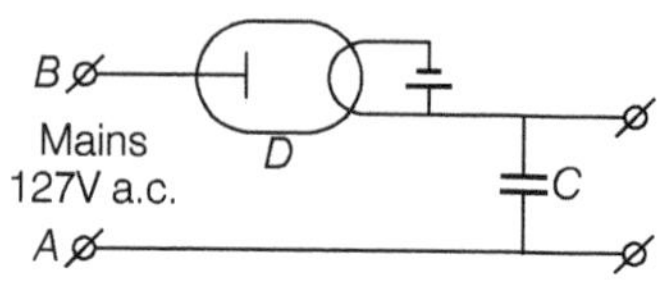

Fig. 215

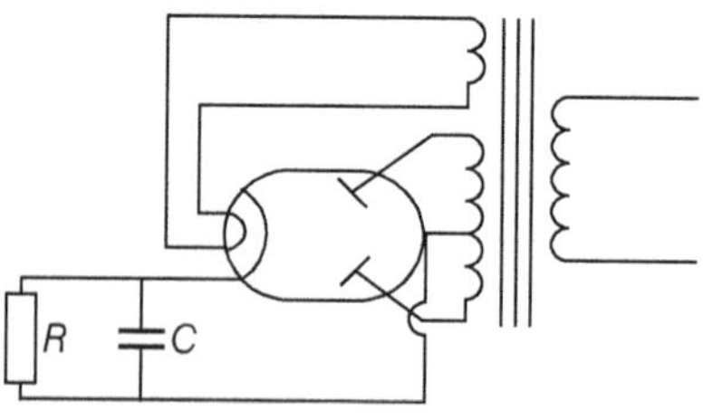

Fig. 216

602. Show that the proportion $\frac{I_1}{I_2} = \frac{N_2}{N_1}$, where I_1 and I_2 are the currents in the windings and N_1 and N_2 the numbers of turns in them, exists, if the no-load current of a transformer and the ohmic resistance of its windings are neglected.

Consider the windings as coils with the same cross section.

603. What puncturing voltages should capacitor C and diode D be calculated for if the rectifier (Fig. 215) can operate both with or without load?

604. An alternating voltage with an amplitude of $U = 600$ V is excited at the ends of the secondary winding of a transformer supplying a biphase rectifier (Fig. 216). The capacitance of the capacitor C is so high that the current I flowing through the resistance $R = 5\ k\ \Omega$ can be considered as approximately constant ($I = 40$ mA).

Assuming that each of the diodes passes no current in the opposite direction, find the share of the period T during which no current flows through the valve.

3-6. Electrical Machines

605. The resistance in the load circuit of an a-c generator increases. How should the power of the motor revolving the generator change for the frequency of the alternating current to remain the same?

606. The force acting on a moving charged particle from the side of a magnetic field (the Lorentz force) is always perpendicular to the velocity. Therefore, this force performs no work (see Problem 561).

Why does an electric motor operate in this case? We know that the force acting on a conductor carrying a current is caused by the action of the field on separate particles whose motion produces a current.

607. Can a d-c series motor connected to mains with a voltage of $U = 120$ V develop a power of $P = 200$ watts if the resistance of its windings is $R = 20\ \Omega$?

608. Determine the efficiency of a series and a shunt-wound motors if they develop the maximum power. The voltages across the terminals U and the resistances of the windings of the rotor R_1 and the stator R_2 are the same in both motors and are known.

609. The rotor of a model of a d-c motor consists of one turn in the form of a rectangle. The intensity of the magnetic field H produced by a permanent magnet (north at the left and south at the right) is directed along the radius because the gap between the pole shoes and iron cylinder C is very small (Fig. 217). A potential difference U is applied to the turn, whose area is A and resistance is R.

Find the power of the motor as a function of the angular velocity ω. At what angular velocity ω will the power be maximum, and what will the current be at this power?

610. Determine how the rotational moment (torque) M depends on the angular velocity, using the condition of the previous problem.

611. Find the nature of the relation between the power of a d-c motor (see Problem 609) and the intensity of the magnetic field H at a given speed. At what value of H does the power reach its maximum?

612. Determine the intensity of the magnetic field in a d-c motor (see Problem 609) at which the torque M is maximum. The speed of the armature is known.

613. A d-c shunt-wound motor develops a mechanical power of $P = 160$ watts with a voltage across the terminals of $U = 120$ V. The armature rotates at $n = 10$ rev/s. Determine the maximum possible speed of the motor at this voltage. The resistance of the armature is $R = 20\ \Omega$.

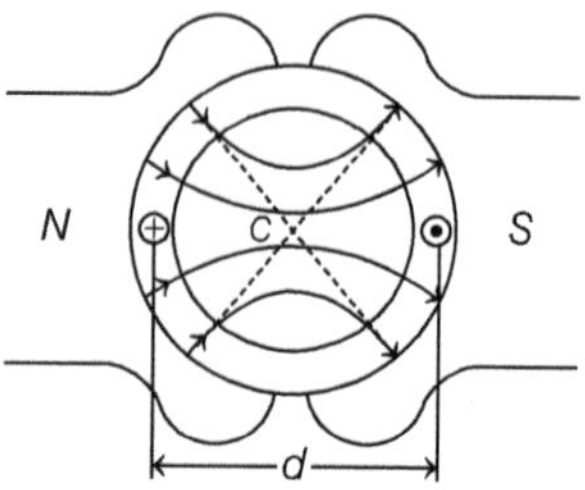

Fig. 217

614. A d-c shunt-wound motor has an angular velocity of the armature rotation of $\omega = 100$ rad/s at a voltage of $U = 120$ V across the terminals. The resistance of the motor armature winding is $R = 20\ \Omega$. What electromotive force will this motor develop when used as a generator if it is rotated with the same angular velocity? The voltage in the stator windings is kept constant and equal to 120 V. At the velocity indicated, the mechanical moment on the motor shaft is $M = 1.6 \times 10^7$ dyne·cm.

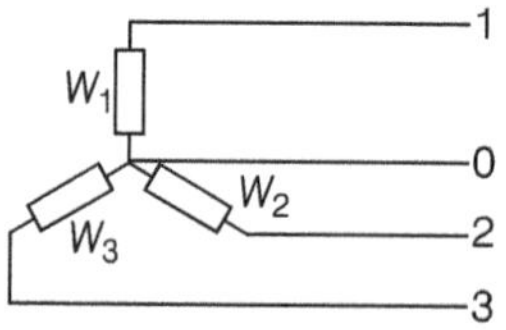

Fig. 218

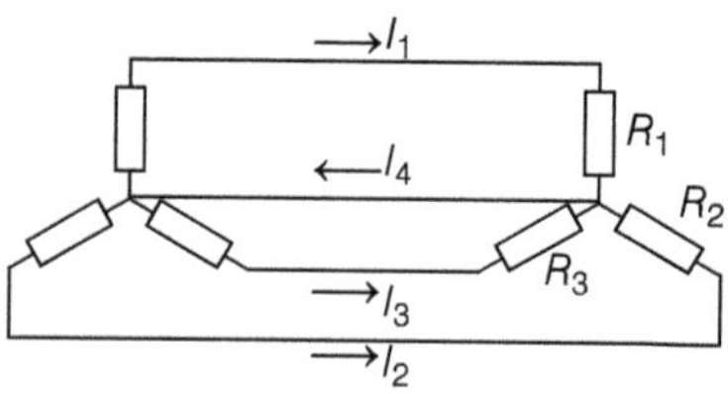

Fig. 219

615. How will the speed n of a shunt-wound motor change when the current in the stator windings grows if the voltage across the armature U and the mechanical moment M applied to the armature axis remain constant?

616. What parameters of mains would determine the power of a d-c series motor connected to them if the winding of the motor were made of a superconductor?

617. Prove that if the windings of a three-phase generator are star-connected (Fig. 218), the voltages between linear conductors U_{12}, U_{18} and U_{23} are $\sqrt{3}$ times greater than the phase voltages U_{01}, U_{02} and U_{03}.

618. Prove that when the windings of a three-phase generator and the load resistances are star-connected (Fig. 219), the current I_4 flowing through the neutral conductor is zero if $R_1 = R_2 = R_3 = R$.

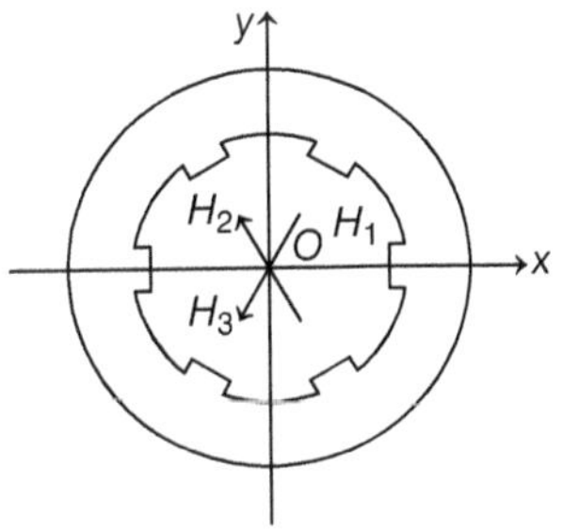

Fig. 220

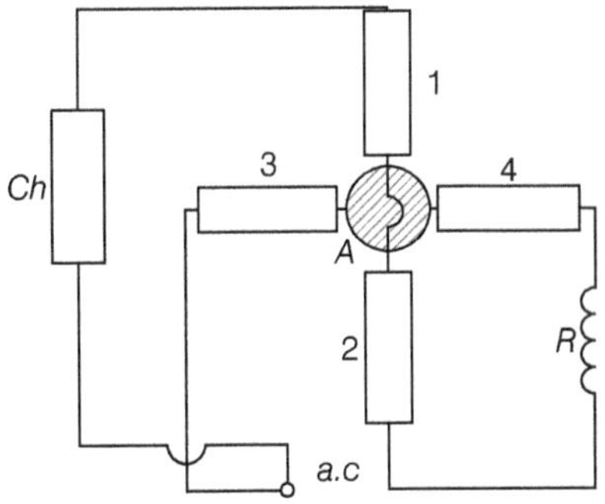

Fig. 221

619. Prove that if the intensities of the magnetic field generated by three pairs of electromagnets are equal in amplitude and shifted in phase through $\frac{2}{3}\pi$ (Fig. 220), the resultant magnetic field can be described by a vector rotating with a constant angular velocity ω about point O. Each pair of electromagnets creates magnetic fields directed along the respective diameters of the ring: H_1, H_2 and H_3. The electromagnets are fed with an alternating current having the frequency ω.

620. Two identical coils perpendicular to each other are divided in half and connected to a circuit as shown in Fig. 221.

The inductance of the choke Ch and the ohmic resistance R are so selected that the intensities of the currents in the coils are the same. The ohmic resistance and inductive reactance of the coils are much less than the inductive reactance of the choke.

What will occur if an aluminium cylinder A secured on an axis is introduced into the space between the poles of the coils?

4

OSCILLATIONS AND WAVES

4-1. Mechanical Oscillations

621. A weight suspended from a long string oscillates in a vertical plane and is deflected through an angle α from the vertical (a mathematical pendulum). The same weight can rotate over a circumference so that the string describes a cone (a conical pendulum). When will the tension of the string deflected through an angle α from the vertical be greater?

622. A clock with an oscillation period of the pendulum of 1 second keeps accurate time on the surface of the Earth. When will the clock go slower in a day: if it is raised to an altitude of 200 metres or lowered into a mine to a depth of 200 metres?

623. Two small spheres each with a mass of $m = 1$ g are secured to the ends of a weightless rod with a length $d = 1$ metre. The rod is so suspended from a hinge that it can rotate without friction around a vertical axis passing through its middle. Two large spheres with masses $M = 20$ kg are fastened on one line with the rod. The distance between the centres of a large sphere and a small one $L = 16$ cm (Fig. 222). Find the period of small oscillations of this torsion pendulum.

624. What is the period of oscillations of a mathematical pendulum in a railway carriage moving horizontally with an acceleration a?

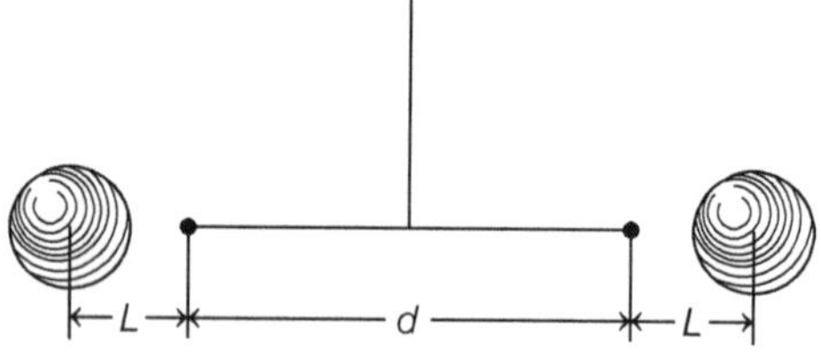

Fig. 222

625. Find the period of oscillations of a pendulum in a lift moving vertically with an acceleration a.

626. A block performs small oscillations in a vertical plane while moving without friction over the internal surface of a spherical cup. Determine the period of oscillations of the block if the internal radius of the cup is R and the face of the block is much smaller than R.

627. How will the period of oscillations of the block in the cup change (see Problem 626) if, besides the force of gravity, the cup is acted upon by a force F directed vertically upward? The mass of the cup M is much greater than that of the block m.

628. How will the period of oscillations of the block in the cup change (see Problem 626) if the cup is placed onto a smooth horizontal surface over which it can move without friction?

629. A hoop with a mass m and a radius r can roll without slipping over the internal surface of a cylinder with a radius R (Fig. 223). Determine the period of motion of the hoop centre if the angle φ is small.

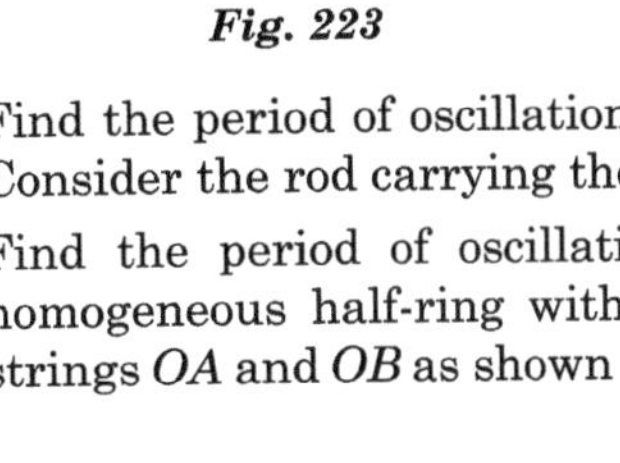

Fig. 223

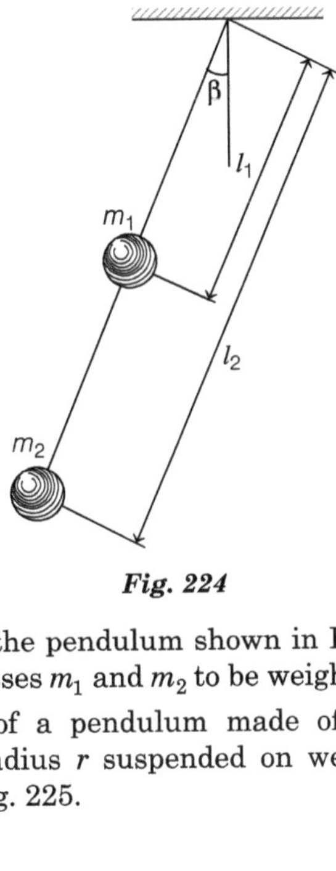

Fig. 224

630. Find the period of oscillations of the pendulum shown in Fig. 224. Consider the rod carrying the masses m_1 and m_2 to be weightless.

631. Find the period of oscillations of a pendulum made of a thin homogeneous half-ring with a radius r suspended on weightless strings OA and OB as shown in Fig. 225.

632. Figure 226 shows a mechanical system consisting of a weight with a mass m, spring A with a coefficient of elasticity k and a pulley with a mass M. A string passed over the pulley connects the weight to the spring. Find the period of oscillations of the weight if the pulley is a thin-walled cylinder.

633. With what frequency will a bottle with a mass of $m = 200$ g and a cross-sectional area of $A = 50\ \text{cm}^2$ oscillate if it floats on the surface of water in a vertical position?

Note. Remember that the period of oscillations of a weight on a spring is $T = 2\pi\sqrt{\frac{m}{k}}$, where k is the coefficient of elasticity of the spring.

634. Mercury is poured into communicating cylindrical vessels. Find the oscillation period of the mercury if the cross-sectional area of each vessel is $A = 0.3\ \text{cm}^2$ and the mass of the mercury $m = 484$ g. The specific weight of mercury is $\gamma = 13.6\ \text{gf/cm}^3$.

635. A mine pierces the Earth along one of its diameters. In what time will a body thrown into the mine reach the centre of the Earth? There is no resistance to motion.

636. A string secured at its ends is stretched with a force f. A point weight with a mass m is attached to the middle of the string (Fig. 227). Determine the period of small oscillations of the weight. The mass of the string and the force of gravity can be neglected.

637. How will the period of vertical oscillations of a weight hanging on two identical springs change if the series connection of the springs is changed to a parallel one?

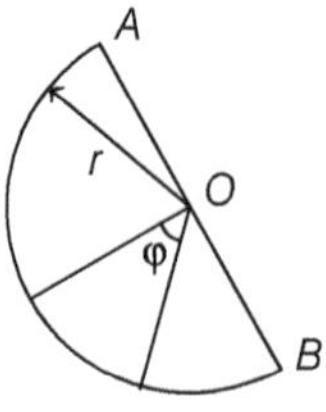

Fig. 225

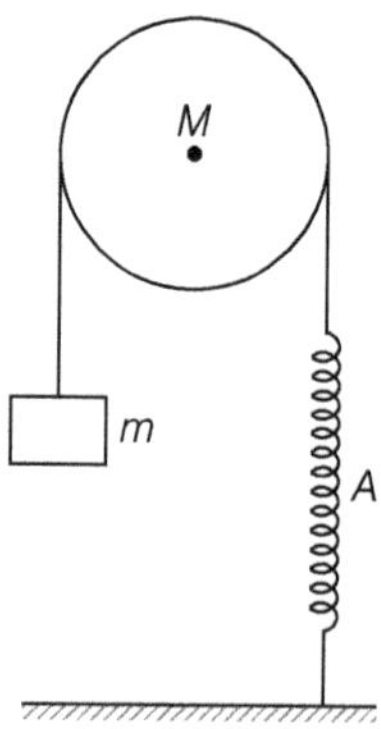

Fig. 226

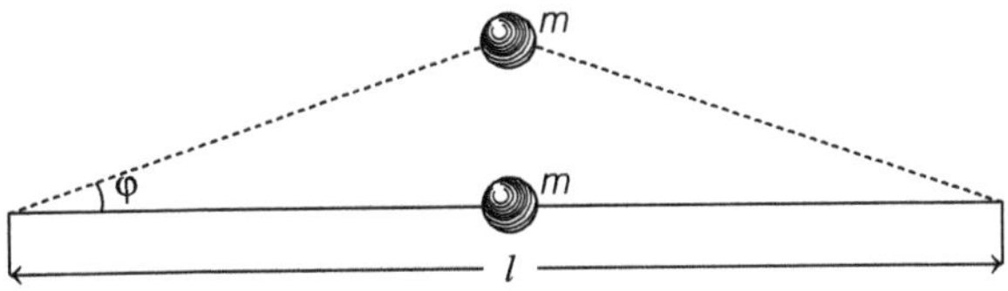

Fig. 227

638. Two mathematical pendulums each with a length l are connected by a weightless spring as shown in Fig. 228. The coefficient of elasticity of the spring is k. In equilibrium the pendulums occupy a vertical position and the spring is not deformed. Determine the frequencies of small oscillations of the two linked pendulums when they are deflected in one plane through equal angles in one direction (oscillations in phase) and in opposite directions (oscillations in antiphase).

639. A force of 5 kgf should be applied to the handle of an open swinging door to keep it in equilibrium (the door is returned to its usual closed position of equilibrium by springs). Can the door be opened by a force of 10 gf applied to the same handle? Disregard friction in the door hinges.

640. A weightless rod with a length l is rigidly attached to a weightless pulley with a radius r. The end of the rod carries a mass m (Fig. 229). A string to whose free end a mass M is fastened is passed over the pulley. In what condition will the motion of the system be oscillatory if the angle α between the rod and the vertical is zero at the initial moment?

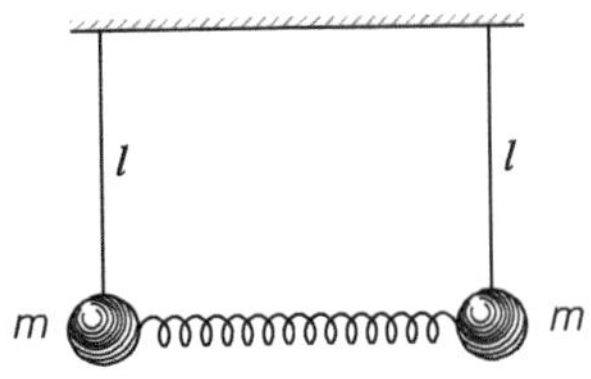

Fig. 228

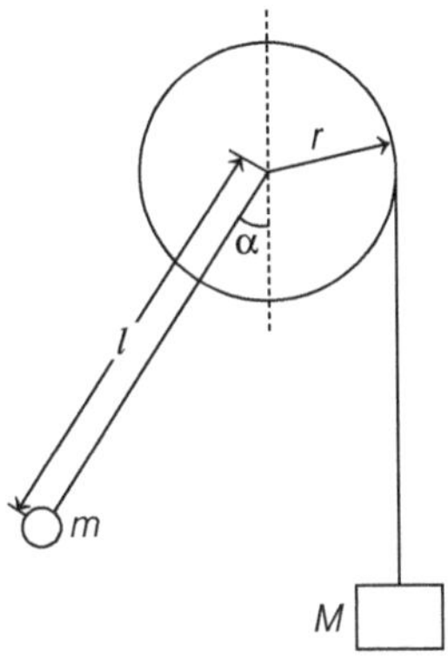

Fig. 229

4-2. Electrical Oscillations

641. Why is a permanent magnet needed in a telephone receiver? Why must the intensity of the magnetic field of this magnet be greater than the maximum intensity of the magnetic field generated by the current flowing through the winding of the telephone coil?

642. Find the frequency of natural oscillations in a circuit consisting of a solenoid with a length of $l = 3$ cm, a cross-sectional area of $S_1 = 1\ \text{cm}^2$ and a plane capacitor with a plate area of $S_2 = 30\ \text{cm}^2$, the distance between them being $d = 0.1$ cm.. The number of solenoid turns is $N = 1{,}000$.

643. An electric circuit consists of a capacitor with a constant capacitance and a coil into which a core can move. One core is made of ferrite and is used as an insulator, and the other is made of copper. How will the frequency of natural oscillations of the circuit change if (a) the copper core, (b) the ferrite core, is moved into the coil?

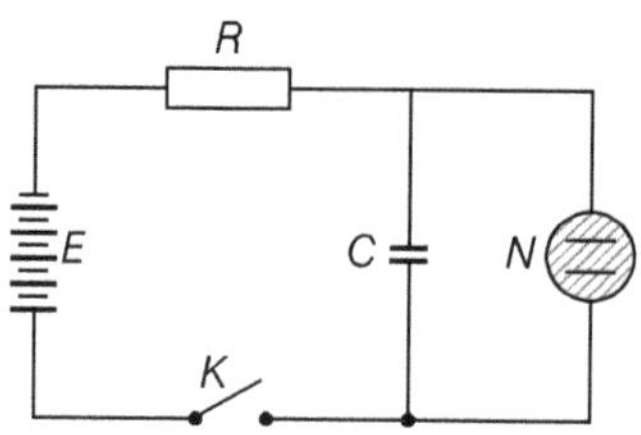

Fig. 230

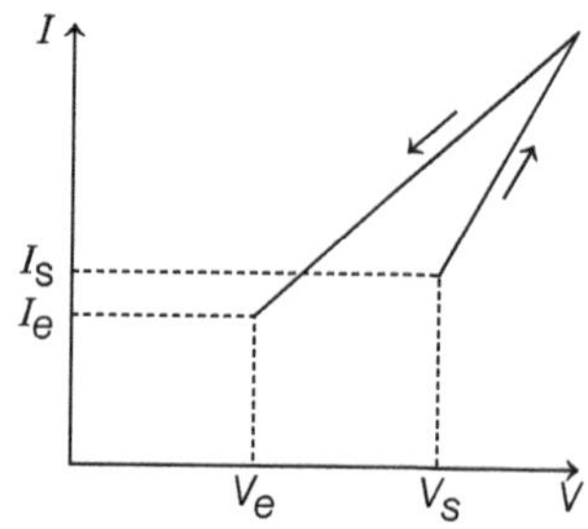

Fig. 231

644. The capacitors described in Problem 594 are connected by a superconductor. No heat is liberated. How can the decrease in the energy of the electric field be explained in this case from a qualitative point of view?

645. A voltage of $V_1 = V_{10} \cos \omega t$ is applied to the vertically deflecting plates of an oscillograph and a voltage of $V_2 = V_{20} \cos(\omega t - \varphi)$ to the horizontally deflecting plates. Find the trajectory of the cathode beam on the oscillograph screen if the phase difference between the voltage on the plates is $\varphi_1 = \frac{\varphi}{2}$ and $\varphi_2 = \pi$.

646. Figure 230 shows a circuit consisting of a battery E, a neon tube N, a capacitor C and a resistor R.

The characteristic of the neon tube (current in the tube versus voltage) is shown in Fig. 231. Current does not flow through the tube when the voltage is low. When the potential in the tube reaches V_s (striking potential), the tube ignites, the current jumps to I_s and then grows in proportion to V. If the voltage drops, the current drops at a slower rate than that at which it increased. The tube goes out at an extinguishing potential V_e.

Draw the approximate relation between the change of voltage across the capacitor and the time when the key K is closed.

647. How will the period of relaxation oscillations change in a circuit with a neon tube (see Problem 646) if the capacitance of the capacitor C and the resistance R change?

648. A plane capacitor incorporated into an oscillatory circuit is so designed that its plates can move relative to each other. How can the circuit be made to oscillate parametrically by moving the plates?

4-3. Waves

649. A thin string is replaced by another made of the same material, but with a double diameter. How many times should the tension of the string be changed to retain the previous frequency of oscillations?

650. Find the frequencies of natural oscillations of a steel string with a length of $l = 50$ cm and a diameter of $d = 1$ mm if the tension of the string is $T = 2{,}450$ dynes. The density of steel is $\rho = 7.8\ \text{g/cm}^3$.

651. Find the frequencies of the natural oscillations of an air column in a pipe with a length of $l = 3.4$ metres closed at both ends.

652. A tuning fork with a frequency of natural oscillations of $\upsilon = 340\ \text{s}^{-1}$ sounds above a cylindrical vessel one metre high. Water is slowly poured into the vessel. At what level of the water in the vessel will the sound of the tuning fork be appreciably intensified?

653. What is the shape of the front of the shock wave produced in air when a bullet flies at a supersonic velocity?

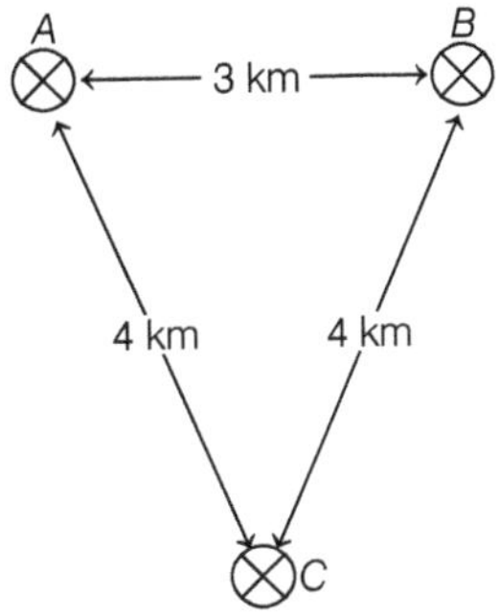

Fig. 232

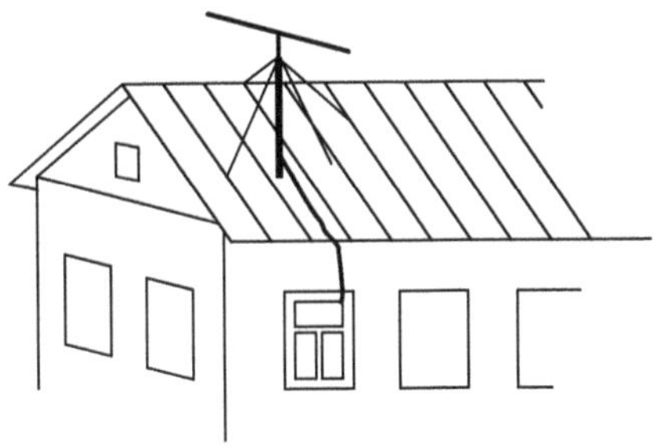

Fig. 233

654. A jet airplane flies at a velocity of 500 m/s at a distance of 6 km from a man. At what distance from the man was the plane when the man heard its sound?

655. If a source of sound and a man are about at the same altitude the sound is heard better in the direction of the wind than in the opposite direction. How can you explain this phenomenon?

656. Why can TV programmes be seen only within the range of direct visibility?

657. A radio direction finder operates in pulse duty. The pulse frequency is $f = 1{,}700$ cps and its duration is $\tau = 0.8\,\mu s$. Determine the maximum and minimum range of this finder.

658. Besides the wave transmitted directly from the station (point A), a TV aerial (point C in Fig. 232) receives the wave reflected from the iron roof of a building (point B). As a result, a double image appears. How many centimetres are the images shifted with respect to each other if the aerial and the roof are located as shown in Fig. 232? The width of the TV screen is $l = 50$ cm.

Note. Remember that the image is resolved into 625 lines, and 25 frames are transmitted a second.

659. A vibrator with a length of $l = 0.5$ metre is immersed into a vessel with kerosene ($\varepsilon_r = 2$). What is the length of the electromagnetic wave radiated by the vibrator in a vacuum as it emerges from the vessel?

660. Figure 233 depicts a TV aerial. How is the plane of oscillations of the magnetic vector of the wave coming from the TV centre orientated?

5

GEOMETRICAL OPTICS

5-1. Photometry

661. A round hall with a diameter of $D = 30$ metres is illuminated by a lamp secured in the centre of the ceiling. Find the height h of the hall if the minimum illumination of the wall is double that of the floor.

662. A lamp rated at $I_1 = 100$ cd hangs above the middle of a round table with a diameter of $D = 3$ metres at a height of $H = 2$ metres. It is replaced by a lamp with $I_2 = 25$ cd and the distance to the table is changed so that the illumination of the middle of the table remains as before. How will the illumination of the edge of the table change?

663. Sources of light S_1 and S_2 of equal intensity are arranged at the vertices of an isosceles right triangle (Fig. 234). How should a small plate A be positioned for its illumination to be maximum? The sides of the triangle $AS_1 = AS_2 = a$.

664. An attempt to use a photometer to measure the luminous intensity of a certain source of light failed since the luminous intensity was very high and the illumination of the photometer fields with the aid of a standard source could not be equalized even when the source being investigated was placed on the very edge of the photometer bench. Then a third source was employed with a luminous intensity lower than that of the one being investigated. At a distance of $r_1 = 10$ cm from the photometer the standard source produced the same

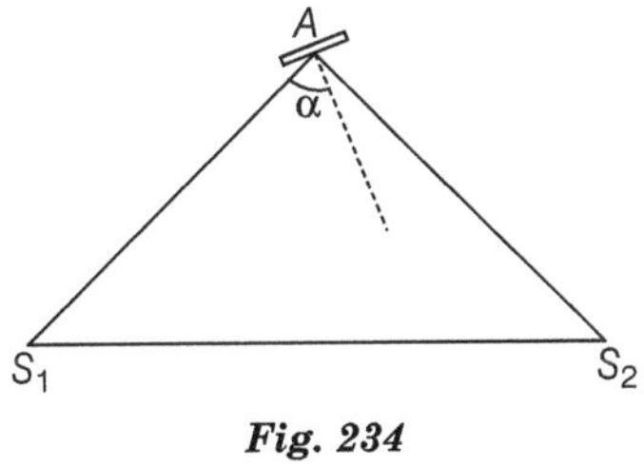

Fig. 234

illumination of the fields as the third one that was placed at a distance of $r_2 = 50$ cm. After that, the standard source was replaced by the one being investigated and equal illuminations were obtained at distances from the photometer $r_3 = 40$ cm (source being investigated) and $r_4 = 10$ cm (auxiliary source). Find how many times the luminous intensity of the source being investigated is greater than that of the standard source.

665. The ray of a searchlight falls on the wall of a house and produces a bright spot with a radius of $r = 40$ cm. How many times will the illumination of the wall of a remote house be smaller if the radius of the spot on it is 2 metres?

666. A lamp with a luminous intensity of $I = 100$ cd is fastened to the ceiling of a room. Determine the total luminous flux falling onto all the walls and the floor of the room.

667. What part of the energy radiated by the Sun reaches the Earth? The radius of the Earth is 6,400 km and the average distance from the Earth to the Sun is 149,000, 000 km.

668. A hot glowing wire is placed on the axis of a hollow cylinder with a radius R. The length of the wire is much greater than the height of the cylinder. How many times will the illumination of the internal surface of the cylinder change if its radius is R_2 (assume that $R_2 < R_1$) ?

669. At what height should a lamp be hung above the centre of a round table to obtain the maximum illumination on its edges?

670. Why can a text be read through tissue paper only if the paper is placed directly on the page?

5-2. Fundamental Laws of Optics

671. Why is the shadow of a man's legs on the ground sharp and that of his head diffused? In what conditions will the shadow be equally distinct everywhere?

672. How should a pencil be held above a table to obtain a sharp shadow if the source of light is a daylight lamp in the form of a long tube secured on the ceiling?

673. In autumn, when the trees have shed all their leaves, one can frequently see shadows from two parallel branches. The lower branch produces a sharp dark shadow and the upper branch a broader and lighter one. If two such shadows are accidently superimposed, one can see a bright light stripe in the middle of the darker shadow, so that the shadow seems to be a double one (Fig. 235). How do you explain this phenomenon?

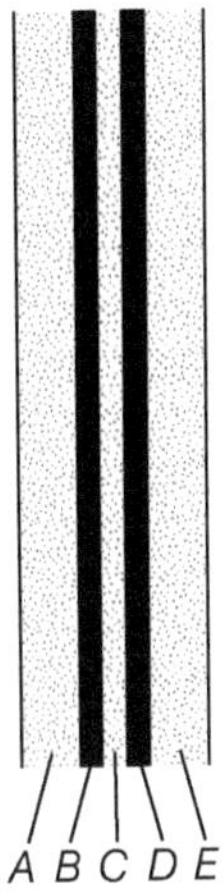

Fig. 235

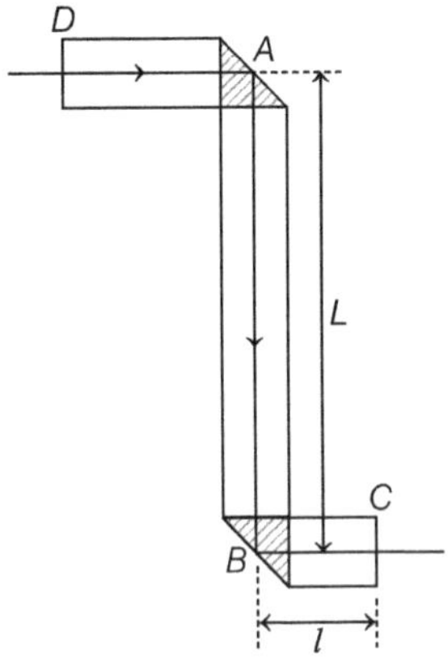

Fig. 236

674. When sunrays pass through a small opening In the foliage at the top of a high tree, they produce an elliptical spot on the ground. The major and minor axes of the ellipses are $a = 12$ cm and $b = 10$ cm, respectively. What is the height of the tree H? The angular dimensions of the Sun's disk are $\beta = 1/108$ rad.

675. A periscope is designed with two reflecting prisms. Determine the ratio of the widths of these prisms if the distance between them $AB = L$ and the distance from the lower prism to the eye of the observer $BC = l$ (Fig. 236). The objects viewed through the periscope are at a great distance from it.

676. What minimum height should a flat mirror secured vertically on a wall have for a man to see his reflection full size without changing the position of his head? Determine also the required distance between the floor and the lower edge of the mirror.

677. Sunrays are reflected from a horizontal mirror and fall on a vertical screen. An oblong object is placed on the mirror (Fig. 237). Describe the shadow on the screen.

678. In what conditions does the shape of the reflection of a sunbeam from a small mirror not depend on the shape of the mirror?

679. How can a real landscape on a photograph be distinguished from its reflection in quiet water?

680. Find graphically the positions of an observer's eye that will allow him to see in a mirror of finite dimensions the image of a straight line arranged as shown in Fig. 238.

681. A flat mirror is arranged parallel to a wall at a distance l from it. The light produced by a point source fastened on the wall falls on the mirror, is reflected and produces a light spot on the wall. With what velocity will the light spot move along the wall if the mirror is brought up to the wall with a velocity v? How will the dimensions of the light spot change?

682. Using the condition of Problem 681, find whether the illumination of the wall at the point where the light spot is will change when the mirror is moved. The dimensions of the mirror are much smaller than the distance from the mirror to the source of light.

683. A flat mirror revolves at a constant angular velocity, making $n = 0.5$ revolution per second. With what velocity will a light spot move along a spherical screen with a radius of 10 metres if the mirror is at the centre of curvature of the screen?

684. When the Russian scientist A. A. Belopolsky investigated the Doppler optical phenomenon he observed light repeatedly reflected from moving mirrors (Fig. 239). The mirrors were placed on disks revolving in different directions.

(a) The angular velocity ω of rotation of the disks being known, find the angular velocity Ω of rotation of a beam that is consecutively reflected n times from the mirrors.

(b) Determine the linear velocity of the n-th image at the moment when the mirrors are parallel to each other and their reflecting portions move with a velocity v in different directions.

685. Solve Problem 684 if the disks rotate in the same direction.

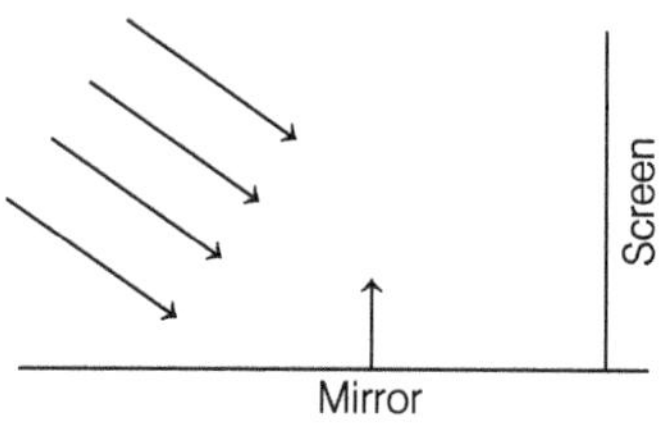

Fig. 237

Fig. 238

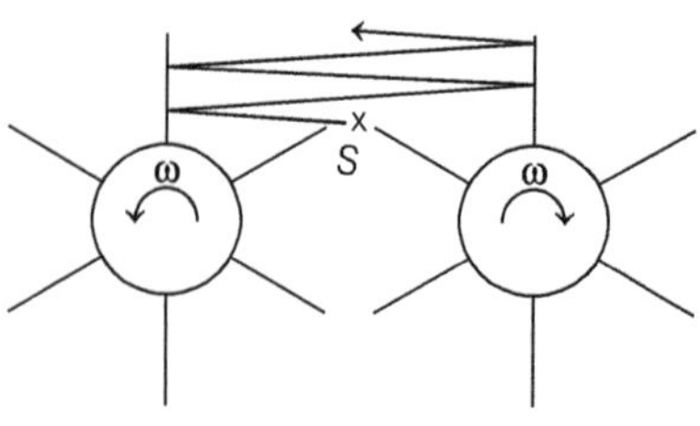

Fig. 239

686. A narrow beam of light S is incident on a dihedral angle $\alpha = 60°$ formed by identical flat mirrors OM and ON secured on axis O (Fig. 240). After being reflected from the mirrors, the light is focussed by lens L and gets into stationary receiver R. The mirrors rotate with a constant angular velocity.

What part of the light energy of the beam will reach the receiver during a time that greatly exceeds the period of rotation if the beam passes at a distance a from an axis equal to half the length of mirror OM?

687. Can a flat mirror be used instead of an ordinary cinema screen?

688. A projecting camera standing near a wall in a room produces an image with an area of $A = 1\ \text{m}^2$ on the opposite wall. What will the area of the image be if a flat mirror is so hung on the wall opposite the camera that the image is obtained on the wall near which the camera stands?

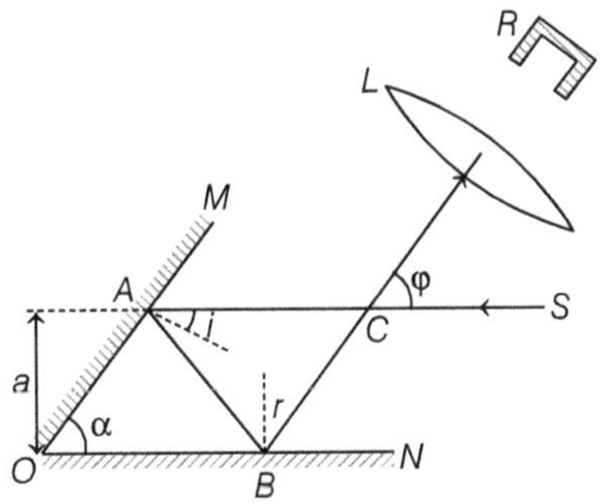

Fig. 240

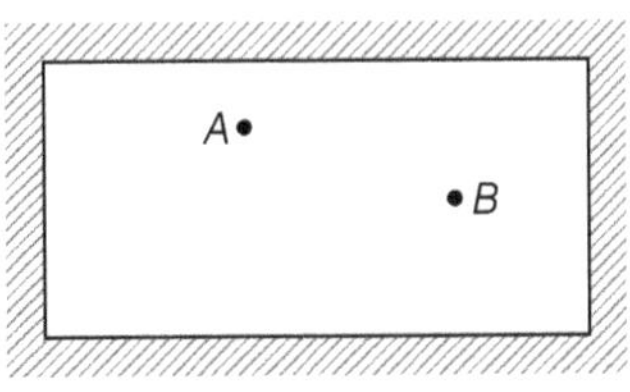

Fig. 241

689. Two flat mirrors AO and OB form a dihedral angle $\phi = \frac{2\pi}{n}$, where n is an integer. A point source of light S is placed between the mirrors at the same distance from both of them. Find the number of images of the source in the mirrors.

690. Two flat mirrors AO and OB form an arbitrary dihedral angle $\phi = \frac{2\pi}{a}$, where a is any number greater than 2. A point source of light S is between the mirrors at equal distances from them. Find the number of images of the source in the mirrors.

691. In what direction should a beam of light be sent from point A (Fig. 241) contained in a mirror box for it to fall onto point B after being reflected once from all four walls?

Points A and B are in one plane perpendicular to the walls of the box (i.e., in the plane of the drawing).

692. Why does the water seem much darker directly below an airplane flying over a sea than at the horizon?

693. Over what distance will a beam passing through a plane-parallel plate be displaced if the thickness of the plate is d, the refraction index is n and the angle of incidence is t?

Can the beam be displaced by more than the thickness of the plate?

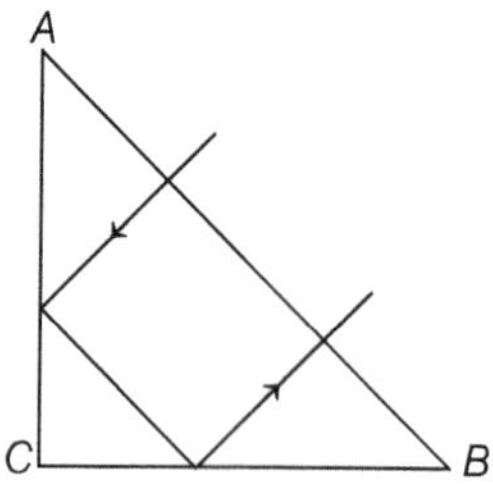

Fig. 242

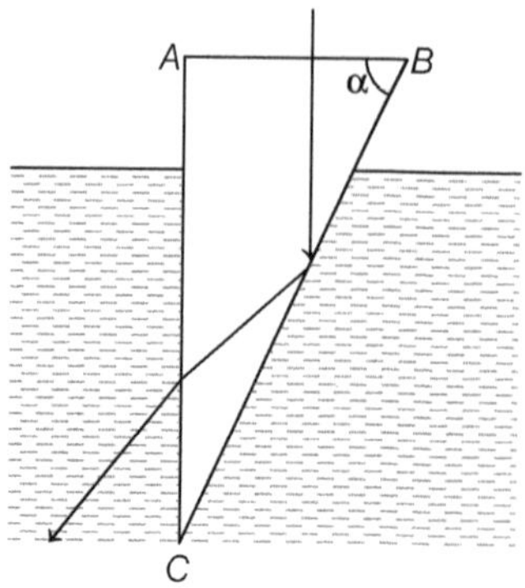

Fig. 243

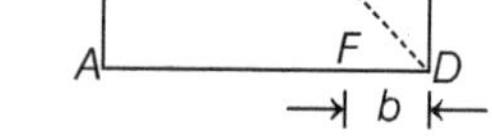

Fig. 244

694. At what values of the refraction index of a rectangular prism can a ray travel as shown in Fig. 242? The section of the prism is an isosceles triangle and the ray is normally incident onto AB.

695. A rectangular glass wedge is lowered into water. The refraction index of glass is $n_1 = 1.5$. At what angle α (Fig. 243) will the beam of light normally incident on AB reach AC entirely?

696. On bright sunny days drivers frequently see puddles on some parts of asphalt country highways at a distance of 80-100 metres ahead of the car. As the driver approaches such places, the puddles disappear and reappear again in other places approximately at the same distance away. Explain this phenomenon.

697. A thick plate is made of a transparent material whose refraction index changes from n_1 on its upper edge to n_2 on its lower edge. A beam enters the plate at the angle α. At what angle will the beam leave the plate?

698. A cubical vessel with non-transparent walls is so located that the eye of an observer does not see its bottom, but sees all of the wall CD (Fig. 244).

What amount of water should be poured into the vessel for the observer to see an object F arranged at a distance of $b = 10$ cm from corner D? The face of the vessel is $a = 40$ cm.

699. A man in a boat is looking at the bottom of a lake. How does the seeming depth of the lake h depend on the angle i formed by the line of vision with the vertical? The actual depth of the lake is everywhere the same and equal to H.

700. The cross section of a glass prism has the form of an equilateral triangle. A ray is incident onto one of the faces perpendicular to it. Find the angle ϕ between the incident ray and the ray that leaves the prism. The refraction index of glass is $n = 1.5$.

701. The cross section of a glass prism has the form of an isosceles triangle. One of the equal faces is coated with silver. A ray is normally incident on another unsilvered face and, being reflected twice, emerges through the base of the prism perpendicular to it. Find the angles of the prism.

702. A ray incident on the face of a prism is refracted and escapes through an adjacent face. What is the maximum permissible angle of refraction of the prism α if it is made of glass with a refraction index of $n = 1.5$?

703. A beam of light enters a glass prism at an angle α and emerges into the air at an angle β. Having passed through the prism, the beam is reflected from the original direction by an angle γ. Find the angle of refraction of the prism ϕ and the refraction index of the material which it is made of.

704. The faces of prism $ABCD$ made of glass with a refraction index n form dihedral angles: $\angle A = 90°$, $\angle B = 75°$, $\angle C = 135°$ and $\angle D = 60°$ (the Abbe prism). A beam of light falls on face AB and after complete internal reflection from face BC escapes through face AD. Find the angle of incidence α of the beam onto face AB if a beam that has passed through the prism is perpendicular to the incident beam.

705. If a sheet of paper is covered with glue or water the text typed on the other side of the sheet can be read. Explain why?

5-3. Lenses and Spherical Mirrors

706. Find the refraction index of the glass which a symmetrical convergent lens is made of if its focal length is equal to the radius of curvature of its surface.

707. A plano-convex convergent lens is made of glass with a refraction index of $n = 1.5$. Determine the relation between the focal length of this lens f and the radius of curvature of its convex surface R.

708. Find the radii of curvature of a convexo-concave convergent lens made of glass with a refraction index of $n = 1.5$ having a focal length of $f = 24$ cm. One of the radii of curvature is double the other.

709. A convexo-convex lens made of glass with a refraction index of $n = 1.6$ has a focal length of $f = 10$ cm. What will the focal length of this lens be if it is placed into a transparent medium with a refraction index of $n_1 = 1.5$? Also find the focal length of this lens in a medium with a refraction index of $n_2 = 1.7$.

710. A thin glass lens has an optical power of $D = 5$ diopters. When this lens is immersed into a liquid with a refraction index n_2, the lens acts as a divergent one with a focal length of $f = 100$ cm. Find the refraction index n_2 of the liquid if that of the lens glass is $n_1 = 1.5$.

711. The distance between an object and a divergent lens is m times greater than the focal length of the lens. How many times will the image be smaller than the object?

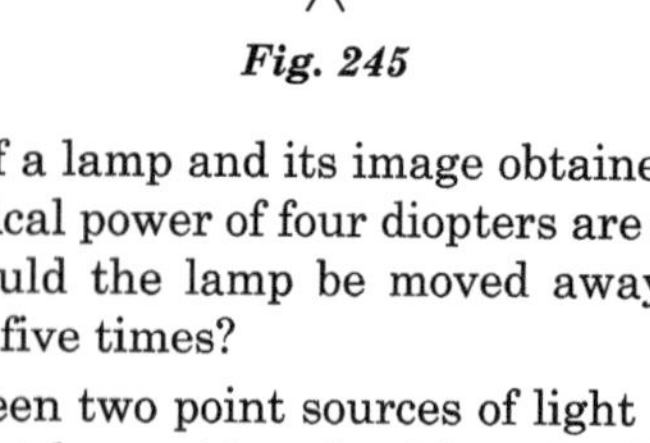

Fig. 245

712. The hot filament of a lamp and its image obtained with the aid of a lens having an optical power of four diopters are equal in size. Over what distance should the lamp be moved away from the lens to decrease its image five times?

713. The distance between two point sources of light is $l = 24$ cm. Where should a convergent lens with a focal length of $f = 9$ cm be placed between them to obtain the images of both sources at the same point?

714. The height of a candle flame is 5 cm. A lens produces an image of this flame 15 cm high on a screen. Without touching the lens, the candle is moved over a distance of $l = 1.5$ cm. away from the lens, and a sharp image of the flame 10 cm high is obtained again after shifting the screen. Determine the main focal length of the lens.

715. A converging beam of rays is incident onto a divergent lens so that the continuations of all the rays intersect at a point lying on the optical axis of the lens at a distance of $b = 15$ cm from it.

Find the focal length of the lens in two cases:

(1) after being refracted in the lens, the rays are assembled at a point at a distance of $a_1 = 60$ cm from the lens;

(2) the continuations of the refracted rays intersect at a point at a distance of $a_2 = 60$ cm in front of the lens.

716. The distance between an electric lamp and a screen is $d = 1$ metre. In what positions of a convergent lens with a focal length of $f = 21$ cm will the image of the lamp filament be sharp?

Can an image be obtained if the focal length is $f' = 26$ cm?

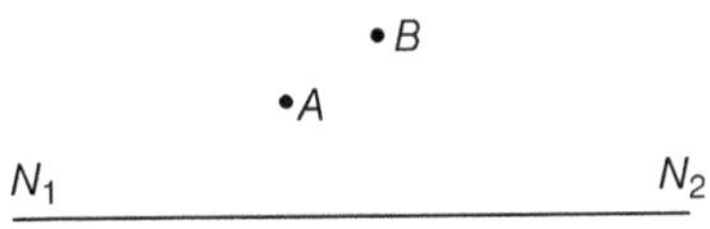

Fig. 246

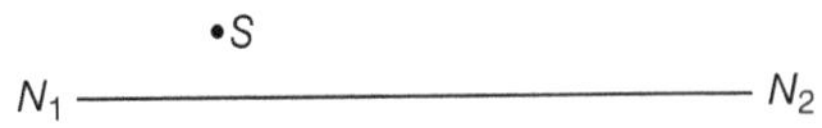

Fig. 2.47

717. A thin convergent lens produces the image of a certain object on a screen. The height of the image is h_1. Without changing the distance between the object and the screen, the lens is shifted, and it is found that the height of the second sharp image is h_2.

Determine the height of the object H.

718. What is the radius R of a concave spherical mirror at a distance of $a = 2$ metres from the face of a man if he sees in it his image that is one and a half times greater than on a flat mirror placed at the same distance from the face?

719. Figure 245 shows ray AB that has passed through a divergent lens. Construct the path of the ray up to the lens if the position of its foci F is known.

720. Figure 246 shows a luminescent point and its image produced by a lens with an optical axis N_1N_2. Find the position of the lens and its foci.

721. Find by construction the optical centre of a lens and its main foci on the given optical axis N_1N_2 if the positions of the source S and the image S' are known (Fig. 247).

722. The position of the optical axis $N_1 N_2$, the path of ray AB incident upon a lens and the refracted ray BC are known (Fig. 248). Find by construction the position of the main foci of the lens.

723. A convergent lens produces the image of a source at point S' on the main optical axis. The positions of the centre of the lens O and its foci F are known, and $OF < OS'$. Find by construction the position of source S.

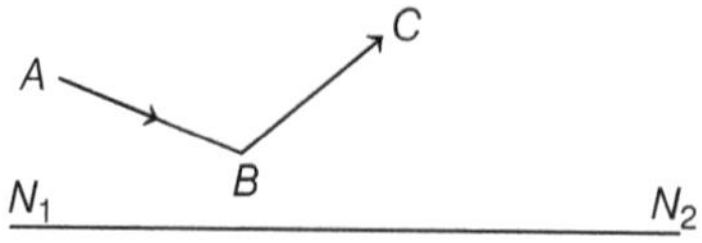

Fig. 248

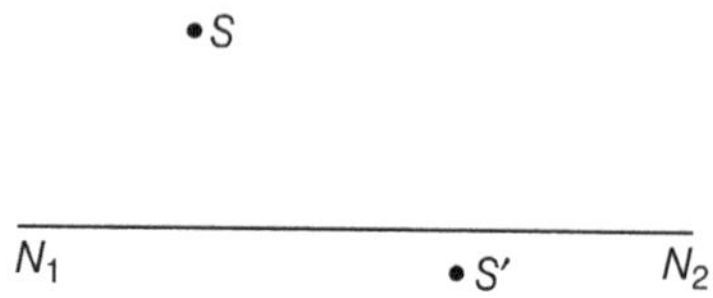

Fig. 249

724. Point S' is the image of a point source of light S in a spherical mirror whose optical axis is N_1N_2 (Fig. 249). Find by construction the position of the centre of the mirror and its focus.

725. The positions of optical axis N_1N_2 of a spherical mirror, the source and the image are known (Fig. 250). Find by construction the positions of the centre of the mirror, its focus and the pole for the cases: (a) A — source, B—image; (b) B—source, A— image.

726. A point source of light placed at a distance from a screen creates an illumination of 2.25 lx at the centre of the screen. How will this illumination change if on the other side of the source and at the same distance from it we place:

(a) an infinite flat mirror parallel to the screen?

(b) a concave mirror whose centre coincides with the centre of the screen?

(c) a convex mirror with the same radius of curvature as the concave mirror?

727. A man wishing to get a picture of a zebra photographed a white donkey after fitting a glass with black streaks onto the objective of his camera. What will be on the photograph?

728. The layered lens shown in Fig. 251 is made of two kinds of glass. What image will be produced by this lens with a point source arranged on the optical axis? Disregard the reflection of light on the boundary between layers.

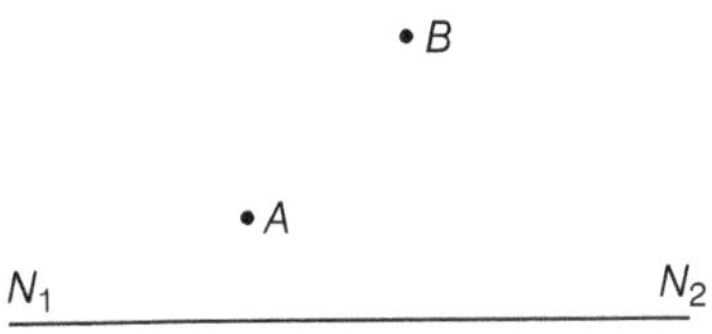

Fig. 250

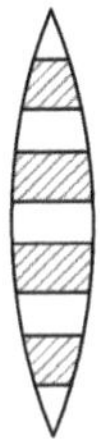

Fig. 251

729. The visible dimensions of the disks of the Sun and the Moon at the horizon seem magnified as compared with their visible dimensions at the zenith. How can it be proved experimentally with the aid of a lens that this magnification is apparent?

5-4. Optical Systems and Devices

730. A source of light is located at double focal length from a convergent lens. The focal length of the lens is $f = 30$ cm. At what distance from the lens should a flat mirror be placed so that the rays reflected from the mirror are parallel after passing through the lens for the second time?

731. A parallel beam of rays is incident on a convergent lens with a focal length of 40 cm. Where should a divergent lens with a focal length of 15 cm be placed for the beam of rays to remain parallel after passing through the two lenses?

732. An object is at a distance of 40 cm from a convex spherical mirror with a radius of curvature of 20 cm. At what distance from the object should a steel plate be placed for the image of the object in the spherical mirror and the plate to be in one plane?

733. At what distance from a convexo-convex lens with a focal length of $f = 1$ metre should a concave spherical mirror with a radius of curvature of $R = 1$ metre be placed for a beam incident on the lens parallel to the major optical axis of the system to leave the lens, remaining parallel to the optical axis, after being reflected from the mirror? Find the image of the object produced by the given optical system.

734. An optical system consists of two convergent lenses with focal lengths $f_1 = 20$ cm and $f_2 = 10$ cm. The distance between the lenses is $d = 30$ cm. An object is placed at a distance of $a_1 = 30$ cm from the first lens. At what distance from the second lens will the image be obtained?

735. Determine the focal length of an optical system consisting of two thin lenses: a divergent one with a focal length f_1 and a convergent one with a focal length f_2. The lenses are fitted tightly against each other, so that the distance between them can be neglected. The optical axes of the lenses coincide.

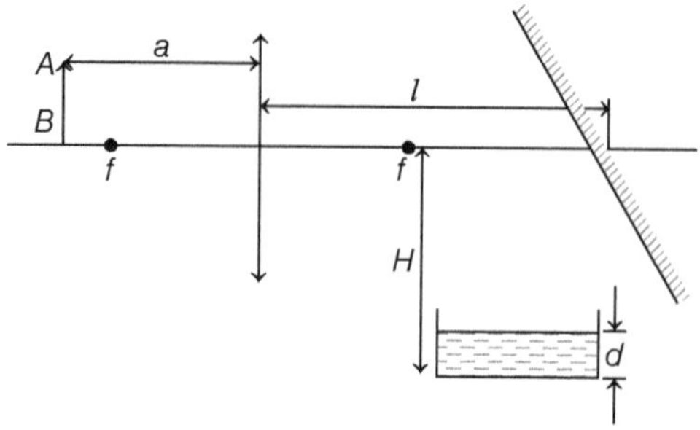

Fig. 252

736. A parallel beam of light is incident on a system consisting of three thin lenses with a common optical axis. The focal lengths of the lenses are equal to $f_1 = +10$ cm, $f_2 = -20$ cm and $f_3 = +9$ cm, respectively. The distance between the first and the second lenses is 15 cm and between the second and the third 5 cm. Find the position of the point at which the beam converges when it leaves the system of lenses.

737. A lens with a focal length of $f = 30$ cm produces on a screen a sharp image of an object that is at a distance of $a = 40$ cm from the lens. A plane-parallel plate with a thickness of $d = 9$ cm is placed between the lens and the object perpendicular to the optical axis of the lens. Through what distance should the screen be shifted for the image of the object to remain distinct? The refraction index of the glass of the plate is $n = 1.8$.

738. An object AB is at a distance of $a = 36$ cm from a lens with a focal length of $f = 30$ cm. A flat mirror turned through 45° with respect to the optical axis of the lens is placed behind it at a distance of $l = 1$ metre (Fig. 252).

At what distance H from the optical axis should the bottom of a tray with water be placed to obtain a sharp image of the object on the bottom? The thickness of the water layer in the tray is $d = 20$ cm.

739. A glass wedge with a small angle of refraction α is placed at a certain distance from a convergent lens with a focal length f, one surface of the wedge being perpendicular to the optical axis of the lens. A point source of light is on the other side of the lens in its focus. The rays reflected from the wedge produce, after refraction in the lens, two images of the source displaced with respect to each other by d. Find the refraction index of the wedge glass.

740. A concave mirror has the form of a hemisphere with a radius of $R = 55$ cm. A thin layer of an unknown transparent liquid is poured into this mirror, and it was found that the given optical system produces, with the source in a certain position, two real images, one of which coincides with the source and the other is at a distance of $l = 30$ cm from it. Find the refraction index n of the liquid.

741. A convexo-convex lens has a focal length of $f_1 = 10$ cm. One of the lens surfaces having a radius of curvature of $R = 10$ cm is coated with silver. Construct the image of the object produced by the given optical system and determine the position of the image if the object is at a distance of $a = 15$ cm from the lens.

742. A recess in the form of a spherical segment is made in the flat surface of a massive block of glass (refraction index n). The piece of glass taken out of the recess is a thin convergent lens with a focal length f. Find the focal lengths f_1 and f_2 of the spherical surface obtained.

743. A narrow parallel beam of light rays is incident on a transparent sphere with a radius R and a refraction index n in the direction of one of the diameters. At what distance f from the centre of the sphere will the rays be focussed ?

744. Find the position of the main planes of a transparent sphere used as a lens.

745. An object is at a distance of $d = 2.5$ cm from the surface of a glass sphere with a radius of $R = 10$ cm. Find the position of the image produced by the sphere. The refraction index of the glass is $n = 1.5$.

746. A spherical flask is made of glass with a refraction index n. The thickness of the flask walls ΔR is much less than its radius R.

Taking this flask as an optical system and considering only the rays close to the straight line passing through the centre of the sphere, determine the position of the foci and the main planes of the system.

747. A beam of light is incident on a spherical drop of water at an angle i. Find the angle θ through which the beam is deflected from the initial direction after a single reflection from the internal surface of the drop.

748. A parallel beam of rays is incident on a spherical drop of water.

(1) Calculate the angles θ through which the rays are deflected from the initial position for various angles of incidence: 0°, 20°, 40°, 50°, 55°, 60°, 65°, and 70°.

(2) Plot a diagram showing θ versus i and use it to find the approximate value of the angle of minimum deflection θ_{min}.

(3) Determine the values of the angle θ near which the rays issuing from the drop are approximately parallel. The refraction index of water is $n = 1.333$. (This value of n is true for red rays.)

749. What magnification can be obtained with the aid of a projecting camera having a lens with a main focal length of 40 cm if the distance from the lens to the screen is 10 metres?

750. Calculate the condenser of a projecting camera, i.e., find its diameter D and focal length f, if the source of light has dimensions of about $d = 6$ mm, and the diameter of the lens is $D_0 = 2$ cm. The distance between the source of light and the lens is $l = 40$ cm. The size of a slide is 6×9 cm.

751. In some photographic cameras ground glass is used for focussing. Why is transparent glass not used for this purpose?

752. Two lanterns of the same luminance are at different distances from an observer.

(1) Will they appear to the observer as equally bright?

(2) Will their images on photographs be equally bright if the lanterns are photographed on different frames so that their images are focussed?

753. An object is photographed from a small distance by two cameras with identical lens speeds, but different focal lengths. Should the exposures be the same?

754. It can be noticed that a white wall illuminated by the setting Sun seems brighter than the surface of the Moon at the same altitude above the horizon as the Sun. Does this mean that the surface of the Moon consists of dark rock?

755. Why does a swimmer see only hazy contours of objects when he opens his eyes under water, while they are distinctly visible if he is using a mask?

756. A short-sighted man, the accommodation of whose eye is between $a_1 = 12$ cm, and $a_2 = 60$ cm, wears spectacles through which he distinctly sees remote objects. Determine the minimum distance a_3 at which the man can read a book through his spectacles.

757. Two men, a far-sighted and a short-sighted ones, see objects through their spectacles as a man with normal eyesight. When the far-sighted man accidently put on the spectacles of his short-sighted friend, he found that he could see distinctly only infinitely far objects.

At what minimum distance a can the short-sighted man read small type if he wears the spectacles of the far-sighted man?

758. An object is examined by a naked eye from a distance D. What is the angular magnification if the same object is viewed through a magnifying glass held at a distance r from the eye and so arranged that the image is at a distance L from the eye?

The focal length of the lens is f.

Consider the cases :

(1) $L = \infty$

(2) $L = D$.

759. The objective is taken out of a telescope adjusted to infinity and replaced by a diaphragm with a diameter D. A screen shows a real image of the diaphragm having a diameter d at a certain distance from the eyepiece. What was the magnification of the telescope?

760. The double-lens objective of a photographic camera is made of a divergent lens with a focal length of $f_1 = 5$ cm installed at a distance of $l = 45$ cm from the film. Where should a convergent lens with a focal length of $f_2 = 8$ cm be placed to obtain a sharp image of remote objects on the film?

761. Calculate the diameter D of the Moon's image on a negative for the three different positions of the lenses described in Problem 760.

The diameter of the Moon is seen from the Earth at an angle of $\phi = 31'5'' \cong 0.9 \times 10^{-2}$ rad.

762. The main focal length of the objective of a microscope is $f_{ab} = 3$ mm and of the eyepiece $f_{eye} = 5$ cm. An object is at a distance of $a = 3.1$ mm from the objective. Find the magnification of the microscope for a normal eye.

6

PHYSICAL OPTICS

6-1. Interference of Light

763. Two light waves are superposed in a certain section of space and extinguish each other. Does this mean that a quantity of light is converted into other kinds of energy?

764. Two coherent sources of light S_1 and S_2 are at a distance l from each other. A screen is placed at a distance $D \gg l$ from the sources (Fig. 253). Find the distance between adjacent interference bands, near the middle of the screen (point A) if the sources send light with a wavelength λ.

765. Two flat mirrors form an angle close to 180° (Fig. 254). A source of light S is placed at equal distances b from the mirrors. Find the interval between adjacent interference bands on screen MN at a distance $OA = a$ from the point of intersection of the mirrors. The length of the light wave is known and equal to λ. Shield C does not allow the light to pass directly from the source to the screen.

766. Lloyd's interference experiment consisted in obtaining on a screen a pattern from source S and from its virtual image S' in mirror AO (Fig. 255). How will the interference pattern obtained from sources S and S' differ from the pattern considered in Problem 764?

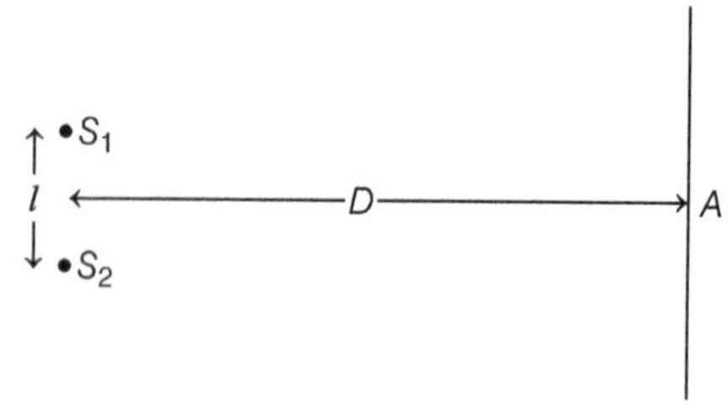

Fig. 253

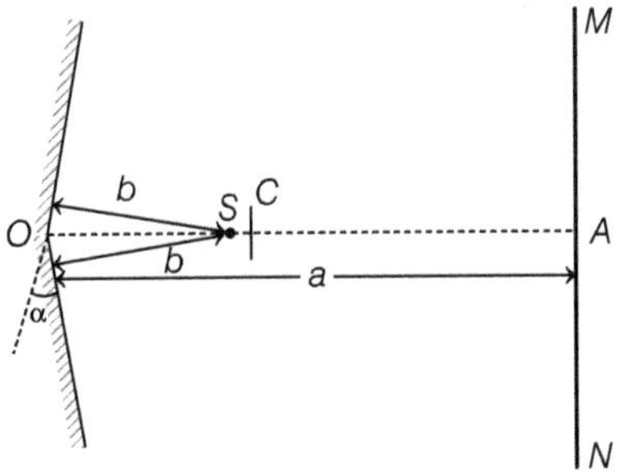

Fig. 254

767. Two point sources with the same phases of oscillation are on a straight line perpendicular to a screen. The nearest source is at a distance of $D \gg \lambda$ from the screen. What shape will the interference bands have on the screen? What is the distance on the screen from the perpendicular to the nearest bright band if the distance between the sources is $l = n\lambda \gg \lambda$ (n is an integer)?

768. Find the radius r_k of the k-th bright ring (see Problem 767) if $D = l = n\lambda$, $n \gg l$, and $k = n$, $n - 1$, $n - 2$, etc.

769. How can the experiment described in Problem 767 be carried out in practice?

770. Light from source S is incident on the Fresnel biprism shown in Fig. 256. The light beams refracted by the different faces of the prism partly overlap and produce an interference pattern on a screen on its section AB.

Find the distance between adjacent interference bands if the distance from the source to the prism is $a = 1$ metre and from the prism to the screen $b = 4$ metres. The angle of refraction of the prism is $\alpha = 2 \times 10^{-3}$ rad.

The glass which the prism is made of has a refraction index of $n = 1.5$. The length of the light wave $\lambda = 6{,}000$ Å.

771. How many interference bands can be observed on a screen in an installation with the biprism described in the previous problem?

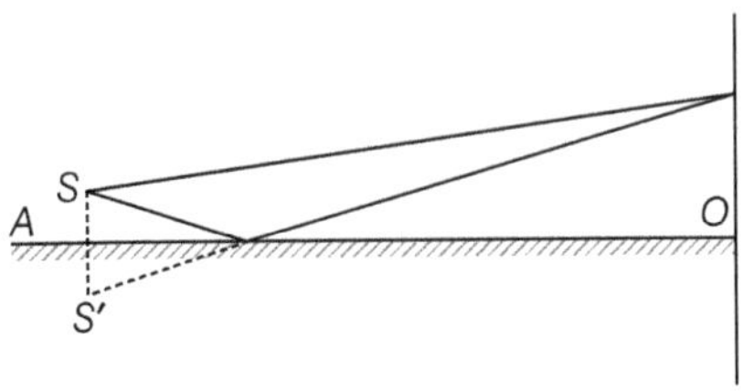

Fig. 255

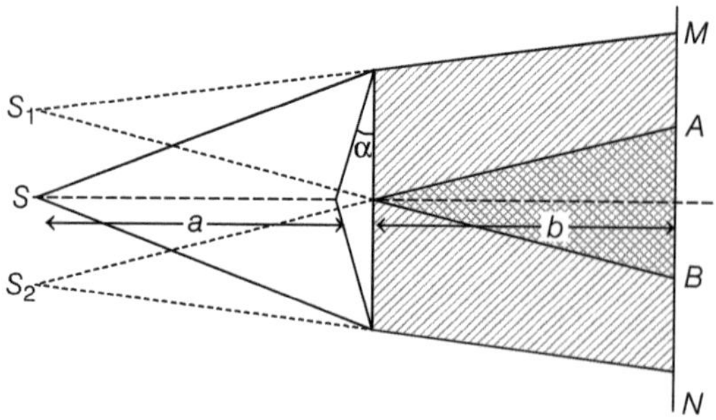

Fig. 256

772. The following method is used to facilitate the fabrication of a biprism with an angle close to 180° (see Problem 770). A biprism with an angle β that appreciably differs from 180° is placed into a vessel filled with a liquid having a refraction index n_1, or serves as one of the walls of this vessel (Fig. 257.).

Calculate the angle δ of an equivalent biprism in air. The refraction index of the prism substance is n_2.

Perform the calculations for $n_1 = 1.5$ (benzene), $n_2 = 1.52$ (glass), $\beta = 170°$.

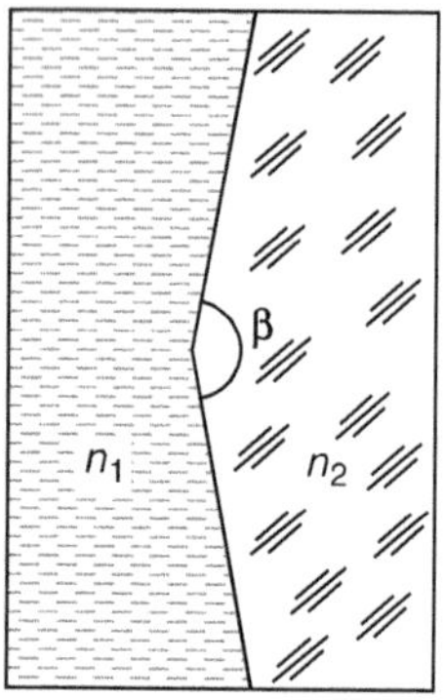

Fig. 257

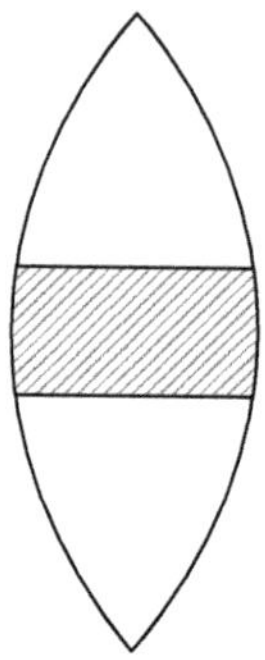

Fig. 258

773. A convergent lens with a focal length of $f = 10$ cm is cut into two halves that are then moved apart to a distance of $d = 0.5$ mm (a double lens). Appraise the number of interference bands on a screen at a distance of $D = 60$ cm behind the lens if a point source of monochromatic light ($\lambda = 5{,}000$ Å) is placed in front of the lens at a distance of $a = 15$ cm from it.

774. A central portion with a width of $d = 0.5$ mm is cut out of a convergent lens having a focal length of $f = 10$ cm, as shown in Fig. 258. Both halves are tightly fitted against each other. The lens receives monochromatic light ($\lambda = 5{,}000$ Å) from a point source at a distance of $a = 5$ cm from it. At what distance should a screen be fixed on the opposite side of the lens to observe three interference bands on it?

What is the maximum possible number of interference bands that can be observed in this installation?

775. Find the distance between the neighbouring bands of an interference pattern produced by a lens with a radius of $R = 1$ cm, described in Problem 774, if this distance does not depend on the position of the screen.

At what position of the screen will the number of interference bands be maximum?

The source sends monochromatic light with a wavelength of $\lambda = 5{,}000$ Å.

776. What will occur with the interference pattern in the installation described in Problem 775 if a plane-parallel glass plate with a thickness of $d_1 = 0.11$ cm is introduced into a light beam that has passed through the upper half of the lens and a plate $d_2 = 0.1$ cm thick into a light beam that has passed through the lower half of the lens? The refraction index of glass is $n = 1.5$. The plates are arranged normally to the light beams passing through them.

777. Why are Newton's rings formed only by the interference of rays 2 and 3 reflected from the boundaries of the air layer between the lens and the glass (Fig. 259), while ray 4 reflected from the flat face of the lens does not affect the nature of the interference pattern?

778. Will the nature of the interference pattern change in the installation described in Problem 765 if shield C is removed?

Consider the distance a to be great (one metre). The waves radiated by the source are not monochromatic.

779. Will Newton's rings be seen more distinctly in reflected light or in transmitted light?

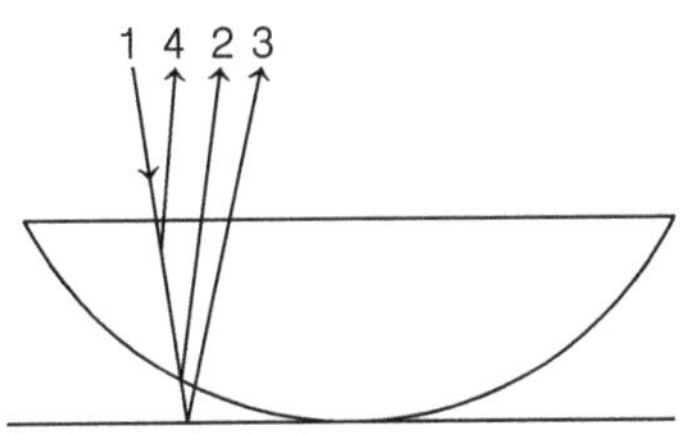

Fig. 259

780. Dust prevents contact between a plane-convex lens and the glass plate on which it is placed. The radius of Newton's fifth dark ring is $r_1 = 0.08$ cm.

If the dust is removed the radius of this ring will increase to $r_2 = 0.1$ cm. Find the thickness of the dust layer d if the radius of curvature of the convex surface of the lens is $R = 10$ cm.

781. A plane-convex lens with a radius of curvature $R_2 < R_1$ is placed with its convex side onto the surface of a plane-concave lens having a radius of curvature of R_1. Find the radii of Newton's rings that appear around the point of contact of the lenses if monochromatic light with a wavelength λ is normally incident onto the system.

782. A thin coat of a transparent substance with a refraction index n lower than that of glass is applied to the surface of optical glass (coated glass) to reduce its reflection factor (reflectance).

Estimate the thickness of this coat if light rays are incident almost normally onto the optical glass.

783. A normal eye can discern various colour tints with a difference of wavelengths of 100 Å. Bearing this in mind, estimate the maximum thickness of a thin air layer at which an interference pattern caused by superposition of the rays reflected from the boundaries of this layer can be observed in white light.

784. A monochromatic flux from a remote source with a wavelength λ is almost normally incident on a thin glass wedge. A screen is placed at a distance d from the wedge. A lens with a focal length f projects the interference pattern produced in the wedge onto the screen. The distance between the interference bands on the screen Δl is known. Find the angle α of the wedge if the refraction index of glass is n.

6-2. Diffraction of Light

785. Calculate the radii of the Fresnel zones of a spherical wave with a radius a for point B removed by a distance of $a + b$ from a source of monochromatic waves with a length of λ, bearing in mind that $a \gg \lambda$ and $b \gg \lambda$.

786. Calculate the radii of the Fresnel zones of a plane wave for point B removed from the wave front by a distance of $b \gg \lambda$, where λ is the wavelength of the source.

787. A point source of monochromatic light with a wavelength of $\lambda = 5{,}000$ Å is at a distance of $a = 6.75$ metres from a shield having an aperture with a diameter $D = 4.5$ mm. A screen is placed at a distance of $b = a$ from the shield (Fig. 260). How will the illumination change at point B of the screen lying on the axis of the beam if the diameter of the aperture is increased to $D_1 = 5.2$ mm?

788. How can the fact that an increase of the aperture (see Problem 787) may reduce the illumination on the axis of a beam be agreed with the law of conservation of energy? Indeed, the total luminous flux penetrating behind the shield increases when the aperture grows.

789. A plane light wave ($\lambda = 6{,}000$ Å) is incident on a shield with a circular diaphragm. A screen is placed at a distance of $b = 2$ metres behind the diaphragm. At what diameter D of the diaphragm will the illumination of the screen at point B lying on the axis of the light beam be maximum?

790. Assuming the distances from a source to a shield and from the shield to a screen to be about the same and equal to a, find the conditions in which the diffraction of light waves with a length λ on the aperture in the shield will be sufficiently distinct (the intensity on the axis of the beam will depend on the diameter of the aperture).

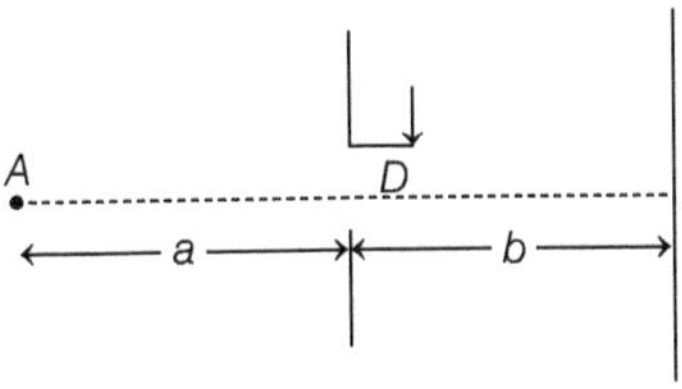

Fig. 260

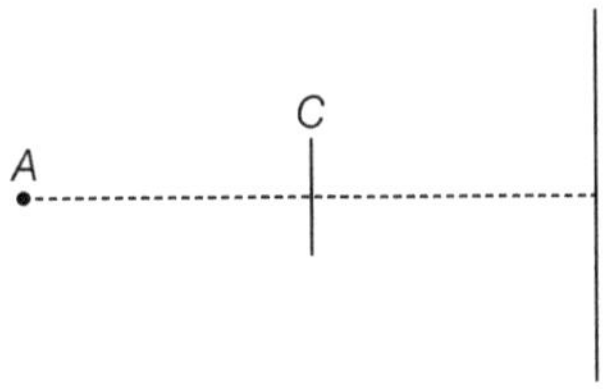

Fig. 261

791. Prove that a bright spot will be observed at point B behind circular screen C (Fig. 261) if its dimensions are sufficiently small.

792. At what distance from each other should two men stand for an eye to distinguish them from a distance of about 11 km? The resolving power of a normal eye is approximately 1′.

793. A plane light wave (with a length λ) is normally incident on a narrow slit with a width b. Determine the directions to the illumination minima.

794. Find the optimum dimensions of the aperture in a pin hole camera depending on the wavelength, i.e., the radius r of the aperture at which a point source will appear on the camera wall as a circle of minimum diameter, if the distance from the source of light to the camera is great as compared with its depth d.

Note. The directions to the illumination minima are determined in the order of their magnitude by the same formula as in the case of a slit (see Problem 793), the diameter of the aperture $2r$ being used instead of the width of the slit b.

795. A monochromatic wave is normally incident on a diffraction grating with a period of $d = 4 \times 10^{-4}$ cm. Find the wave length λ if the angle between the spectra of the second and third orders is $\alpha = 2°30'$. The angles of deflection are small.

796. A plane monochromatic wave ($\lambda = 5 \times 10^{-5}$ cm) is incident on a diffraction grating with 500 lines. Determine the maximum order of the spectrum k that can be observed when the rays are normally incident on the grating.

797. Find the constant d of a grating that can analyse infrared radiation with wavelengths up to $\lambda = 2 \times 10^{-2}$ cm. The radiation is normally incident on the grating.

798. A monochromatic wave is normally incident on a diffraction grating with a period of $d = 4 \times 10^{-4}$ cm. A lens with a focal length of $f = 40$ cm that produces an image of the diffraction pattern on a screen is arranged behind the grating.

Find the wavelength λ if the first maximum is obtained at a distance of $l = 5$ cm from the central one.

799. A source of white light, a diffraction grating and a screen are immersed into water. What will the changes in the diffraction pattern be if the angles by which the light rays are deflected by the grating are small?

800. Light passed through a light filter is normally incident on a diffraction grating with a period of $d = 2 \times 10^{-4}$ cm. The filter passes wavelengths from $\lambda_1 = 5{,}000$ Å to $\lambda_2 = 6{,}000$ Å. Will the spectra of different orders be superposed?

801. Solve Problem 796, assuming that a plane wave ($\lambda = 5 \times 10^{-5}$ cm) is incident on the grating at an angle of 30°.

802. Solve Problem 797, assuming that the rays may fall on the grating diagonally.

803. Find the condition that determines the directions to the principal maxima if light waves fall diagonally on a grating with a period $d \gg k\lambda$ (k is the order of the spectrum).

6-3. Dispersion of Light and Colours of Bodies

804. A beam of white light falls at an angle of $\alpha = 30°$ on a prism with a refraction angle of $\phi = 45°$. Determine the angle θ between the extreme rays of the spectrum when they emerge from the lens if the refraction indices of the prism for the extreme rays of the visible spectrum are $n_r = 1.62$ and $n_v = 1.67$.

805. White light is incident from a point source on the optical axis of a convexo-convex lens at a distance of $a = 50$ cm from it. The radii of curvature of the lens are $R_1 = R_2 = 40$ cm. A diaphragm with a diameter of $D = 1$ cm that restricts the cross section of the light beam is placed tightly in front of the lens.

The refraction indices for the extreme rays of the visible spectrum are $n_r = 1.74$ and $n_v = 1.8$, respectively. What pattern can be observed on a screen arranged at a distance of $b = 50$ cm from the lens and perpendicular to its optical axis?

806. Using the results of Problem 748, construct the elementary theory of the rainbow, i.e., show that the centre of a rainbow lies on a straight line drawn from the Sun through the eye of an observer and that the arc of the rainbow is a part of a circle all of whose points can be seen at an angle of 42° (for red light) with reference to the straight line connecting the eye of the observer and the centre of the rainbow.

807. Explain from a qualitative point of view the appearance of a double rainbow. How do the colours alternate in the primary and the reflection rainbows?

808. Can a rainbow be observed at midday in Moscow during the summer solstice (on June 22)?

Note. At this time the Sun is the highest above the horizon in the northern hemisphere.

809. The length of a wave in water diminishes n times, n being the refraction index. Does this mean that a diver cannot see surrounding objects in their natural colours?

810. The word "excellent" is written on a sheet of white paper with a red pencil and the word "good" with a green pencil. A green and a red pieces of glass are available. Through which glass can the word "excellent" be seen?

811. Why do coated lenses (see Problem 782) have a purple-violet (lilac) tint?

812. Why do the colours of thin films (for example, oil films on water) and the colours of a rainbow have different tints?

813. A thin soapy film is stretched over a vertical frame. When the film is illuminated by white light it shows three bands coloured purple (crimson), yellow and light-blue (blue-green). Find the arrangement and the order of the bands.

814. Why does the Moon, purely white in the daytime, have a yellowish hue after sunset?

815. Why does a column of smoke rising above the roofs of houses seem blue against the dark background of the surrounding objects, and yellow or even reddish against the background of a bright sky?

816. Why do the colours of moist objects seem deeper and richer than those of dry ones?

ANSWERS AND SOLUTIONS

1. MECHANICS

1-1. Kinematics of Uniform Rectilinear Motion

1. During the first hour after the meeting, the boat travelled away from the rafts. During the next 30 minutes, when the engine was being repaired, the distance between the boat and the rafts did not increase. The boat overtook the rafts in one hour because its speed with respect to the water and hence to the rafts was constant. Thus the velocity of the current

$$v = \frac{s}{t} = \frac{7.5}{1 + 0.5 + 1} \text{ km/h} = 3 \text{ km/h}$$

2. As can be seen from Fig. 262, $S = \frac{H}{H - h} s$. Since the man moves uniformly, $s = vt$. Hence $S = \frac{H}{H - h} vt$. The shadow moves with a constant velocity $\frac{Hv}{H - h}$, greater than the speed of the man. For this reason the velocity diagram has the form of a straight line parallel to the axis of abscissas.

3. The time of the meeting was 8 a. m. The man's speed was 4 km/h. The other questions can easily be answered with the aid of the chart in Fig. 263—the man met the twelfth bus at a distance of 10.7 km from the mill; the cyclist was overtaken by four buses.

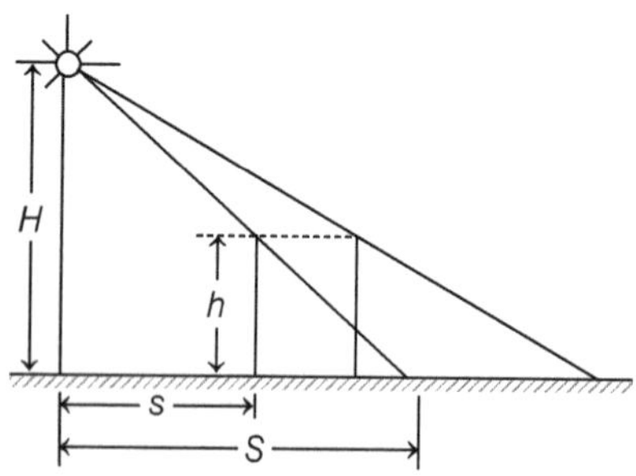

Fig. 262

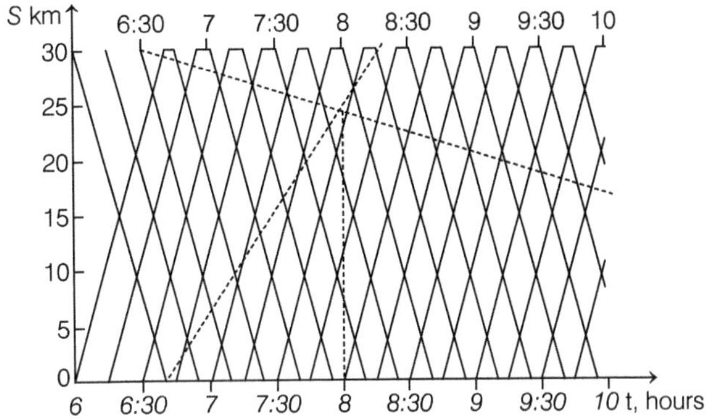

Fig. 263

4. The distance between the trains is $s = vt$; on the other hand, $s = v\tau + u\tau$.

Hence $u = \dfrac{v(t-\tau)}{\tau} = 45$ km/h.

5. In Fig. 264, AMN shows the usual trip of the car, SC—the engineer's walk until he met the car at point C, CB— the motion of the car after it met the engineer.

According to the conditions, $BN = KM = 10$ minutes. The time during which the engineer had walked before he met the car is

$$SD = SM - DM = SM - \frac{KM}{2} = 55 \text{ minutes}$$

6. The chart in Fig. 265 shows the movement of the launches between the landing-stages M and K. It follows from the chart that the landing-stages are served by eleven launches. A launch travelling from M to K meets eight launches, as does a launch travelling from K to M.

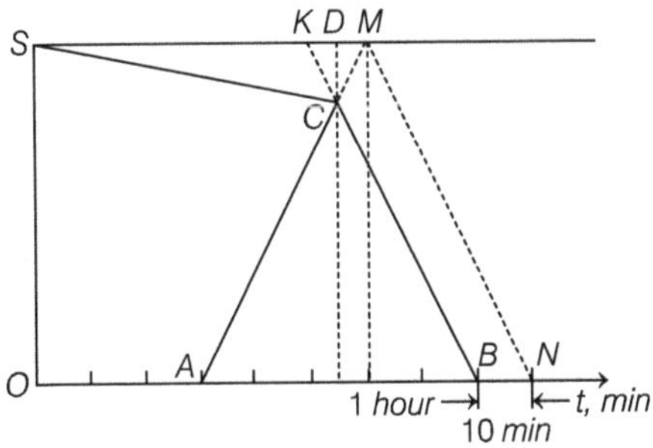

Fig. 264

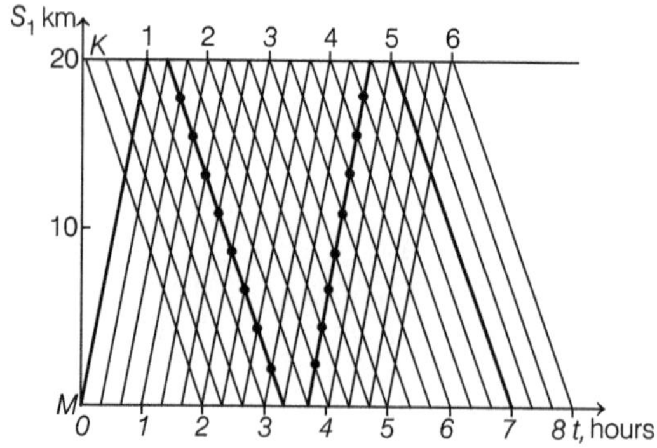

Fig. 265

7. Each tourist will walk half the distance and ride the bicycle the other half. The entire distance will be covered in $t = \frac{s}{2v_1} + \frac{s}{2v_2}$

$$= 5 \text{ hours } 20 \text{ minutes.}$$

Hence, the mean speed is $v = \frac{s}{t} = 7.5 \text{ km/h}$. The bicycle remains unused during half the time of motion, i.e., during 2 hours 40 minutes.

8. Assume that the first candle burns down by the amount Δh_1 and the other by Δh_2 during the time Δt (Fig. 266). The shadow on the left wall (from the first candle) will then lower over the distance

$$\Delta x = \Delta h_1 + (\Delta h_1 - \Delta h_2)$$
$$= 2\Delta h_1 - \Delta h_2$$

and that on the right wall over the distance

$$\Delta y = \Delta h_2 - (\Delta h_1 - \Delta h_2) = 2\Delta h_2 - \Delta h_1$$

Remembering that $\Delta h_1 = \frac{h}{t_1} \Delta t$ and $\Delta h_2 = \frac{h}{t_2} \Delta t$, we get

$$v_1 = \frac{\Delta x}{\Delta t} = \frac{2h}{t_1} - \frac{h}{t_2} = \frac{h}{t_1 t_2} (2\, t_2 - t_1)$$

$$v_2 = \frac{\Delta y}{\Delta t} = \frac{2h}{t_2} - \frac{h}{t_1} = \frac{h}{t_1 t_2} (2\, t_1 - t_2)$$

If $t_2 > t_1$, then $v_1 > 0$ and v_2 may be negative, i.e., the shadow on the right wall may move upward.

9. The bus is at point A and the man at point B (Fig. 267). Point C is the spot where the man meets the bus, α is the angle between the direction towards the bus and the direction in which the man should run, $AC = v_1 t_1$, $BC = v_2 t_2$, where t_1 and t_2 are the times required for the bus and the man, respectively, to reach point C.

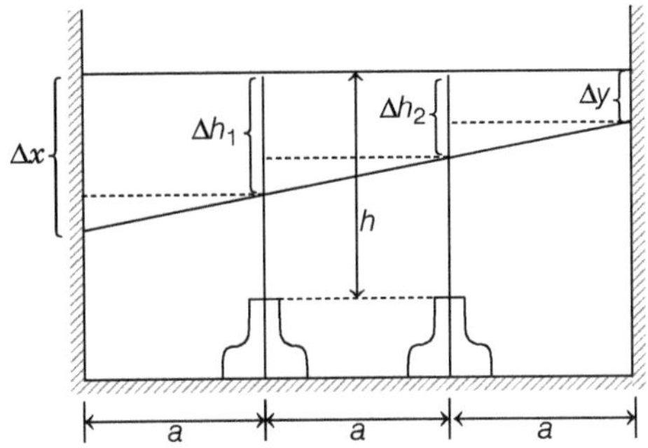

Fig. 266

A glance at triangle ABC shows that $AC = \frac{b\sin\alpha}{\sin\beta}$, where $\sin\beta = \frac{a}{BC}$.

Consequently, $\sin\alpha = \frac{a}{b}\frac{v_1t_1}{v_2t_2}$. According to the condition, $t_1 \ge t_2$ and therefore $\sin\alpha \ge \frac{av_1}{bv_2} = 0.6$. Hence $36°45' \le \alpha \le 143°15'$.

The directions in which the man can move are within the limits of the angle DBE. Upon running in the directions BD or BE, the man will reach the highway at the same time as the bus. He will reach any point on the highway between D and E before the bus.

10. The minimum speed can be determined from the conditions

$$t_1 = t_2,\ \sin\alpha = \frac{av_1}{bv_2} = 1$$

Hence $\quad v_2 = \frac{a}{b}v_1 = 2.4 \text{ m/s}.$

Here $\alpha = 90°$. Therefore, the man should run in a direction perpendicular to the initial line (AB, Fig. 267) between him and the bus.

11. Since the man's speed in water is lower than that along the shore, the route AB will not necessarily take the shortest time.

Assume that the man follows the route ADB (Fig. 268). Let us determine the distance x at which the time will be minimum.

The time of motion t is

$$t = \frac{\sqrt{d^2 + x^2}}{v_1} + \frac{s - x}{v_2}$$

$$= \frac{v_2\sqrt{d^2 + x^2} - v_1x + v_1s}{v_1v_2}$$

This time will be minimum if $y = v_2\sqrt{d^2 + x^2} - v_1 x$ has the smallest value. Obviously, the value of x that corresponds to the minimum time t does not depend on the distances S. To find the value of x corresponding to the minimum value of y, let us express x through y and obtain the quadratic equation

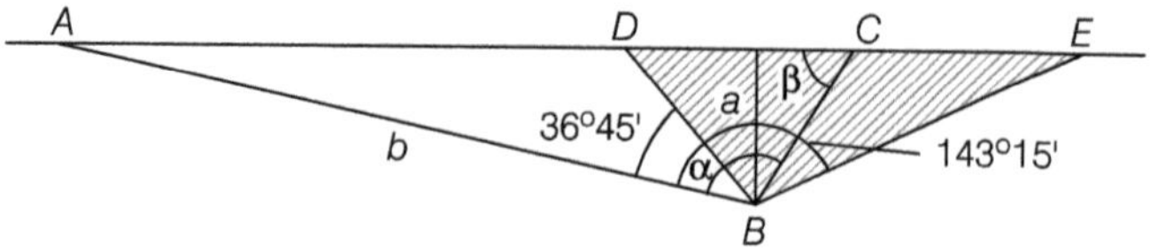

Fig. 267

$$x^2 - \frac{2yv_1}{v_2^2 - v_1^2}x + \frac{v_2^2 d^2 - y^2}{v_2^2 - v_1^2} = 0$$

Its solution leads to the following expression

$$x = \frac{v_1 y \pm v_2\sqrt{y^2 + d^2 v_1^2 - v_2^2 d^2}}{v_2^2 - v_1^2}$$

Since x cannot be a complex number, $y^2 + d^2 v_1^2 \geq v_2^2 d^2$.

The minimum value of y is equal to $y_{min} = d\sqrt{v_2^2 - v_1^2}$, and $x = \dfrac{dv_1}{\sqrt{v_2^2 - v_1^2}}$ corresponds to it.

If $s \leq \dfrac{dv_1}{v_2^2 - v_1^2}$, the man should immediately swim to point B along AB.

Otherwise, the man should run along the shore over the distance $AD = s - \dfrac{dv_1}{\sqrt{v_2^2 - v_1^2}}$, and then swim to B.

Let us note that $\sin\alpha = \dfrac{v_1}{v_2}$ for the route corresponding to the shortest time.

12. (1) Graphically, it is easier to solve the problem in a coordinate system related to the water. The speed of a raft equal to the velocity of the river current is zero in this system, and the speed of the ship upstream and downstream will be the same in magnitude. For this reason $\tan\alpha_1 = \tan\alpha_2 = v_1$ on the chart showing the motion of the motor-ship (Fig. 269). When the ship stops at the landing-stage, its speed with respect to the water will be equal to the river current velocity v_2. Hence $\tan\alpha = v_2$.

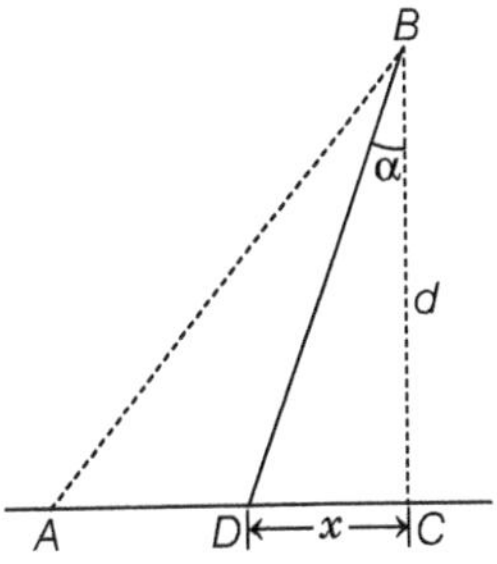

Fig. 268

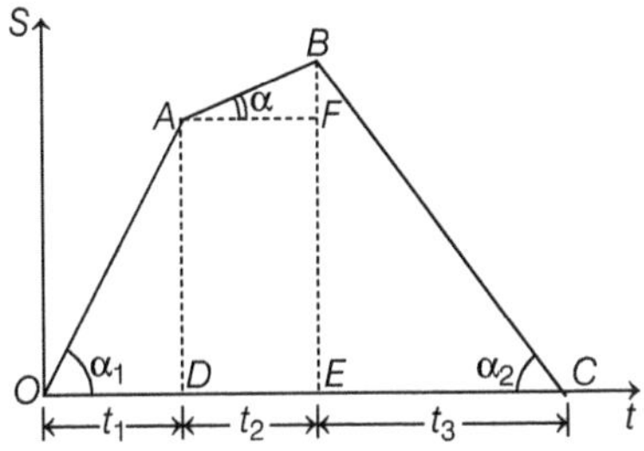

Fig. 269

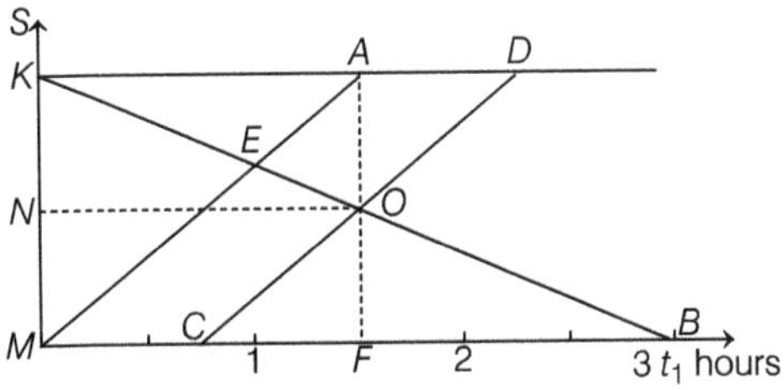

Fig. 270

It can be seen from Fig. 269 that

$$v_2 = \tan\alpha = \frac{BF}{AF} = \frac{\tan\alpha_2 \cdot t_3 - \tan\alpha_1 \cdot t_1}{t_2}$$

$$= \frac{v_1(t_3 - t_1)}{t_2} = 2.5 \text{ km/h}$$

(2) From the moment the ship meets the rafts to the moment it overtakes them, the rafts will cover a distance equal to

$$s = v_2(t_1 + t_2 + t_3)$$

On the other hand, this distance is equal to the difference between the distances travelled by the ship upstream and downstream:

$$s = t_3(v_1 + v_2) - t_1(v_1 - v_2)$$

Hence,

$$v_2(t_1 + t_2 + t_3) = t_3(v_1 + v_2) - t_1(v_1 - v_2)$$

and
$$v_2 = \frac{v_1(t_3 - t_1)}{t_2} = 2.5 \text{ km/h}$$

13. The motion of the launches leaving their landing-stages at the same time is shown by lines *MEB* and *KEA*, where *E* is their meeting point (Fig. 270). Since the speeds of the launches relative to the water are the same, *MA* and *KB* are straight lines.

Both launches will travel the same time if they meet at the middle between the landing-stages. Point *O* where they meet lies on the intersection of line KB with a perpendicular erected from the middle of distance KM. The motion of the launches is shown by lines KOD and COB. It can be seen from Fig. 270 that ΔMAF is similar to ΔCOF, and, therefore, the sought time $MC = 45$ minutes.

14. The speed of the launches with respect to the water v_1 and the velocity of the river current v_2 can be found from the equations $s = t_1(v_1 + v_2)$ and $s = t_2(v_1 - v_2)$, where t_1 and t_2 are the times of motion of the launches downstream and upstream. It follows from the condition that $t_1 = 1.5$ hours and $t_2 = 3$ hours.

Hence

$$v_1 = \frac{s(t_1 + t_2)}{2\,t_1 t_2} = 15 \text{ km/h}$$

$$v_2 = \frac{s(t_2 - t_1)}{2\,t_1 t_2} = 5 \text{ km/h}$$

The point of the meeting is at a distance of 20 km from landing-stage M.

15. Let us assume that the river flows from C to T with a velocity of v_0. Since the duration of motion of the boat and the launch is the same, we can write the equation

$$\frac{s}{v_1 + v_0} = 2\left(\frac{s}{v_2 + v_0} + \frac{s}{v_2 - v_0}\right)$$

where s is the distance between the landing-stages. Hence,

$$v_0^2 + 4v_2v_0 + 4v_2v_1 - v_2^2 = 0$$

and therefore

$$v_0 = -2v_2 \pm \sqrt{5v_2^2 - 4v_1v_2} = -20 \pm 19.5$$

The solution $v_0 = -39.5$ km/h should be discarded, since with such a current velocity neither the boat nor the launch could go upstream. For this reason, $v_0 = -0.5$ km/h. i. e., the river flows from T to C.

16. The speed of the boat v with respect to the bank is directed along AB (Fig. 271). Obviously, $v = v_0 + u$. We know the direction of vector v and the magnitude and direction of vector v_0. A glance at the drawing shows that vector u will be minimum when $u \perp v$.

Consequently,

$$u_{min} = v_0 \cos\alpha, \text{ where } \cos\alpha = \frac{b}{\sqrt{a^2 + b^2}}$$

17. Let the speed u be directed at an angle α to the bank (Fig. 272). Hence

$$t(u\cos\alpha - v) = BC = a, \text{ and } tu\sin\alpha = AC = b$$

where t is the time the boat is in motion.

By excluding α from these equations, we get

$$t^2(u^2 - v^2) - 2vat - (a^2 + b^2) = 0$$

whence $t = 15/21$ hour. It is therefore impossible to cover the distance AB in 30 minutes.

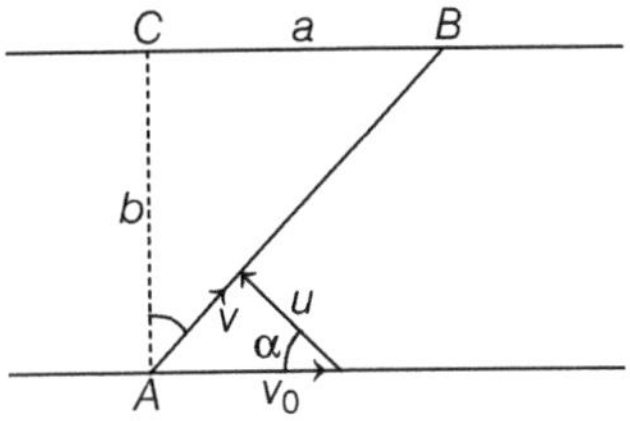

Fig. 271

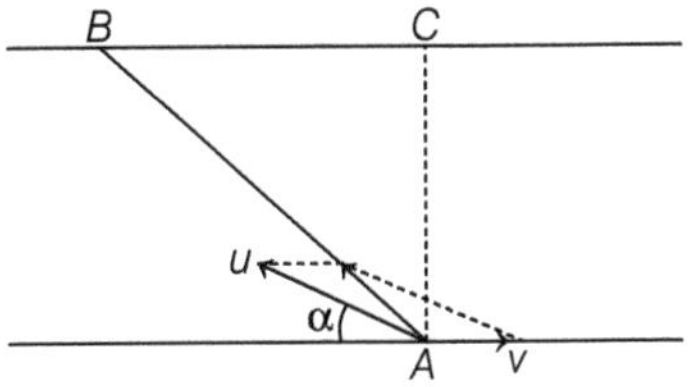

Fig. 272

18. Let u_0 be the velocity of the wind relative to the launch. Hence the flag on the mast will be directed along u_0. If v is the speed of the launch with respect to the bank, then u = $u_0 + v$ (Fig. 273).

In triangle FCD the angle $DCF = \beta + \alpha - \frac{\pi}{2}$ and the angle $FDC = \pi - \beta$. According to the sine theorem,

$$\frac{v}{\sin\left(\alpha + \beta - \frac{\pi}{2}\right)} = \frac{u}{\sin(\pi - \beta)}$$

and therefore $v = u\dfrac{\sin\left(\alpha + \beta - \frac{\pi}{2}\right)}{\sin(\pi - \beta)}$.

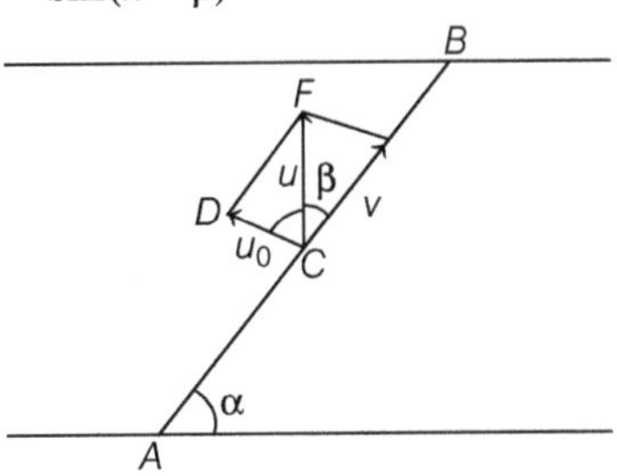

Fig. 273

It is impossible to determine the velocity of the current from the known speed of the launch with respect to the bank, since we do not know the direction of the moving launch with respect to the water.

19. (1) If the speed of the plane relative to the air is constant and equal to v, then its speed with respect to the Earth with a tail wind (along side BC) is $v_{BC} = v + u$, with a head wind $v_{DA} = v - u$ and with a side wind $v_{AB} = v_{CD} = \sqrt{v^2 - u^2}$ (Fig. 274 *a* and *b*).

Hence, the time required to fly around the square is

$$t_1 = \frac{a}{v+u} + \frac{a}{v-u} + \frac{2a}{\sqrt{v^2-u^2}} = 2a\,\frac{v+\sqrt{v^2-u^2}}{v^2-u^2}$$

(2) If the wind blows along the diagonal of the square from A to C, then (see Fig. 274*c*)

$$v^2 = v_{AB}^2 + u^2 - 2\,uv_{AB}\cos 45°$$

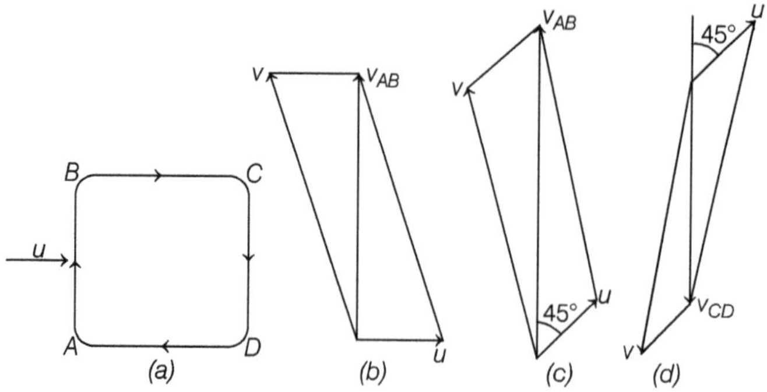

Fig. 274

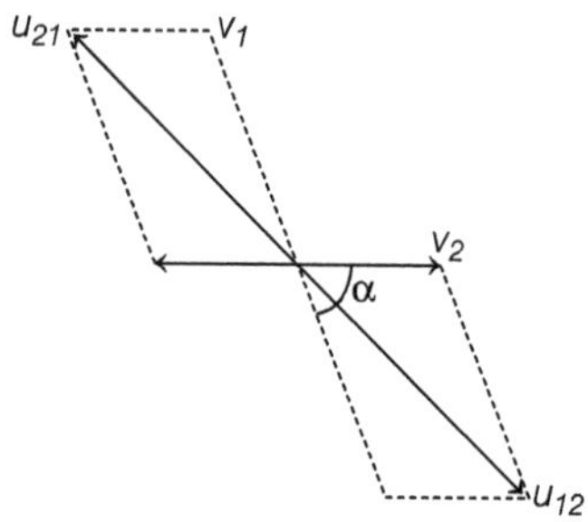

Fig. 275

The speed along sides AB and BC is

$$v_{AB} = v_{BC} = \frac{\sqrt{2}}{2}\,u + \sqrt{v^2 - \frac{u^2}{2}}$$

The speed along sides CD and AD (Fig. 274*d*)

$$v_{CD} = v_{DA} = -\frac{\sqrt{2}}{2}\,u + \sqrt{v^2 - \frac{u^2}{2}}$$

Let us leave only the plus sign before the root in both solutions to preserve a clockwise direction of the flight. The time required to fly around the square is

$$t_2 = \frac{4a\sqrt{v^2 - \frac{u^2}{2}}}{v^2 - u^2}$$

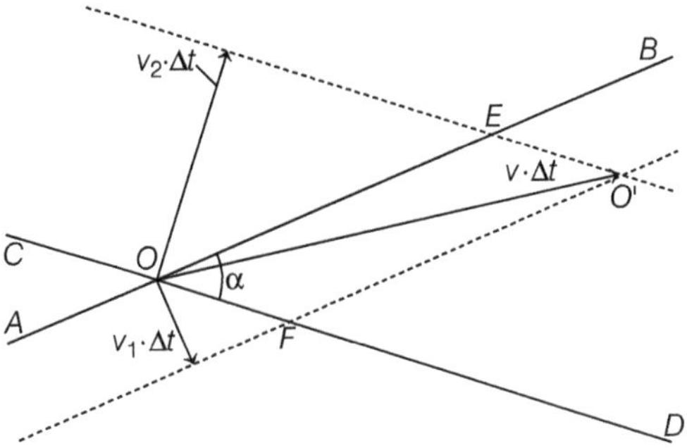

Fig. 276

20. Let us use the following notation: u_{12} = speed of the second vehicle with respect to the first one; u_{21} = speed of the first vehicle with respect to the second one.
Obviously, $u_{12} = u_{21}$ and $u_{12}^2 = v_1^2 + v_2^2 + 2v_1v_2\cos\alpha$ (Fig. 275). The time sought is $t = \frac{s}{u_{12}}$.

21. During the time Δt the straight line AB will travel a distance $v_1\Delta t$ and the straight line CD a distance $v_2\Delta t$. The point of intersection of the lines will travel to position O' (Fig. 276). The distance OO' can be found from triangle OFO' or OEO'. where $OF = \frac{v_1\Delta t}{\sin\alpha} = EO'$ and

$$OE = \frac{v_2\,\Delta t}{\sin\alpha} = FO' \text{ i.e.,}$$

$$OO' = \sqrt{OF^2 + OE^2 + 2OF\cdot OE\cos\alpha} = v\,\Delta t$$

whence, $$v = \frac{1}{\sin\alpha}\sqrt{v_1^2 + v_2^2 + 2v_1v_2\cos\alpha}$$

1-2. Kinematics of Non-Uniform and Uniformly Variable Rectilinear Motion

22. The mean speed over the entire distance $v_m = \frac{s}{t_1 + t_2 + t_3}$, where t_1, t_2 and t_3 are the times during which the vehicle runs at the speeds v_1, v_2 and v_3 respectively. Obviously,

$$t_1 = \frac{s}{3v_1},\ t_2 = \frac{s}{3v_2},\ \text{and } t_3 = \frac{s}{3v_3}$$

Consequently,

$$v_m = \frac{3v_1v_2v_3}{v_1v_2 + v_1v_3 + v_2v_3} = 18 \text{ km/h}$$

23. The path s travelled by the point in five seconds is equal numerically to the area enclosed between *Oabcd* and the time axis (see Fig. 6): $s_1 = 10.5$ cm.

The mean velocity of the point in five seconds is $v_1 = s_1/t_1 = 2.1$ cm/s and the mean acceleration of the point during the same time is

$$a_1 = \frac{\Delta v}{t_1} = 0.8 \text{ cm/s}^2$$

The path travelled in 10 seconds is $s_2 = 25$ cm.

Therefore, the mean velocity and the mean acceleration are

$$v_2 = \frac{s_2}{t_2} = 2.5 \text{ cm/s}, \ a_2 = 0.2 \text{ cm/s}^2$$

24. During a small time interval Δt the bow of the boat will move from point A to point B (Fig. 277). The distance $AB = v_1\Delta t$, where v_1 is the speed of the boat. A rope length of $OA - OB = CA = v\Delta t$ will be taken up during this time. The triangle ABC may be considered as a right one, since $AC \ll OA$.

Therefore $v_1 = \dfrac{v}{\cos\alpha}$.

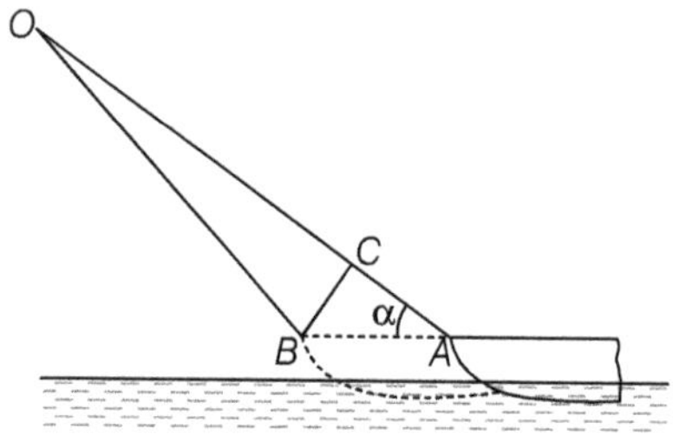

Fig. 277

25. Assume that at the initial moment $t = 0$ the object was at point S (Fig. 278), and at the moment t occupied the position CD. The similarity of the triangles SCD and SBA allows us to write the equation

$$AB = \frac{hl}{SD} = \frac{hl}{v_1 t}$$

The velocity of point B at a given moment of time is $v_2 = \dfrac{BB'}{\Delta t}$ if the time Δt during which the edge of the shadow is shifted by the distance BB' tends to zero.

$$\text{Since } BB' = AB - AB' = \frac{hl}{v_1}\left(\frac{1}{t} - \frac{1}{t+\Delta t}\right)$$

$$= \frac{hl\Delta t}{t(t+\Delta t)},$$

then

$$v_2 = \frac{hl}{v_1 t(t+\Delta t)},$$

or, remembering that $\Delta t \ll t$, we have $v_2 = \dfrac{hl}{v_1 t^2}$

26. For uniformly accelerated motion $x = x_0 + v_0 t + \dfrac{at^2}{2}$. Therefore $v_0 = 35$ cm/s, $a = 82$ cm/s^2, and $x_0 = 11$ cm is the initial coordinate of the point.

27. It follows from the velocity chart (see Fig. 8) that the initial velocity $v_0 = 4$ cm/s ($OA = 4$ cm/s). The acceleration $a = \dfrac{OA}{OB} = 1$ cm/s^2. First the velocity of the body decreases. At the moment $t_1 = 4$ s it is zero and then grows in magnitude.

The second chart (see Fig. 9) also shows uniformly variable motion. Before the body stops, it travels a distance of $h = 10$ cm. According to the first chart, the distance to the stop equal to the area of triangle OAB is 8 cm. Therefore, the charts show different motions.

A different initial velocity $v' = \dfrac{2h}{t_1} = 5$ cm/s and a different acceleration $a' = \dfrac{2h}{t_1^2} = 1.25$ cm/s^2 correspond to the second chart.

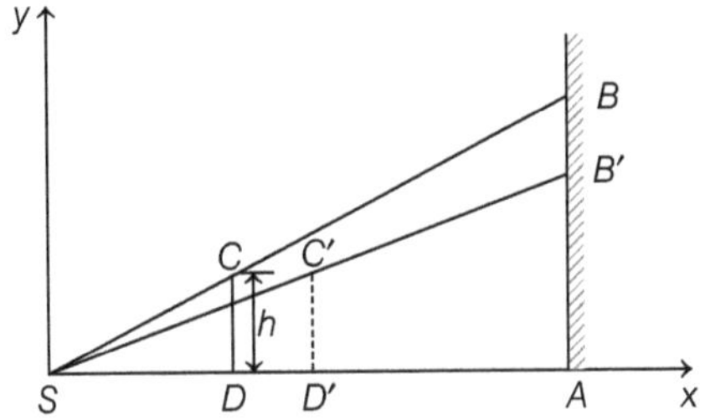

Fig 278

28. The mean speeds of both the motor vehicles are the same and equal to

$$v_m = \frac{2v_1 v_2}{v_1 + v_2} = 36 \text{ km/h}$$

Therefore, the distance between points A and B is 72 km. The first vehicle travelled half of this distance in $t' = 6/5$ h, and the other half in $t'' = 4/5$ h. The second vehicle travelled all the time with the acceleration

$$a = \frac{2s}{t_0^2} = 36 \text{ km/h}^2$$

and reached a speed of $v_f = at_0 = 72$ km/h at the end of its trip.
It acquired a speed of 30 km/h in

$$t_1 = \frac{v_{30}}{a} = \frac{5}{6} \text{ h}$$

and a speed of 45 km/h in $t_2 = \dfrac{v_{45}}{a} = 5/4$ h after the moment of departure.

At these moments the first vehicle moved at the same speed as the second.

At the moment when one vehicle overtook the other, both of them travelled the same distance, and therefore the following equalities should be true

$$v_1 t = \frac{at^2}{2} \text{ for } t \le \frac{6}{5} \text{ h}$$

and $$v_1 t' + v_2(t - t') = \frac{at^2}{2}$$

for $$\frac{6}{5} \text{ h} \le t \le 2 \text{ h}$$

In the first case $t = 0$ (the vehicles run side by side at the initial moment) or $t = 5/3$ h, which disagrees with the condition that $t < 6/5$ h. In the second case $t = 2$ h (the vehicles arrive simultaneously at point B), and $t = 1/2$ h. This does not satisfy the condition that $t > 6/5$ h. Hence, neither vehicle overtakes the other.

29. The maximum velocity of the ball when it touches the support is $v_{max} = \sqrt{2gH}$.

During the impact the velocity of the ball is reversed, remaining the same in absolute magnitude. The velocity chart has the form shown in Fig. 279*a*.

Figure 279*b* shows how the coordinate changes with time.

30. The time during which the first ball falls is $t_1 = \sqrt{\dfrac{2h_1}{g}} = 0.3$ s.

The ratio of the maximum velocities of the balls is $\dfrac{v_2}{v_1} = \sqrt{\dfrac{h_2}{h_1}} = \dfrac{1}{2}$.

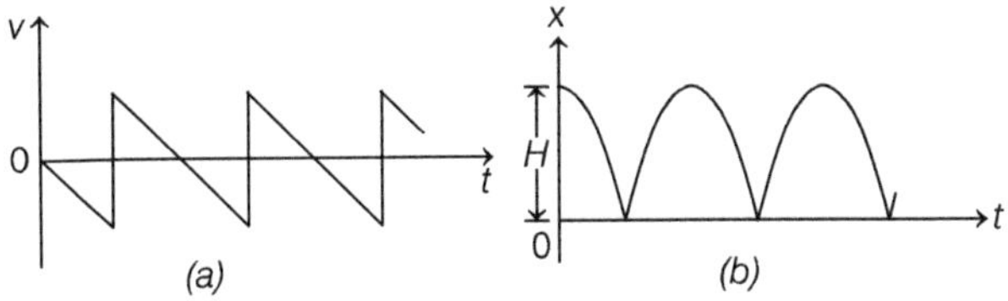

Fig. 279

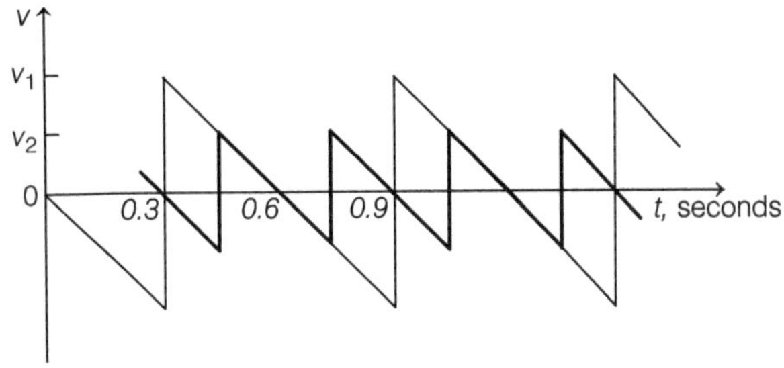

Fig. 280

It follows from the velocity chart (Fig. 280) that the minimum time $\tau = 0.3$ s. Besides, the second ball may begin to fall in 0.6, 0.9, 1.2 s, etc., after the first ball begins to drop.

The time t during which the velocities of the two balls are the same is 0.3 s. The process periodically repeats every 0.6 second.

31. The initial equations are

$$\frac{gt^2}{2} = n, \ \frac{g(t-\tau)^2}{2} = n - 1$$

where τ is the duration of motion of the body over the n-th centimetre of its path.

Hence,

$$t = \sqrt{\frac{2n}{g}}, \ t - \tau = \sqrt{\frac{2(n-1)}{g}}$$

$$\tau = \sqrt{\frac{2}{g}} \ (\sqrt{n} - \sqrt{n-1})$$

32. Upon denoting the coordinate and the velocity of the first body with respect to the tower by x_1 and v_1 and those of the second by x_2 and v_2, we can write the following equations

$$x_1 = v_0 t - \frac{gt^2}{2}; \ v_1 = v_0 - gt;$$

$$x_2 = -v_0(t-\tau) - \frac{g(t-\tau)^2}{2}; v_2 = -v_0 - g(t-\tau)$$

(The upward direction is considered to be positive here.)

The velocity of the first body relative to the second is $u = v_1 - v_2 = 2v_0 - g\tau$ and it does not change with time.
The distance between the bodies is

$$s = x_1 - x_2 = (2v_0 - g\tau)\, t - v_0\tau + \frac{g\tau^2}{2}$$

The bodies move uniformly with respect to each other and therefore the distance between them changes in proportion to the time.

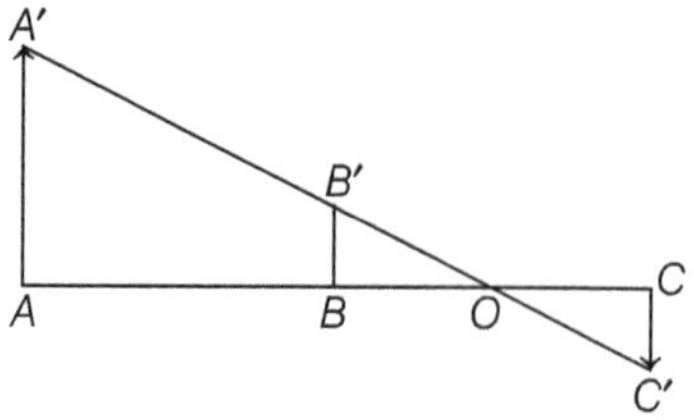

Fig. 281

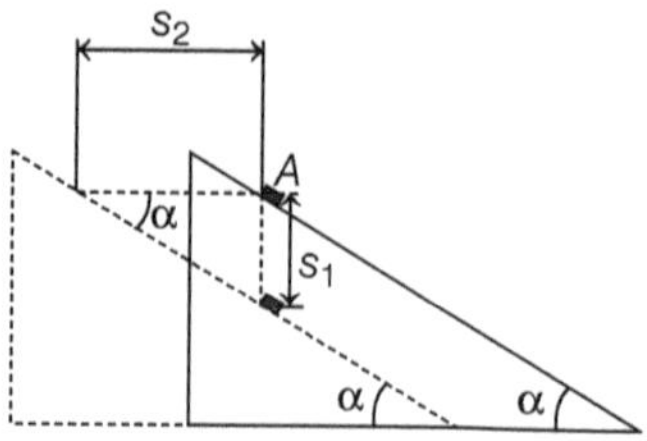

Fig. 282

33. According to the condition, $AA' = vt$ and $CC' = \frac{at^2}{2}$ (Fig. 281). From the similarity of the triangles $AA'O$, $BB'O$ and $CC'O$ we have

$$\frac{AA'}{AO} = \frac{BB'}{BO} = \frac{CC'}{CO}$$

A glance at Fig. 281 shows that $AO = AB + BO$ and $CO = BC - BO$. These ratios allow us to determine

$$BB' = \frac{AA' - CC'}{2} = \frac{vt}{2} - \frac{at^2}{4}$$

Hence, point B moves with the initial velocity $\frac{v}{2}$ directed upward and a constant acceleration $\frac{a}{2}$ directed downward. After reaching the height $h = \frac{v^2}{4a}$, the point will move downward.

34. Let us denote the speed of the left-hand truck at a certain moment of time by v_1, of the right-hand one by v_2, and of the towed one by v_3. Then, during the time t the left-hand truck will cover the distance

$$s_1 = v_1 t + \frac{a_1 t^2}{2}$$

the right-hand truck the distance, $s_2 = v_2 t + \frac{a_2 t^2}{2}$

and the towed truck the distance, $s_3 = v_3 t + \frac{a_3 t^2}{2}$

At the same time it is easy to see that $s_3 = \frac{s_1 + s_2}{2}$

Since this equality must be true at any value of t, then

$$v_3 = \frac{v_1 + v_2}{2} \text{ and } a_3 = \frac{a_1 + a_2}{2}$$

35. The acceleration of the book with respect to the floor of the lift depends on the direction of acceleration of the lift and not on the direction of its motion (the direction of its velocity).

If the acceleration of the lift is directed upward, the acceleration of the book will be $g + a$. If it is directed downward, the acceleration of the book will be $g - a$.

36. The acceleration of the stone with respect to the Earth is g and with respect to the railway carriage $\sqrt{a^2 + g^2}$.

37. If the speed of the lift did not change, the ball would jump up to a height H.

In a system of reading having a constant speed equal to that of the lift at the moment the ball begins to fall, the lift will rise during the time t to the height $h_1 = \frac{at^2}{2}$, and during the next time interval t to the height $h_2 = at^2 - \frac{at^2}{2}$. Its total height of rising is $h = h_1 + h_2 = at^2$.

The sought height which the ball will jump up to above the floor of the lift is $x = H - h = H - at^2$.

38. The time during which the load is lifted to the height h is $t = \sqrt{\frac{2h}{a_1}}$.

The speed of the load relative to the crane in a vertical direction at this moment is $v_1 = a_1 t$ and in a horizontal one $v_2 = a_2 t$.

The total speed of the load with respect to the ground is

$$v = \sqrt{v_1^2 + v_2^2 + v_0^2}$$

39. In free falling, body A will travel a vertical distance $s_1 = \frac{gt^2}{2}$ during the time t. During the same time the wedge should move over a distance $s_2 = \frac{at^2}{2}$. If the body is constantly in contact with the wedge, then $\frac{s_2}{s_1} = \cot\alpha$, as can be seen from Fig. 282.

Therefore, the acceleration sought $a = g\cot\alpha$.

If the acceleration of the wedge in a horizontal direction is greater than $g\cot\alpha$, the body will move away from the wedge.

1-3. Dynamics of Rectilinear Motion

40. The force F applied to the sphere determines, according to Newton's second law, the magnitude and direction of acceleration of the sphere, but does not determine its velocity. For this reason the sphere can move in any direction under the force F, and may also have a velocity equal to zero.

41. The resultant of all the forces is 6 kgf and coincides in direction with the force 5 kgf. Therefore, the acceleration of the sphere is $a = \frac{F}{m} = 14.7$ m/s^2 and is directed towards the force 5 kgf.

Nothing can be said about the direction of motion (see the solution to Problem 40).

42. In the MKS (or SI) system the weight of the body $G = mg = 9.8\text{N}$. The unit of force in the technical system is 1 kgf, i. e., the force with which the Earth attracts a body with a mass of 1 kg. In the CGS system the weight of the body is 980,000 dynes. When Newton's second law is used to find the force in the technical system of units, it should be borne in mind that the mass should also be expressed in technical units.

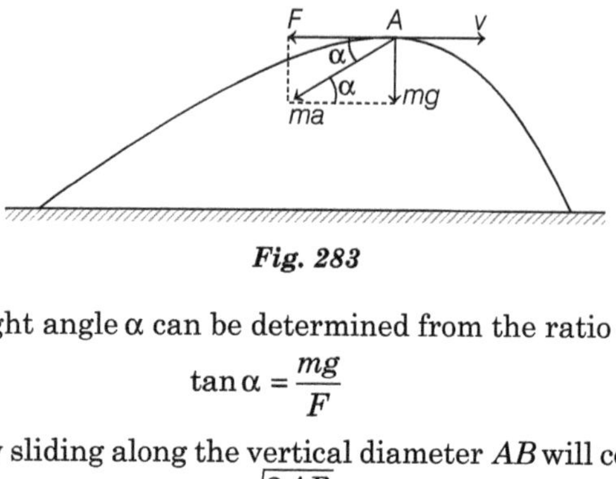

Fig. 283

43. The sought angle α can be determined from the ratio (Fig. 283):

$$\tan\alpha = \frac{mg}{F}$$

44. The body sliding along the vertical diameter AB will cover the entire distance in the time $t_{AB} = \sqrt{\frac{2AB}{g}}$. For an arbitrary groove arranged at an angle α to the diameter AB, the time of motion is

$$t_{AC} = \sqrt{\frac{2AC}{g\cos\alpha}}.$$

Since $AC = AB\cos\alpha$, then $t_{AC} = t_{AB}$. All the bodies will reach the edge of the disk simultaneously.

45. The force of air resistance F will reach its minimum after the parachutist's speed becomes constant, and we have $F = mg = 75\,\text{kgf}$.

46. According to Newton's second law, $N - mg = \pm ma$. Therefore, $N = mg + ma$ if the acceleration of the lift is directed upward, and $N = mg - ma$ if downward, irrespective of the direction of the speed. When $a = g$, then $N = 0$. (Here and below N denotes the force of normal pressure, or the force of normal reaction.)

47. According to Newton's second law, $ma = kmg$. Hence, the coefficient of friction $k = \frac{a}{g}$. Since in an elastic impact only the direction of the velocity changes, irrespective of the angle, then $a = \frac{v^2}{2s}$, where $s = 12.5$ metres is the total path travelled by the puck before it stops. Therefore, $k = \frac{v^2}{2gs} = 0.102$.

48. Assuming the acceleration of a motor vehicle to be constant, we can write $a = \frac{v^2}{2s}$. Since the maximum force of friction in braking is kmg, then, according to Newton's second law, $m\frac{v^2}{2s} = kmg$, where m is the mass of the motor vehicle. Hence, $k = \frac{v^2}{2gs}$. Upon inserting the values of v and s from the table in this formula, we can find the coefficients of friction for various pavements.

ice	$k = 0.1$
dry snow	$k = 0.2$
wet wood block	$k = 0.3$
dry wood block	$k = 0.5$
wet asphalt	$k = 0.4$
dry asphalt	$k = 0.6$
dry concrete	$k = 0.7$

The coefficient of friction does not depend on the speed if the accuracy to the first digit after the decimal point is wanted.

49. When the motor vehicle accelerates, the rear wall of the fuel tank imparts an acceleration u/t to the petrol. According to Newton's second law, the force F required for this acceleration is $Al\rho\frac{v}{t}$, where A is the area of the rear wall of the fuel tank. In conformity with Newton's third law, the petrol will act with the same force on the wall. The hydrostatic pressure of the petrol on both walls is the same. Hence, the difference of pressures exerted on the walls is $\Delta\rho = \frac{F}{A} = l\rho\frac{v}{t}$.

50. The mass of the left-hand part of the rod $m_1 = \frac{M}{L} l$ and of the right-hand part $m_2 = \frac{M}{L}(L - l)$, where M is the mass of the entire rod. Under the action of the applied forces each part of the rod moves with the same acceleration a. Therefore,

$$F_1 - F = m_1 a$$
$$F - F_2 = m_2 a$$

Hence the force F is

$$F = \frac{F_1 m_2 + F_2 m_1}{m_1 + m_2} = F_1 \frac{L - l}{L} + F_2 \frac{l}{L}$$

51. The motion of the ball will be uniform. The images of the ball on the film appear at intervals of $t = 1/24$ s.

The distance between the positions A and B of the ball in space that correspond to the positions C and D of the images on the film is $AB = CD\frac{OE}{OF}$, as shown in Fig. 284. The focal length of the lens $OF = 10$ cm, $OE = 15$ metres, and $CD = 3$ mm. The velocity of the ball $v_1 = \frac{AB}{t} = 10.8$ m/s.

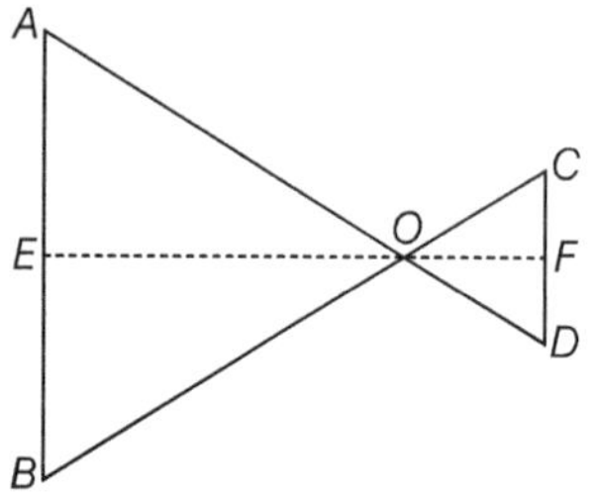

Fig. 284

When the ball is in uniform motion, $mg = kv_1^2$. In the second case $4mg = kv_2^2$. Hence, $v_2^2 = 4v_1^2$ and $v_2 = 21.6$ m/s.

$k = \frac{G}{v_1^2} \approx 3.9 \times 10^{-5}\,\text{kgf}\cdot\text{s}^2/\text{m}^2$.

52. Figure 285 shows the forces that act on the weights. The equations of motion for the weights can be written as follows:

$$m_1 a = T - m_1 g \quad \text{and} \quad m_2 a = m_2 g - T$$

where T is the tension of the string and a is the acceleration. (The accelerations of the weights are the same since the string is considered to be unstretchable. The weightlessness of the string and the pulley determine the constancy of T.)

Therefore,

$$a = \frac{m_2 - m_1}{m_1 + m_2} g = 327 \text{ cm/s}^2$$

$$T = m_1(a + g)$$
$$= 130{,}700 \text{ dynes} = 133 \text{ gf}$$

The time of motion $t = \sqrt{\frac{2H}{a}} \cong 1$ s.

53. If the mass of the pulleys and the rope is negligibly small, then (Fig. 286) $2F - T = 0$ and $T - G = ma$.

Hence, $F = \frac{G}{2}\left(1 + \frac{a}{g}\right)$. When $a = 0$, we have $F = \frac{G}{2}$.

54. If the mass of the second weight is much greater than 200 g, both weights will move with an acceleration somewhat below g, the acceleration of the lighter weight being directed upwards. To make a weight of mass m move upward with an acceleration g, a force of $2mg$ should be applied to it.

For this reason the string should withstand a tension of about 400 gf.

55. The dynamometer first shows $F = 3$ kgf. If the reading of the dynamometer does not change, the weight 2 kgf is acted upon by the upward force of the string tension equal to 3 kgf. Therefore, this weight rises with an acceleration of $a = \frac{g}{2}$. The other weight lowers with the same acceleration.

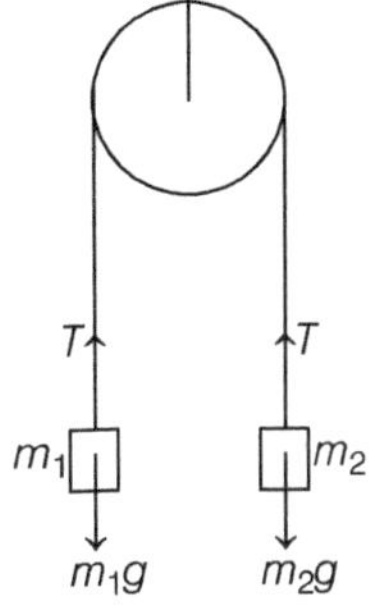

Fig. 285

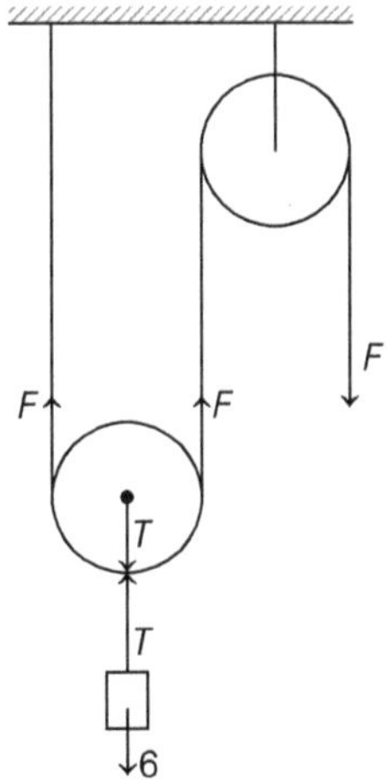

Fig. 286

The additional weight on the second pan can be found from the equation

$$\frac{G+G_1}{g}\times\frac{g}{2}=(G+G_1-F)$$

Hence $G_1 = 3$ kgf.

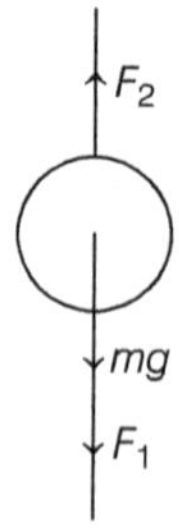

Fig. 287

56. The sphere is acted upon by three forces: the force of gravity, the force of tension of the upper rope and the force applied to the lower one when it is pulled (Fig. 287).

The acceleration imparted to the sphere by the pull can be found from the equation $ma = F_1 + mg - F_2$.

For the lower rope to break, the force applied to it should be greater than the tension of the upper one, i.e., $F_1 > F_2$.

For this condition, the acceleration imparted to the sphere is greater than that of gravity, i.e., $a > g$.

57. According to Newton's second law,

$(m_1 + m_2)\,a = m_1 g\sin\alpha - m_2 g\sin\beta - km_1 g\cos\alpha - km_2 g\cos\beta$

The weights will be at the same height after travelling the distance s, which conforms to the following equations: $s\sin\alpha = h - s\sin\beta$ and $s=\dfrac{a\tau^2}{2}$.

Upon eliminating s and a from the three equations we obtain

$$\frac{m_1}{m_2}=\frac{g\tau^2(\sin\alpha+\sin\beta)(k\cos\beta+\sin\beta)+2h}{g\tau^2(\sin\alpha+\sin\beta)(\sin\alpha-k\cos\alpha)-2h}$$

58. The equations of motion give the following formulas for the acceleration of the stone:

$a_1 = g(\sin\alpha + k\cos\alpha)$ in upward motion

$a_2 = g(\sin\alpha - k\cos\alpha)$ in downward motion

The kinematic equations can be written as follows:

$$l = v_0 t_1 - \frac{a_1 t_1^2}{2};\; l=\frac{a_2 t_2^2}{2};$$

$$v_0 - a_1 t_1 = 0$$

We find from these five equations that

$$k = \frac{2l - gt_1^2 \sin\alpha}{gt_1^2 \cos\alpha} \cong 0.37$$

$$t_2 = t_1 \sqrt{\frac{l}{gt_1^2 \sin\alpha - l}} = 4.2 \text{ s}$$

59. For this case the equations of dynamics can be written as

$$mg - T = ma, \text{ and } T = Ma$$

where T is the tension of the string.

Hence,

$$a = \frac{m}{M+m} g = \frac{2}{7} g$$

The equations of kinematics give $x = v_0 t - \frac{at^2}{2}$, and $v_t = v_0 - at$. Upon solving this system of equations, we find that in 5 seconds the cart will stay at the same place ($x = 0$) and will have a speed $v_t = 7$ m/s directed to the right. The cart will cover the distance

$$s = 2\left\{v_0 \frac{t}{2} - \frac{a\left(\frac{t}{2}\right)^2}{2}\right\} = 17.5 \text{ metres}$$

60. The ice-boat can move only in the direction of its runners. When the speed of the boat exceeds that of the wind, the velocity of the latter with respect to the boat has a component directed backward. If the velocity of the wind with respect to the boat also has a component perpendicular to the direction of motion, the sail can be so set that the force F acting on it will push the boat forward (Fig. 288).

Therefore, the speed of the boat can exceed that of the wind. In practice it can be two or three times greater.

61. (1) At the initial moment the acceleration is $a_0 = \frac{F}{M} \approx 13.1 \text{ m/s}^2$. It changes with time according to the law $a = \frac{F}{M - \mu t}$, where $\mu = 200$ kg/s is the mass of the fuel consumed by the rocket in a unit of time. A diagram of the acceleration is shown in Fig. 289. In 20 seconds the velocity is numerically equal to the hatched area, $v \cong 300$ m/s.

(2) Newton's second law can be written as

$$(M - \mu t)\, a = F - (M - \mu t)\, g - f$$

According to the initial conditions, $t = 20$ s and $a = 0.8$ g. Hence, the force of air resistance is

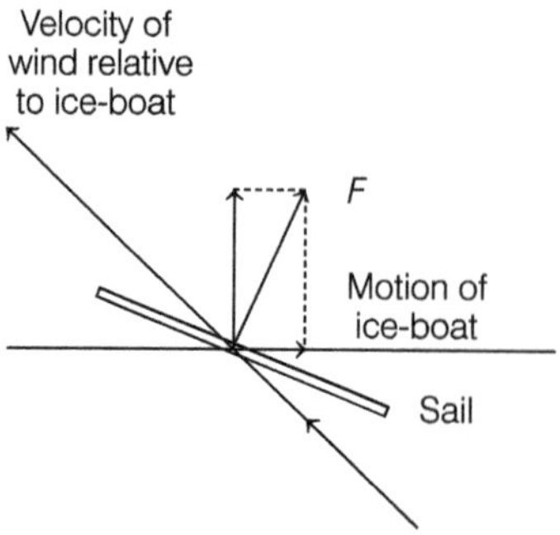

Fig. 288

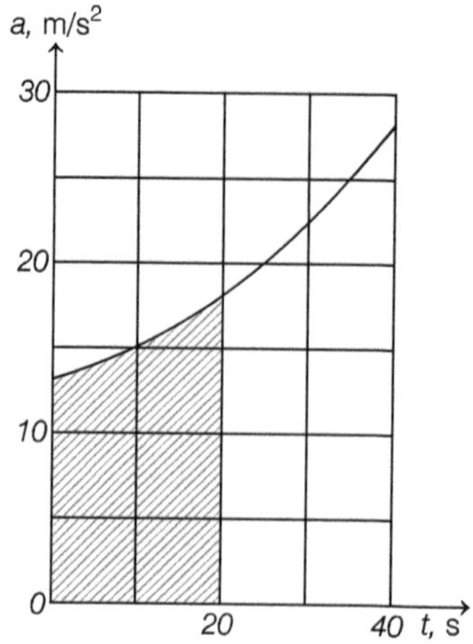

Fig. 289

$$f = F - (M - \mu t)g - (M - \mu t)0.8\ g = 12{,}800 \text{ kgf}$$

(3) Newton's equation for the weight gives $m_1 a = kx - m_1 g$, where m_1 is the mass of the weight at the end of the spring, c is the acceleration of the rocket, k is the coefficient of elasticity of the spring, x is the elongation of the spring. According to the condition, $m_1 g = kl_0$. Therefore, $x = \frac{l_0}{g}(a + g)$. The scale of the device should be graduated uniformly (Fig. 290). An acceleration of g corresponds to a division of one centimetre.

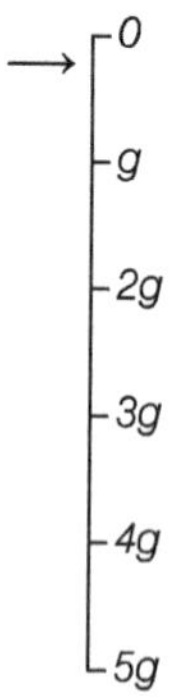

Fig. 290

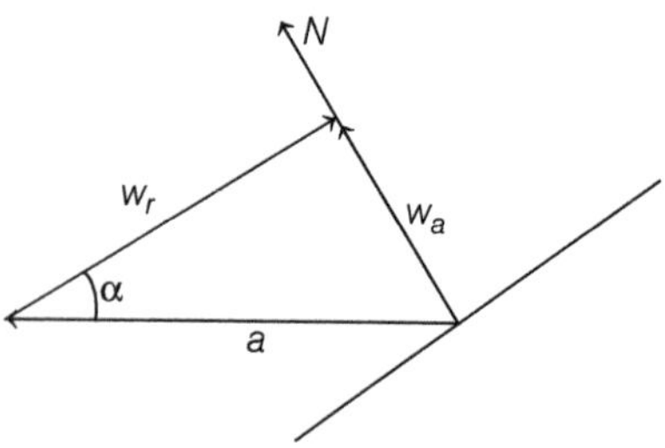

Fig. 291

62. The only force acting on the bead is the reaction force of the rod N directed at right angles to the rod. The absolute acceleration w_a of the bead (relative to a stationary observer) will be directed along the line of action of the reaction force N. The relative acceleration w_r is directed along the rod (Fig. 291)

$$w_a = a + w_r$$

It follows from the triangle of accelerations that $w_r = a\cos\alpha$ and $w_a = a\sin\alpha$.

From Newton's second law, the reaction force is $N = ma\sin\alpha$.

The time τ during which the bead moves along the rod can be found from the equation $l = \frac{a\cos\alpha \cdot \tau^2}{2}$. Hence, $\tau = \sqrt{\frac{2l}{a\cos\alpha}}$.

63. When the bead moves it is acted upon by the friction force kN and the reaction force N.

The absolute acceleration will be directed along the resultant force F. It follows from Fig. 292 that

$N = ma\sin\alpha$, and $w_r = a\cos\alpha - \frac{kN}{m} = a(\cos\alpha - k\sin\alpha)$

Hence (see Problem 62),

$$\tau = \sqrt{\frac{2l}{a(\cos\alpha - k\sin\alpha)}}$$

If $k \geq \cot\alpha$, the bead will not move with respect to the rod, and the force of friction is $ma\cos\alpha$.

64. The equations of motion of the block and the body have the form:

$$ma = f \qquad (1)$$

$$Mb = F - f \qquad (2)$$

where f is the force of friction; a and b are accelerations.

Let us assume that there is no sliding, then $a = b$. The acceleration and the force of friction can be found from the equations of motion. The force of friction is $f = m\frac{F}{M+m}$. For the body not to slide, the force of friction should satisfy the inequality $f \leq kmg$, i.e., $\frac{F}{M+m} \leq kg$. If $F > k(M+m)g$, the body will begin to slide. Here equations (1) and (2) will take the form :

$$ma = kmg, \text{ and } Mb = F - kmg$$

These equations can be used to find a and b:

$$a = kg, \text{ and } b = \frac{F - kmg}{M}$$

Obviously, $b > a$. The acceleration of the body relative to the block will be directed oppositely to the motion and will be equal in magnitude to $\frac{F - kmg}{M} - kg$.

The time during which the body moves over the block is

$$\tau = \sqrt{\frac{2lM}{F - kg(M + m)}}.$$

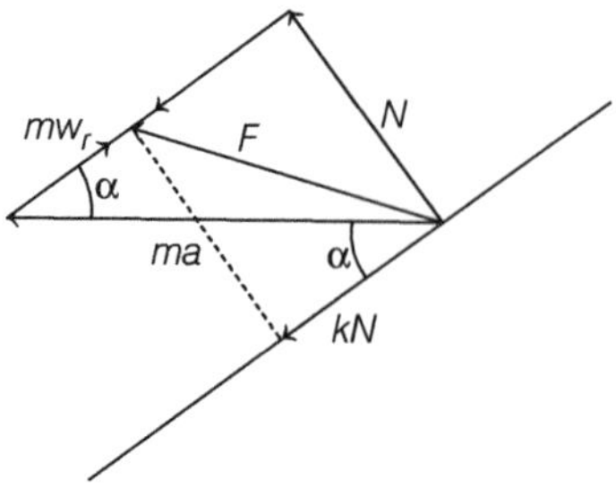

Fig. 292

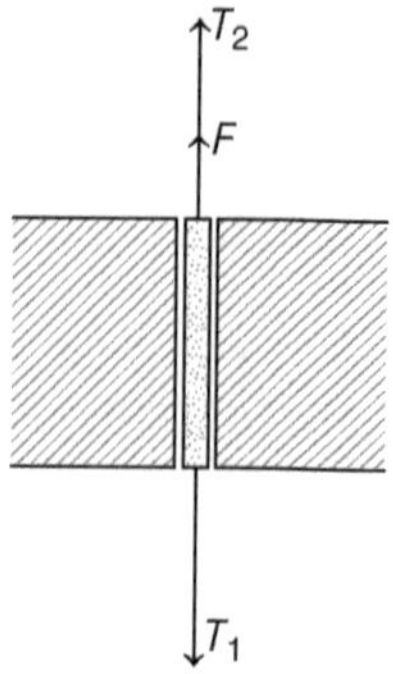

Fig. 293

65. Initially the wagon is in uniformly retarded motion at a speed of $v = v_0 - \frac{f}{M}t$, where f is the force of friction equal to kmg. The body is in uniformly accelerated motion at a speed of $u = \frac{f}{m}t$.

If the wagon is long, the speeds of the body and the wagon may become equal. This will occur at the moment of time $\tau = \frac{v_0}{\frac{f}{m} + \frac{f}{M}}$.

After this, both the body and the wagon will begin to move at a constant speed equal to $\frac{Mv_0}{M + m}$. By this time the wagon will have covered a distance

$$s = v_0\tau - \frac{f}{2M}\tau^2$$

and the body a distance

$$s = \frac{f}{2m}\tau^2$$

The distance covered by the body relative to the cart is $S - s$. It should be shorter than l. Thus, the body will not slip off the wagon if $S - s \leq l$, i.e.,

$$\frac{Mv_0^2}{2gk(M + m)} \leq l$$

68. Let us consider the string element in the slit. Assume that the string moves downward. Then, the string element will be acted upon by the force of string tension on both sides and the force of friction (Fig. 293).

Since the mass of this string element is neglected, $T_1 - F - T_2 = 0$.

The equations of dynamics can be written as follows:

$$m_1 g - T_1 = m_1 a$$
$$m_2 g - T_2 = -m_2 a$$

Hence

$$a = \frac{(m_1 - m_2)g - F}{m_1 + m_2}$$

67. Since the weights move uniformly, the tension of the string is equal to the weight m_1. Therefore, the force with which the pulley acts on the bar is $2m_1 g$, i.e., in the first case it is 2 kgf and in the second 6 kgf. In both cases the balance will show the sum of the first and second weights, i.e., 4 kgf. The force of friction equal to 2 kgf is applied to the bar at the side of the second weight. In the first case it is added to the force of pressure exerted by the pulley on the bar and in the second it is subtracted from it.

68. Since the masses of the pulleys and the string may be neglected, the tension of the string will be the same everywhere (Fig. 294). Therefore,

$$m_1 g - T = m_1 a_1$$
$$m_2 g - 2T = m_2 a_2$$
$$m_3 g - T = m_3 a_3$$
$$a_2 = -\frac{a_1 + a_3}{2}$$

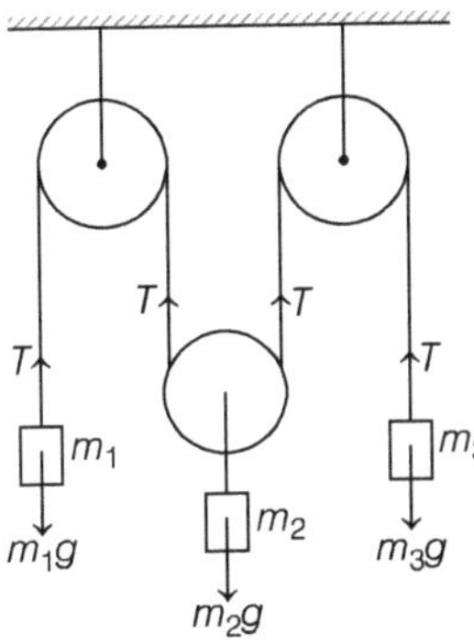

Fig. 294

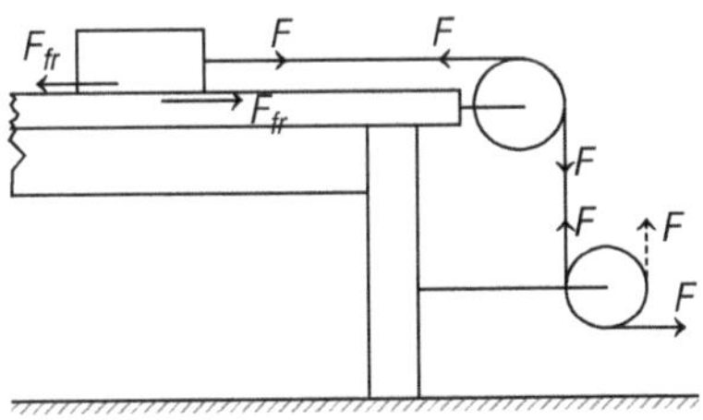

Fig. 295

(see Problem 34). Hence,

$$a_1 = \frac{4m_1 m_3 - 3m_2 m_3 + m_1 m_2}{4m_1 m_3 + m_2 m_3 + m_1 m_2} g$$

$$a_2 = \frac{m_1 m_2 - 4m_1 m_3 + m_2 m_3}{4m_1 m_3 + m_2 m_3 + m_1 m_2} g$$

$$a_3 = \frac{4m_1 m_3 - 3m_1 m_2 + m_2 m_3}{4m_1 m_3 + m_2 m_3 + m_1 m_2} g$$

69. The second monkey will be at the same height as the first. If the mass of the pulley and the weight of the rope are disregarded, the force T tensioning both ends of the rope will be the same and, therefore, the forces acting on each monkey will equal $F = T - mg$. Both monkeys have the same accelerations in magnitude and direction and will reach the pulley at the same time.

70. Since the mass of the pulleys and the string is negligibly small, the tension of the thread is the same everywhere.

Therefore,

$$m_1 g - T = m_1 a_1$$
$$m_2 g - 2T = m_2 a_2$$
$$2T - T = 0$$

Hence, $T = 0$ and $a_1 = a_2 = g$.

Both weights fall freely with an acceleration g. Pulleys B and C rotate counterclockwise and pulley A clockwise.

71. (1) The forces acting on the table and the weight are shown in Fig. 295. The equations of horizontal motion have the following form:

for the table with the pulley

$$F - F + F_{fr} = \frac{G_1}{g} a_1$$

and for the weight

$$F - F_{fr} = \frac{G_2}{g} a_2$$

Let us assume that the force F is so small that the weight does not slide over the table. Hence $a_1 = a_2$ and $F_{fr} = F \frac{G_1}{G_1 + G_2}$.

By gradually increasing the force F we shall thereby increase the force of friction F_{fr}. If the table and the weight are immovable relative to each other, however, the force of friction between them cannot exceed the value $F_{fr \cdot max} = kG_2$.

For this reason the weight will begin to slide over the table when

$$F > F_{fr \cdot max} \frac{G_1 + G_2}{G_1}$$

$$= k \frac{G_2}{G_1}(G_1 + G_2) = 10 \text{ kgf}$$

In our case $F = 8$ kgf, consequently, the weight will not slide, and

$$a_1 = a_2 = \frac{F}{G_1 + G_2} g = \frac{8}{25} g \cong 314 \text{ cm/s}^2$$

(2) In this case the equations of motion to the table with the pulleys and the weight will have the form:

$$-F + F_{fr} = \frac{G_1}{g} a_1$$

$$F - F_{fr} = \frac{G_2}{g} a_2$$

The accelerations of the table and the weight are directed oppositely, and the weight will be sure to slide.

Hence, $F_{fr} = kG_2$

The acceleration of the table is

$$a_1 = \frac{-F + kG_2}{G_1} g = -\frac{2}{15} g = -131 \text{ cm/s}^2$$

and it will move to the left.

72. According to Newton's second law, the change in the momentum of the system "cannon-ball" during the duration of the shot τ should be equal to the impulse of the forces acting on the system.

Along a horizontal line

$$mv_0 \cos\alpha - Mv_1 = F_{fr}\tau$$

where $F_{fr}\tau$ is the impulse of the forces of friction.

Along a vertical line

$$mv_0 \sin\alpha = N\tau - (Mg + mg)\tau$$

were $N\tau$ is the impulse of forces of normal pressure (reactions of a horizontal area),$(Mg + mg)\tau$ is the impulse of the forces of gravity. Remembering that $F_{fr} = kN$, we obtain

$$v_1 = \frac{m}{M} v_0 \cos\alpha - k\frac{m}{M} v_0 \sin\alpha - k\frac{M + m}{M} g\tau$$

or since $g\tau \ll v_0$, $v_1 \cong \frac{m}{M} v_0(\cos\alpha - k\sin\alpha)$

This solution is suitable for $k \le \cot\alpha$. When $k > \cot\alpha$ the cannon will remain in place.

1-4. The Law of Conservation of Momentum

73. The momentum of the meteorite is transmitted to the air molecules and, in the final run, to the Earth.

74. Let us divide the mass of the disk into pairs of identical elements lying on one straight line at equal distances from the centre. The momentum of each pair is zero, since the momenta of both masses are equal but oppositely directed. Therefore, the momentum of the entire disk is zero.

75. The propeller of a conventional helicopter is rotated by an engine mounted inside the fuselage. According to Newton's third law, oppositely directed forces are applied from the propeller to the engine. These forces create a torque that tends to rotate the fuselage in a direction opposite to rotation of the propeller. The tail rotor is used to compensate for this torque. In a jet helicopter the forces from the propeller are applied to the outflowing gases and for this reason do not create any torque.

76. The velocity of the boat u can be found with the aid of the law of conservation of momentum. In a horizontal direction

$$Mu = mv\cos\alpha$$

Hence, $u = 8$ cm/s.

77. At the highest point the rocket reaches, its velocity is zero. The change in the total momentum of the rocket fragments under the action of external forces (the force of gravity) is negligible since the impulse of these forces is very small in view of the instantaneous nature of the explosion. For this reason the total momentum of the rocket fragments before and immediately after the explosion remains constant and equal to zero. At the same time, the sum of the three vectors $(m_1\mathrm{v}_1, m_2\mathrm{v}_2, m_3\mathrm{v}_3)$ may be zero only when they are in one plane. Hence it follows that the vectors v_1, v_2 and v_3 also lie in one plane.

78. Let the mass of the man be m and that of the boat M. If the man moves with a speed v relative to the boat the latter will move at a speed $-u$ with respect to the bank, and therefore the man moves with a speed $v_1 = v - u$ with respect to the bank.

According to the law of conservation of momentum,

$$m(v-u) - Mu = 0$$

Hence, $u = \dfrac{m}{m+M}v$ and the speed of the man relative to the bank is

$$v_1 = \frac{M}{m+M}v.$$

Since the signs of v_1 and v coincide, the distance between the man and the bank will increase whatever the ratio between the masses m and M.

79. The speed of the boat u with respect to the shore is related to the speed of the man v with respect to the boat by the equation $u = \frac{m}{m+M} v$ (see Problem 78). The relation between the speeds remains constant during motion. For this reason the relation between the distances travelled will be equal to the relation between the speeds

$$\frac{s}{l} = \frac{m}{m+M}$$

where s is the distance travelled by the boat and l is its length (the distance covered by the man in the boat).

Therefore, for the boat to reach the shore its length should be atleast

$$l = \frac{m+M}{m} s = 2.5 \text{ metres.}$$

80. When the spring extends, it will act on both weights. The weight at the wall will first remain stationary while the second weight will begin to move. When the spring extends completely(i.e., is no longer deformed), the second weight will have a certain velocity. Therefore, the system will acquire momentum in a horizontal direction that will be retained in the following period, since the external forces will not act in this direction. Thus, the system as a whole will move away from the wall, the weights alternately converging and diverging.

81. The speed of the cart will not depend on the point of impact. The momentum of the revolving cylinder is zero, irrespective of its direction and velocity of rotation (see Problem 74). For this reason the bullet will impart to the "cylinder-cart" system the same momentum as it would to a cylinder rigidly secured on the cart.

82. Let us denote the velocity of the rocket at the end of the k-th second by v_k. Gas with a mass m is ejected from the rocket at the end of the $(k + 1)$-th second, and it carries away a momentum equal to

$$m(-u + v_k)$$

It follows from the law of conservation of momentum that

$$(M - km)v_k = [M - (k+1)m]v_{k+1} + m(-u + v_k)$$

The change in the velocity of the rocket per second is

$$v_{k+1} - v_k = \frac{mu}{M - (k+1)m}$$

If we know how the velocity changes in one second, we can write the expression for the velocity at the end of the n-th second

$$v_n = v_0 + u\left(\frac{m}{M-m} + \frac{m}{M-2m} + \ldots + \frac{m}{M-nm}\right)$$

83. The velocity of the rocket will grow. This becomes obvious if we pass over to a reading system with respect to which the rocket is at rest at the given moment. The pressure of the ejected gases will push the rocket forward.

84. Let the mass of the boat be M, that of a sack m and the velocity of the boats v_0. When a sack is thrown out of a boat the latter is acted upon by a certain force in a direction perpendicular to v_0. It should be noted, however, that the boat does not change its velocity, since the resistance of the water prevents lateral motion of the boats. The velocity of a boat will change only when a sack is dropped into it.

Applying the law of conservation of momentum to the "sack-boat" system, we can write in the first case:

for one boat

$$(M + m)v_0 - mv_0 = (M + 2m)v_1$$

for the other one $- Mv_0 + mv_1 = (M + m)v_2$

Here v_1 and v_2 are the final velocities of the boats. From these simultaneous equations

$$v_1 = - v_2 = \frac{M}{M+2m} v_0.$$

When the sacks are exchanged simultaneously, the final velocities of the boats v'_1 and v'_2 can be found from the equations:

$$Mv_0 - mv_0 = (M + m)v'_1 ;$$

$$- Mv_0 + mv_0 = (M + m)v'_2$$

Hence, $v'_1 = - v'_2 = \dfrac{M-m}{M+m} v_0.$

Thus, the final velocity of the boats will be higher in the first case.

85. External forces do not act in a horizontal direction on the "hoop-beetle" system. For this reason the centre of gravity of the system (point C in Fig. 296) will not move in a horizontal plane. The distance from the centre of gravity of the system to the centre of the hoop is

$$CO = \frac{m}{m+M} R.$$

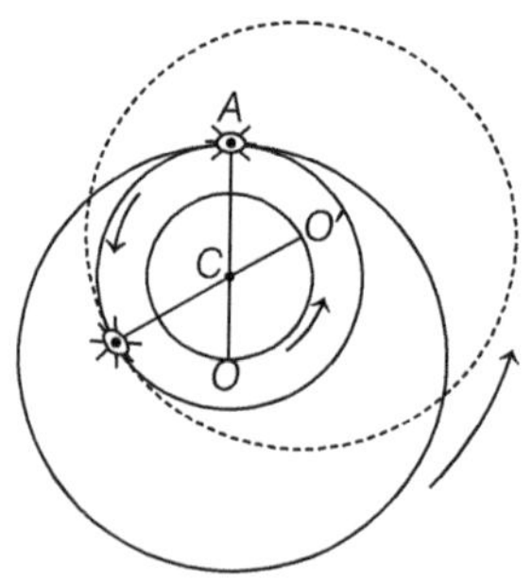

Fig. 296

Since this distance is constant, the centre of the hoop O will describe a circle with the radius CO about the stationary point C. It is easy to see that the trajectory of the beetle is a circle with the radius

$$AC = \frac{M}{m+M} R.$$

The mutual positions and the direction of motion of the beetle and the hoop are shown in Fig. 296.

86. Since no external forces act on the system in a horizontal direction, the projection of the total momentum of the "wedge-weights" system onto the horizontal direction must remain constant (equal to zero). It thus follows that the wedge will begin to move only if the weights move.

For the weight m_2 to move to the right, the condition

$$m_2 g \sin\alpha \geq m_1 g + k m_2 g \cos\alpha$$

should be observed.

Therefore, $\frac{m_1}{m_2} \leq \sin\alpha - k\cos\alpha$. Here the wedge will move to the left.

For the weight m_2 to move to the left, the following condition should be observed :

$$m_1 g \geq m_2 g \sin\alpha + k m_2 g \cos\alpha$$

or

$$\frac{m_1}{m_2} \geq \sin\alpha + k\cos\alpha$$

Here the wedge will move to the right.

Hence, for the wedge to be in equilibrium, the ratio between the masses of the weights should satisfy the inequality

$$\sin\alpha - k\cos\alpha \leq \frac{m_1}{m_2} \leq \sin\alpha + k\cos\alpha$$

1-5. Statics

87. $l_1 = l\frac{k}{k+1}$.

88. In the position of equilibrium (Fig. 297) $mg - 2mg\cos\alpha = 0$. Therefore, $\alpha = 60°$. The sought distance $h = l\cot\alpha = \frac{l}{\sqrt{3}}$. Equilibrium will set in after the oscillations caused by the weight being lowered attenuate.

89. The equality of the projections of the forces onto the direction of the vertical (Fig. 298) gives the equation

$$2N\sin\frac{\alpha}{2} - 2F_{fr}\cos\frac{\alpha}{2} = 0$$

where N is the force of normal pressure and $F_{fr} \le kN$ is the force of friction.

The weight of the wedge may usually be neglected.

Hence, $\tan\frac{\alpha}{2} \le k$ and $\alpha \le 2$ arc tan k.

90. If the weight G_1 lowers over the height h, then point F will lower by $h/3$. The weight G_2 will rise by $2h/3$. Applying the "golden" rule of mechanics, we have $G_1 h = G_2 \frac{2}{3} h$.

Hence, $\quad G_1 = \frac{2}{3} G_2.$

91. If the box is not overturned, the moment of the force F rotating the box in one direction, say counterclockwise, about a bottom edge is less than or equal to the moment of the force of gravity rotating the box clockwise. For the box to slide, the force should be greater than the maximum force of friction applied to it. Therefore,

$$Fh \le mg\frac{l}{2} \text{ and } F \ge kmg$$

whence $k \le \frac{l}{2h}$.

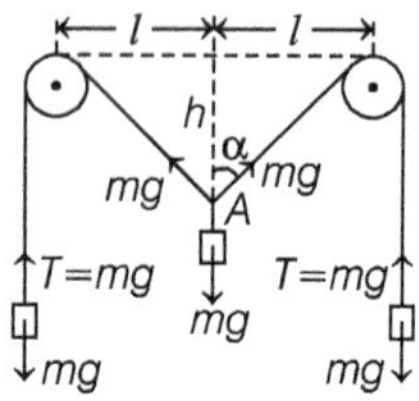

Fig. 297

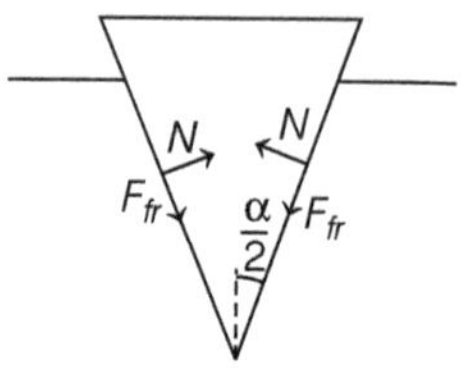

Fig. 298

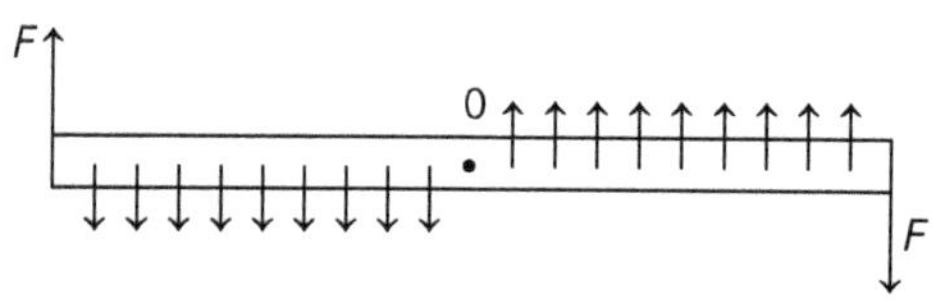

Fig. 299

92. To turn the beam, the moment of the forces applied to its ends should be greater than the moment of the forces of friction when they reach their maximum.

The forces of friction are distributed uniformly along the beam (Fig. 299). The mean arm of the forces of friction acting on the left- or right-hand part of the beam is $l/4$, if the length of the entire beam is l.

The moment of all the forces of friction with respect to the beam centre is $2\frac{kG}{2}\frac{l}{4}$. Consequently, to turn the beam around, the applied forces F should satisfy the inequality $2F\frac{l}{2} > \frac{kGl}{4}$ whence $F > \frac{kG}{4}$.

To move the beam translationally, $2F$ should be greater than kG. Therefore, it is easier to turn the beam.

93. The equation of motion of the load is $\frac{G_0}{g}a = F - G_0$ (Fig. 300). The sum of the forces acting on the crane vertically is zero. Therefore, $G_1 + G_2 = G + F$. Since the sum of the moments of the forces relative to point A is zero, we have $Fl + G\frac{L}{2} = LG_2$.

Solution of these simultaneous equations gives

$$G_1 \cong 2.23 \text{ tonf, and } G_2 \cong 1.77 \text{ tonf}$$

94. For the lever to be in equilibrium, the force applied at point D should produce a moment equal to $G \cdot AB$. The force will be minimum when the maximum arm is equal to BD.

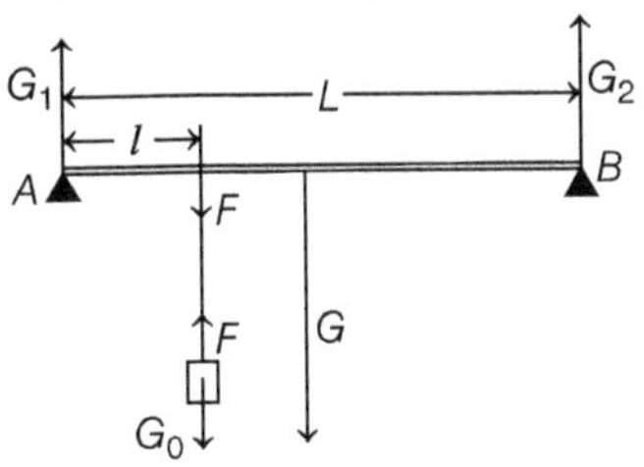

Fig.300

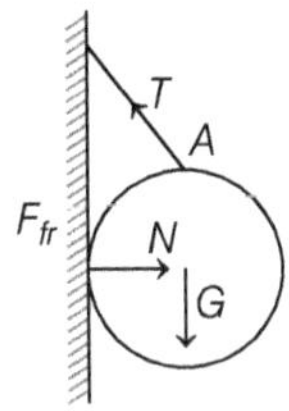

Fig.301

Hence, $F = G\frac{AB}{BD} = \frac{G}{\sqrt{2}}$, and it is directed at right angles to BD.

95. If there is no friction between the floor and the boxes, the latter will move simultaneously. If the coefficient of friction is not zero, the right-hand box will move first (see Fig. 35), since the force applied to it by the rod will be greater than the force applied to the left-hand box. Indeed, the rod is acted upon from the side of the right-hand box by the force F_1 directed opposite to F, and from the side of the left-hand box by the force F_2 directed along F. The sum of the forces in equilibrium is zero. Therefore, $F_1 = F + F_2$, and the force F_1 will reach the maximum force of friction of rest before F_2.

96. The equality to zero of the sum of the moments of the forces acting on the sphere with respect to point A (Fig. 301) gives us the equation

$$F_{fr}R - NR = 0$$

Since $F_{fr} \le kN$, then $k \ge 1$.

97. For a body to be at rest, the total moment of the forces that tends to turn the body clockwise should be equal to the moment of the forces that tends to turn the body counterclockwise around a point (around the centre of gravity, for example). In our case the moment of the forces of friction that turns the brick clockwise should be equal to the moment of the forces of the pressure exerted by the plane on the brick. It follows that the force of pressure exerted by the plane on the right half of the brick should be greater than on the left one. According to Newton's third law, the force with which the right half of the brick presses against the plane should be greater than that of the left half.

98. To lift the roller onto the step, the moment of the forces turning the roller around point A (Fig. 302) counterclockwise should at least be equal to the moment of the forces turning it clockwise, i.e.,

$$G(R-h) = G\sqrt{R^2 - (R-h)^2}$$

Hence, $$h = \frac{2 \pm \sqrt{2}}{2} R.$$ Since $h < R$, then

$$h = \left(1 - \frac{\sqrt{2}}{2}\right) R \cong 0.29R$$

99. Since the force of friction is zero on one of the planes, it is also zero on the other one. Otherwise, the sphere would rotate around its centre, for the moment of all the other forces relative to this centre is zero (because the arm of each of these forces with reference to the centre of the sphere is zero).

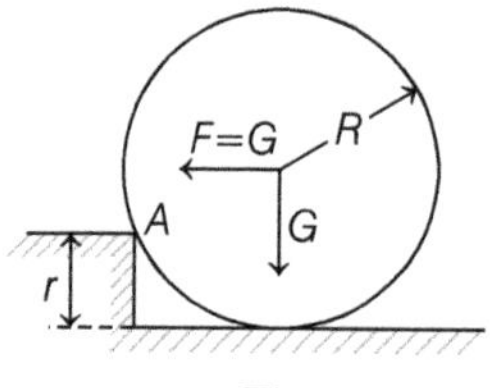

Fig. 302

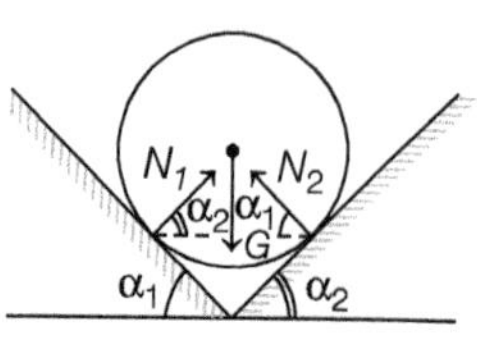

Fig. 303

The sums of the projections of the forces on the vertical and horizontal directions are equal to zero (Fig. 303). For this reason

$$N_1 \cos\alpha_2 - N_2 \cos\alpha_1 = 0, \quad G - N_1 \sin\alpha_2 - N_2 \sin\alpha_1 = 0$$

where N_1 and N_2 are the sought pressure forces. Hence,

$$N_1 = \frac{G}{\sin\alpha_2 + \cos\alpha_2 \cdot \tan\alpha_1} \cong 2.6 \text{ kgf}$$

$$N_2 = \frac{G}{\sin\alpha_1 + \cos\alpha_1 \cdot \tan\alpha_2} \cong 1.5 \text{ kgf}$$

100. Let us denote the force applied to one handle by F. The force F will turn the drawer and induce elastic forces N_1 and N_2 at points A and B (Fig. 304) which will act on the drawer from the side of the cabinet. These forces are equal: $N_1 = N_2 = N$. Since the moment of all the acting forces relative to the centre of the drawer C is zero, $N = F\dfrac{l}{2a}$.

The drawer can be pulled out if the applied force F is greater than the maximum force of friction of rest: $F > f_1 + f_2 = 2kN$.

For the last inequality to be satisfied, k should be smaller than $\dfrac{a}{l}$.

101. A board on a rough log forming an angle α with a horizontal plane is similar to a body retained by the forces of friction on an inclined plane with an angle α at its base. Therefore, in equilibrium, $F_{fr} = mg\sin\alpha$. Bearing in mind that $F_{fr} \le kmg\cos\alpha$, we have $\tan\alpha \le k$.

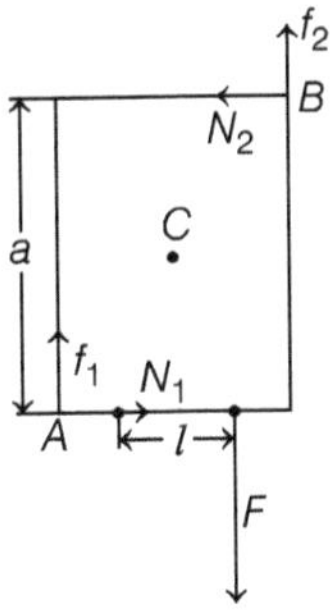

Fig. 304

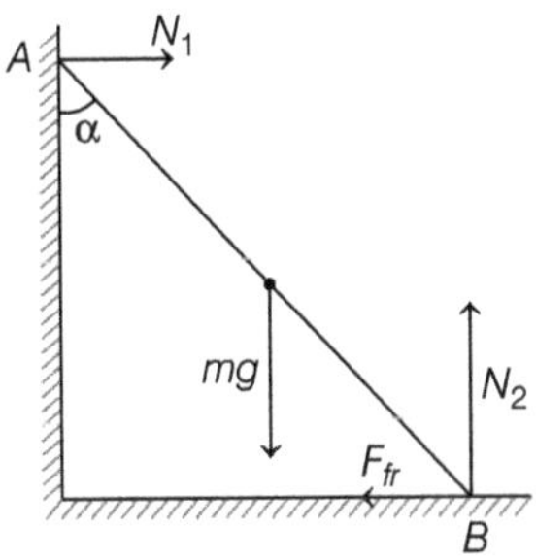

Fig. 305

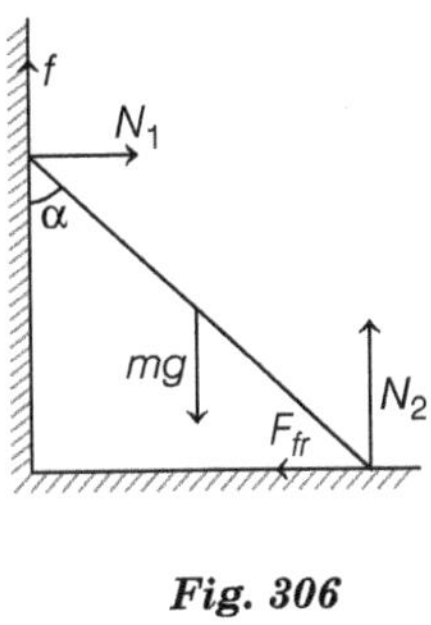

Fig. 306

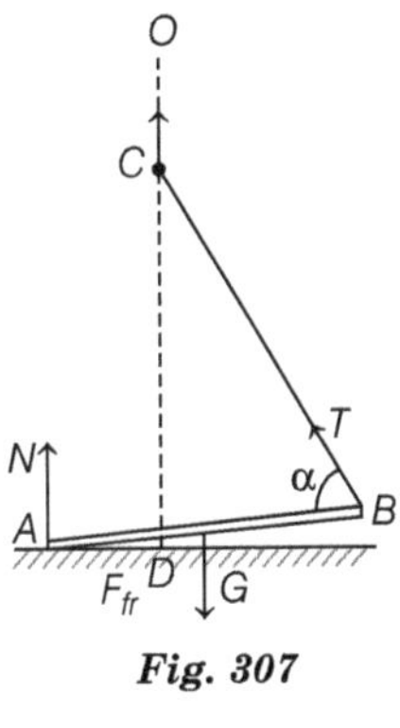

Fig. 307

102. The forces applied to the ladder are shown in Fig. 305. In equilibrium the sums of the projections of the forces along a vertical and a horizontal lines are equal to zero.

Therefore, $\qquad N_1 = F_{fr}$ and $N_2 = mg$.

Since the sum of the moments of the forces relative to point B is zero, we can write another equation

$$N_1 \cos\alpha = mg\frac{\sin\alpha}{2}$$

Hence, $F_{fr} = mg\frac{\tan\alpha}{2}$. Since the force of friction satisfies the inequality $F_{fr} \le kN_2$, the ladder will be in equilibrium if

$$\tan\alpha \le 2k$$

103. The forces applied to the ladder are shown in Fig. 306. Since the sum of the forces and the sum of the moments of the forces are equal to zero:

$$f + N_2 = mg \quad (1)$$

$$N_1 = F_{fr} \quad (2)$$

$$f\sin\alpha + N_1\cos\alpha = mg\frac{1}{2}\sin\alpha \quad (3)$$

The forces of friction f and F_{fr} satisfy the inequalities $f \le kN_1$ and $F_{fr} \le kN_2$. By using the first inequality and equations (1) and (3), we get : $\cos\alpha \ge \frac{N_2}{2N_1} - \frac{k}{2}$. Since $k \ge \frac{N_1}{N_2}$, then $\cot\alpha \ge \frac{1-k^2}{2k}$. If we assume that $k = \tan\beta$, the inequality can be written in a more convenient form for calculations:

$$\cot\alpha \ge \cot 2\beta \text{ or } \alpha \le 2\beta$$

104. If at the moment when end B of the rod begins to rise, the force of friction $F_{fr} \le kN$ proves sufficient for end A not to slip, the rod will begin to rotate around point A. Otherwise, end A will begin to slip until the force of friction $F_{fr} = kN$ can keep the rod in equilibrium (Fig. 307). After this the rod will begin to rotate around end A.

Let us find the coefficient of friction k at which slipping stops with a definite angle α between the rod and the string.

The equality of the forces at the moment when the rod is almost horizontal gives us the equations:

$$F_{fr} = T\cos\alpha$$

$$G = N + T\sin\alpha$$

The equality of the moments of the forces with respect to point A can be written as

$$G\frac{l}{2} = Tl\sin\alpha$$

By using this system of equations, we find that

$$k = \frac{F_{fr}}{N} = \cot\alpha$$

For the rod not to slip at all it is necessary that $k \ge \cot 60° = \frac{1}{\sqrt{3}}$.

105. The sum of the moments of the forces acting on the man relative to his centre of gravity is zero. For this reason the force F acting from the Earth is always directed towards the man's centre of gravity C (Fig. 308). The horizontal component of this force cannot be greater than the maximum force of friction of rest: $F \sin\alpha \leq kF \cos\alpha$. Hence, $k \geq \tan\alpha$.

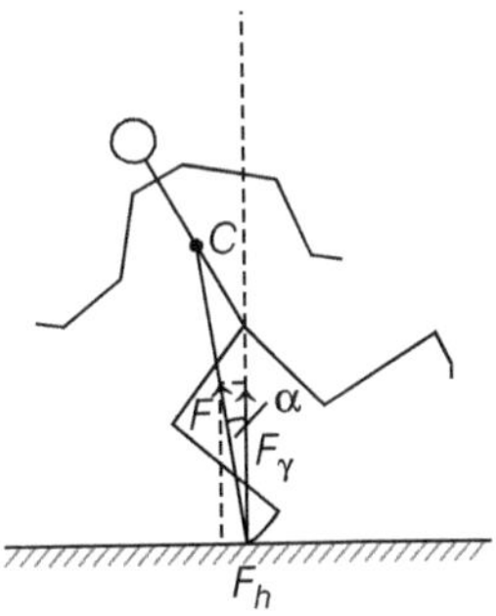

Fig. 308

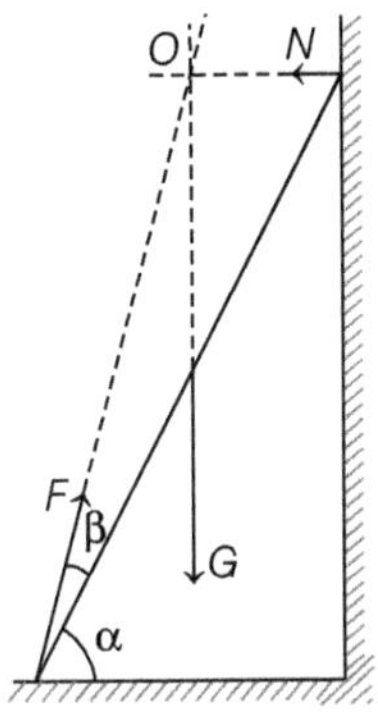

Fig. 309

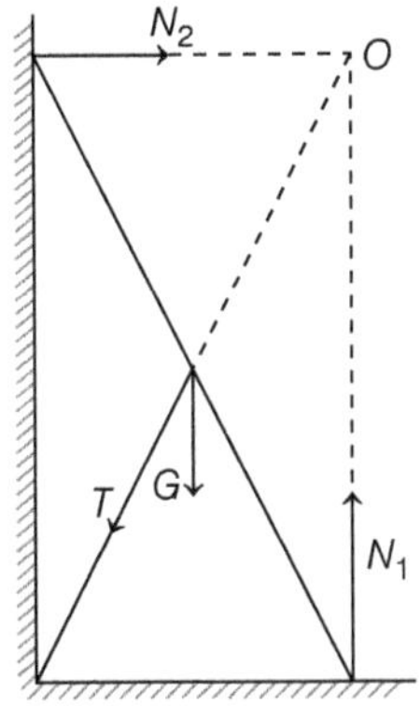

Fig. 310

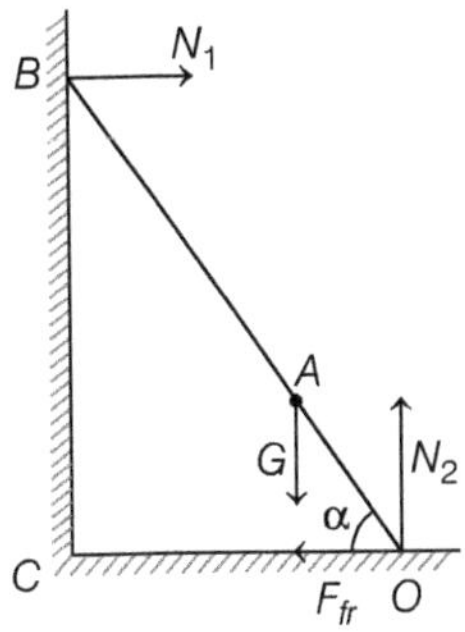

Fig. 311

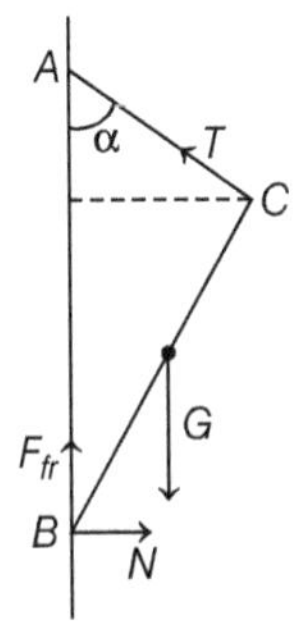

Fig. 312

106. The ladder is acted upon by three forces: its weight G, the force from the Earth F and the reaction of the support N (Fig. 309). Since the wall is smooth, the force N is perpendicular to it.

It will be the easiest to determine the force F if we find the point with respect to which the moments of the forces G and N are equal to zero. This will be the point at which the straight lines ON and OG intersect. The moment of the force F relative to this point should also be zero. Therefore, the force should be directed so that it continues past point O.

Figure 309 shows that the direction of the force F forms with the ladder the angle $\beta = 30° — \text{arc tan}\,\frac{1}{2\sqrt{3}} \cong 14°10'$.

The force which acts on the ladder from the Earth will be directed along the ladder only if all the other forces are applied to the centre of the ladder masses or act along it.

107. The ladder cannot be prevented from falling down by means of a rope tied to its middle. The moments of the forces of reaction of the floor and the wall as well as the moments of the force with which the rope is tensioned with respect to point O are zero whatever the tension T (Fig. 310). The moment of the force of gravity with respect to the same point differs from zero. For this reason the ladder will fall down without fail.

108. From the wall the ladder is acted upon by the reaction of the support N_1 perpendicular to the wall. The bottom end of the ladder is acted upon by the forces N_2 (reaction of the support) and F_{fr} (force of friction) (Fig. 311). If, for the sake of simplicity, we disregard the weight of the ladder, these forces will be supplemented by the weight of the man G. The equality of the projections of the forces along a horizontal and a vertical lines gives us:

$$N_2 = G \text{ and } N_1 = F_{fr}.$$

Let the man first stand on the lower part of the ladder (point A). The equality of the moments of the forces with respect to point O gives us the equation $N_1 CB = G \cdot \cos\alpha \cdot AO$.

Hence, the higher the man, the greater will be the force N_1. But $F_{fr} = N_1$. For this reason the force of friction retaining the ladder grows as the man climbs up. As soon as F_{fr} reaches its maximum value equal to kG the ladder will begin to slip.

109. In equilibrium the sum of the forces acting on the picture is zero (Fig. 312). Therefore, $G = F_{fr} + T\cos\alpha$, and $N = T\sin\alpha$. The force of friction should satisfy the inequality

$$F_{fr} \le kN \text{ or } k \ge \frac{F_{fr}}{N}$$

The equality of the moments relative to point B gives us the equation

$$\frac{G}{2} l \sin\alpha = T(l\cos\alpha + \sqrt{d^2 - l^2\sin^2\alpha})\sin\alpha$$

Hence,

$$\frac{F_{fr}}{N} = \frac{l\cos\alpha + 2\sqrt{d^2 - l^2\sin^2\alpha}}{l\sin\alpha}$$

and

$$k \ge \frac{l\cos\alpha + 2\sqrt{d^2 - l^2\sin^2\alpha}}{l\sin\alpha}$$

110. Let us first find the direction of the force f with which rod BC acts on rod CD. Assume that this force has a vertical component directed upward. Then, according to Newton's third law, rod CD acts on rod BC with a force whose vertical component is directed downward. This contradicts the symmetry of the problem, however. Therefore, the vertical component of the force f should be equal to zero. The force acting on rod CD from rod DE will have both a horizontal and a vertical components, as shown in Fig. 313*a*.

Since all the forces acting on CD are equal to zero, $F = mg$ and $f = f'$. The equality to zero of the moment of the forces with respect to D gives us:

$$f \sin\beta CD = mg\frac{\cos\beta}{2}CD$$

or

$$\tan\beta = \frac{mg}{2f}$$

Figure 313*b* shows the forces acting on rod DE. Since the moment of the forces with respect to E is equal to zero, it follows that

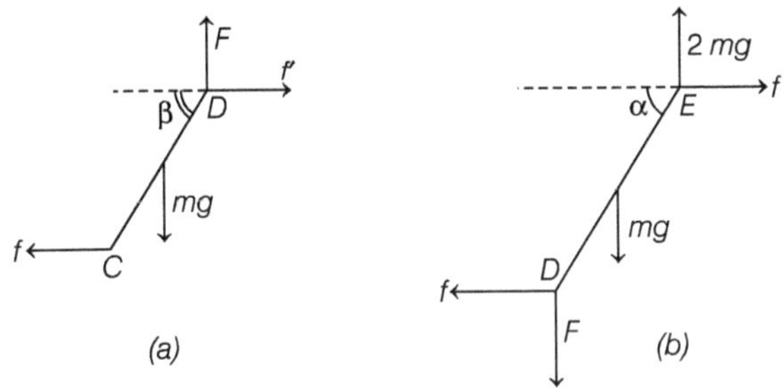

Fig. 313

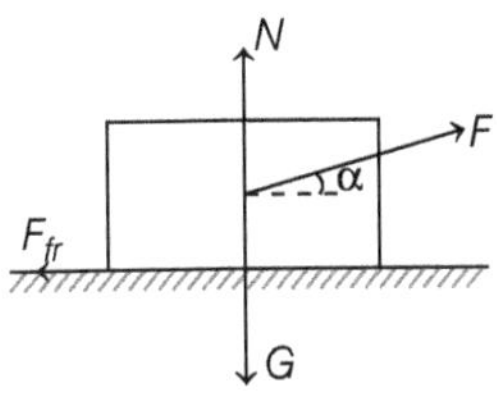

Fig.314

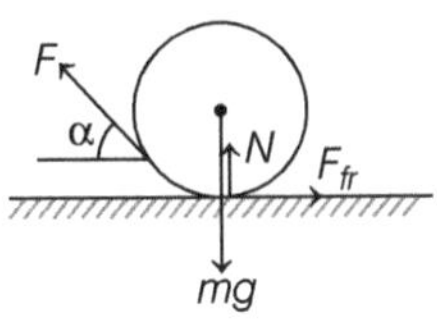

Fig. 315

$$f \sin\alpha\ DE = F\cos\alpha\ DE + mg\frac{\cos\alpha}{2}DE$$

or

$$\tan\alpha = \frac{3mg}{2f}$$

Consequently, $\tan\alpha = 3\tan\beta$.

111. The forces acting on the box are shown in Fig. 314. The conditions of equilibrium have the form: $F\cos\alpha = F_{fr}$, and $G = N + F\sin\alpha$.

At the moment equilibrium is disturbed, the force of friction reaches its maximum: $F_{fr} = kN$. Hence, $F = \dfrac{kG}{\cos\alpha + k\sin\alpha}$. The value of F will be minimum at an angle α corresponding to the maximum denominator. To find the maximum let us transform the denominator, introducing a new quantity φ instead of k so that $\tan\varphi = k$.

Then,

$$\cos\alpha + k\sin\alpha = \frac{\cos(\alpha - \varphi)}{\cos\varphi}$$

or

$$\cos\alpha + k\sin\alpha = \sqrt{1 + k^2}\cos(\alpha - \varphi)$$

Since the maximum value of $\cos(\alpha - \varphi)$ is unity, then

$$F_{\min} = \frac{kG}{\sqrt{1 + k^2}}$$

Hence,

$$k = \frac{F}{\sqrt{G^2 - F^2}} = 0.75.$$

112. The forces acting on the cylinder are shown in Fig. 315. Since the cylinder does not move translationally,

$$F_{fr} - F\cos\alpha = 0$$

$$F\sin\alpha - mg + N = 0$$

The force of friction $F_{fr} = kN$.

Hence,

$$F = \frac{kmg}{\cos\alpha + k\sin\alpha}$$

The denominator of this expression can be written as $A\sin(\alpha + \phi)$, where $A = \sqrt{1 + k^2}$ (see Problem 111).

Therefore, the minimum force with which the string should be pulled is $F_{min} = \dfrac{kmg}{\sqrt{1 + k^2}}$.

The angle α_1 can be found from the equation $\cos\alpha_1 + k\sin\alpha_1 = \sqrt{1 + k^2}$, and $\tan\alpha_1 = k$.

113. The forces acting on the piston and the rear lid of the cylinder are $F_1 = F_2 = pA$ (Fig. 316*a*). Point C of the wheel is also acted upon in a horizontal direction by the force F_2 transmitted from the piston through the crank gear.

The sum of the moments of the forces acting on the wheel with respect to its axis is zero. (The mass of the wheel is neglected.)

Therefore, $F_{fr}R = F_2 r$, where F_{fr} is the force of friction. Since the sum of the forces which act on the wheel is also zero, the force F_3 applied to the axis by the bearings of the locomotive is $F_3 = F_{fr} + F_2$. According to Newton's third law, the force $F_4 = F_3$ acts on the locomotive from the side of the axis. Hence, the tractive effort

$$F = F_4 - F_1 = F_{fr}$$
$$= pA\frac{r}{R}.$$

In the second position of the piston and the crank gear, the forces we are interested in are illustrated in Fig. 316*b*. For the same reason as in the previous case, $F_{fr} = F_2\frac{r}{R}$.

The tractive effort $F = F_1 - F_4 = F_{fr}$

$$= pA\frac{r}{R}.$$

As could be expected, the tractive effort is equal to the force of friction, for the latter is the only external force that acts on the locomotive.

114. The maximum length of the extending part of the top brick is $l/2$. The centre of gravity of the two upper bricks C_2 is at a distance of $l/4$ from the edge of the second brick (Fig. 317). Therefore, the second brick may extend by this length relative to the third one.

The centre of gravity of the three upper bricks C_3 is determined by the equality of the moments of the forces of gravity with respect to C_3; namely, $G\left(\frac{l}{2} - x\right) = 2Gx$. Hence $x = \frac{l}{6}$, i.e., the third brick may extend over the fourth one by not more than $l/6$. Similarly it can be found that the fourth brick extends over the fifth one by $l/8$, etc. The nature of the change in the length of the extending part with an increase in the number of bricks is obvious. The maximum distance over which the right-hand edge of the upper brick can extend over the right-hand edge of the lowermost brick can be written as the series

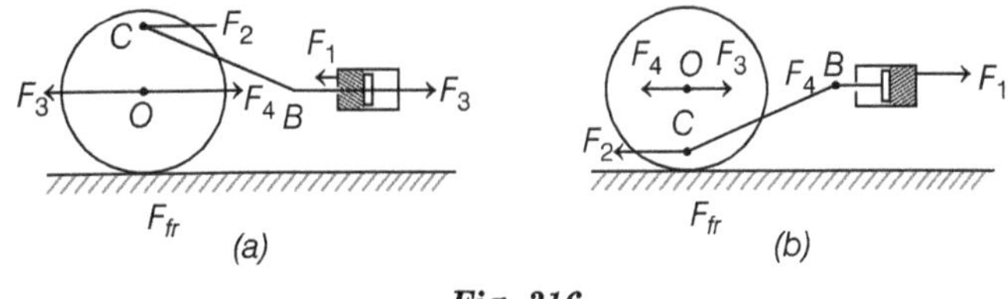

Fig. 316

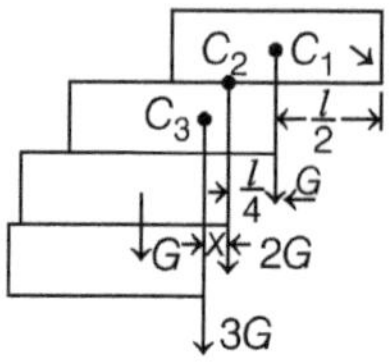

Fig. 317

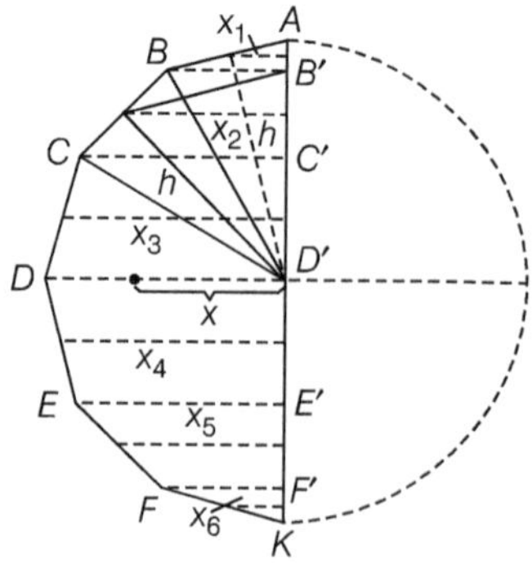

Fig. 318

$$L = \frac{l}{2}\left(1 + \frac{1}{2} + \frac{1}{3} + \frac{1}{4} + \ldots\right)$$

When the number of the bricks is increased infinitely this sum tends to infinity.

Indeed, the sum of the series

$$1 + \frac{1}{2} + \frac{1}{3} + \frac{1}{4} + \frac{1}{5} + \frac{1}{6} + \frac{1}{7} + \frac{1}{8} + \ldots$$

is greater than that of the series

$$1 + \frac{1}{2} + \overbrace{\frac{1}{4} + \frac{1}{4}}^{1/2} + \overbrace{\frac{1}{8} + \frac{1}{8} + \frac{1}{8} + \frac{1}{8}}^{1/2} + \ldots$$

and the latter sum will be infinitely great if the number of terms is infinite. The centre of gravity of all the bricks passes through the right-hand edge of the lowermost brick. Equilibrium will be unstable. The given example would be possible if the Earth were flat.

115. Let us inscribe a right polygon into a circle with a radius r (Fig. 318). Let us then find the moment of the forces of gravity (with respect to the

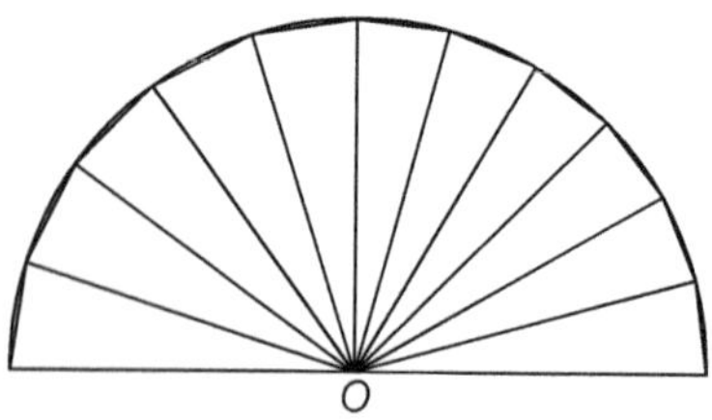

Fig. 319

axis AK) applied to the middles of the sides of the polygon AB, BC, CD, DE, etc., assuming that the force of gravity acts at right angles to the drawing. This moment is equal to $\rho g(ABx_1 + BCx_2 + CDx_3 + DEx_4 + EFx_5 + FKx_6)$, where ρ is the mass of a unit of the wire length.

From the similarity of the corresponding triangles it can be shown that the products ABx_1, BCx_2, CDx_3, etc., are equal respectively to $AB'h$, $B'C'h$, $C'D'h$, etc., where h is the apothem of the polygon.

Therefore, the moment is equal to

$$\rho gh(AB' + B'C' + C'D' + D'E' + E'F' + F'K) = \rho gh2r.$$

If the number of sides increases infinitely, the value of h will tend to r and the moment to $2r^2\rho g$. On the other hand, the moment is equal to the product of the weight of the wire $\pi r\rho g$ and the distance x from the centre of gravity to axis AK. Therefore, $2r^2\rho g = \pi r\rho gx$,

whence $x = \frac{2}{\pi}r$.

116. Let us divide the semicircle into triangles and segments, as shown in Fig. 319. The centre of gravity of a triangle, as is known, is at the point of intersection of its medians. In our case the centre of gravity of each triangle is at a distance of $\frac{2}{3}h$ from point O (h is the apothem).

When the sides are increased infinitely in number the centres of gravity of the triangles will lie on a circle with a radius of $\frac{2}{3}r$, while the areas of the segments will tend to zero.

Thus, the problem consists in determining the centre of gravity of a semicircle with a radius of $\frac{2}{3}r$.

It follows from the solution of Problem 115 that x, which is the distance between the centre of gravity of the semicircle and point O, is equal to $x = \frac{2}{\pi}\frac{2}{3}r = \frac{4}{3\pi}r$.

117. By applying the method used to solve Problems 115 and 116, it can be shown that the centre of gravity is at point C at a distance $CO = \frac{2\sin\frac{\alpha}{2}}{\alpha}r$ from the centre of curvature of the arc (see Fig. 45).

118. From the solutions of Problems 115, 116 and 117, it can be shown that the centre of gravity is at point C at a distance of $CO = \frac{4}{3}\frac{\sin\frac{\alpha}{2}}{\alpha}r$ from point O.

119. When the centre of gravity is determined, the plate with a cut-out portion can formally be considered as a solid one if we consider that a semicircle with a negative mass equal in magnitude to the mass of the cut-out portion is superposed on it.

The moment of the gravity forces of the positive and negative masses with respect to axis AB is equal to

$$\rho g\left(2r^2\frac{r}{2}-\frac{\pi r^2}{2}\frac{4}{3\pi}r\right)=\frac{1}{3}r^3\rho g$$

if the force of gravity acts at right angles to the drawing (see Fig. 47), where ρ is the mass of a unit area of the plate (see the solution to Problem 116). On the other hand, this moment is equal to the product of the weight of the plate and the distance $x = OC$ from its centre of gravity to axis AB.

Hence, $$x\rho g\left(2r^2-\frac{\pi r^2}{2}\right)=\frac{1}{3}r^3\rho g.$$

Therefore, $$x=\frac{2}{3(4-\pi)}r.$$

1-6. Work and Energy

120. The work of a force does not depend on the mass of the body acted upon by the given force. A force of 3 kgf will perform the work $W = Fh = 15$ kgf - m. This work is used to increase the potential energy (5 kgf-m) and the kinetic energy (10 kgf-m) of the load.

121. $k = 0.098$ J/kgf · cm.

122. First let us find the force with which the air presses on one of the hemispheres. Assume that its base is covered with a flat lid in the form of a disk with a radius R. If the air is pumped out from this vessel the force of pressure on the flat cover will be $F_1 = pA = p\pi R^2$. Obviously, the same pressure will be exerted on the hemisphere by the air, otherwise the forces will not be mutually balanced and the vessel will perpetually move in the direction of the greater force. The number of horses should be F_1 / F, since the other hemisphere may simply be tied to a post. The tensioned rope will produce the same force as the team of horses pulling at the other side.

123. The change in the momentum of a body is equal to the impulse of the force of gravity. Since the forces acting on the stone and the Earth are the same and act during the same time, the changes in the momenta of these bodies are also the same.

The change in the kinetic energy of a body is equal to the work of the forces of gravitational attraction. The forces are equal, but the paths traversed by the stone and the Earth are inversely proportional to their masses. This is why the law of conservation of energy may be written in a form which disregards the change in the kinetic energy of the Earth: $\frac{mv^2}{2} + E_p = \text{const}$, where m is the mass of the stone and Ep the potential energy of interaction.

124. According to the law of conservation of energy,

$$m_1 gh = \frac{m_1 v_1^2}{2}$$

where m_1 is the mass of the pile driver, h the height from which it drops, and v_1 its velocity before the impact.

Since the impact is instantaneous, the force of resistance cannot appreciably change the total momentum of the system.

Seeing that the impact is inelastic

$$m_1 v_1 = (m_1 + m_2) v_2$$

where m_2 is the mass of the pile, v_2 the velocity of the driver and the pile at the first moment after the impact.

The mechanical energy of the driver and the pile is spent on work done to overcome the resistance of the soil F:

$$\frac{(m_1 + m_2) v_2^2}{2} + (m_1 + m_2) gs = Fs$$

where s is the depth which the pile is driven to into the soil.

Hence,

$$F = \frac{m_1}{m_1 + m_2} \cdot \frac{h}{s} m_1 g + m_1 g + m_2 g = 32{,}500 \text{ kgf}$$

125. As a result of the inelastic impact, the linear velocity of the box with the bullet at the first moment will be equal to $u = \frac{mv}{M + m}$, where v is the velocity of the bullet. On the basis of the law of conservation of energy, the angle of deflection α is related to the velocity v by the expression.

$$\frac{(M + m)u^2}{2} = \frac{m^2 v^2}{2(M + m)} = (M + m) L (1 - \cos\alpha) g$$

whence

$$v = 2 \sin\frac{\alpha}{2} \frac{(M + m)}{m} \sqrt{Lg}$$

126. Since the explosion is instantaneous, the external horizontal forces (forces of friction) cannot appreciably change the total momentum of the system during the explosion. This momentum is zero both before and directly after the explosion.

Therefore, $m_1 v_1 + m_2 v_2 = 0.$

Hence, $\frac{v_1}{v_2} = -\frac{m_2}{m_1}$.

Since the carts finally stop, their initial kinetic energies are spent on work against the forces of friction:

$$\frac{m_1 v_1^2}{2} = km_1 g s_1, \text{ and } \frac{m_2 v_2^2}{2} = km_2 g s_2$$

Hence, $\frac{v_1^2}{v_2^2} = \frac{s_1}{s_2}$, and, therefore, $s_2 = 2$ metres.

127. Let us denote the speed of the body and the wagon after they stop moving with respect to each other by u. According to the law of conservation of momentum,

$$(M + m)\, u = Mv_0 \tag{1}$$

The wagon loses its kinetic energy in view of the fact that the force of friction f acting on it performs negative work

$$\frac{Mv_0^2}{2} - \frac{Mu^2}{2} = fS$$

where S is the distance travelled by the wagon.

The body acquires kinetic energy because the force of friction acting on it performs positive work

$$\frac{mu^2}{2} = fs$$

Here s is the distance travelled by the body.

It is easy to see that the change in the kinetic energy of the system

$$\frac{Mv_0^2}{2} - \left[\frac{Mu^2}{2} + \frac{mu^2}{2}\right] = f(S - s) \tag{2}$$

is equal to the force of friction multiplied by the motion of the body relative to the cart.

It follows from equations (1) and (2) that

$$S - s = \frac{mMv_0^2}{2f(M + m)}$$

Since $S - s \le l$, then $l \ge \frac{mMv_0^2}{2f(M + m)}$

Bearing in mind that $f = kmg$, we have $l \ge \frac{Mv_0^2}{2kg(M + m)}$.

128. The combustion of the second portion increases the velocity of the rocket v by Δv. Since combustion is instantaneous, then according to the law of conservation of momentum,

$$(M + m)\, v = M(v + \Delta v) + m(v - u)$$

where m is the mass of a fuel portion, M the mass of the rocket without fuel, u the outflow velocity of the gases relative to the rocket. The velocity increment of the rocket $\Delta v = \frac{m}{M} u$ does not depend on the velocity v before the second portion of the fuel burns. On the contrary, the increment in the kinetic energy of the rocket (without fuel)

$$\Delta E_k = \frac{M(v+\Delta v)^2}{2} - \frac{Mv^2}{2} = mu\left(\frac{m}{2M}u + v\right)$$

will be the greater, the higher is v.

The maximum altitude of the rocket is determined by the energy it receives. For this reason the second portion of the fuel can be burnt to the greatest advantage when the rocket attains its maximum velocity, i.e., directly after the first portion is ejected. Here thc greatest part of the mechanical energy produced by the combustion of the fuel will be imparted to the rocket, while the mechanical energy of the combustion products will be minimum.

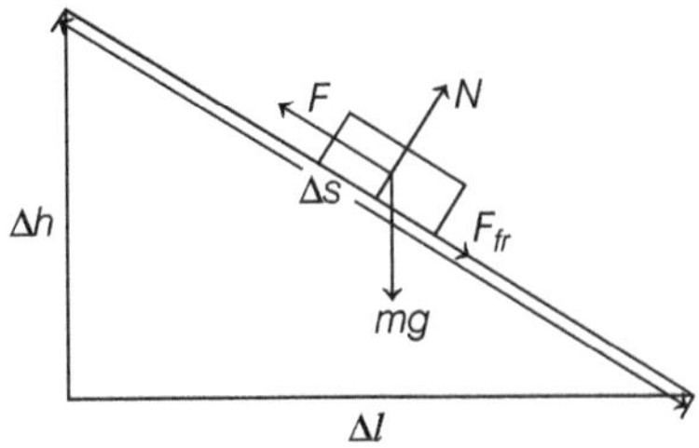

Fig. 320

129. It will be sufficient to consider the consecutive combustion of two portions of fuel. Let the mass of the rocket with the fuel first be equal to $M + 2m$.

After combustion of the first portion, the velocity of the rocket $v = \frac{mu_1}{M+m}$, where u_1 is the velocity of the gases with respect to the rocket. The initial velocity of the rocket is assumed to be zero.

The increment in the velocity of the rocket after the second portion burns is $\Delta v = \frac{mu_2}{M}$, where u_2 is the new velocity of the gases with respect to the rocket.

Combustion of the first portion produces the mechanical energy $\Delta E_1 = \frac{(M+m)v^2}{2} + \frac{mu_1^2}{2}$, and of the second portion the energy

$$\Delta E_2 = \frac{M(v+\Delta v)^2}{2} + \frac{m(v-u_2)^2}{2} - \frac{(M+m)v^2}{2}$$

According to the initial condition, $\Delta E_1 = \Delta E_2$.

Hence, $$u_1^2\left(\frac{m^2}{2(M+m)} + \frac{m}{2}\right) = u_2^2\left(\frac{m^2}{2M} + \frac{m}{2}\right)$$

Therefore, $u_1 > u_2$; the velocity of the gases with respect to the rocket diminishes because the mass of the rocket decreases as the fuel burns.

130. Both slopes may be broken into any arbitrary number of small inclined planes with various angles of inclination. Let us consider one of them (Fig. 320).

The work done to lift the body up such an inclined plane is equal to the work against the forces of gravity $mg\Delta h$ plus the work against the forces of friction $F_{fr}\Delta s$.

But $F_{fr} = kmg\cos\alpha$ and $\Delta s = \dfrac{\Delta l}{\cos\alpha}$. Therefore, $F_{fr}\Delta s = kmg\Delta l$. The total work $\Delta W = mg(\Delta h + k\Delta l)$. If we consider all the inclined surfaces and sum up the elementary works, the total work will be

$$W = \Sigma\Delta W = mg\,(\Sigma\Delta h + k\Sigma\Delta l) = mgh + kmgl$$

The work is determined only by the height of the mountain h and the length l of its foot.

131. The force applied to the handle will be minimum if it forms a right angle with the handle. Denoting the force sought by F, we shall have from the golden rule of mechanics: $2\pi RF = Gh$.

Hence,
$$F = \frac{Gh}{2\pi R}.$$

132. According to the definition, the efficiency $\eta = \dfrac{W_1}{W_1 + W_2}$, where $W_1 = GH$ is the work done to lift the load G to a height H, and W_2 is the work done against the forces of friction. Since the force of friction is capable of holding the load in equilibrium, the work of this force cannot be less than the work W_1. The minimum work of the forces of friction is $W_2 = W_1$. Therefore, $\eta \le 50\%$.

133. As the man climbs the ladder the balloon will descend by a height h. Therefore, the work done by the man will be spent to increase his potential energy by the amount $mg(l - h)$ and that of the balloon by mgh (the balloon without the man will be acted upon by the lifting force mg directed upwards). Hence,

$$W = mg(l - h) + mgh = mgl$$

This result can be obtained at once in calculating the work done by the man in the system related to the ladder.

If the man climbs with a velocity v with respect to the ladder, he moves at $v - v_1$ with respect to the Earth, where v_1 is the velocity of the ascending balloon.

According to the law of conservation of momentum, $(v - v_1)m = Mv_1$.

Hence,
$$v_1 = \frac{m}{M + m}v$$

134. To deliver twice as much water in a unit of time, a velocity two times greater should be imparted to the double mass of the water. The work of the motor is spent to impart a kinetic energy $\frac{mv^2}{2}$ to the water. Therefore, the power of the motor should be increased eight times.

135. (1) The work done to raise the water out of the pit is

$$W_1 = \rho g \frac{H}{2} A \times \frac{3}{4} H = \frac{3}{8} \rho g A H^2$$

where ρ is the density of water.

The work

$$W_2 = \frac{1}{2} \rho \frac{H}{2} A v^2$$

is required to impart kinetic energy to the water.

The velocity v with which the water flows out of the pipe onto the ground can be found from the ratio $\frac{H}{2} A = \pi R^2 v \tau$.

The total work is

$$W = \frac{3}{8} \rho g A H^2 + \frac{1}{16} \rho \frac{H^3 A^3}{\pi^2 R^4 \tau^2}$$

(2) In the second case the work required to raise the water is less than W_1 by $\Delta W'_1 = \rho g A_1 h \left(H - \frac{h}{2} \right)$. The work required to impart kinetic energy to the water is

$$W'_2 = \frac{1}{2} \rho \frac{\left(\frac{HA}{2} - hA_1 \right)^3}{\pi^2 R^4 \tau^2}$$

The total work $W' = W_1 - \Delta W'_1 + W'_2$.

136. It is the simplest to solve the problem in a coordinate system related to the escalator. The man will walk a distance $l = \frac{h}{\sin\alpha} + v\tau$ relative to the escalator, where $v\tau$ is the distance covered by the escalator. He should perform the work $W = \left(\frac{h}{\sin\alpha} + v\tau \right) mg \sin\alpha$, since during the ascent the force mg is applied over the distance l and forms an angle of $90° - \alpha$ with it.

Part of the work mgh is spent to increase the potential energy of the man, and the remainder $mgv\tau \sin\alpha$ together with the work of the motor that drives the escalator is spent to overcome the forces of friction.

137. The relation between the elastic force and the deformation is shown in Fig. 321. The work done to stretch (or compress) the spring by a small amount Δx is shown by the area of the hatched rectangle $\Delta W = F\Delta x$. The total work in stretching or compressing the spring by the amount l, equal to its potential energy E_p, is shown by the area of triangle $OBC: W = E_p = \frac{kl^2}{2}$.

Let us recall that an expression for the distance covered in uniformly accelerated motion can usually be obtained by similar reasoning.

138. The man acting with force F on the spring does the work $W_1 = -FL$. At the same time, the floor of the railway carriage is acted upon from the side of the man by the force of friction F. The work of this force $W_2 = FL$. Therefore, the total work performed by the man in a coordinate system related to the Earth is zero, in the same way as in a system related to the train.

139. In the system of the train the work done is equal to the potential energy of the stretched spring (see Problem 137) $W = \frac{kl^2}{2}$, since the force of friction between the man and the floor of the railway carriage does not do work in this system.

In the system related to the Earth, the work the man does to stretch the spring is equal to the product of the mean force $\frac{kl}{2}$ and the distance $L - l$, i.e., $W_1 = \frac{kl}{2}(L - l)$. The man acts on the floor of the carriage with the same mean force $\frac{kl}{2}$. The work of this force $W_2 = \frac{kl}{2}L$. The total work in the given coordinate system $W = W_1 + W_2 = \frac{kl^2}{2}$ is the same as in the system of the carriage.

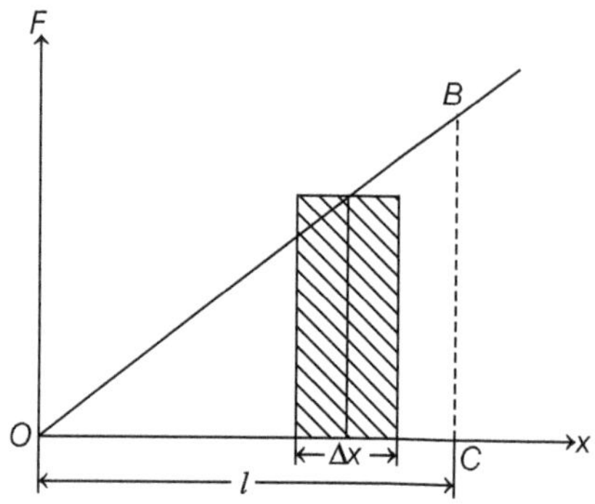

Fig. 321

140. On the basis of the laws of conservation of momentum and energy we can write the following equations:

$$m_1v_1 + m_2v_2 = m_1v'_1 + m_2v'_2$$

$$\frac{m_1v_1^2}{2} + \frac{m_2v_2^2}{2} = \frac{m_1v'^2_1}{2} + \frac{m_2v'^2_2}{2}$$

where v'_1 and v'_2 are the velocities of the spheres after the collision. Upon solving these simultaneous equations, we obtain

$$v'_1 = \frac{(m_1 - m_2)v_1 + 2m_2v_2}{m_1 + m_2};\ \text{and}\ v'_2 = \frac{(m_2 - m_1)v_2 + 2m_1v_1}{m_1 + m_2}$$

(1) If the second sphere was at rest before the collision ($v_2 = 0$), then

$$v'_1 = \frac{(m_1 - m_2)v_1}{m_1 + m_2};\ v'_2 = \frac{2m_1v_1}{m_1 + m_2}$$

If $m_1 > m_2$, the first sphere continues to move in the same direction as before the collision, but at a lower velocity.

If $m_1 < m_2$, the first sphere will jump back after the collision. The second sphere will move in the same direction as the first sphere before the collision.

(2) If $m_1 = m_2$, then $v'_1 = \frac{2mv_2}{2m} + v_2$ and $v'_2 = \frac{2mv_1}{2m} = v_1$. Upon collision the spheres exchange their velocities.

141. The elastic impact imparts a velocity v to the left-hand block. At this moment the right-hand block is still at rest since the spring is not deformed.

Let us denote the velocities of the left-and right-hand blocks at an arbitrary moment of time by u_1 and u_2, and the absolute elongation of the spring at this moment by x.

According to the laws of conservation of momentum and energy, we have

$$m(u_1 + u_2) = mv$$

$$\frac{mu_1^2}{2} + \frac{mu_2^2}{2} + \frac{kx^2}{2} = \frac{mv^2}{2}$$

or

$$kx^2 = m[v^2 - (u_1^2 + u_2^2)]$$

Upon substituting $u_1 + u_2$ for v in the last equation, we obtain

$$kx^2 = 2mu_1u_2$$

Therefore, $u_1u_2 = \frac{kx^2}{2m}$

and $\quad u_1 + u_2 = v.$

It can be seen from the last two equations that u_1 and u_2 will have the same sign and both blocks will move in the same direction.

The quantity x^2 will be maximum when the product of the velocities u_1 and u_2 is also

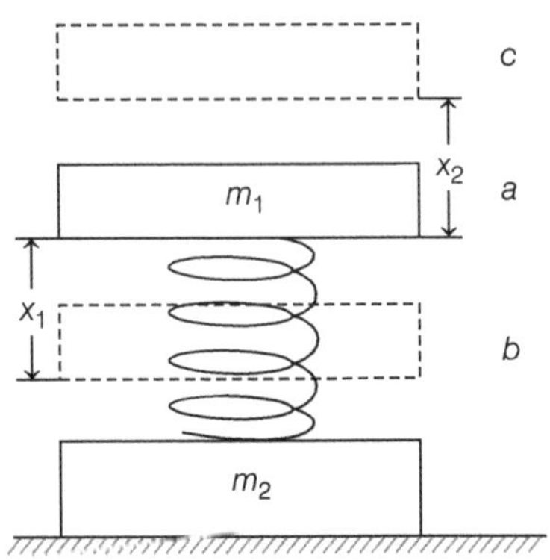

Fig.322

maximum. Hence, to find the answer to the second question, it is necessary to determine the maximum product $u_1 u_2$ assuming that the sum $u_1 + u_2$ is constant and equal to v.

Let us consider the obvious inequality $(u_1 - u_2)^2 \geq 0$, or $u_1^2 - 2u_1 u_2 + u_2^2 \geq 0$. Let us add $4u_1 u_2$ to the right and left sides of the inequality. Hence, $u_1^2 + 2u_1 u_2 + u_2^2 \geq 4u_1 u_2$ or $(u_1 + u_2)^2 \geq 4u_1 u_2$. Since $u_1 + u_2 = v$, then $4u_1 u_2 \leq v^2$.

Therefore, the maximum value of $u_1 u_2$ is equal to $v^2/4$ and it is attained when $u_1 = u_2 = \frac{v}{2}$.

At this moment the distance between the blocks is

$$l \pm x_{max} = l \pm v\sqrt{\frac{m}{2k}}$$

142. The lower plate will rise if the force of elasticity acting on it is greater than its weight: $kx_2 > m_2 g$. Here x_2 is the deformation of the spring stretched to the maximum (position c in Fig. 322). Position a shows an undeformed spring.

For the spring to expand over a distance of x_2 it should be compressed by x_1 (position b in Fig. 322), which can be found on the basis of the law of conservation of energy :

$$\frac{kx_1^2}{2} = \frac{kx_2^2}{2} + m_1 g(x_1 + x_2)$$

Hence, $$x_1 > \frac{2m_1 g}{k} + \frac{m_2 g}{k}$$

To compress the spring by x_1, the weight of the plate should be supplemented with a force which satisfies the equation $F + m_1 g = kx_1$.

Therefore, the sought force $F > m_1 g + m_2 g$.

143. In a reading system related to the wall, the velocity of the ball is $v + u$. After the impact the velocity of the ball will be $-(v + u)$ in the same reading system. The velocity of the ball after the impact with respect to a stationary reading system will be

$$-(v + u) - u = -(v + 2u)$$

The kinetic energy after the impact is $\frac{m}{2}(v + 2u)^2$ and before the impact $\frac{m}{2}v^2$.

The change in the kinetic energy is equal to $2mu(u + v)$.

Let us now calculate the work of the elastic forces acting on the ball during the impact. Let the collision continue for τ seconds. For the sake of simplicity we assume that during the impact the elastic force is constant (generally speaking the result does not depend on this assumption). Since the impact changes the momentum by $2m(v+u)$, the elastic force is

$$F = \frac{2m(v+u)}{\tau}$$

The work of this force is

$$W = Fs = Fu\tau = \frac{2m(v+u)u\tau}{\tau} = 2m(v+u)u$$

It is easy to see that this work is equal to the change in the kinetic energy.

144. (1) Up to the moment when the rope is tensioned the stones will fall freely

$$s_1 = \frac{gt^2}{2} \text{ and } s_2 = \frac{g(t-\tau)^2}{2}$$

The moment of tensioning of the rope can be determined from the condition $l = s_1 - s_2$. Hence, $t = 3$ seconds, $s_1 = 44.1$ metres, $s_2 = 4.9$ metres. The time is counted from the moment the first stone begins to fall. When the rope is tensioned, there occurs an elastic impact and the stones exchange their velocities (see Problem 140). At the moment of impact $v_1 = gt = 29.4$ m/s, and $v_2 = g(t-\tau) = 9.8$ m/s.

The duration of falling of the first stone t_1 (after the rope is tensioned) can be found from the condition

$$h - s_1 = v_2 t_1 + \frac{g t_1^2}{2}$$

and of the second stone t_2 from the condition

$$h - s_2 = v_1 t_2 + \frac{g t_2^2}{2}$$

Therefore $t_1 \cong 1.6$ seconds, and $t_2 \cong 1.8$ seconds.

The first stone falls during 4.6 seconds and the second during 2.8 seconds.

(2) If the rope is inelastic, the velocities of the stones after it is tensioned are equalized (inelastic impact): $v = \frac{v_1 + v_2}{2} = 19.6$ m/s. The duration of falling of the stones with the rope tensioned is determined from the equations :

$$h - s_1 = vt'_1 + \frac{gt'^2_1}{2} \text{ and } h - s_2 = vt'_2 + \frac{gt'^2_2}{2}$$

s_1 and s_2 are the same as in the first case.

Hence, $t'_1 \cong 1.2$ seconds, and $t'_2 \cong 3.3$ seconds.

The first stone falls during 4.2 seconds and the second during 4.3 seconds.

145. If only the right-hand ball is moved aside, the extreme left-hand ball will bounce off after the impact through an angle equal to that by which the right-hand ball was deflected.

If two right-hand balls are simultaneously moved aside and released, then two extreme left-hand balls will bounce off, and so on.

When the first ball strikes the second one, the first ball stops and transmits its momentum to the second one (see the solution to Problem 140), the second transmits the same momentum to the third, etc. Since the extreme left-hand ball has no "neighbour" at its left, it will bounce off (provided there is no friction or losses of energy) through the same angle by which the extreme right-hand ball was deflected.

When the left-hand ball, after deviation through the maximum angle, strikes the ball at its right, the process of transmission of momentum will take the reverse course.

When two right-hand balls are deflected at the same time, they will transmit their momenta to the row in turn after a very small period of time. In this way- the other balls will receive two impulses that will be propagated along the row at a certain time interval. The extreme left-hand ball will bounce off after it receives the "first portion" of the momentum. Next, its "neighbour" will bounce off after receiving the next portion of the momentum from the extreme right-hand ball.

If three right-hand balls are deflected, the row of balls will receive portions of momentum from the third, second and first ball following each other in very small intervals of time. If four balls are deflected and released at the same time, four balls will bounce off at the left while the other two will remain immobile.

146. The striking ball will jump back and the other balls up to the steel one will remain in place. The steel ball and all the others after it will begin to move to the left with different velocities.

The fastest velocity will be imparted to the extreme left-hand ball. The next one will move slower, etc. The balls will move apart (see the solutions to Problems 140 and 145).

147. Assume that the weight 2 m lowers through a distance of H. The weights m will accordingly rise to the height h (Fig. 323).

On the basis of the law of conservation of energy, $2mgh + \frac{2mv_1^2}{2} + \frac{2mv_2^2}{2} = 2mgH$, or $v_1^2 + v_2^2 = 2g(H - h)$, where v_1 is the velocity of the weights m and v_2 that of the weight $2m$.

When the weight $2m$ lowers, its velocity v_2 approaches the velocity v_1, since the angles between the parts of the string thrown over the pulleys tend to zero. In the limit, $v_2 \cong v_1$. At the same time, $H - h \cong l$. Therefore, the maximum velocity of the weights is

$$v = \sqrt{gl}$$

148. The velocities of the weights are the same if they cover identical distances Δs in equally small intervals of time. These distances will be the same at such an angle ANB at which lowering of the weight m_1 through $\Delta s = NK$ (Fig. 324) is attended by an increase in part ANB of the string also by the amount Δs. Therefore, when the velocities are equal, $HK = BK - BN = \frac{\Delta s}{2}$ and $FK = AK - AN = \frac{\Delta s}{2}$.

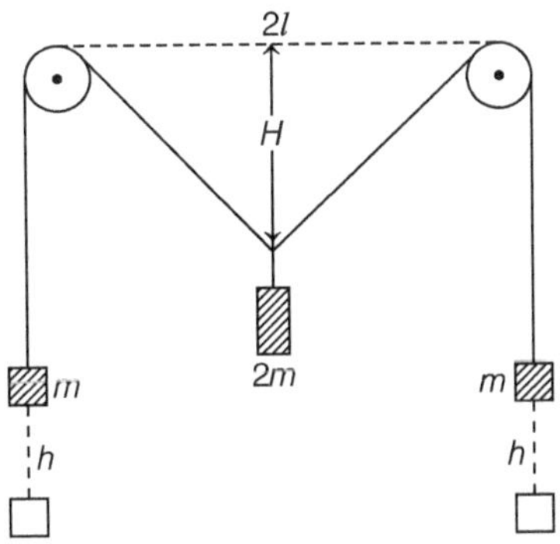

Fig. 323

The triangles NHK and NFK will be the closer to right ones, the smaller is the distance Δs. When $\Delta s \to 0$, the angles NHK and NFK tend to be right ones, while the angles KNH and KNF tend to $30°$. Therefore, the velocities will be equal when $\angle ANB = 120°$.

Let us use the law of conservation of energy to find these velocities

$$m_1 gh = 2(2 - \sqrt{3})\, m_2 gh + \frac{m_1 + m_2}{2} v^2$$

Hence,
$$v^2 = 2gh \frac{m_1 - 2(2 - \sqrt{3})m_2}{m_1 + m_2} \cong 0$$

The weights will oscillate near the position of equilibrium, which corresponds to the angle $ANB = 2 \arccos \frac{m_1}{2m_2} \cong 149°$.

The maximum deviation from the position of equilibrium corresponds to the angle $ANB = 120°$.

149. Since there is no slipping of the board on the rollers and of the rollers on the horizontal surface, the distance between the axes of the

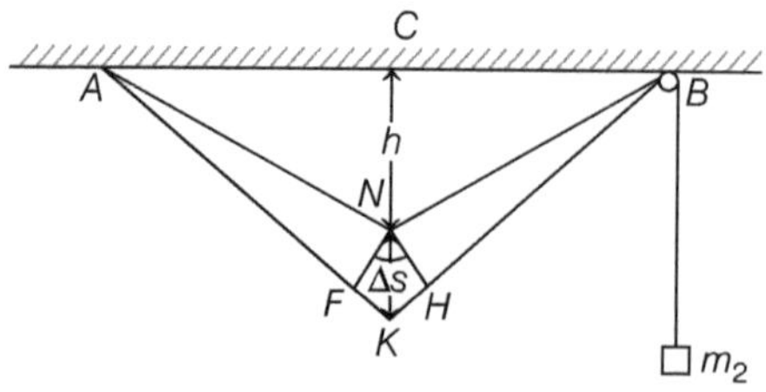

Fig. 324

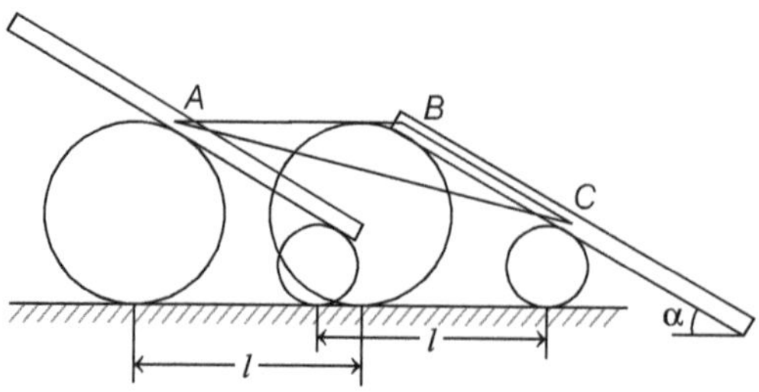

Fig. 325

rollers in motion remains constant. For this reason the board will move transnationally in a horizontal direction and at the same time down along the rollers. When the rollers move through a certain distance l, each point on the board (in particular, its centre of gravity A) will move in a horizontal direction through the same distance l and will also move the same distance along the rollers: $AB = BC = l$ (Fig. 325). (This is particularly obvious if we consider the motion of the rollers in a coordinate system travelling together with the rollers.) As a result, the centre of gravity of the board will move along straight line AC inclined to the horizon at an angle $\alpha/2$, since ABC is an isosceles triangle. The motion will be uniformly accelerated. The board acquires kinetic energy owing to the reduction of its potential energy $\frac{mv^2}{2} = mgl\sin\alpha$, or $v^2 = 2gl\sin\alpha$. On the other hand, in uniformly accelerated motion $v^2 = 2as$,

where $s = AC = 2l\cos\frac{\alpha}{2}$.

Hence, the acceleration

$$a = \frac{v^2}{2s} = g\sin\frac{\alpha}{2}.$$

150. Let us calculate the difference between the potential energies for the two positions of the chain—when it lies entirely on the table and when a part of it x hangs from the table. This difference is equal to the weight $\frac{M}{2l}xg$ of the hanging part multiplied by $x/2$, since the chain is homogeneous and the centre of gravity of the hanging end is at a distance of $x/2$ from the edge of the table.

On the basis of the law of conservation of energy, $\frac{Mv^2}{2} = \frac{Mg}{4l}x^2$ or $v = \sqrt{\frac{gx^2}{2l}}$. At this moment of time the acceleration can be found from Newton's second law:

$$Ma = \frac{M}{2l}gx. \text{ Therefore, } a = \frac{gx}{2l}.$$

To calculate the reaction of the table edge, let us first find the tension of the chain at the point of its contact with the table. It is equal to the change in the momentum of the part of the chain lying on the table

$$F = \frac{M}{2l}(2l - x)\,a - \frac{M}{2l}v^2 = \frac{Mg}{2l^2}(l - x)x$$

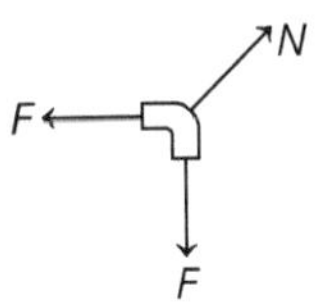

Fig. 326

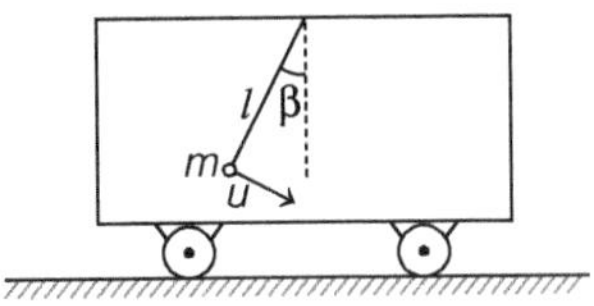

Fig. 327

Let us now consider a very small element of the chain in contact with the table edge. This element is acted upon by the three forces (Fig. 326). Since this element is infinitely small, the sum of the three forces which act on it should be zero.

Therefore, the force of reaction is

$$N = F\sqrt{2} = \sqrt{2}\,\frac{M(l - x)x}{2l^2}\,x$$

When $x > l$, the chain no longer touches the edge of the table.

151. Let us denote the velocity of the wagon by v. The horizontal component of the velocity of the pendulum with respect to the wagon is $u\cos\beta$ (Fig. 327). Since the wagon moves, the velocity of the pendulum with respect to the rails is $v + u\cos\beta$. The external forces do not act on the system in a horizontal direction. Therefore, on the basis of the law of conservation of momentum, we have

$$m(v + u\cos\beta) + Mv = 0 \qquad (1)$$

since the system was initially at rest.

The vertical component of the pendulum velocity with respect to the wagon and the rails is $u \sin\beta$.

According to the Pythagorean theorem, the square of the pendulum velocity relative to the rails is $(v + u\cos\beta)^2 + u^2\sin^2\beta$.

With the aid of the law of conservation of energy, we can obtain a second equation interrelating the velocities v and u:

$$\frac{m}{2}[(u\cos\beta + v)^2 + u^2\sin^2\beta] + \frac{M}{2}v^2 = mgl(\cos\beta - \cos\alpha) \quad (2)$$

It can be found from equations (1) and (2) that

$$v^2 = \frac{2m^2gl}{(M+m)} \cdot \frac{(\cos\beta - \cos\alpha)\cos^2\beta}{(M + m\sin^2\beta)}$$

In a particular case,

when $\beta = 0 \left(\text{assuming} \frac{m}{M} \ll 1\right)$

$$v^2 = 2\frac{m^2}{M^2}gl(1-\cos\alpha)$$

or $$v = 2\frac{m}{M}\sin\frac{\alpha}{2}\sqrt{gl}$$

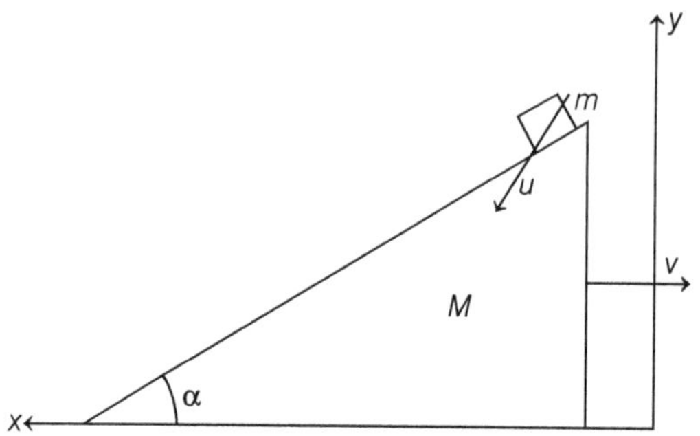

Fig. 328

152. Let us denote the velocity of the wedge by v, and the horizontal and vertical components of the velocity u of the block with reference to a stationary reading system by u_x and u_y (Fig. 328).

On the basis of the laws of conservation of momentum and energy we can write

$$-Mv + mu_x = 0,$$

and $$\frac{Mv^2}{2} + \frac{m}{2}(u_x^2 + u_y^2) = mgh$$

It should be noted that the angle α with the horizontal surface is formed by the relative velocity u_{rel}, i.e., the velocity of the block with respect to the moving wedge, and not the absolute velocity of the block u, by which is meant the velocity relative to a stationary horizontal surface.

It follows from the velocity diagram (Fig. 329) that $\frac{u_y}{v+u_x} = \tan\alpha$.

Upon solving these equations with respect to v, we obtain

$$v = \sqrt{\frac{2mgh}{M + m\left[\left(\frac{M}{m}\right)^2 + \left(\frac{M}{m}+1\right)^2 \tan^2\alpha\right]}}$$

$$= \sqrt{\frac{2gh}{\frac{M}{m} + \left(\frac{M}{m}\right)^2 + \left(\frac{M}{m}+1\right)^2 \tan^2\alpha}}$$

At the same moment of time the absolute velocity of the block is

$$u = \sqrt{u_x^2 + u_y^2} = \sqrt{2gh}\sqrt{1 - \frac{1}{1 + \frac{M}{m} + \frac{m}{M}\left(1 + \frac{M}{m}\right)^2 \tan^2\alpha}}$$

When the mass of the wedge is much greater than that of the block, u tends, as should be expected, to $\sqrt{2gh}$.

153. The velocity of the rod with respect to the moving wedge is directed at an angle α to the horizon. If the velocity of the wedge is added to this relative velocity, the result will be the absolute velocity u of the rod

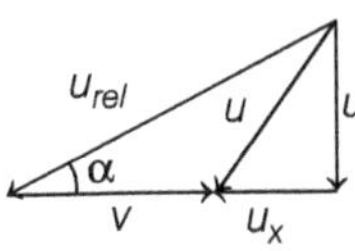

Fig. 329

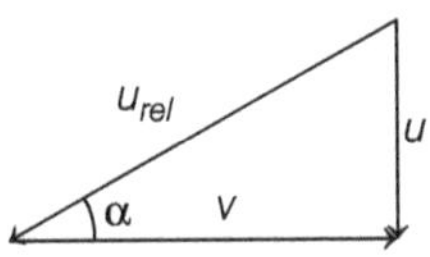

Fig. 330

(Fig. 330). The relation between the velocities is obviously equal to

$$\frac{u}{v} = \tan\alpha$$

It follows from the law of conservation of energy that

$$\frac{Mv^2}{2} + \frac{mu^2}{2} = mgh.$$

Upon cancelling u from these two equations, we get an expression for v:

$$v = \sqrt{\frac{2mgh}{M + m\tan^2\alpha}}$$

We can then write for the relative velocity of the rod:

$$u_{rel} = \frac{1}{\cos\alpha}\sqrt{\frac{2mgh}{M + m\tan^2\alpha}}$$

The velocity of the rod is $u = \tan\alpha\sqrt{\frac{2mgh}{m + M\tan^2\alpha}}$

$$= \sqrt{2\frac{mg\tan^2\alpha}{m + M\tan^2\alpha}h}$$

It can be seen from this formula that the velocity of the rod changes with the path h travelled according to the law of uniformly accelerated motion $u = \sqrt{2ah}$. Therefore, the acceleration of the rod is

$$a = \frac{m\tan^2\alpha}{m + M\tan^2\alpha}$$

1-7. Kinematics of Curvilinear Motion

154. The driving pulley rotates with an angular velocity of $\omega_1 = 2\pi n_1$ and the driven one with a velocity of $\omega_2 = 2\pi n_2$. The velocity of the drive belt is $v = \omega_1 r_1 = \omega_2 r_2$.

Hence, $$\frac{r_1}{r_2} = \frac{\omega_2}{\omega_1} = \frac{n_2}{n_1}.$$

The sought diameter is $D_1 = D_2\frac{n_2}{n_1} = 100\text{ mm}.$

155. (1) Let us denote the length of the crawler by $L = na$.

Hence, $l = \frac{L - 2\pi R}{2}$ is the distance between the axes of the wheels.

The number of links taking part in translational motion is $n_1 = \frac{l}{a} = \frac{L - 2\pi R}{2a}$.

The same number of links is at rest relative to the Earth. In rotary motion there are $n_2 = \frac{2\pi R}{a}$ links.

(2) The time during which the tractor moves is $t_0 = \frac{s}{v}$. During a full revolution of the crawler the link will move translationally a

distance $2l$ at a speed of $2v$. The duration of motion of one link during a revolution is $\frac{2l}{2v}$.

Altogether, the crawler will make $N = \frac{s}{L}$ revolutions. Therefore, the duration of translational motion of the link is $t_1 = \frac{Nl}{v}$. This is the time during which the link is at rest.

The link will participate in rotary motion during the time

$$t_2 = t_0 - \frac{2Nl}{v} = \frac{s + 2\pi RN - NL}{v}$$

If $s \gg L$, the number of revolutions may be assumed as a whole number, neglecting the duration of an incomplete revolution.

156. The duration of flight of a molecule between the cylinders is $t = \frac{R - r}{v}$. During this time the cylinders will turn through the angle ωt and, therefore,

$$l = R\omega t = \omega R \frac{R - r}{v}$$

Hence,

$$v = \frac{\omega R(R - r)}{l}.$$

157. Let us denote the sought radius by R and the angular speed of the tractor over the arc by ω. Hence (Fig. 331),

$$v_1 = \omega\left(R - \frac{d}{2}\right), \text{ and } v_0 = \omega\left(R + \frac{d}{2}\right)$$

Thus

$$\frac{v_1}{v_0} = \frac{R - \frac{d}{2}}{R + \frac{d}{2}} \text{ and } R = \frac{d}{2}\frac{v_0 + v_1}{v_0 - v_1} = 6m$$

158. First the observer is at the pole (point O in Fig. 332). The axis of the Earth passes through point O perpendicular to the drawing. Line OA (parallel to BC) is directed toward the star.

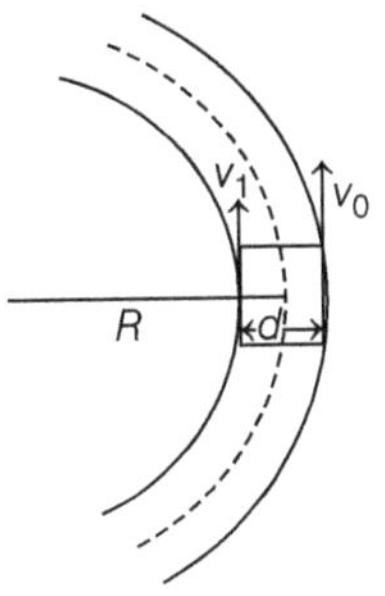

Fig. 331

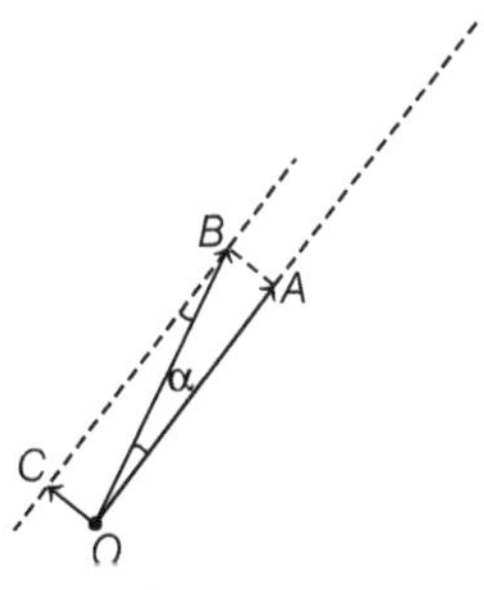

Fig. 332

The mountain is at the right of point A. The angle $\alpha = \omega\Delta t$ is the angle through which the Earth rotates during the time Δt with the angular velocity ω. To see the star, the observer should run a distance $OC \cong OA\omega\Delta t$.

The observer's speed $v = \frac{OC}{\Delta t} = OA\omega = 0.7$ m/s.

159. Let us take point A from which the boat departs as the origin of the coordinate system. The direction of the axes is shown in Fig. 333. The boat moves in a direction perpendicular to the current at a constant velocity u. For this reason the boat will be at a distance y from the bank in the time $t = \frac{y}{u}$ after departure. Let us consider the motion of the boat up to the middle of the river $\left(y \le \frac{c}{2}\right)$. The current velocity is $v = \frac{2v_0}{c}y$ at a distance y from the bank.

Upon inserting $y = ut$ into the expression for the current, we get $v = \frac{2v_0 ut}{c}$.

It follows from this relationship that the boat moves parallel to the banks with a constant acceleration $a = \frac{2v_0 u}{c}$.

The boat will reach the middle of the river in the time $T = \frac{c}{2u}$ and will be carried downstream during the same time over a distance of $s = \frac{aT^2}{2} = \frac{v_0 c}{4u}$. When moving from the middle of the river (point D) to the opposite bank, the boat will again be carried away over the same distance s. Thus, the sought distance is $\frac{v_0 c}{2u}$. When the boat moves to the middle of the river $x = \frac{at^2}{2} = \frac{v_0 u}{c}t^2$ and $y = ut$.

Let us use these ratios to determine the trajectory of the boat from A to D.

We get $y^2 = \frac{cu}{v_0}x$ (a parabola). The other half of the trajectory (DB) is of the same nature as the first one.

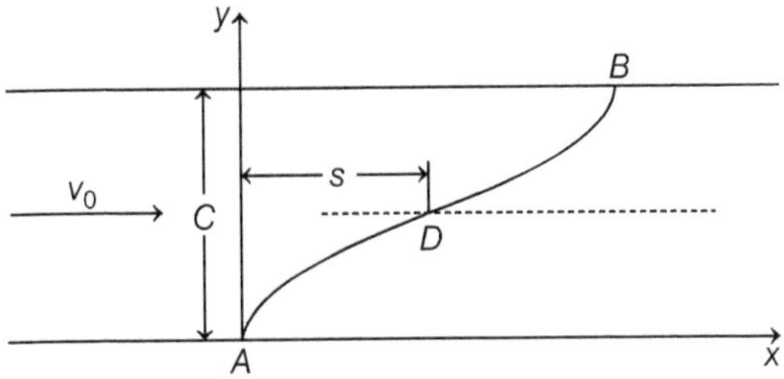

Fig. 333

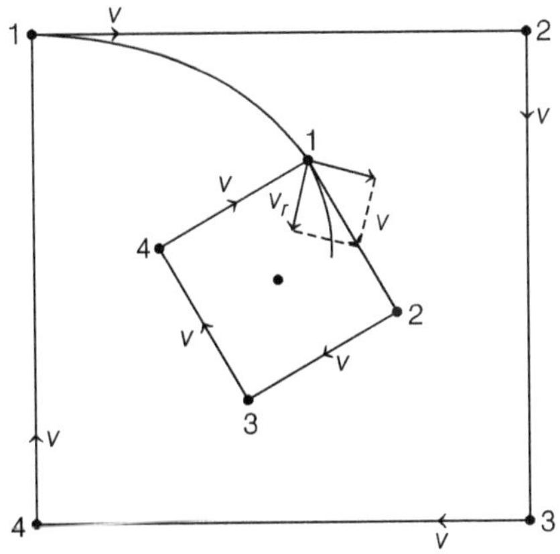

Fig. 334

160. It is obvious from considerations of symmetry that at any moment of time the tortoises will be at the corners of a square whose side gradually diminishes (Fig. 334). The speed of each tortoise can be resolved into a radial (directed towards the centre) and a perpendicular components. The radial speed will be equal to $v_r = \dfrac{v}{\sqrt{2}}$.

Each tortoise has to walk a distance of $l = \dfrac{a}{\sqrt{2}}$ to the centre.

Therefore, the tortoises will meet at the centre of the square after the time $t = \dfrac{l}{v_r} = \dfrac{a}{v}$ elapses.

161. Ship B moves toward ship A with the speed v. At the same time ship A sails away from ship B with the speed $v\cos\alpha$ (Fig. 335). Therefore, the distance AB reduces with a speed of $v(1 - \cos\alpha)$. Point C (the projection of point B onto the trajectory of ship A) moves with a speed of $v(1 - \cos\alpha)$. For this reason distance AC increases with a speed of $v\cos\alpha$. Therefore, the sum of the distances $s = AB + AC$ remains constant as the ships move. At the initial moment point C coincides with A, and therefore $s = AB = a$.

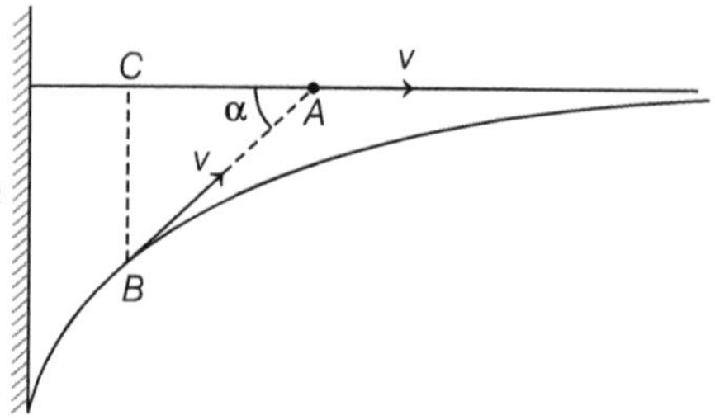

Fig.335

After a sufficiently great interval of time point C will coincide with B, and $AB = AC = \frac{s}{2} = \frac{a}{2}$, and the ships will move at a distance of 1.5 km from each other.

162. With respect to the reading system shown in Fig. 336, the coordinates and the velocities of the body can be determined at any moment of time from the following formulas :

$$x = v_{0x}t \tag{1}$$

$$y = v_{0y}t - \frac{gt^2}{2} \tag{2}$$

$$v_x = v_{0x} \tag{3}$$

$$v_y = v_{0y} - gt \tag{4}$$

Here $v_{0x} = v_0 \cos\alpha$ and $v_{0y} = v_0 \sin\alpha$ are the projections of the initial velocities on the axes x and y. Equations (1), (2), (3) and (4) provide an answer to all the questions stipulated in the problem.

The duration of flight T can be found from equation (2). When $y = 0$, we have $v_0 \sin\alpha T - \frac{gT^2}{2} = 0$. Hence, $T = \frac{2v_0 \sin\alpha}{g}$.

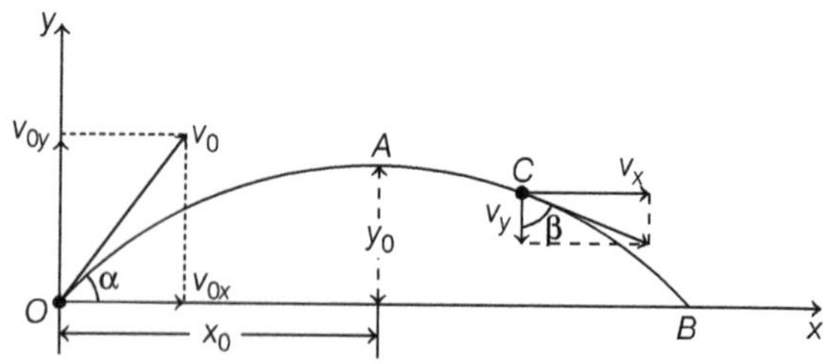

Fig. 336

The distance of the flight is $L = v_0 \cos\alpha\, T = \frac{v_0^2 \sin 2\alpha}{g}$. This distance will be maximum when $\alpha = 45°$:

$$L_{max} = \frac{v_0^2}{g}$$

The height at which the body will be after the time τ elapses is equal to

$$h = v_0 \sin\alpha\tau - \frac{g\tau^2}{2}.$$

The velocity of the body at the moment τ is equal to $v = \sqrt{v_x^2 + v_y^2}$, where $v_x = v_0 \cos\alpha$ and $v_y = v_0 \sin\alpha - g\tau$.
Hence, $v = \sqrt{v_0^2 + g^2\tau^2 - 2v_0 g\tau \sin\alpha}$ and it forms an angle of β with the vertical that can be found from the equation $\tan\beta = \frac{v_0 \cos\alpha}{v_0 \sin\alpha - g\tau}$.

163. The coordinates of the body x and y change with time according to the law

$$y = v_0 \sin\alpha t - \frac{gt^2}{2}$$

$$x = v_0 \cos\alpha t$$

Upon excluding the time from these expressions, we obtain an equation for the trajectory $y = -\frac{g}{2v_0^2 \cos^2\alpha} x^2 + \tan\alpha \cdot x$. This is an equation of a parabola. By denoting the coordinates of the vertex of the parabola (point A in Fig. 336) by x_0 and y_0, the equation of the trajectory can be written as $y - y_0 = k(x - x_0)^2$, where

$$k = -\frac{g}{2v_0^2 \cos^2\alpha};\ y_0 = \frac{v_0^2 \sin^2\alpha}{2g} \text{ and } x_0 = \frac{v_0^2 \sin 2\alpha}{2g}$$

164. The trajectory of the ball takes the form of a parabola passing through a point with the coordinates h and s. Therefore (see the solution to Problem 163),

$$h = -\frac{g}{2v_0^2 \cos^2\alpha} s^2 + \tan\alpha\, s$$

Hence,

$$v_0^2 = \frac{gs^2}{2\cos^2\alpha(\tan\alpha\, s - h)} = \frac{gs^2}{(s\sin 2\alpha - h\cos 2\alpha) - h}$$

$$= \frac{gs^2}{\sqrt{s^2 + h^2}\sin(2\alpha - \varphi) - h}$$

where $\tan\varphi = h/s$.
The minimum velocity

$$v_0 = \sqrt{\frac{gs^2}{\sqrt{s^2+h^2}-h}} = \sqrt{g(h+\sqrt{h^2+s^2})}$$

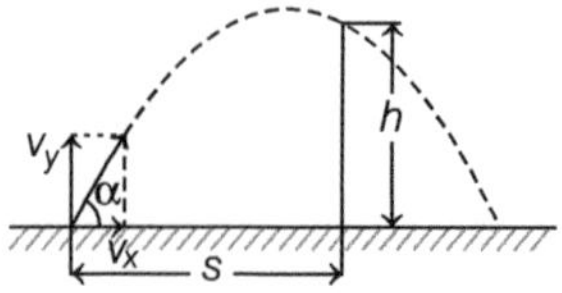

Fig. 337

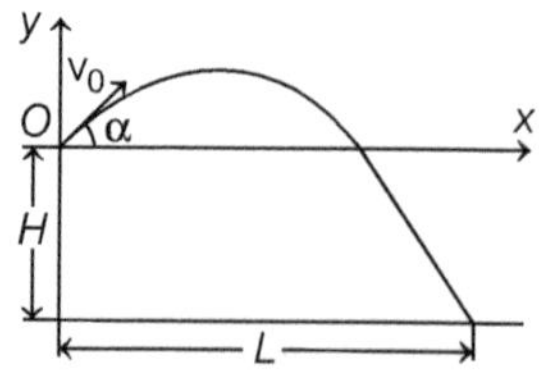

Fig. 338

is attained when

$$\alpha = \frac{\psi}{2} + \frac{\pi}{4} = \arctan\frac{h+\sqrt{s^2+h^2}}{s}$$

$$= \arctan\frac{s}{\sqrt{h^2+s^2}-h} \text{ (Fig. 337.)}$$

165. The coordinates and the velocities of the body at any moment of time with respect to the reading system shown in Fig. 338 are determined by the same equations as in Problem 162.

At the moment when the body falls into the water its coordinate

$$y = -H.$$

For this reason the duration of flight T can be found from the equation

$$-H = v_0 \sin\alpha T - \frac{gT^2}{2}$$

Hence,

$$T = \frac{v_0\sin\alpha \pm \sqrt{v_0^2\sin^2\alpha + 2gH}}{g}$$

Since $T > 0$, we shall retain the plus sign. The distance from the bank is

$$L = v_0\cos\alpha T = \frac{v_0^2\sin 2\alpha}{2g} + \frac{v_0\cos\alpha}{g}\sqrt{v_0^2\sin^2\alpha + 2gH}$$

The body will be at a height h above the water after the time

$$\tau = \frac{v_0\sin\alpha \pm \sqrt{v^2\sin^2\alpha + 2g(H-h)}}{g}$$

If $|h| < |H|$, only the plus sign has a physical meaning. When $h \geq H$, both solutions have a meaning. During its motion the body will be twice at the same height above the water.

It is the simplest to find the final velocity v with the aid of the law of conservation of energy

$$\frac{mv_0^2}{2} + mgH = \frac{mv^2}{2}$$

Therefore, $$v = \sqrt{v_0^2 + 2gH}$$

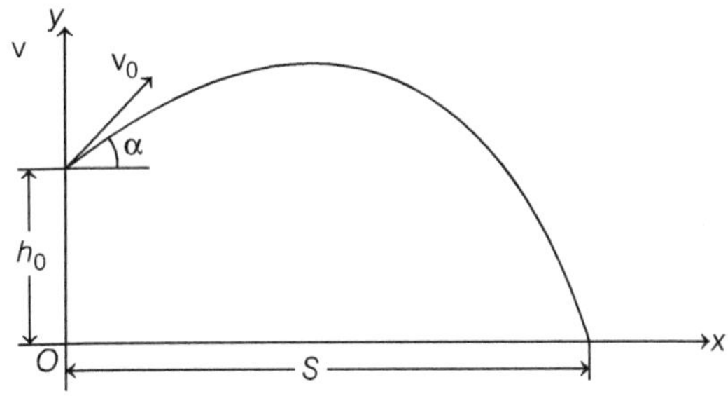

Fig. 339

166. In the reading system depicted in Fig. 339, the coordinates of the stone are determined at any moment of time by the following equations:

$$x = v_0 \cos \alpha t$$

$$y = h_0 + v_0 \sin \alpha t - \frac{gt^2}{2}$$

At the moment when the stone falls, $y = 0$ and $x = s$, where s is the distance covered by the stone.

Upon solving these equations with respect to the angle α, we obtain

$$\tan \alpha = \frac{v_0^2}{gs}\left(1 \pm \sqrt{1 + \frac{2gh_0}{v_0^2} - \frac{g^2 s^2}{v_0^4}}\right)$$

This expression has a meaning when

$$1 + \frac{2gh_0}{v_0^2} - \frac{g^2 s^2}{v_0^4} \geq 0$$

Hence, $$s \leq \frac{v_0\sqrt{v_0^2 + 2gh_0}}{g}.$$

Its maximum value is

$$s_{max} = \frac{v_0\sqrt{v_0^2 + 2gh_0}}{g}.$$

When s is smaller, two values of the angle α correspond to each of its values, the difference between which is the less, the nearer s is to its maximum value.

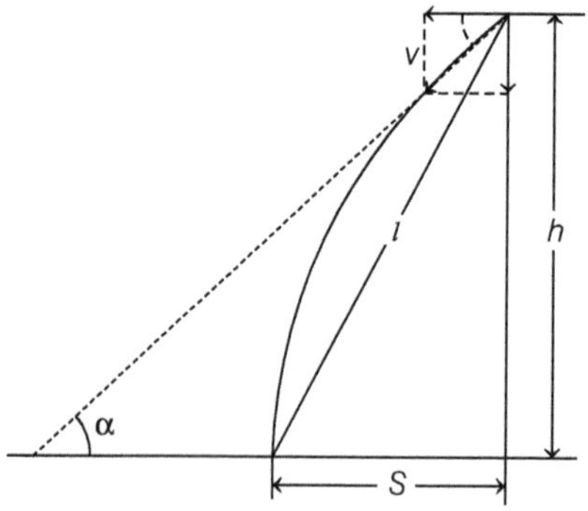

Fig. 340

Therefore, for the maximum distance of flight,

$$\tan\alpha = \frac{v_0^2}{gs_{max}} = \frac{v_0}{\sqrt{v_0^2 + 2gh_0}} = \frac{1}{\sqrt{3}};$$

and $\alpha = 30°$

167. The components of the velocities of the bodies along x and y at any moment of time are determined as

$$v_{1y} = v_0 \sin\alpha_1 - gt;$$
$$v_{2y} = v_0 \sin\alpha_2 - gt;$$
$$v_{1x} = v_0 \cos\alpha_1;$$

and $$v_{2x} = -\, v_0 \cos\alpha_2$$

Let u be the velocity of the second body with respect to the first one. Hence

$$u_y = v_0 \sin\alpha_1 - gt - v_0 \sin\alpha_2 + gt$$
$$= v_0(\sin\alpha_1 - \sin\alpha_2)$$
$$u_x = v_0(\cos\alpha_1 + \cos\alpha_2)$$

Therefore, the velocity u is equal to

$$u = \sqrt{u_x^2 + u_y^2} = 2\cos\left(\frac{\alpha_1 + \alpha_2}{2}\right)v_0$$

The bodies move with respect to each other at a constant velocity. After the time τ the distance between them will be

$$s = 2v_0 \cos\left(\frac{\alpha_1 + \alpha_2}{2}\right)\tau$$

168. The horizontal path of the bomb $s = \sqrt{l^2 - h^2} = v\cos\alpha t$, where t is the duration of falling of the bomb. The vertical path is $h = v\sin\alpha t + \frac{gt^2}{2}$ (Fig. 340).

Upon excluding the time from these equations, we find

$$\tan\alpha = -\frac{v^2}{gs} \pm \sqrt{\left(\frac{v^2}{gs}\right)^2 + \frac{2hv^2}{gs^2} - 1}$$

The solution with the plus sign has a meaning. The minus sign corresponds to $\alpha < 0$, i.e., the bomb is dropped when the dive-bomber is flying upward.

169. It will be convenient to solve this problem in a reading system related to the uniformly moving vehicles.

In this system, the highway moves back with the speed of $v = 50$ km/h, the vehicles are at rest with respect to each other and their wheels rotate. The linear speed of points on the circumference of the wheel and that of the stuck stone is also v. The stone will fly the maximum distance if it flies out when its speed forms an angle of 45° with the horizon. Let us find this distance. Neglecting the fact that the stone is somewhat above the level of the highway when it is thrown out, we obtain $l = \frac{v^2 \sin 2\alpha}{g} = \frac{v^2}{g} = 19.6$ metres. The minimum distance between the vehicles should be 19.6 metres.

170. It will be much easier to solve the problem if the axes of coordinates are directed along the inclined plane and perpendicular to it (Fig. 341).

In this case the components of the acceleration of the ball on the axes x and y will be respectively equal to $a_x = g_x = g\sin\alpha$ and $a_y = g_y = -g\cos\alpha$.

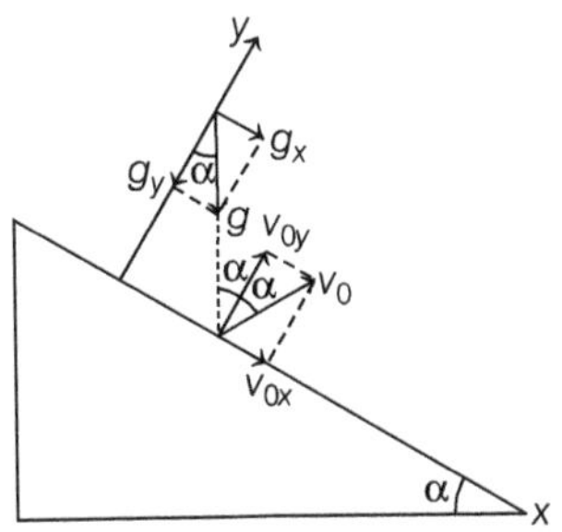

Fig. 341

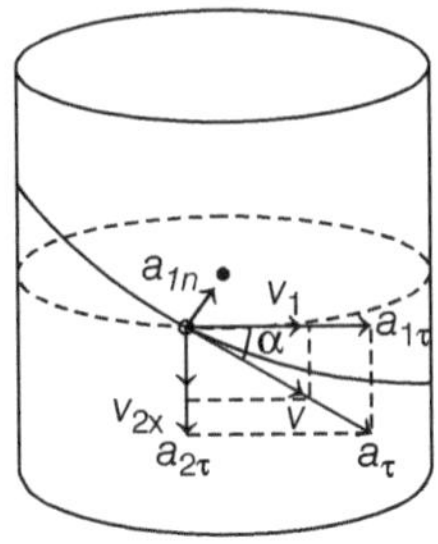

Fig. 342

Upon the first impact with the inclined plane, the velocity of the ball will be $v_0 = \sqrt{2gh}$. The initial velocity of the ball after the first impact is v_0 and forms an angle α with the y-axis (Fig. 341).

The distance between the points of the first and second impacts is $l_1 = v_0 \sin\alpha t_1 + \frac{g\sin\alpha t^2}{2}$, where t_1 is the duration of flight and is determined by the equation

$$v_0 \cos\alpha t_1 - \frac{g\cos\alpha t_1^2}{2} = 0$$

Hence, $t_1 = \frac{2v_0}{g}$ and $l_1 = 8h\sin\alpha$. The velocity of the ball at the second impact can be found from the equations

$$v_{1x} = v_{0x} + a_x t_1 = v_0 \sin\alpha + g\sin\alpha t_1 = 3v_0 \sin\alpha$$

$$v_{1y} = v_{0y} + a_y t_1 = v_0 \cos\alpha - g\cos\alpha t_1 = -\, v_0 \cos\alpha$$

After the impact these velocities are equal to

$$v_{2x} = v_{1x}, \text{ and } y_{2y} = -\, v_{1y}$$

The distance between the points of the second and third impacts is equal to

$$l_2 = 3v_0 \sin\alpha t_2 + \frac{g\sin\alpha t_2^2}{2}$$

where t_2 is the time during which the ball is in flight. Since the initial velocity along the y' axis is the same as during the first impact, $t_2 = t_1$.

Therefore, $l_2 = 16h\sin\alpha$.

Similarly, it can be shown that the distance between the next points

$$l_3 = 24h\sin\alpha.$$

Consequently, $l_1 : l_2 : l_3 \ldots = 1:2:3$, etc.

171. The motion of the body can be considered as superposition of movement along a circumference with a radius R in a horizontal plane and vertical falling. Accordingly, the velocity of the body v at the given moment can be represented as the geometrical sum of two components: $v_1 = v\cos\alpha$ directed horizontally and $v_2 = v\sin\alpha$ directed vertically (Fig. 342). Here α is the angle formed by the helical line of the groove with the horizon.

In curvilinear motion the acceleration of a body is equal to the geometrical sum of the tangential and normal accelerations. The normal acceleration that corresponds to movement along the circumference is

$$a_{1n} = \frac{v_1^2}{R} = \frac{v^2 \cos^2\alpha}{R}$$

The vertical motion is rectilinear, and therefore $a_{2n} = 0$.

The sought acceleration $a = \sqrt{a_1^2\tau + a_2^2\tau + a_1^2 n}$, where $a_{1\tau}$ and $a_{2\tau}$ are the tangential accelerations that correspond to motion along the circumference and along the vertical. The total tangential acceleration a_τ is obviously equal to $a_\tau = \sqrt{a_{1\tau}^2 + a_{2\tau}^2}$.

The value of a_τ can be found by mentally developing the surface of the cylinder with the helical groove into a plane. In this case the groove will become an inclined plane with a height nh and a length of its base $2\pi Rn$.

Apparently, $a_\tau = g\sin\alpha = g\dfrac{h}{\sqrt{h^2 + 4\pi^2R^2}}$.

To determine a_{1n}, let us find v from the law of conservation of energy: $\dfrac{mv^2}{2} = mghn$. Consequently, $v^2 = 2ghn$ and $a_{1n} = \dfrac{8\pi^2 nhgR}{h^2 + 4\pi^2R^2}$.

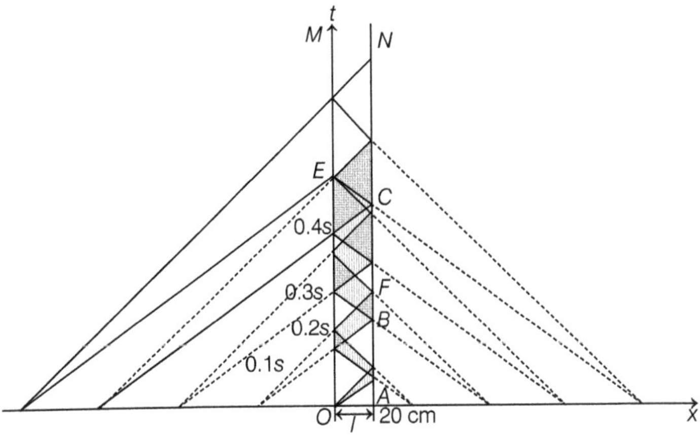

Fig. 343

Upon inserting the found accelerations a_τ and a_{1n}, into the expression for the sought acceleration, we get

$$a = \frac{gh\sqrt{h^2 + 4\pi^2R^2 + 64\pi^4n^2R^2}}{h^2 + 4\pi^2R^2}$$

172. As usual, the motion of the ball can be considered as the result of summation of vertical (uniformly accelerated) and horizontal (uniform) motions.

The simplest method of solution is to plot a diagram showing how the coordinates of the ball along the horizontal depend on the time for the limiting velocities 267 cm/s and 200 cm/s (Fig. 343). The lower broken line corresponds to the maximum velocity and the upper one to the minimum velocity. In the course of time, as can be seen from the diagram, the indefiniteness of the ball coordinate x shown by the section of the horizontal straight line between the lines of the diagram increases. The vertical hatching in Fig. 343 shows the movement of the ball from M to N, and the horizontal hatching—from N to M. The cross-hatched areas correspond to indefiniteness in the direction of the horizontal velocity.

(1) The diagram shows that after the ball bounces once from slab N the direction of its horizontal velocity will be indefinite when the duration of falling $OK \le t \le OL$ or $t > AB$ (where $OK = 0.15s$, $OL = 0.2s$ and $AB = 0.225s$).

Hence, $10 \text{ cm} \le H \le 20 \text{ cm}$ or $h \ge \frac{gt^2}{2} \approx 26 \text{ cm}$.

(2) The ball may strike any point on the base supporting the slabs if the duration of falling of the ball $t \ge AF = 0.3$ s.

Therefore, $H_{min} = 44$ cm.

173. (1) During the time T of a complete revolution the disk will cover a distance equal to the length of its circumference, i.e., $s = 2\pi r$, where r is the radius of the disk. Therefore, the translational velocity: of any point on the disk $v_{tr} = \frac{2\pi r}{T} = v$. On the other hand, the linear velocity of rotation of points on the disk rim with respect to the centre O is $v_{lin} = \omega r$, where ω is the angular velocity of rotation. Since $\omega = \frac{2\pi}{T}$, then $v_{lin} = \frac{2\pi r}{T} = v_{tr}$.

(2) The velocity of points on the disk rim with respect to a standing observer will be the sum of the translational and rotational velocities. The total velocity for point A will be equal to $2v$.

For points B and D the velocities being added are equal in absolute magnitude and their sum is $\sqrt{2}v$ (Fig. 344*a*).

For point C the total velocity with respect to a standing observer is zero, since the translational and rotational velocities are equal in absolute magnitude and oppositely directed.

(3) The instantaneous velocities of points on diameter AC increase in direct proportion to the distance from point C. For this reason the motion of the disk may be considered at the given moment of time as rotation around the point where the disk touches the path. The axis passing through point C perpendicular to the plane of the disk is known as the instantaneous axis of rotation. When the disk moves, this axis constantly passes through the point of contact between the disk and the path.

Therefore, all the points on the disk equidistant from point C at the given moment of time will have the same total velocity with respect to a standing observer. The points which are at a distance equal to the disk radius from the instantaneous axis (point C) will have the same velocity (in absolute magnitude) as that of the axis, i.e., v (Fig. 344*b*).

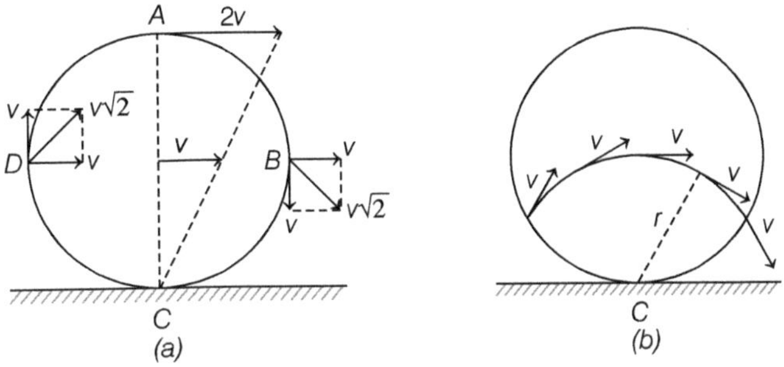

Fig. 344

174. The angle between adjacent spokes of the front wheel is $\phi = \frac{2\pi}{N_1}$. The wheel will seem stationary on the screen if it turns through the angle $\alpha = k\varphi$ during the time between the filming of two successive frames $\tau = 1/24$ s. Here k is a positive integer. On the other hand, the angle of rotation of the wheel during the time τ is equal to $\alpha = \omega\tau$, where ω is the angular speed of the wheel. Therefore, the front wheel will seem stationary if $\omega = \frac{2\pi k}{N_1\tau}$, and the speed of the cart $v = \omega r = \frac{2\pi kr}{N_1\tau}$. The minimum speed of the cart $v_{min} = \frac{2\pi r}{N_1} = 8.8$ m/s.

The rear wheel will also seem stationary if

$$\frac{2\pi k_1 r}{N_1\tau} = \frac{2\pi k_2 R}{N_2\tau}$$

Hence, $$N_2 = \frac{N_1 R}{r} = 9 \text{ when } k_1 = k_2 = 1.$$

175. (1) The spokes will seem to rotate counterclockwise if during the time τ (see Problem 174) the wheel turns through the angle β_1 which satisfies the condition $k\varphi > \beta_1 > k - \frac{\varphi}{2}$, where $k = 1, 2, 3, \ldots$. The consecutive positions of the wheel spokes are shown for this case in Fig. 345*a*. It seems to the audience that each spoke turns through the angle $\alpha < \frac{\varphi}{2}$ counterclockwise. The possible angular velocities lie within the interval

$$\frac{k\varphi}{\tau} > \omega_1 > \frac{(2k-1)\varphi}{2\tau}$$

Since the front and rear wheels have the same number of spokes, the wheels will seem to revolve counterclockwise if the speed of the cart is

$$\frac{k\varphi r}{\tau} > v > \frac{k\varphi r}{\tau} - \frac{\varphi r}{2\tau} \tag{1}$$

$$\frac{k\varphi R}{\tau} > v > \frac{k\varphi R}{\tau} - \frac{\varphi R}{2\tau} \tag{2}$$

Since $R = 1.5r$, the second inequality can be rewritten as follows:

$$1.5\frac{k\varphi r}{\tau} > v > \frac{1.5k\varphi r}{\tau} - \frac{1.5\varphi r}{2\tau}$$

Both inequalities, which are congruent only when $k = 1$, give the permissible speeds of the cart in the form

$$\frac{\varphi r}{\tau} > v > 0.75\frac{\varphi r}{\tau}$$

Or, since $\varphi = \frac{2\pi}{6}$, we have $8.8\ \text{m/s} > v > 6.6\ \text{m/s}$.

(2) The spokes of the rear wheel will seem to revolve clockwise if during the time τ the wheel turns through the angle β_2 which satisfies the condition $(2k-1)\frac{\varphi}{2} > \beta_2 > (k-1)\varphi$ (Fig. 345*b*). Hence, the following inequality is true for the speed of the cart:

$$1.5\frac{(2k-1)\varphi r}{2\tau} > v > \frac{1.5(k-1)\varphi r}{\tau}$$

At the same time inequality (1) should be complied with. When $k = 1$, both inequalities are congruent if $0.75\frac{\varphi r}{\tau} > v > 0.5\frac{\varphi r}{\tau}$. When $k = 2$ they are congruent if $\frac{\varphi r}{\tau} > v > 1.5\frac{\varphi r}{\tau}$. If $k > 2$ the inequalities are incongruent.

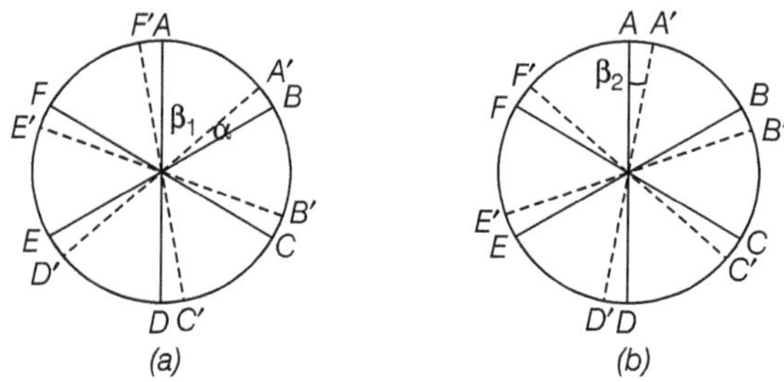

Fig. 345

Therefore, $6.6\ \text{m/s} > v > 4.4\ \text{m/s}$

or $17.6\ \text{m/s} > v > 14.2\ \text{m/s}$

176. The instantaneous axis of rotation (see Problem 173) passes through point C (Fig. 346). For this reason the velocity of point A relative to the block is $v_A = v\frac{R+r}{r}$.

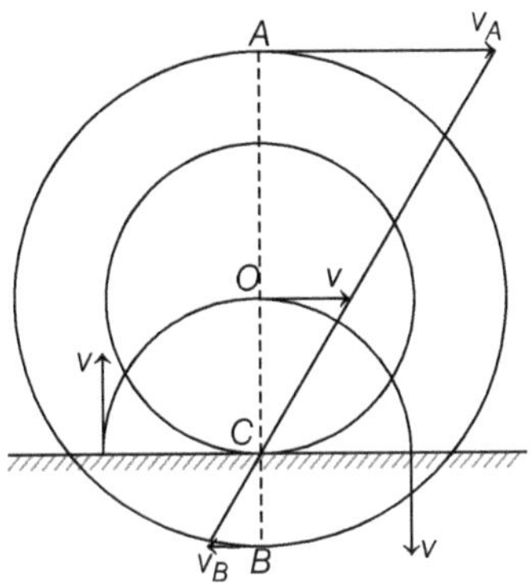

Fig. 346

Point B has the velocity $v_B = v\dfrac{R-r}{r}$.

Points on a circle with the radius r whose centre is point C have an instantaneous velocity equal to that of the spool core.

177. The trajectories of points A, B and C are shown in Fig. 347. Point B describes a curve usually called an ordinary cycloid. Points A and C describe an elongated and a shortened cycloids.

178. The linear velocity of points on the circumference of the shaft $v_1 = \omega\dfrac{d}{2}$ and that of points on the race $v_2 = \Omega\dfrac{D}{2}$. Since the balls do not slip, the same instantaneous velocities will be imparted to the points on the ball bearing that at this moment are in contact with the shaft and the race. The instantaneous velocity of any point on the ball can be regarded as the sum of two velocities: the velocity of motion of its centre v_0 and the linear velocity of rotation around the centre. The ball will rotate with a certain angular velocity ω_0 (Fig. 348).

Therefore,

$$v_1 = v_0 - \omega_0 r$$

$$v_2 = v_0 + \omega_0 r$$

Hence, $$v_0 = \frac{1}{2}(v_1 + v_2) = \frac{1}{4}(\omega d + \Omega D)$$

In this expression, each angular velocity may be positive (clockwise rotation) and negative (counterclockwise rotation).

When $\Omega = 0$, $v_0 = \dfrac{\omega d}{4}$.

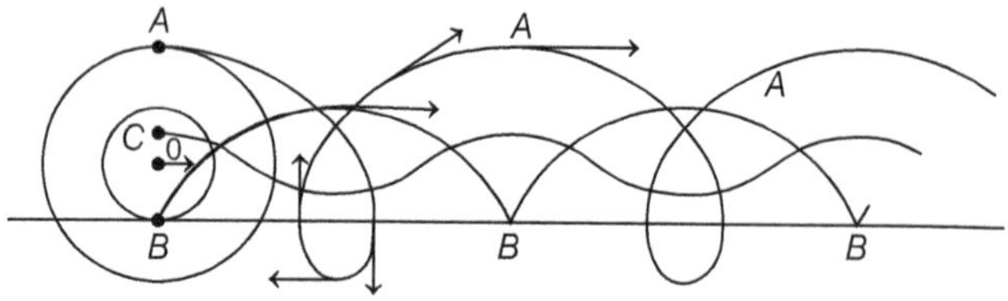

Fig. 347

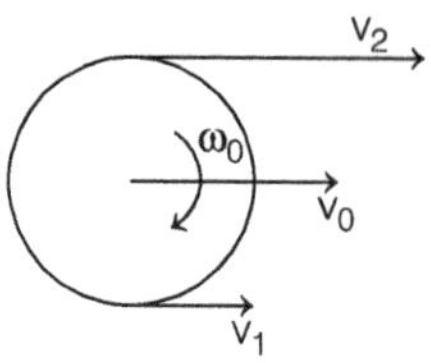

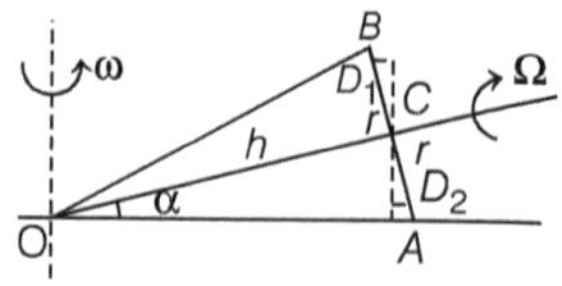

Fig.348 *Fig.349*

179. Since the cone rolls without slipping, the points on the generatrix OA (Fig. 349) should be stationary. This fact is used to determine the velocity Ω with which the cone rotates around its axis.

For point A we have $\dfrac{\omega h}{\cos\alpha} = \Omega h \tan\alpha$. Hence, $\Omega = \dfrac{\omega}{\sin\alpha}$. The velocity of an arbitrary point D_1 on diameter AB of the cone base is the sum of two velocities: $v_1 = \omega(h\cos\alpha - r\sin\alpha) + \dfrac{r\omega}{\sin\alpha}$

where r is the distance from the centre of the base C to the given point. For point D_2 below centre C we have

$$v_2 = \omega(h\cos\alpha + r\sin\alpha) - \frac{r\omega}{\sin\alpha}$$

The velocity of the lowermost point is zero and of the uppermost point $v = 2\omega h\cos\alpha$.

180. The linear velocities should be the same where bevel gears E and C, and E and D, mesh. Since gears E rotate with the velocity ω around axle A, and the axle rotates in another plane with the velocity Ω, the following equation is true for gears E and C to mesh:

$$r_1\omega_1 = r\omega + r_1\Omega$$

For meshing of gears E and D, a similar equation has the form

$$r_1\omega_2 = -r\omega + r_1\Omega$$

Hence,

$$2\Omega = \omega_1 + \omega_2$$

$$2\omega = \frac{r_1}{r}(\omega_1 - \omega_2)$$

When wheel B rotates at a speed of Ω, the angular velocities of the driving wheels may differ from each other from zero to 2Ω.

1-8. Dynamics of Curvilinear Motion

181. On the basis of Newton's second law,

$$\frac{(M+m)u^2}{L} = T - (M+m)g, \text{ where } u = 2\sin\frac{\alpha}{2}\sqrt{Lg} \text{ (see Problem 125).}$$

Hence, $\quad T = (M+m)g\left(4\sin^2\frac{\alpha}{2} + 1\right)$

182. $T_1 = 10m\omega^2 l$; $T_2 = 9m\omega^2 l$; $T_3 = 7m\omega^2 l$, and $T_4 = 4m\omega^2 l$.

183. The distances from the centre of gravity to the masses m_1 and m_2 are equal, respectively, to

$$x = \frac{m_2}{m_1 + m_2} l \text{ and } y = \frac{m_1}{m_1 + m_2} l$$

Let us denote the velocity of the centre of gravity by u and the angular velocity of rotation by ω. Thus $u + \omega x = v_1$ and $u - \omega y = v_2$.

Hence, $\omega = \frac{v_1 - v_2}{l}$, and $u = \frac{m_1 v_1 + m_2 v_2}{m_1 + m_2}$

184. The velocity of rotation will be retarded. The platform imparts to the shell an additional momentum along a tangent to the trajectory of the end of the cannon barrel. According to Newton's third law, the shell ejected from the barrel will press against its inner part in a direction opposite to rotation.

185. When the body touches the horizontal plane, the vertical and horizontal components of its velocity will be $v_{ver} = \sqrt{2gH}\sin\alpha$ and $v_{hor} = \sqrt{2gH}\cos\alpha$. If the impact is absolutely elastic, the vertical component will change its sign, while the horizontal component will remain the same. The trajectory of the body will take the form of sections of parabolas (Fig. 350), with $h = H\sin^2\alpha$ and $l = 2H\sin 2\alpha$.

If the impact is absolutely inelastic, the vertical component of the velocity will be zero, and the body will move uniformly over the horizontal plane with the velocity $v = \sqrt{2gH}\cos\alpha$.

186. The Earth acts on the motor-cycle with two forces (Fig. 351), namely, N – the reaction of the support, and f – the force of friction. The sum of these forces T is directed along the motor-cycle. (Otherwise, the moment of the forces tending to tip the motor-cycle would be acting with respect to the centre of gravity O).

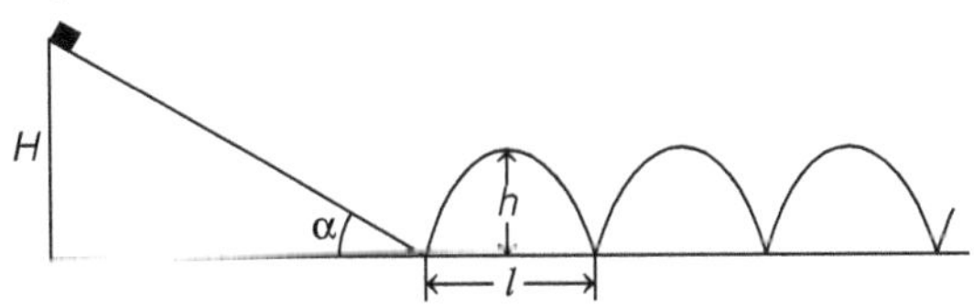

Fig. 350

Thus, the centre of gravity is acted upon by the resulting force $F = T + G$, where G is the weight of the motor-cyclist. Since $F = T\cos\alpha = f$, a centripetal acceleration will be imparted to the motor-cycle only by the friction force f.

According to Newton's second law, $f = \dfrac{mv^2}{R}$ and $f \le kmg$. Figure 351 shows that $mg = f\tan\alpha$.

The minimum value of R from this system of equations is $R_{min} = \dfrac{v^2}{kg} = 147$ metres, when $\tan\alpha = \dfrac{Rg}{v^2} \cong 33.3$, and therefore $\alpha \cong 73°20'$.

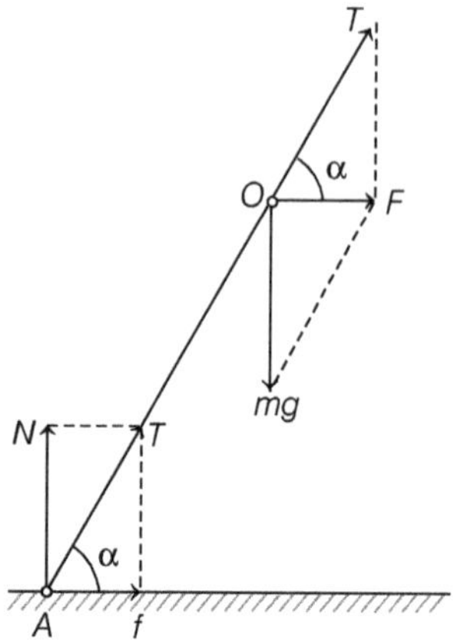

Fig. 351

187. Let us consider the intermediate position of the rod when it has moved through the angle α from the vertical. In accordance with the law of conservation of energy, $MgR = MgR\cos\alpha + \dfrac{M\omega^2R^2}{2}$, where R is the distance from the end of the rod to the centre of gravity of the sphere.

Therefore, the angular velocity ω can be expressed as $\omega = 2\sin\dfrac{\alpha}{2}\sqrt{\dfrac{g}{R}}$

With the given value of α, it will be the smaller, the greater is R. Hence, the rod will fall faster if it is placed on end B.

188. According to Newton's second law, $m\omega^2R = mg\cos\alpha - N$, where N is the force which the deformed rod acts on the sphere with.

The deformation of the rod will disappear when the rod no longer presses on the floor, and $N = 0$. As shown in Problem 187,

$$\omega = 2\sqrt{\frac{g}{R}}\sin\frac{\alpha}{2}.$$

Upon inserting this value of ω in the equation of motion, we find that $\cos\alpha = \frac{2}{3}$. Hence, $\alpha = 48°10'$.

For the rod not to slip, the condition $N\sin\alpha \le kN\cos\alpha$ should be fulfilled (Fig. 352). Therefore, $k \ge \tan\alpha$.

Hence $k \ge \frac{\sqrt{5}}{2}$.

189. If $k > \frac{\sqrt{5}}{2}$, the rod will not slip until N is zero, i.e., until $\alpha > \text{arc cos}\,\frac{2}{3}$.

When $\alpha > \text{arc cos}\,\frac{2}{3}$, the equation $m\omega^2 R = mg\cos\alpha - N$ will give us $N < 0$. This means that if the end of the rod had been attached to the ground, the rod would have been stretched. If the rod is not secured, the sphere will begin to fall freely from the moment when the angle α reaches the value $\alpha_0 = \text{arc cos}\,\frac{2}{3}$.

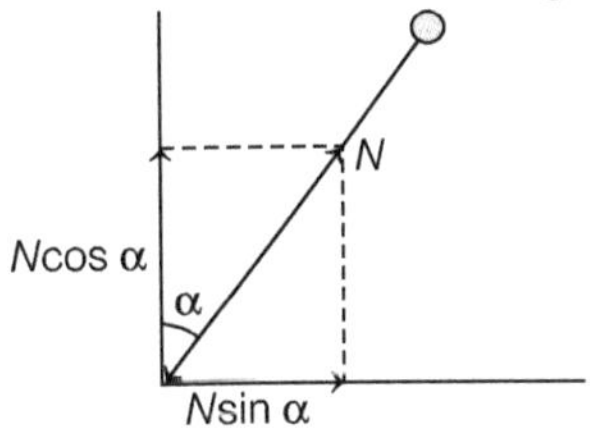

Fig. 352

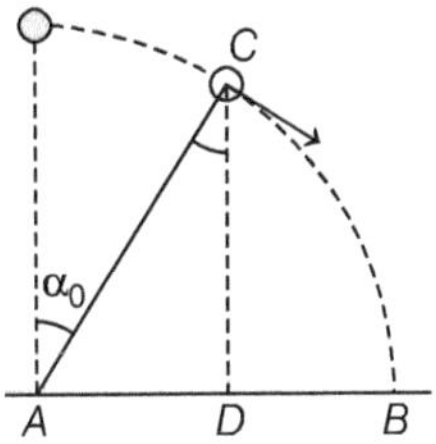

Fig. 353

At this moment $v = \omega R = \sqrt{\frac{2}{3}gR}$ forms an angle α_0 with the horizon and the height of the sphere above the ground is $CD = \frac{2}{3}R$ (Fig. 353).

Let us find the distance sought by using the laws of free falling:

$$AB = AD + DB = R\frac{5\sqrt{5} + 4\sqrt{23}}{27} \cong 1.12R$$

190. The bead moves along section ADB under the action of the force of gravity (Fig. 354). For the bead to reach point B after it leaves point A, the horizontal path which it travels should be $2R\sin\alpha$. For this purpose the velocity of the bead at point A should satisfy the condition $\frac{2u^2\sin\alpha\cos\alpha}{g} = 2R\sin\alpha$ (see Problem 162). Hence, $u^2 = \frac{gR}{\cos\alpha}$.

At point A the bead will have the velocity u if at point O it is given a velocity v equal, according to the law of conservation of energy, to:

$$v = \sqrt{u^2 + 2gR(1 + \cos\alpha)} = \sqrt{gR\left(2 + 2\cos\alpha + \frac{1}{\cos\alpha}\right)}$$

191. Assume that no segment has been cut out. Hence at point C (Fig. 355)

$$mg = \frac{mv^2}{R} \qquad (1)$$

According to the law of conservation of energy,

$$mgh = mg \cdot 2R + \frac{mv^2}{2} \qquad (2)$$

From equations (1) and (2) we get $h = \frac{5}{2}R$. The velocity at point A can be found from the law of conservation of energy

$$mg\frac{5}{2}R = \frac{mv_A^2}{2} + mgR(1 + \cos\alpha) \qquad (3)$$

The body thrown at an angle α to the horizon will fly a horizontal distance of

$$AB = \frac{v_A^2 \sin 2\alpha}{g} \qquad (4)$$

On the other hand,

$$AB = 2R \cdot \sin\alpha \qquad (5)$$

It follows from equations (4) and (5) that

$$v_A^2 = \frac{Rg}{\cos\alpha}$$

Upon inserting this value into equation (3), we obtain :

$$mg\frac{5}{2}R = \frac{mgR}{2\cos\alpha} + mgR + mgR\cos\alpha$$

Therefore, $\cos\alpha = \frac{3 \pm 1}{4}$ and, consequently,

$$\alpha_1 = 0, \text{ and } \alpha_2 = 60°.$$

It is easy to see that if $\alpha > 60°$, the body will fall inside the loop; if $\alpha < 60°$ it will fly out.

192. Let us consider the forces that act on the string thrown over the left nail (Fig. 356). The vertical components of the forces of tension T acting on the weights are mg if the string is secured on the nail. According to Newton's third law, the knot (point O) is acted upon by the same forces T. Their sum is directed vertically downward and is equal to $2mg$.

If only one weight is rotating, the vertical component of the string tension T' is equal to $2mg$ (if the weight does not move downward). The tension of the string itself, however, is $T' > 2mg$ (Fig. 356). Therefore, the system will not be in equilibrium. The right-hand weight will have a greater pull.

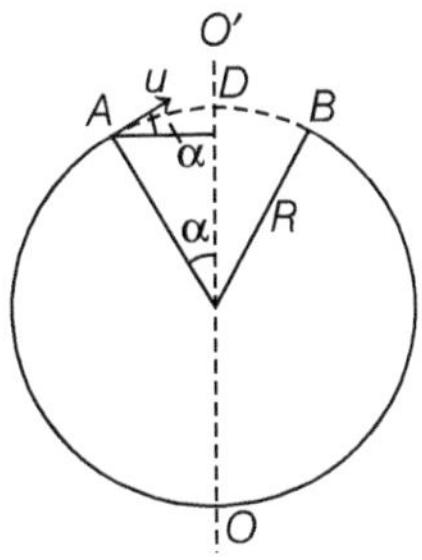

Fig. 354

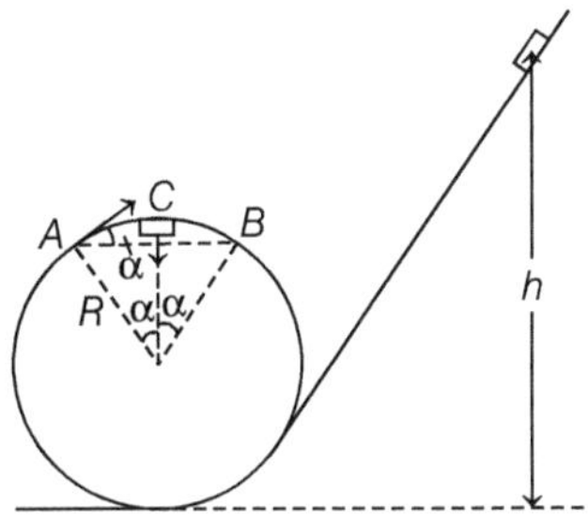

Fig. 355

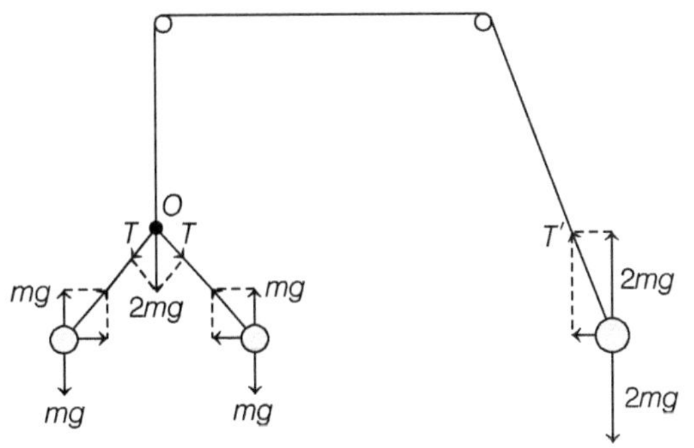

Fig. 356

193. The direction of the acceleration coincides with that of the resultant force. The acceleration is directed downward when the ball is in its two extreme upper positions B and C (Fig. 357). The acceleration will be directed upward if the ball is in its extreme bottom position A and horizontally in positions D and L determined by the angle α.

Let us find α. According to Newton's second law, the product of the mass and the centripetal acceleration is equal to the sum of the projections of the forces on the direction of the radius of rotation:

$$\frac{mv^2}{l} = T - mg\cos\alpha$$

On the other hand, as can be seen from Fig. 357, we have $T = \dfrac{mg}{\cos\alpha}$.

On the basis of the law of conservation of energy:

$$\frac{mv^2}{2} = mgl\cos\alpha$$

We can find from these equations that $\cos\alpha = \dfrac{1}{\sqrt{3}}$, and therefore $\alpha \cong 54°45'$.

194. Let us denote the angular velocity of the rod by ω at the moment when it passes through the vertical position. In conformity with the law of conservation of energy :

$$\frac{\omega^2}{2}(m_1r_1^2 + m_2r_2^2) = g(1 - \cos\alpha)(m_1r_1 + m_2r_2)$$

or
$$\omega = 2\sin\frac{\alpha}{2}\sqrt{g\frac{m_1r_1 + m_2r_2}{m_1r_1^2 + m_2r_2^2}}$$

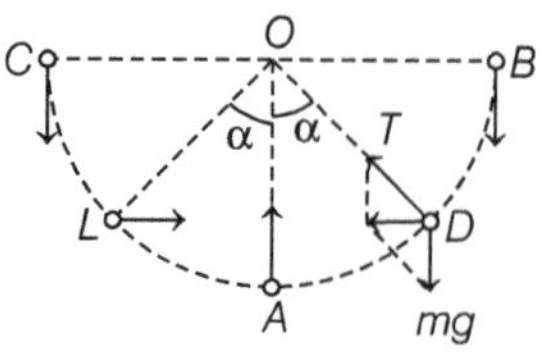

Fig. 357

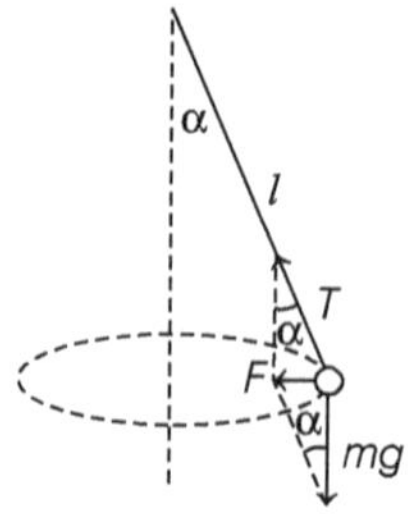

Fig. 358

whence

$$v_1 = \omega r_1 = 2r_1\sin\frac{\alpha}{2}\sqrt{g\frac{m_1r_1 + m_2r_2}{m_1r_1^2 + m_2r_2^2}}$$

$$v_2 = \omega r_2 = 2r_2\sin\frac{\alpha}{2}\sqrt{g\frac{m_1r_1 + m_2r_2}{m_1r_1^2 + m_2r_2^2}}$$

195. The resultant of the forces applied to the ball $F = mg\tan\alpha$ should build up a centripetal acceleration $a = \omega^2 r$, where $r = l\sin\alpha$ (Fig. 358). Hence, $mg\tan\alpha = m\omega^2 l\sin\alpha$

This equation has two solutions :

$$\alpha_1 = 0$$

$$\alpha_2 = \arccos\frac{g}{\omega^2 l}$$

Both solutions are valid in the second case : $\alpha_1 = 0$ (here the ball is in a state of unstable equilibrium) and $\alpha_2 = 60°$.

In the first case the only solution is $\alpha_1 = 0$.

196. Let us resolve the force F acting from the side of the rod on the weight m into mutually perpendicular components T and N (Fig. 359).

Let us project the forces onto a vertical and a horizontal lines and write Newton's equations for these directions

$$m\omega^2 l\sin\varphi = T\sin\varphi - N\cos\varphi$$

$$mg = T\cos\varphi + N\sin\varphi$$

Let us determine T and N from these equations

$$T = m(\omega^2 l \sin^2\varphi + g\cos\varphi)$$
$$N = m(g - \omega^2 l\cos\varphi)\sin\varphi$$

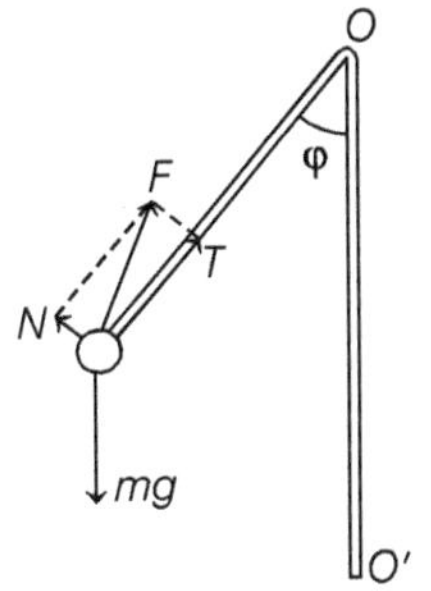

Fig. 359

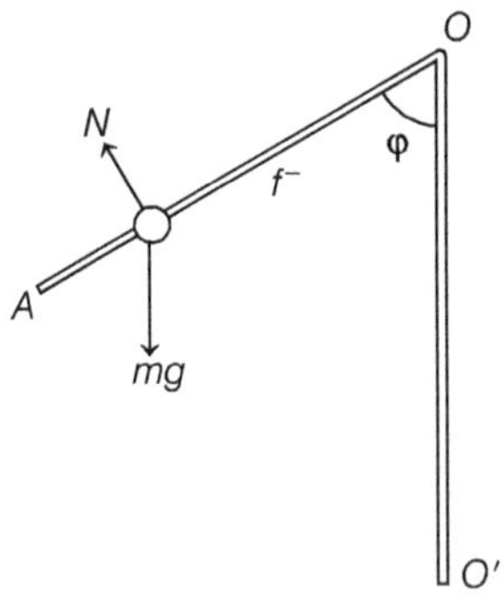

Fig. 360

Therefore, $$F = \sqrt{T^2 + N^2} = m\sqrt{g^2 + \omega^4 l^2 \sin^2\varphi}$$

197. The forces acting on the bead are shown in Fig. 360: f is the force of friction, mg the weight and N the force of the normal reaction. Newton's equations for the projection of the forces on a horizontal and a vertical directions will have the form

$$f\sin\varphi \mp N\cos\varphi = m\omega^2 l\sin\varphi$$
$$f\cos\varphi \pm N\sin\varphi - mg = 0$$

The upper sign refers to the case shown in Fig. 360 and the lower one to the case when the force N acts in the opposite direction. We find from these equations that

$$f = m\omega^2 l\sin^2\varphi + mg\cos\varphi$$
$$N = \pm\,(mg\sin\varphi - m\omega^2 l\sin\phi\cos\varphi)$$

In equilibrium $$f \le kN$$

or $$l \le \frac{k\sin\phi - \cos\varphi}{\sin\varphi(k\cos\varphi + \sin\varphi)}\cdot\frac{g}{\omega^2}$$

when $k \ge \cot\phi$

and $$l \ge \frac{k\sin\varphi + \cos\varphi}{\sin\varphi\,(k\cos\varphi - \sin\varphi)}\cdot\frac{g}{\omega^2} \text{ when } k \ge \tan\varphi$$

198. Figure 361 shows the forces acting on the weights. Here T_1 and T_2 are the tensions of the string. Let us write Newton's equations for the projections onto a horizontal and a vertical directions.

For the first weight

$$T_1\sin\varphi - T_2'\sin\psi = m\omega^2 l\sin\varphi$$

$$T_1 \cos\varphi - T_2 \cos\psi - mg = 0 \qquad (1)$$

For the second weight

$$m\omega^2 l(\sin\varphi + \sin\psi) = T_2 \sin\psi \qquad (2)$$

$$T_2 \cos\psi = mg$$

Upon excluding T_1 and T_2 from the system of equations (1) and (2), we obtain the equations

$$a \sin\varphi = 2\tan\varphi - \tan\psi$$

$$a(\sin\varphi + \sin\psi) = \tan\psi$$

where $a = \dfrac{\omega^2 l}{g}$.

From these equations we get $2\tan\phi - \tan\psi < \tan\psi$ and, therefore, $\varphi < \psi$.

199. The forces acting on the weights are shown in Fig. 362. Here T_1, N_1 and T_2, N_2 are the components of the forces acting from the side of the rod on the weights m and M.

The forces N_1 and N_2 act in opposite directions, since the sum of the moments of the forces acting on the rod with respect to point O is zero because the rod is weightless: $N_1 b - N_2(b + a) = 0$. The equations of motion of the masses m and M for projections on a horizontal and a vertical directions have the form

$$m\omega^2 \sin\varphi = T_1 \sin\varphi - N_1 \cos\varphi;$$

$$T_1 \cos\varphi + N_1 \sin\varphi = mg$$

$$M\omega^2(b + a)\sin\varphi = T_2 \sin\phi + N_2 \cos\varphi;$$

$$T_2 \cos\varphi - N_2 \sin\varphi = Mg$$

Upon excluding the unknown quantities T_1, T_2, N_1 and N_2 from the system, we find that

(1) $\varphi = 0$, and

(2) $\cos\varphi = \dfrac{g}{\omega^2} \dfrac{mb + M(a + b)}{mb^2 + M(a + b)^2}$

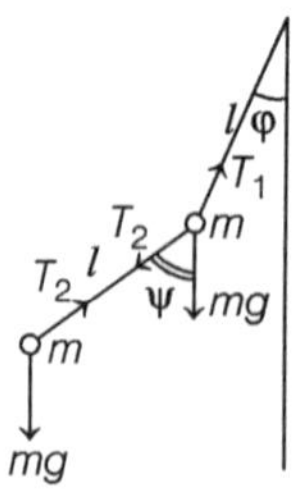

Fig. 361

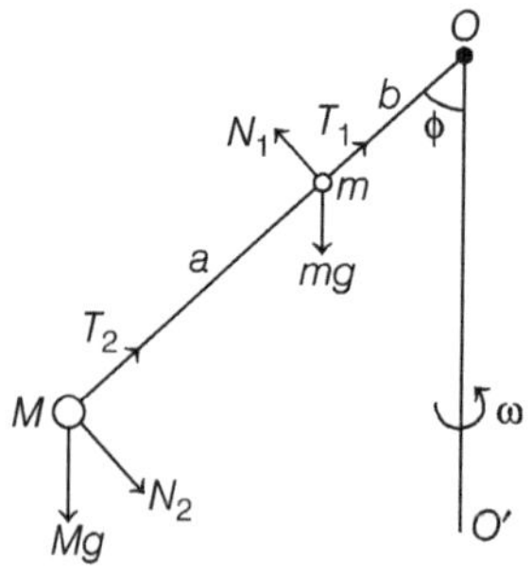

Fig. 362

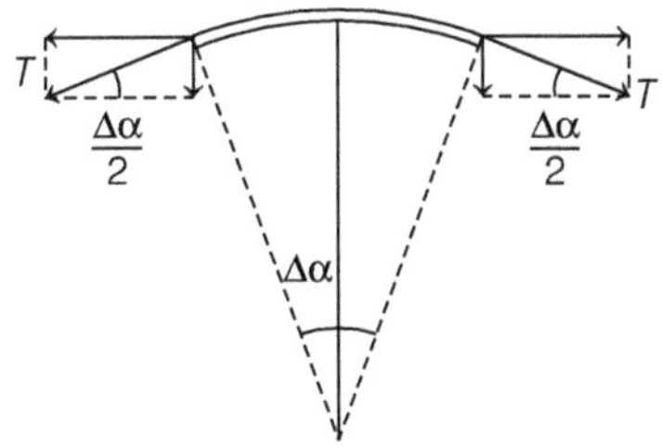

Fig. 363

The first solution is true for any angular velocities of rotation, and the second when $\omega \geq \sqrt{g \frac{mb + M(a + b)}{mb^2 + M(a + b)^2}}$ (see the solution to Problem 195).

200. In the state of equilibrium $m\omega^2 x = kx$, where x is the distance from the body to the axis.

It is thus obvious that with any value of x the spring imparts the centripetal acceleration necessary for rotation to the body. For this reason the latter will move after the impetus with a constant velocity up to stop A or as long as the law of proportionality between the force acting on the spring and its deformation is valid.

201. Let us write Newton's second law for a small portion of the chain having the mass $\frac{m}{l} R\Delta\alpha$ and shown in Fig. 363 :

$$\frac{m}{l} R\Delta\alpha (2\pi n)^2 R$$
$$= 2T \sin \frac{\Delta\alpha}{2}$$

Since the angle $\Delta\alpha$ is small, $\sin \frac{\Delta\alpha}{2} \approx \frac{\Delta\alpha}{2}$, whence $T = mln^2 = 9.2$ kgf.

202. Let us take a small element of the tube with the length $R\Delta\alpha$ (Fig. 364). The stretched walls of the tube impart an acceleration $a = \frac{v^2}{R}$ to the water flowing along this element. According to Newton's third law, the water will act on the element of the tube with the force

$$\Delta F = \rho \frac{\pi d^2}{4} R\Delta\alpha \frac{v^2}{R}$$

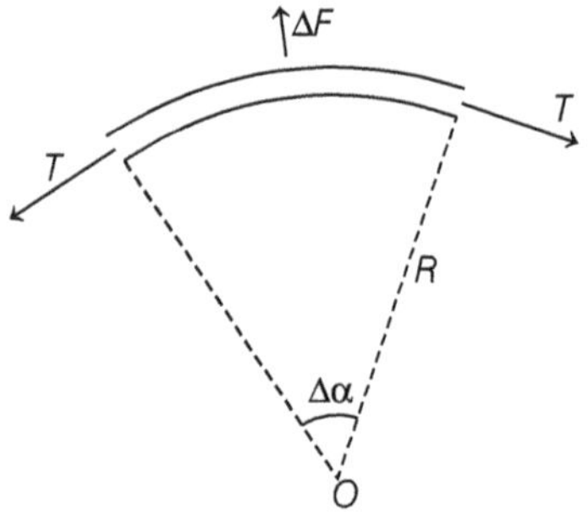

Fig. 364

where ρ is the density of the water. The force ΔF is balanced by the tension forces of the ring T. From the condition of equilibrium, and remembering that $\Delta\alpha$ is small,

we have

$$\Delta F = 2T\sin\frac{\Delta\alpha}{2} \cong T\Delta\alpha$$

Therefore, the sought force is

$$T = \frac{\rho\pi d^2}{4}v^2.$$

203. Let us divide the rod into n sections of equal length and consider an arbitrary section with the number i (Fig. 365). The acceleration of the various points in this section will be different, since the distances from the points to the axis of rotation are not the same. If the difference $r_{i+1} - r_i$ is small, however, the acceleration of the i-th section may be assumed as equal to $\omega^2\frac{r_{i+1}+r_i}{2}$, and this will be the more accurate, the smaller is the length of the section.

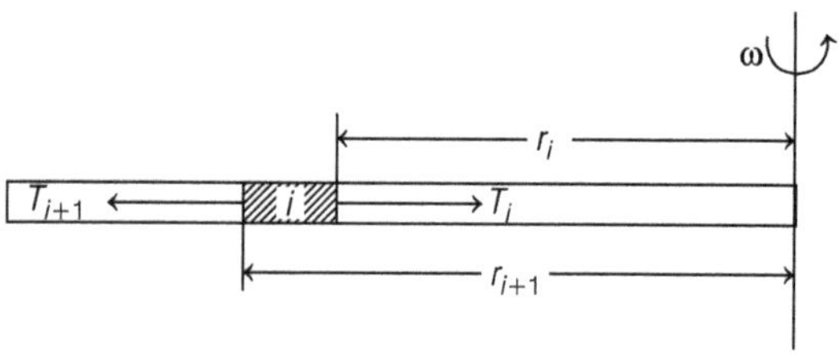

Fig. 365

The i-th section is acted upon by the elastic force T_{i+1} from the side of the deformed section $i + 1$ and the force T_i from the side of the section $i - 1$. Since the mass of the i-th section is $\frac{m}{l}(r_{i+1} - r_i)$, on the basis of Newton's second law we can write that

$$T_i - T_{i+1} = \frac{m}{l}(r_{i+1} - r_i)\omega^2 \frac{r_{i+1} + r_i}{2}$$

or $$T_{i+1} - T_i = -\frac{m\omega^2}{2l}(r_{i+1}^2 - r_i^2)$$

Let us write the equations of motion for the sections from k to n, inclusive, assuming that $r_{n+1} = l$ and $r_k = x$:

$$-T_n = -\frac{m\omega^2}{2l}(l^2 - r_n^2)$$

$$T_n - T_{n-1} = -\frac{m\omega^2}{2l}(r_n^2 - r_{n-1}^2)$$

. . . .

. . . .

$$T_{k+2} - T_{k+1} = -\frac{m\omega^2}{2l}(r_{k+2}^2 - r_{k+1}^2)$$

$$T_{k+1} - T_x = -\frac{m\omega^2}{2l}(r_{k+1}^2 - x^2)$$

The first equation in this system takes into account the fact that the elastic force does not act on the end of the rod, i.e., $T_{n+1} = 0$. Upon summing up the equations of the system, we find that the sought tension is equal to

$$T_x = \frac{m\omega^2}{2}(l^2 - x^2).$$

The closer the sections of the rod are to the axis of rotation, the more will they be stretched.

204. In the reading system that is stationary with respect to the axis, the force of tension of the rod does not perform any work, since this force constantly remains perpendicular to the velocity of the ball. In the moving system this force performs work other than zero, and this changes the kinetic energy of the ball.

205. Section AB of the hoop with a mass m at the highest point has an energy of $mg\,2R + \frac{m(2v)^2}{2}$. During motion the kinetic and potential energies of section AB begin to decrease owing to the work of the forces of the elastic deformation of the hoop whose resultant produces a centripetal force always directed towards the centre. The velocity of section AB forms an obtuse angle α with the force F (Fig. 366). For this reason the work of the force $W_1 = F\Delta S \cos\alpha$ is negative, and, consequently, the energy of the section with the mass m diminishes.

After section AB passes through the lowermost position, it is easy to see that the work of the force F becomes positive and the energy of section AB will grow.

206. Let us draw a tangent to the inner circumference of the spool (Fig. 367) from point A, which is the instantaneous axis of rotation (see Problem 173). If the directions of the thread and the tangent AC coincide, the moment of the forces that rotate the spool around the instantaneous axis will be zero. Therefore, if the spool is at rest, it will not rotate around the instantaneous axis and, consequently, it will not move translationally.

The angle α at which the motion of the spool is reversed can be found from triangle AOB; namely, $\sin\alpha = \frac{r}{R}$. If the thread is inclined more than α, the spool will roll to the right, if α is smaller it will move to the left on condition that it does not slip. If the tension of the thread T satisfies the condition $Tr \leq fR$, where f is the force of friction, the spool will remain in place. Otherwise, when $\sin\alpha = \frac{r}{R}$ it will begin to rotate counterclockwise around point O.

207. Break the hoop into equal small sections each with a mass of Δm. Consider two symmetrical sections (relative to the centre). All the particles of the hoop simultaneously participate in translational motion with the velocity v and in rotational motion with the velocity $v_1 = \omega R$. The resultant velocity v_2 of the upper section of the hoop can be found as the geometrical sum of the velocities v and v_1 (Fig. 368):

$$v_2^2 = v_1^2 + v^2 + 2vv_1 \cos\alpha$$

For a symmetrical section

$$v_3^2 = v^2 + v_1^2 - 2vv_1 \cos\alpha$$

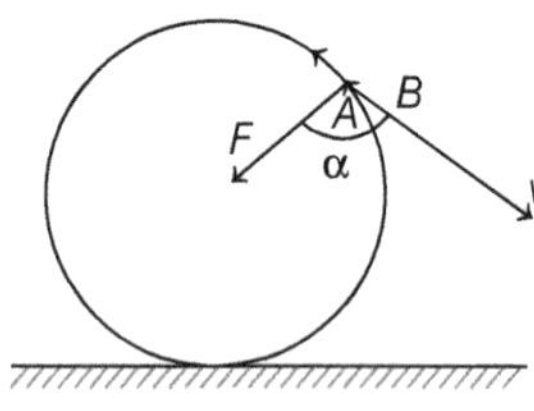

Fig. 366

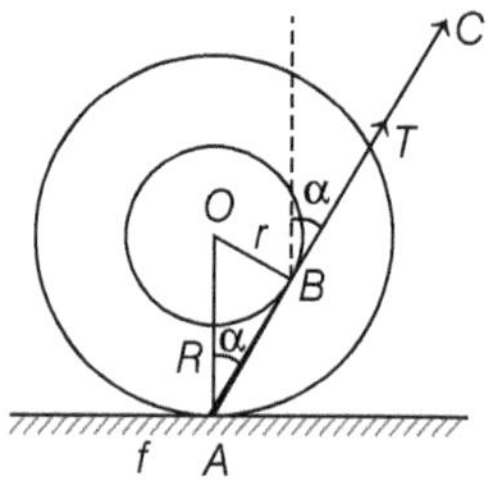

Fig. 367

The total kinetic energy of both sections is

$$\Delta E_k = \frac{\Delta m v_2^2}{2} + \frac{\Delta m v_3^2}{2} = \Delta m v^2 + \Delta m \omega^2 R^2$$

Since this expression is valid for any two sections, we can write for the entire hoop

$$E_k = \frac{Mv^2}{2} + \frac{MR^2\omega^2}{2}$$

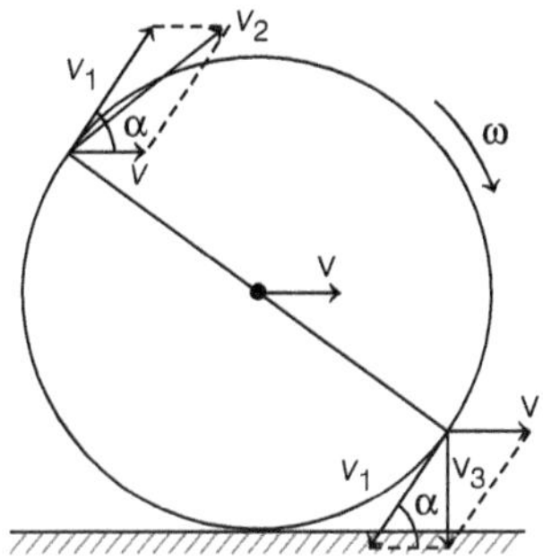

Fig. 368

If the hoop rolls without slipping, $v = \omega R$ and, therefore, $E_k = Mv^2$.

208. $E_k = \dfrac{2Gv^2}{g}(\pi r + l)$.

209. The cylinder made of a denser material will obviously be hollow. At the same translational velocities without slipping the kinetic energy of rotation will be greater in the hollow cylinder, since the particles of its mass are further from the centre and, therefore, have higher velocities.

For this reason the hollow cylinder will roll down an inclined plane without slipping slower than the solid one. At the end of the plane, the total kinetic energies of both cylinders are the same. This is possible only when the velocities are different, since when the velocities are the same, the energies of translational motion are identical, while the energy of rotational motion of a solid cylinder will always be smaller than that of a hollow one.

210. When the drum moves, the force of friction performs no work, since the cable and the drum do not slip. Hence, the energy of the system does not change:

$$\frac{G}{g}v^2 + GR = \frac{G - \rho x}{g}u^2 + (G - \rho x)R$$

where u is the sought velocity. Therefore,

$$u = \sqrt{\frac{Gv^2 + \rho Rgx}{G - \rho x}}.$$

It is equal to infinity when $G = \rho x$ because the mass of the drum is neglected.

The momentum is diminished by the action of the force of friction, which is opposite to the direction of motion.

211. Since the force of friction is constant, motion will be uniformly retarded. The power developed by the friction force is fv, where $v = \omega r$ is the instantaneous velocity of the point on the pulley which the force f is applied to. The work during the time t is equal to the mean power multiplied by the time t :

$$W = f\frac{\omega_0 r + \omega r}{2}t$$

The change in the kinetic energy of the pulley is equal to this work:

$$\frac{mr^2}{2}(\omega_0^2 - \omega^2) = \frac{frt}{2}(\omega_0 + \omega)$$

Hence,
$$\omega = \omega_0 - \frac{ft}{mr}.$$

212. Since the force of friction f is constant, the change in the momentum of the hoop during the time t is $mv = ft$. If the hoop rolls without slipping, the velocity of the point on the hoop which the friction force is applied to is zero.

Upon equating the work of the friction forces to the difference between the kinetic energies, we have

$$\frac{m\omega_0^2 r}{2} - mv^2 = f\frac{\omega_0 r + 0}{2}t$$

(see Problem 207).

Upon solving the equations with respect to v, we find:

$$v = \frac{\omega_0 r}{2}$$

213. The equations which show the change in the momentum and the kinetic energy of the hoop have the form:

$$m(v_0 - v) = ft$$

$$\frac{mv_0^2}{2} - mv^2 = f\frac{v_0 + 0}{2}t$$

where $v = \omega r$ is the velocity of the centre of the hoop when it rolls without slipping.

Upon solving these equations with respect to v, we have:

$$v = \frac{v_0}{2}$$

Therefore, the sought value is $w = \frac{v_0}{2r}$.

214. The equations that show the change in the momentum and the kinetic energy of the hoop have the form:

$$m(v_0 - v) = ft$$

$$\frac{mv_0^2}{2} + \frac{m\omega_0^2 r^2}{2} - \frac{mv^2}{2} - \frac{m\omega^2 r^2}{2} = f\frac{(v_0 + \omega_0 r) + (v + \omega r)}{2}t$$

where v is the velocity of the hoop centre at any subsequent moment of time.

Upon solving this system of equations, we find that

$$v = v_0 - \frac{f}{m}t, \text{ and } \omega = \omega_0 - \frac{ft}{mr}$$

If $v_0 < \omega_0 r$, the hoop will stop at the moment of time $\tau = \frac{mv_0}{f}$ when rotating with the angular velocity $\omega = \omega_0 - \frac{v_0}{r}$. Then the hoop begins to move with slipping in the reverse direction. In a certain time the hoop will stop slipping and will roll without slipping to the left with a translational velocity $v = \frac{\omega_0 r - v_0}{2}$ (see Problem 213).

If $v_0 > \omega_0 r$, then after the time $\tau = \frac{mr\omega_0}{f}$ elapses the hoop will stop rotating and will move to the right with a translational velocity $v = v_0 - r\omega_0$. Next the hoop will rotate in the reverse direction, and in some time it will roll without slipping to the right. Its angular velocity will be $\omega = \frac{v_0 - r\omega_0}{2r}$

Practice shows that the loop will also be braked when it does not slip. We did not get such a result since the specific rolling friction was neglected.

215. Since the hoops do not slip, v_0 (velocity of the centre of gravity of the hoops) and v (velocity of the weight) are related by the expression

$$v_0 = v\frac{R}{R - r}$$

Assume that the weight lowers through the distance h. If the system was at rest at the initial moment, from the law of conservation of energy we have

$$mgh = \frac{mv^2}{2} + Mv_0^2$$

(see Problem 207).

This expression can be used to find the velocity of the weight:

$$v = \sqrt{\frac{2mgh}{m + 2M\left(\frac{R}{R - r}\right)^2}}$$

Hence, the acceleration of the weight is

$$a = \frac{mg}{m + 2M\left(\frac{R}{R - r}\right)^2}$$

The weight lowers with the acceleration a under the action of the force of gravity mg and the tension T of the string.

The sought tension T is

$$T = m(g - a) = \frac{2mMg\left(\frac{R}{R-r}\right)^2}{m + 2M\left(\frac{R}{R-r}\right)^2}$$

Since the centre of gravity of the hoop moves with an acceleration equal to $a\dfrac{R}{R-r}$ under the action of the force T and the force of friction F, Newton's second law gives us the following equation for the force F

$$F = T - Ma\frac{R}{R-r}$$

or

$$F = \frac{Mmg\left(\frac{R}{R-r}\right)^2\left(2 - \frac{R-r}{R}\right)}{m + 2M\left(\frac{R}{R-r}\right)^2} = \frac{Mmg\left(1 + \frac{r}{R}\right)}{m\left(1 - \frac{r}{R}\right)^2 + 2M}$$

The friction force of rest cannot exceed the value kMg. For this reason slipping occurs when

$$\frac{Mmg\left(\frac{R}{R-r}\right)^2\left(1 + \frac{r}{R}\right)}{m + 2M\left(\frac{R}{R-r}\right)^2} > kMg$$

or

$$k < \frac{1 + \frac{r}{R}}{2\frac{M}{m} + \left(1 - \frac{r}{R}\right)^2}$$

216. The centre of gravity of the spool will not be displaced if the tension of the thread satisfies the equation $T = Mg\sin\alpha$

Let us find the acceleration of the weight with a mass m to determine the tension of the thread T. Let the weight lower through a distance h. Since, according to the initial condition, the centre of gravity of the spool should remain at rest, the change in the potential energy is equal to mgh. If v is the velocity with which the weight having a mass m moves, the velocity of the points on the spool at the distance R from the axis of rotation is $v\dfrac{R}{r}$.

Hence, the kinetic energy of the system is

$$E_k = \frac{mv^2}{2} + \frac{Mv^2}{2}\frac{R^2}{r^2}$$

It follows from the law of conservation of energy that

$$\left(m + M\frac{R^2}{r^2}\right)\frac{v^2}{2} = mgh$$

or

$$v = \sqrt{\frac{2mgh}{m + M\dfrac{R^2}{r^2}}}$$

Therefore, the acceleration of the weight is

$$a = \frac{mg}{m + M\dfrac{R^2}{r^2}}$$

If we know the acceleration of the weight, we can find the tension of the thread

$$T = m(g - a) = mg\frac{M\dfrac{R^2}{r^2}}{M\dfrac{R^2}{r^2} + m}$$

Thus, for $\sin\alpha$ we get the expression

$$\sin\alpha = \frac{1}{\dfrac{M}{m} + \dfrac{r^2}{R^2}}.$$

The centre of gravity of the spool can be at rest only if

$$\frac{M}{m} + \frac{r^2}{R^2} \geq 1$$

217. If the velocity of the board is v, the velocity of the centre of gravity of each roll will be $v/2$ (see Problem 173). The kinetic energy of the system (the board and the two rolls) is

$$\frac{Mv^2}{2} + \frac{2mv^2}{4} = \frac{M + m}{2}v^2$$

By equating the kinetic energy to the work of the force Q over the distance s, we obtain

$$\frac{M + m}{2}v^2 = Qs, \text{ and } v = \sqrt{\frac{2Qs}{M + m}}$$

(Fig. 369). (The forces of friction perform no work since there is no slipping.)

It follows from the expression for the velocity of the board that its acceleration is

$$a = \frac{Q}{M + m}$$

To determine the force of friction applied to the board from the side of a roll, let us write the equation of motion of the board $Ma = Q - 2F$. Upon inserting the value of the acceleration a in this equation, we get

$$F = \frac{mQ}{2(M + m)}$$

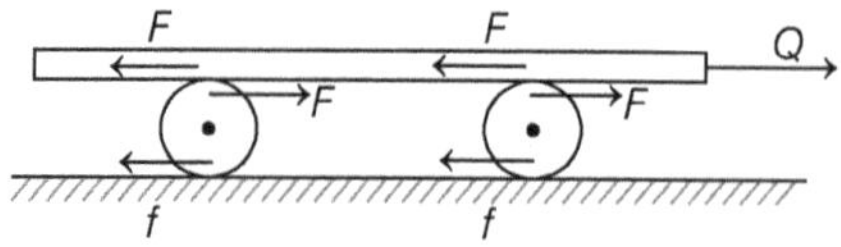

Fig. 369

Since the velocity of the centre of gravity of the roll is half that of the board, the respective accelerations will be in the same ratio. Therefore, the equation of motion of the centre of gravity of a roll will have the form:

$$m\frac{a}{2} = F - f$$

It follows from this equation that $f = 0$.

218. Let us assume, to introduce determinancy, that $m_1R > m_2r$. In this case the first weight will lower and the second one will rise. If the first weight lowers through a distance of h, the other one will rise through $h\frac{r}{R}$.

The decrease in the potential energy will be

$$m_1gh - m_2gh\frac{r}{R} = gh\left(m_1 - m_2\frac{r}{R}\right)$$

If the absolute velocity of the first weight is v, that of the second weight will be $v\frac{r}{R}$.

All the points of the first step of the pulley move with the velocity v, and those of the second step with the velocity $v\frac{r}{R}$. The kinetic energy of the system will be

$$\frac{m_1 + M_1}{2}v^2 + \frac{m_2 + M_2}{2}\frac{r^2}{R^2}v^2$$

It follows from the law of conservation of energy that

$$\frac{m_1 + M_1}{2} v^2 + \frac{m_2 + M_2}{2} \frac{r^2}{R^2} v^2 = \left(m_1 - m_2 \frac{r}{R}\right) gh$$

or

$$v = \sqrt{\frac{2\left(m_1 - m_2 \frac{r}{R}\right) gh}{(m_1 + M_1) + (m_2 + M_2)\frac{r^2}{R^2}}}$$

Therefore, the acceleration of the first weight is

$$a_1 = \frac{m_1 - m_2 \frac{r}{R}}{(m_1 + M_1) + (m_2 + M_2)\frac{r^2}{R^2}} g$$

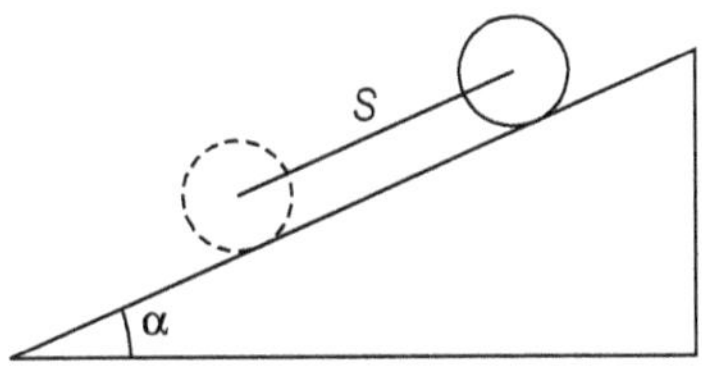

Fig. 370

From the ratio $\frac{a_1}{a_2} = \frac{R}{r}$, where a_2 is the acceleration of the second weight, we can find that

$$a_2 = \frac{\left(m_1 - m_2 \frac{r}{R}\right) g \frac{r}{R}}{(m_1 + M_1) + (m_2 + M_2)\frac{r^2}{R^2}}$$

On the basis of Newton's second law, the tensions of the strings T_1 and T_2 are equal to:

$$T_1 = \frac{M_1 + m_2 \frac{r}{R} + \frac{r^2}{R^2}(m_2 + M_2)}{m_1 + M_1 + (m_2 + M_2)\frac{r^2}{k^2}} m_1 g$$

$$T_2 = \frac{m_1 + M_1 + \frac{r}{R}\left(m_1 + M_2 \frac{r}{R}\right)}{m_1 + M_1 + (m_2 + M_2)\frac{r^2}{R^2}} m_2 g$$

The force F which the system acts on the axis of the pulley with is

$$F = T_1 + T_2 + (M_1 + M_2)g$$

219. Let the path travelled by the centre of gravity of the cylinder during the time t be equal to s, and the velocity of the centre of gravity be v at his moment (see Fig. 370).

Hence, from the law of conservation of energy we have

$$Mv^2 = Mgs\sin\alpha$$

Thus, the velocity is $v = \sqrt{gs\sin\alpha}$ and the acceleration $a = \frac{g\sin\alpha}{2}$.

The velocity of the centre of gravity of the cylinder and the angular velocity of its rotation will be

$$v = \frac{g\sin\alpha}{2}t$$

and
$$\omega = \frac{g\sin\alpha}{2R}t$$

1-9. The Law of Gravitation

220. According to Newton's second law, $m_i g = F$, where m_i is the inertial mass—a quantity that characterizes the ability of bodies to acquire an acceleration under the influence of a definite force.

On the other hand, according to the law of gravitation $F = \gamma\frac{m_g M_g}{R^2}$, where γ is the gravity constant and m_g and M_g are the gravitational masses of the interacting bodies. The gravitational mass determines the force of gravity and, in this sense, can be referred to as a gravitational charge.

It is not obvious in advance that $m_i = m_g$. If this equation (proportionality is sufficient) is satisfied, however, the gravity acceleration is the same for all bodies since, when the gravity force is introduced into Newton's second law, the masses m_i and m_g can be cancelled, and g will be equal to $\gamma\frac{M}{R^2}$.

Identical accelerations are imparted to all bodies irrespective of their masses, only by the force of gravity.

221. The gravitational acceleration $g = \gamma\frac{M}{R^2}$ (see Problem 220). Assuming that $g = 982$ cm/s, we find that $\gamma = 6.68 \times 10^{-8}$ cm^3 g^{-1} s^{-2}.

222. The bodies inside the spaceship will cease to exert any pressure on the walls of the cabin if they have the same acceleration as the spaceship. Only the force of gravity can impart an identical acceleration in this space to all the bodies respective of their mass. Consequently, it is essential that the engine of the spaceship be shut off and there be no resistance of the external medium. The spaceship may move in any direction with respect to that of the force of gravity.

223. The force of gravity imparts the same acceleration to the pendulum and the block. Gravity does not cause any deformations in the system during free falling. For this reason the pendulum will so move with respect to the block as if there is no gravitation (see the solution to Problem 222). The pendulum will move with a constant angular velocity as long as the block falls.

224. On section *BCA* (Fig. 371) the force of gravity performs positive work (the angle θ_1 is acute) and the velocity of the planet will increase, reaching its maximum at point *A*.

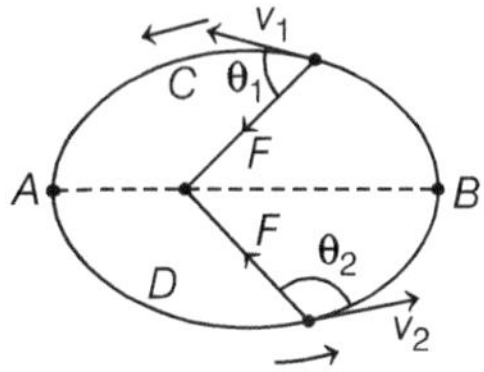

Fig. 371

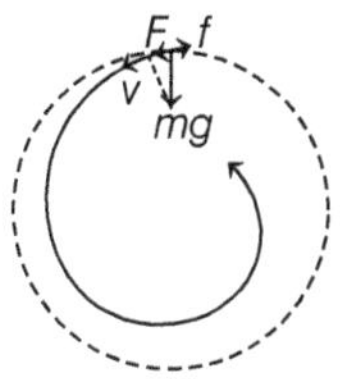

Fig. 372

On section *ADB* the force of gravity performs negative work (the angle θ_2 is obtuse) and the velocity of the planet will decrease and reach its minimum at point *B*.

225. For the satellite to move along a closed orbit (a circle with a radius $R + h$) should be acted upon by a force directed toward the centre. In our case this is the force of the Earth's attraction. According to Newton's second law,

$$\frac{mv^2}{R+h} = \gamma\frac{mM}{(R+h)^2}$$

where M is the mass of the Earth, $R = 6370$ km is the radius of the Earth, and γ is the gravitational constant.

At the Earth's surface

$$\gamma\frac{mM}{R^2} = mg$$

Therefore,

$$v = \sqrt{\frac{gR^2}{R+h}} \cong 7.5 \text{ km/s}$$

226. The resistance of the atmosphere will cause the satellite to gradually approach the Earth and the radius of its orbit diminishes.

Since this resistance is small in the upper layers, the decrease of the radius is insignificant during one revolution. Considering the orbit to be approximately circular, we can write

$$\frac{mv^2}{R} = \gamma\frac{mM}{R^2}$$

where R is the radius of the orbit. Therefore, $v = \sqrt{\frac{\gamma M}{R}}$, i.e., the velocity of the satellite increases with a reduction of R.

This can be illustrated as follows. In view of the atmospheric resistance, a satellite placed, for example, Into a circular orbit (dotted line in Fig. 372) will actually move along a certain helix (solid line in Fig. 372). For this reason the projection of the force of gravity F onto the direction of the satellite velocity v differs from zero. It is the work of the force F (greater than the atmospheric resistance f) that increases the velocity.

When the satellite moves in the atmosphere, its total mechanical energy diminishes but, as the Earth is approached, the potential energy drops faster than the total energy, causing the kinetic energy to grow.

It should be stressed that the high force of resistance in the dense layers of the atmosphere does not allow us to consider, even approximately, the motion of the satellite as rotation along a circle, and our conclusion is not correct.

227. If the container is thrown in the direction opposite to the motion of satellite A. It will begin to move along a certain ellipse 2 inside the orbit of the satellite (Fig. 373). The period of revolution of the container will be slightly less than that of satellite B. Therefore they can meet at the point of contact between the orbits only after a great number of revolutions.

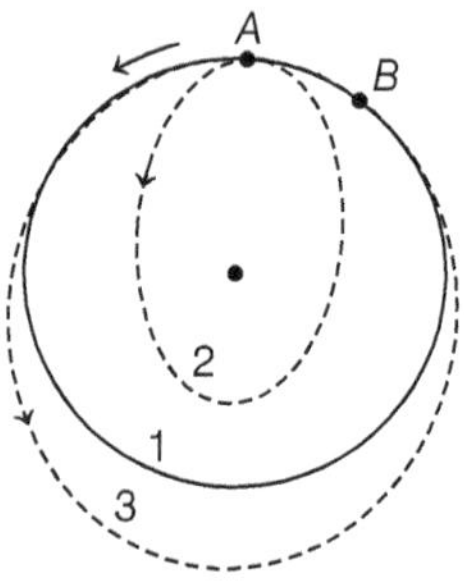

Fig. 373

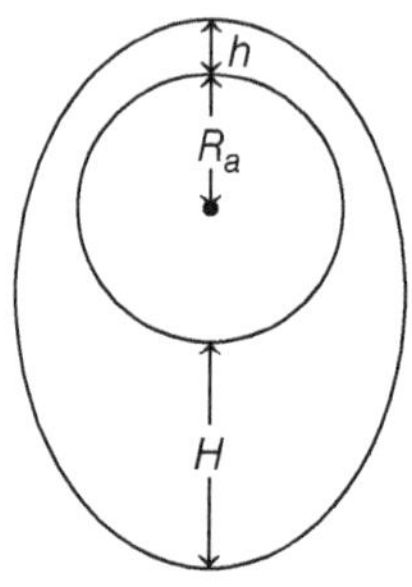

Fig. 374

The container should be thrown in the direction of motion of satellite A. It will begin to move along ellipse 3. The velocity u should be such that during one revolution of the container, satellite B also makes one revolution and in addition covers the distance AB. This is quite possible, since the period of revolution along ellipse 3 is somewhat greater than along circular orbit 1. The container will meet the satellite at the point where orbits 3 and 1 coincide.

228. Assuming the Earth's orbit to be approximately circular, the force of gravity can be determined by the equation $F = m\omega^2 R$, where m is the mass of the Earth, and $\omega = \frac{2\pi}{T}$ is its angular velocity ($T = 365$ days).

On the other hand, according to the law of gravitation, $F = \gamma \frac{mM}{R^2}$, where M is the mass of the Sun. Hence,

$$\gamma \frac{mM}{R^2} = m\omega^2 R \quad \text{or} \quad M = \frac{\omega^2 R^3}{\gamma} \cong 2 \times 10^{33}\,\text{g}$$

229. Since both the Moon and the satellite move in the field of gravity of the Earth, let us use Kepler's third law

$$\frac{T_1^2}{T_2^2} = \frac{(h + H + 2R_0)^3}{8R^3}$$

(Fig. 374). Therefore,

$$h = 2R\left(\frac{T_1}{T_2}\right)^{2/3} - H - 2R_0 = 220\ \text{km}$$

230. Since the mass of the ball is greater than the mass of the water in the same volume, the field of gravity will be greater near the ball than away from it. Therefore, the water near the ball will be compressed additionally. The pressure of the water acting on the bubble from the left will be smaller than the pressure that acts from the right. On the other hand, the force of gravity between the air in the bubble and the ball is greater than the force of attraction between the air and the volume of the water shown by dotted circle a (Fig. 375).

Since the mass of the air in the bubble is very small, the first factor becomes decisive. The bubble will be repulsed from the ball. Conversely, the motion of the iron ball will be determined by the fact that the force of attraction between the air in the bubble and the ball is much less than that between the ball and the volume of the water shown by dotted circle b.

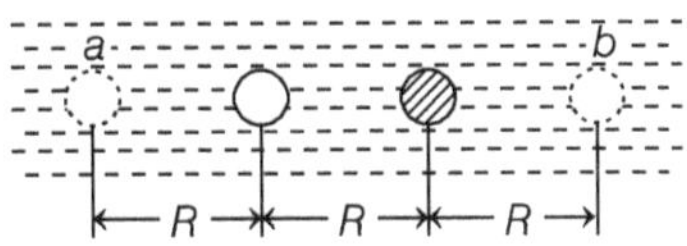

Fig. 375

To calculate the force, let us reason as follows. A homogeneous medium (water) contains a sphere almost devoid of mass (the bubble) and a sphere with an excess mass (the ball). From a formal standpoint, this can be regarded as the presence of negative and positive masses.

The force of interaction between the spheres in the water is equal to the interaction in vacuum of a negative mass equal to the mass of the water in the bubble and *a* positive mass equal to the mass of the iron ball that exceeds the mass of the water in the same volume. Therefore,

$$F = -\gamma \frac{m_1 (m_2 - m_1)}{R^2}$$

Here m_1 is the mass of the water in a sphere with a radius r, and m_2 is the mass of the iron ball.

231. The field of gravity is smaller near the bubble than in a homogeneous liquid, and the liquid is compressed less. For this reason one bubble will move into the volume of liquid near the other one, and the bubbles will be mutually attracted.

Two bubbles in a homogeneous liquid with negligibly small masses can be considered from a formal standpoint as negative masses superimposed upon the positive mass m of the medium in the volume of a bubble:

$$F = \gamma \frac{(-m)(-m)}{R^2} = \gamma \frac{m^2}{R^2}$$

232. If the ball were solid, the force of gravity $F_1 = \gamma \frac{Mm}{l^2}$, where $M = \frac{4}{3}\pi R^3 \rho$ is the mass of the ball without a spherical space. The presence of this space is equivalent to the appearance of a force of repulsion $F_2 = \gamma \frac{m' m}{s^2}$, where $m' = \frac{4}{3}\pi r^3 \rho$, and s is the distance between the centre of the space and the material particle.

The sought force F is the geometrical sum of the forces F_1 and F_2 (Fig. 376).

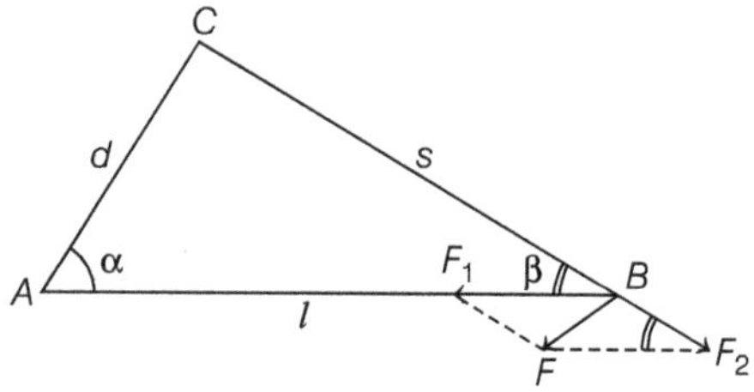

Fig. 376

According to the cosine rule,

$$F = \sqrt{F_1^2 + F_2^2 - 2F_1F_2\cos\beta} = \frac{4}{3}\pi\gamma m\rho$$

$$\sqrt{\frac{R^6}{l^4} + \frac{r^6}{(l^2 - d^2)^2} - \frac{2R^3 r^3 \cos\beta}{l^2(l^2 - d^2)}} \cong 5.8 \times 10^{-4} \text{ gf}$$

233. The sought force of attraction is the geometrical sum of the forces created by separate elements of the sphere. The small elements σ_1, and σ_2 (Fig. 377) are cut out of the sphere as cones with vertices at point A obtained when the generatrix BC revolves around axis S_1S_2. The areas of the elements are $\frac{(AS_1)^2\omega}{\cos\alpha_1}$ and $\frac{(AS_2)^2\omega}{\cos\alpha_2}$, respectively, and their masses are $\frac{(AS_1)^2\omega\rho}{\cos\alpha_1}$ and $\frac{(AS_2)^2\omega\rho}{\cos\alpha_1}$, where ω is the solid angle at which both elements can be seen from point A; ρ is the surface density of the sphere (the mass per unit of surface); $\angle\alpha_1 = \angle\alpha_2$, since S_1OS_2 is an isosceles triangle. The forces of attraction created by the elements are equal, respectively, to:

$$\gamma\frac{m(AS_1)^2\omega\rho}{(AS_1)^2\cos\alpha_1} = \gamma\frac{m\omega\rho}{\cos\alpha_1} \text{ and } \gamma\frac{m(AS_2)^2\omega\rho}{(AS_2)^2\cos\alpha_2} = \gamma\frac{m\omega\rho}{\cos\alpha_2}$$

where m is the mass of the body. These forces act in opposite directions and their resultant is zero.

Reasoning in the same way for the other corresponding elements of the sphere, we convince ourselves that all of them compensate one another in pairs.

Therefore, the force of attraction acting from the sphere on the body inside it is zero. It should be noted that this result is also true for a sphere with a finite thickness; it can be divided into any arbitrary number of thin spherical shells, for each of which, the assumption proved above will be true.

234. The force of attraction will be equal to the force with which the body with the mass m is attracted to a sphere with a radius r and a density ρ. The outer layers of the Earth do not act on the body, as was proved in Problem 233.

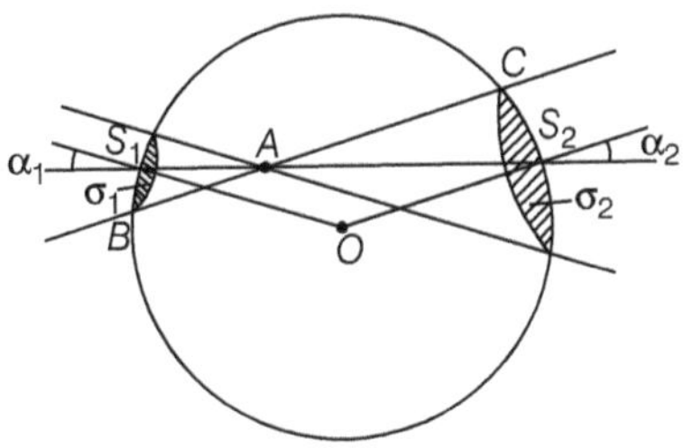

Fig. 377

Therefore, the sought force $F = \gamma\frac{\frac{4\pi}{3}\rho r^3 m}{r^2} = \gamma\frac{4\pi}{3}\rho m r$

This force decreases in proportion to r as the centre of the Earth is approached.

1-10. Hydro- and Aerostatics

235. The level of the water will not change because the quantity of water displaced will remain the same.

236. Equilibrium will not be violated, since according to Pascal's law the pressure on the bottom of the vessel will be the same everywhere.

237. (1) Since the piece of ice floats, the weight of the water displaced by it is equal to the weight of the ice itself or the weight of the water it produces upon melting. For this reason the water formed by the piece of ice will occupy a volume equal to that of the submerged portion, and the level of the water will not change.

(2) The volume of the submerged portion of the piece with the stone is greater than the sum of the volumes of the stone and the water produced by the melting ice. Therefore, the level of the water in the glass will drop.

(3) The weight of the displaced water is equal to that of the ice (the weight of the air in the bubble may be neglected). For this reason, as in case (1), the level of the water will not change.

238. In the first case the weight of the body submerged into the liquid is $G_1 = (\gamma - \gamma_1)V$ and in the second $G_2 = (\gamma - \gamma_2)V$, where V is the volume of the body.

Therefore,

$$\gamma = \frac{G_2\gamma_1 - G_1\gamma_2}{G_2 - G_1}$$

239. The ice can be supported by the edge of the shore only in small ponds. It will always float in the middle of a large lake. The ratio between the densities of the ice and the water is 0.9. Therefore, nine-tenths of the entire thickness of the ice will be in the water. The distance from the surface of the ice to the water is one metre.

240. After the stone is taken out, the match-box becomes lighter by the weight of the stone, and, consequently, the volume of water the box displaces decreases by $V_1 = G/\gamma_1$, where G is the weight of the stone and γ_1 the specific weight of the water. When dropped into the water, the stone will displace a volume of water equal to its own volume $V_2 = G/\gamma_2$, where γ_2 is the specific weight of the material of the stone. Since $\gamma_2 > \gamma_1$, then $V_1 > V_2$. Therefore, the level of the water in the cup will lower.

241. In both cases the pumps perform identical work, since the same amount of pumped in water rises to the same level.

242. The inverted L will be stable on the bottom of the empty vessel, since a perpendicular dropped from the centre of gravity of the figure is within the limits of the supporting area. As water is poured into the vessel, the expulsion force acting on the rectangle will grow in magnitude (it is assumed that the water can flow underneath the figure). When the depth of the water in the vessel is $0.5a$, the sum of the moments of the forces which tend to turn the body clockwise will be equal to the sum of those acting in the opposite direction. As the vessel is filled further, the figure will fall.

243. The length of the tube x can be found from the condition $\gamma x = \gamma_0 (x - h)$, which shows the equality of the pressures at the depth of the lower end of the tube. Here γ_0 is the specific weight of the water.

Hence,

$$x = \frac{\gamma_0 h}{\gamma_0 - \gamma} = 50 \text{ cm}$$

244. Let us separate a column with a height of h inside the water (Fig. 378). The equation of motion of this column has the form: $ma = mg - pA$, where $m = \rho A h$ is the mass of the water and p is the pressure at the depth h.

Therefore, $p = \rho h(g - a)$.

245. In accordance with the solution of Problem 244, the force of expulsion can be written as follows: $F = \rho V(g - a)$, where V is the volume of the submerged portion of the body. The equation of motion of a floating body with a mass M has the form:

$$Ma = Mg - \rho V(g - a).$$

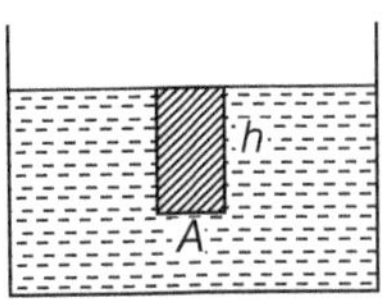

Fig. 378

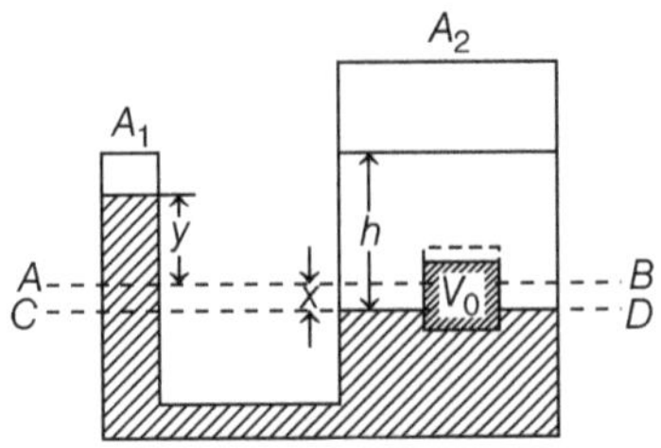

Fig. 379

Hence, $V = \dfrac{M}{\rho}$, as in a stationary vessel, and the body does not rise to the surface.

246. If the tank were at rest or moved uniformly, the pressure at the depth h would be equal to $p_1 = \rho g h$.

On the other hand, if the tank moved with an acceleration and the force of gravity were absent, the pressure at point A would be equal to $p_2 = \rho al$. It is this pressure that would impart, in conformity with Newton's second law, the required acceleration a to the column of water with a length l.

In accelerated motion of the tank, both the pressure p_1 and the pressure p_2 appear in the field of gravity. According to Pascal's law, the pressure in the water is the same in all directions. For this reason the pressures p_1 and p_2 are summated, and the resulting pressure at point A is $p = \rho(gh + al)$.

247. Using the law of conservation of energy and Archimedes' principle, we obtain the following equation

$$mgx = \left(\frac{4}{3}\pi R^3\rho - m\right)gh$$

where ρ is the density of the water and x is the sought height.

Hence,

$$x = \frac{\left(\frac{4}{3}\pi R^3\rho - m\right)h}{m}$$

248. If the level of the water in the vessel is the same, the level of the mercury will also be the same before the piece of wood is dropped.

Dropping of the piece of wood gives the same result as adding of the amount of water that will be displaced by this piece, i. e., the amount of water equal to it by weight. Therefore, if the cross sections of the vessels are the same, the levels of the water and the mercury in both vessels will coincide.

If the cross sections are different, the water will be higher and the mercury lower in the vessel with a smaller cross section. This will occur because the pressure on the surface of the mercury will increase differently when amounts of water equal in weight and volume are added to the vessels with various cross sections.

249. After the block is dropped into the broad vessel, the level of the mercury in both vessels will rise by the amount x and occupy position AB (Fig. 379).

The required height of the water column in the broad vessel is determined by the equality of the pressures, for example, at level CD

$$(y + x)\rho_1 g = h\rho_2 g$$

where ρ_1 is the density of the mercury and ρ_2 that of the water. The value of y can be found by using the condition that the volume of the mercury is constant:

$$(x + y)\,A_1 = V_2$$

where V_2 is the volume of the mercury displaced by the block after the water is poured in.

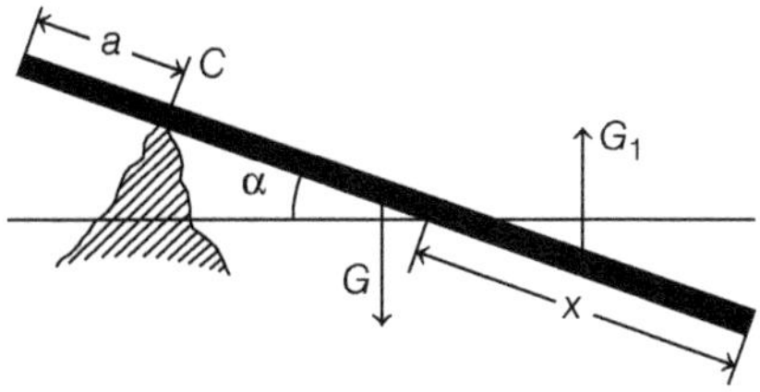

Fig. 380

If the water covers the block entirely, then according to Archimedes' principle, $V_0\rho_0 g = V_2\rho_1 g + (V_0 - V_2)\rho_2 g$, where ρ_0 is the density of iron. Solving the equations, we obtain

$$h = \frac{\rho_1(\rho_0 - \rho_2)V_0}{\rho_2(\rho_1 - \rho_2)A_1}$$

If the water does not cover the block, Archimedes' principle can be written as $V_0\rho_0 g = V_2\rho_1 g + hA\rho_2 g$, where $A = V_0^{2/3}$ is the area of a block face. In this case the sought height

$$h = \frac{\rho_0 V_0}{\rho_2(A_1 + V_0^{2/3})}$$

The first solution is true when $A_1 \le \frac{\rho_1(\rho_0 - \rho_2)}{\rho_2(\rho_1 - \rho_0)} V_0^{2/3}$ and the second one when

$$A_1 \ge \frac{\rho_1(\rho_0 - \rho_2)}{\rho_2(\rho_1 - \rho_0)} V_0^{2/3}.$$

250. It follows from the equality of the moments of the forces acting on the board with respect to point C (Fig. 380) that

$$G_1\left(l - a - \frac{x}{2}\right)\cos\alpha = G\left(\frac{l}{2} - a\right)\cos\alpha$$

where $G_1 = Ax\gamma_0$ and $G = Al\gamma$. Here A is the cross-sectional area of the board and γ_0 the specific weight of the water.

Hence,

$$x = (l - a) \pm \sqrt{(l-a)^2 - \frac{\gamma}{\gamma_0} l(l - 2a)}$$

Since $x < l - a$, only one solution is valid:

$$x = (l - a) - \sqrt{(l-a)^2 - \frac{\gamma}{\gamma_0} l(l - 2a)}$$

251. The pressure on the "bottom" of the vessel is $\rho g h$ and the force with which the hatched portion of the liquid (Fig 381) presses on the table is $\rho g h\pi(2Rh\tan\alpha - h^2\tan^2\alpha)$. According to Newton's third law, an identical force acts on the liquid. The condition of equilibrium of the liquid at the moment when the vessel stops exerting pressure on the table has the form

$$G + G_1 = \rho g h \pi\,(2Rh\tan\alpha - h^2\tan^2\alpha)$$

where G_1 is the weight of the hatched portion of the liquid (the truncated cone minus the cylinder volume)

$$G_1 = \frac{\rho g h}{3}\{\pi R^2 + \pi(R - h\tan\alpha)^2 + \pi R(R - h\tan\alpha)\} - \rho g h \pi (R - h\tan\alpha)^2$$

Therefore,

$$\rho = \frac{G}{\pi g h^2 \tan\alpha\left(R - \frac{h\tan\alpha}{3}\right)}$$

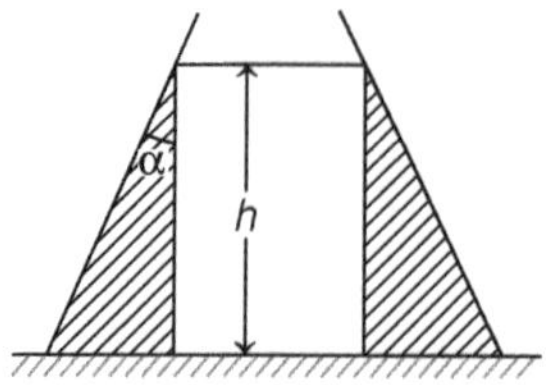

Fig. 381

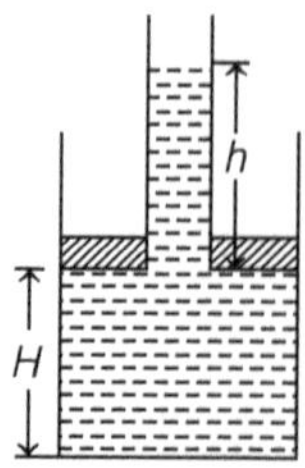

Fig. 381

252. A liquid moves in a syphon by the action of the forces of cohesion between the elements of the liquid. The liquid in the long elbow outweighs the liquid in the short one, and pumps it over. It could be assumed on this basis that water can be pumped over a wall of any height with the aid of a syphon. This is not so, however. At a lifting height of 10 metres the pressure inside the liquid becomes zero. The air bubbles always present in water will begin to expand, and the water column will be broken, thus stopping the action of the syphon.

253. The device will first act as a syphon and the water will flow through the narrow pipe into the reservoir. Then an air bubble will slip through A and divide the liquid in the upper part into two portions. After this the liquid will no longer flow out.

254. The pressure of the water directly under the piston of each pump is less than atmospheric pressure by $\rho g(H + h)$, where ρ is the density of water. Therefore, to keep the piston in equilibrium, it should be pulled upward with the force $F = \rho g(H + h)A$, where A is the area of the piston.

Hence, a greater force should be applied to the pistons with a greater area.

255. The pressure on the bottom is $p = \rho g(H + h)$ (Fig. 382). On the other hand, since the vessel is a cylinder,

$$p = \frac{G + mg}{\pi R^2}.$$

The height h can be determined if we equate the forces acting on the piston

$$\rho g h \pi (R^2 - r^2) = G$$

Hence,

$$H = \frac{1}{\pi R^2 \rho}\left(m - \frac{G}{g}\frac{r^2}{R^2 - r^2}\right) \cong 10\,\text{cm}$$

256. To prevent flowing out of the water, the vessel should be given such an acceleration at which the surface of the water takes the position shown in Fig. 383. The maximum volume of the water is $\frac{bcA}{2l}$ and the mass of the entire system is $M + \frac{bcA}{2l}\rho$.

The required acceleration can be found from the condition that the sum of the forces acting on a small element of the water with a mass Δm near the surface is directed horizontally (Fig. 383).

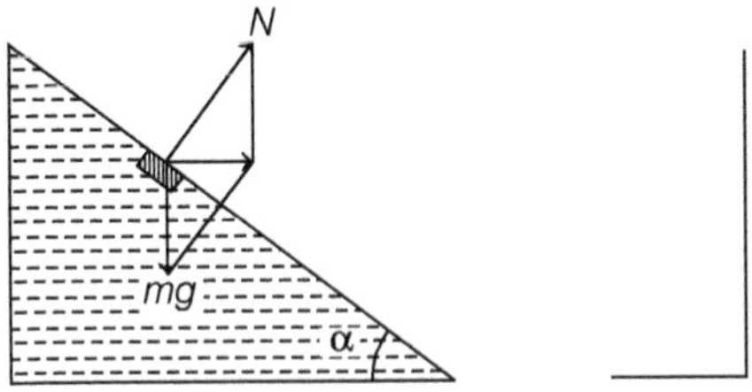

Fig. 383

According to Newton's second law,

$$\Delta m a = \Delta m g \tan \alpha.$$

Therefore, the sought force is

$$F = \left(M + \frac{bcA}{2l}\rho\right) g \frac{b}{c}$$

257. The lower part of a chamber is filled with denser air. The air leaves the chambers in the upper part. The pressure is gradually equalized. The machine will continue to function only as long as the difference in the pressures between the two halves of the vessel is sufficient to raise the water along the tube to the upper portion.

258. Here the disk is not asymmetrical and the air pressure on the right-hand side of the disk will be greater than on the left-hand one. The surplus force of the pressure acting on the right-hand side of the disk is $F = (p_1 - p_2)A$, where A is the cross-sectional area of the chamber. The weight of the chambers filled with water cannot exceed $G = \rho gAh$. Since $h \le \frac{G_1 - G_2}{\rho g}$, then $F \ge G$.

The disk will begin to rotate counterclockwise, and the chambers will rise from the bottom of the vessel to the top filled with air. The disk will rotate counterclockwise until the reduced difference of the pressure can no longer lift the water to the height h.

259. The bottom of the cylindrical vessel will fall off in all three cases, since the pressure exerted on the bottom from the top will always be equal to 1 kgf. In the vessel narrowing upward the bottom will fall off only when oil is poured in, since its level here will be higher than in the cylindrical vessel. In the vessel widening toward the top the bottom will fall off when the mercury is poured in, because its level will be higher than in the cylindrical vessel. This will also occur when the weight is put in. In this case the weight will be distributed over a smaller area than in the other two cases.

260. The reading of the balance will increase if the mean density of the body being weighed is less than the density of the weights. The reading will decrease if the mean density of the body is greater than that of the weights. The equilibrium of the balance will not be disturbed if the weights and the body have the same mean density.

261. The man's aim will not be achieved, since, while increasing the explosive force, he will also considerably increase the weight of the tube (the density of the compressed air in the tube is greater than that of the atmospheric air).

262. The true weight of the body is

$$G = G_1 + \gamma_0\left(V - \frac{G_1}{\gamma_1}\right) \cong 801.16 \text{ gf}$$

The error is equal to

$$\frac{G - G_1}{G} 100\% \cong 0.14\%$$

263. When the atmospheric pressure changes, the Archimedean force acting on the barometers from the side of the air will change owing to the change in the density of the air and in the volume of the barometers when the level of the mercury changes in their open parts.

If all the conditions in the problem are taken into account, the barometers have not only the same weight, but also the same volume. For this reason the change in the expulsion force due to the first reason will be identical for each of them. The change in the volumes will be different. To change the difference in the levels by a certain amount in a U-shaped barometer, the level of the mercury in each elbow should change by only half of this amount. In a cup barometer the level of the mercury in the cup changes negligibly, and in the tube practically by the entire change in the difference of the levels. Here the change of the volume of the mercury in the tube should be the same as in the cup.

Therefore, the change of the volume in the cup barometer will also become less, and it will therefore outweigh the U-shaped barometer.

264. The normal atmospheric pressure is approximately equal to 1 kgf/cm^2. This means that an atmospheric column of air with an area of 1 cm^2 weighs 1 kgf. If we know the surface of the Earth, we can find the weight of its atmosphere.

The Earth's surface $A = 4\pi R^2$, where $R = 6{,}370$ km is the mean radius of the Earth.

The weight of the atmosphere $G \cong 4\pi R^2 \times 1\ \text{kgf} / \text{cm}^2 \cong 5 \times 10^{15}$ ton f.

265. If the man stands on the mattress, his weight will be distributed over a smaller area (that of his feet) than if he lies down. Therefore, the state of equilibrium will set in in the first case at a higher air pressure in the mattress than in the second.

266. Let us first consider the tube inflated with air (Fig. 384*a* shows a cross section of the tube). For sections *AB* and *CD* of the tube to be in equilibrium it is obviously necessary that the tension of the expanded walls of the tube *T* balance the excess pressure inside the tube *p*.

Let us now consider the forces that act on sections *AB* and *CD* when the tube is fitted onto a loaded wheel (Fig. 384*b*). The distribution of the forces acting on *AB* does not appreciably change in the top of the tube. There will be a difference in the bottom. Section *CD* will be acted upon by an elastic force from the side of the rim equal to the load applied to the wheel (the weight of the wheel and one-fourth of the weight of the motor vehicle).

This additional force flattens the tube and the angle between the forces *T* tensioning the rubber increases. The total force of tension acting on *CD* diminishes and the excess pressure of the air in the tube equalizes the force of tension, and also the weight of the wheel and of part of the motor vehicle.

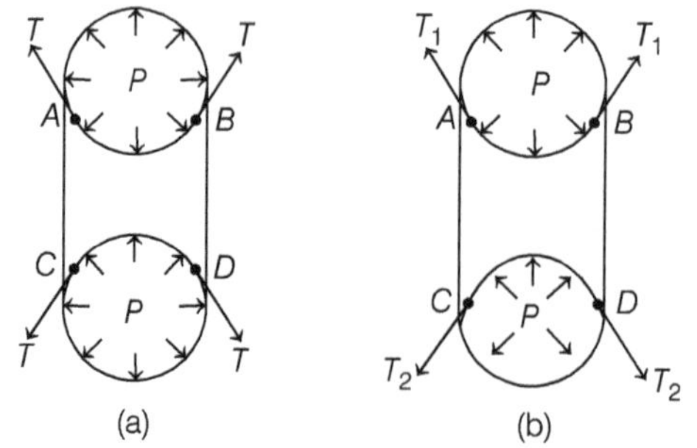

Fig. 384

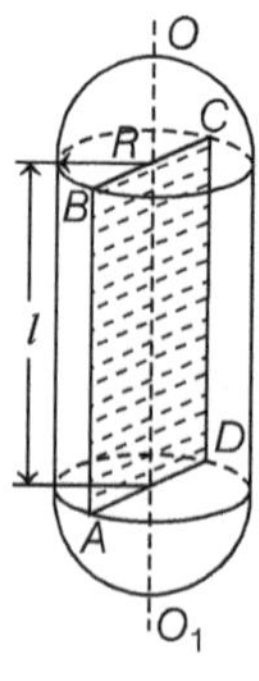

Fig. 385

Thus, the rim does not lower because it is supported by the excess pressure of the air in the tube. In the top of the tube the excess pressure is equalized by the tension of the tube walls, and in the bottom it equalizes both the reduced tension of the walls and the force applied to the wheel.

267. The force per unit of length with which the cylindrical portion of the boiler is stretched in a direction perpendicular to axis OO_1, is

$$f_1 = \frac{2Rl}{2l}p = pR$$

where $2Rl$ is the cross-sectional area $ABCD$ of the boiler and p is the pressure inside it (Fig. 385); $2Rlp$ is the force acting on one half of the cylinder (see Problem 122).

The maximum force per unit length of the hemispherical heads can be found from the formula

$$f_2 = \frac{\pi R^2}{2\pi R}p = \frac{pR}{2} = \frac{f_1}{2}$$

Consequently, the heads can withstand twice the pressure that the cylindrical portion of the boiler does if their walls are equally thick. For the strength of all the parts of the boiler to be identical, the thickness of the head walls may be half that of the cylindrical walls, i.e., 0.25 cm.

268. The shape of the boiler should be such that the force applied per unit length of its cross section is minimum. This force is $f = \frac{pA}{l}$ where A is the cross-sectional area of the boiler, l the perimeter of the section and p the pressure of the steam. The force f will be minimum with the smallest ratio between the cross-sectional area and the perimeter.

As is known, this ratio will be minimum for a circle. It is also known that a circle can be obtained by cutting a sphere with any plane. Therefore, a sphere is the most advantageous shape for the boiler.

269. The ceiling of a stratosphere balloon is determined not by the maximum altitude which it can ascend to, but by the altitude ensuring a safe velocity of landing. The envelope of a stratosphere balloon is filled with a light gas (hydrogen or helium) only partly, since when the balloon ascends, the gas in the envelope expands and forces out the air, making it possible to maintain the lifting force approximately constant. At a certain altitude. the gas will fill the entire envelope. Even after this the lifting force of the balloon continues to increase at the expense of the gas flowing out from the bottom hole in the envelope. The weight of the balloon decreases and it will reach its ceiling only after a certain amount of gas has leaked out.

For the balloon to descend, some gas should additionally be expelled through the upper valve in the envelope, so that the lifting force is only slightly smaller than the weight of the balloon. At a small altitude the velocity of descent will be too high, since the volume of the gas decreases and less of it remains in the balloon than during the ascent. The ballast is dropped to reduce the velocity of descent.

1-11. Hydro and Aerodynamics

270. Let us denote the distance from the level of the water to the upper hole by h, the sought distance from the vessel to the point where the streams intersect by x, and the distance from the level of the water in the vessel to this point by y (Fig. 386).

The point of intersection will remain at the same place if the level of the water in the vessel does not change. This will occur if $Q = Av_1 + Av_2$, where $v_1 = \sqrt{2gh}$ and $v_2 = \sqrt{2g(H + h)}$ are the outflow velocities of the streams from the holes.

On the basis of the laws of kinematics,

$$x = v_1 t_1 = v_2 t_2$$

and

$$y = h + \frac{gt_1^2}{2} = h + H + \frac{gt_2^2}{2}$$

where t_1 and t_2 are the times during which the water falls from the holes to the point of intersection.

Hence,

$$x = \frac{1}{2}\left(\frac{Q^2}{2gA^2} - H^2\frac{2gA^2}{Q^2}\right) = 120 \text{ cm}$$

$$y = \frac{1}{2}\left(\frac{Q^2}{2gA^2} + H^2\frac{2gA^2}{Q^2}\right) = 130 \text{ cm}$$

271. The velocity of water outflow from a hole is $v = \sqrt{2gh}$. The impulse of the force acting from the side of the vessel on the outflowing water $F\Delta t = \Delta m v$, where $\Delta m = \rho A v \Delta t$ is the mass of the water flowing out during the time Δt. Hence, $F = \rho v^2 A = 2\rho g h A$. The pressure at the bottom $p = \rho g h$ and therefore $F = 2pA$. The same force acts on the vessel from the side of the stream.

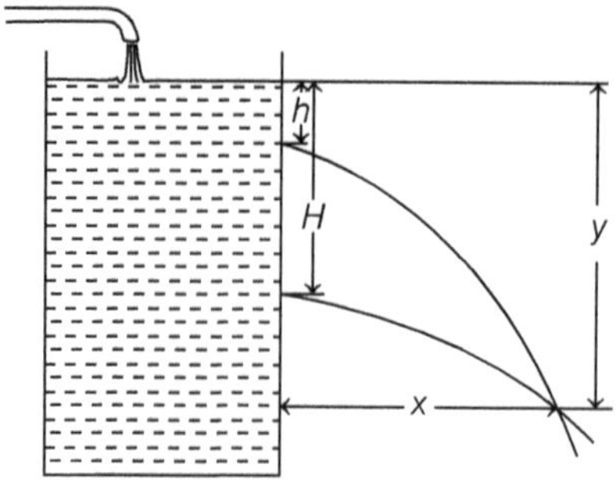

Fig. 386

Thus, the water acts on the wall with the hole with a force $2pA$ smaller than that acting on the opposite wall, and not with a force smaller by pA as might be expected. This is due to a reduction in the pressure acting on the wall with the hole, since the water flows faster at this wall.

The vessel will begin to move if $KG < 2pA$ or $k < \dfrac{2\rho g h A}{G}$

272. According to Newton's second law, the equality $pA_0 = 2pA$ should exist. Therefore, if the liquid flows out through the tube, the cross-sectional area of the stream should be halved

$$A = \frac{A_0}{2}$$

This compression of the stream can be explained as follows.

The extreme streamlets of the liquid approaching the tube from above cannot, in view of inertia, flow around the edge of the tube directly adhering to its walls, and move towards the centre of the stream. Under the pressure of the particles nearer to the centre of the stream, the lines of flow straighten out and a contracted stream of the liquid flows along the tube.

273. By neglecting splashing of the water, we thus assume the impact of the stream against the wall to be absolutely inelastic. According to Newton's second law, the change in the momentum of the water during the time Δt is $\Delta m v = F\Delta t$, where $\Delta m = \rho \dfrac{\pi d^2}{4} v \Delta t$ is the mass of the water flowing during the time Δt through the cross section of the pipe.

Hence, $$F = \frac{\rho\pi d^2}{4}v^2 \cong 8 \text{ gf}$$

274. When the gas flows along the pipe (Fig. 387), its momentum changes in direction, but not in magnitude.

A mass of ρAv passes in a unit of time through cross section I of the vertical part. This mass brings in the momentum $\mathbf{p}_1 = \rho Av\mathbf{v}_1$, where $\mathbf{v}_1$ is the vector of the velocity with which the gas flows in the vertical part, numerically equal to the given velocity v.

During the same time the momentum $\mathbf{p}_2 = \rho Av\mathbf{v}_2$, is carried away through cross section II. Here $\mathbf{v}_2$ is the vector of velocity in the horizontal part, also numerically equal to v.

The change in the momentum is equal to the impulse of the force F that acts from the side of the pipe on the gas: $\mathbf{F} = \rho Av(\mathbf{v}_2 - \mathbf{v}_1)$. In magnitude $F = \rho Av^2\sqrt{2}$.

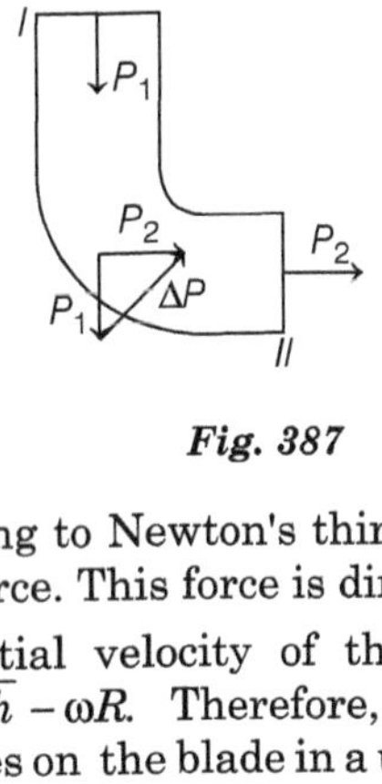

Fig. 387

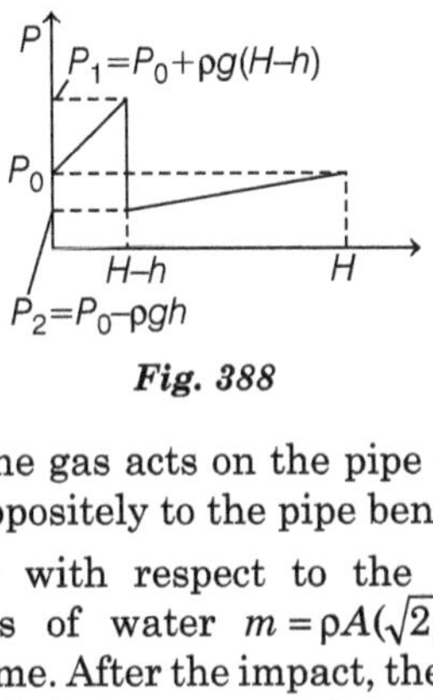

Fig. 388

According to Newton's third law, the gas acts on the pipe with the same force. This force is directed oppositely to the pipe bend.

275. The initial velocity of the water with respect to the blade is $v = \sqrt{2gh} - \omega R$. Therefore, a mass of water $m = \rho A(\sqrt{2gh} - \omega R)$ impinges on the blade in a unit of time. After the impact, the velocity of the water with reference to the blade is zero, and for this reason the change in the momentum of the water in a unit of time is mv. According to Newton's second law, the sought force is

$$F = \rho A(\sqrt{2gh} - \omega R)^2$$

276. At the first moment the ship will begin to move to the right, since the pressure on the starboard side diminishes by $2pA$, where p is the pressure at the depth h of the hole, and A is its area (see Problem 271). As soon as the stream of water reaches the opposite wall, this wall will be acted upon by the force $F = \rho Av^2$, where v is the velocity of the stream with respect to the ship (see Problem 273). The force F is slightly greater than $2pA$, since $v > \sqrt{2gh}$ because the ship moves towards the stream. As a result, the motion will begin to retard.

277. The velocity of the liquid in the pipe is constant along the entire cross section because the liquid has a low compressibility and the stream is continuous. This velocity is $v = \sqrt{2gH}$.

The velocity of the liquid in the vessel is practically zero, since its area is much greater than the cross-sectional area of the pipe.

Therefore, a pressure jump which we shall denote by $p_1 - p_2$ should exist on the vessel-pipe boundary. The work of the pressure forces causes the velocity to change from zero to $\sqrt{2gH}$.

On the basis of the law of conservation of energy,

$$\frac{\Delta m v^2}{2} = (p_1 - p_2)\, A\Delta h$$

where A is the cross-sectional area of the pipe, Δh is the height of a small element of the liquid, and $\Delta m = \rho A\Delta h$ is the mass of this element.

Hence,

$$\frac{\rho v^2}{2} = p_1 - p_2 = \rho g H$$

Since the flow velocity is constant, the pressure in the pipe changes according to the law

$$p = p_0 - \rho g(h - x)$$

as in a liquid at rest. Here p_0 is the atmospheric pressure and x is the distance from the upper end of the pipe.

The change of pressure in height is shown in Fig. 388. The pressure is laid off along the axis of ordinates, and the distance from the surface of the liquid in the vessel along the axis of abscissas.

278. The water flowing out of the pipe during a small interval of time Δt will carry with it the momentum $\Delta p = \rho A v^2 \Delta t$, where $v = \sqrt{2gH}$ is the velocity of the outflowing stream (see Problem 277). According to Newton's law, $F\Delta t = 2\rho g H A \Delta t$. The same force will act on the vessel with the water from the side of the outflowing stream. Therefore, the reading of the balance will decrease by $2\rho g H A$ at the initial moment.

279. At the first moment when the stream has not yet reached the pan, equilibrium will be violated. The pan will swing upwards since the water flowing out of the vessel no longer exerts pressure on its bottom.

When the stream reaches the pan, equilibrium will be restored. Let us consider an element of the stream with the mass Δm. This element, when falling onto the pan, imparts to it an impulse $\Delta m\sqrt{2gh}$ in a vertical direction, where h is the height of the cock above the pan. On the other hand, after leaving the vessel, this element will cease to exert pressure on its bottom and on the pan during the time of falling $t = \sqrt{\frac{2h}{g}}$. This is equivalent to the appearance of an impulse of force acting on the vessel vertically upward when the element of the liquid is falling. The mean value of this impulse during the duration of the fall is

$$\Delta mg\sqrt{\frac{2h}{g}} = \Delta m\sqrt{2gh}$$

Thus, each element of the liquid Δm is accompanied during its falling by the appearance of two equal and oppositely directed impulses of force. Since the stream is continuous, the balance will be in equilibrium.

At the moment when the stream stops flowing, the pan will swing down, since the last elements of the liquid falling on the pan act on it with a force that exceeds the weight of the elements, and there will no longer be a reduction in the pressure on the bottom of the vessel.

280. On the basis of the law of conservation of energy we can write

$$\frac{Mv^2}{2} = mgh$$

where M is the mass of the water in the tube stopped by valve V_2 and m is the mass of the water raised to the height h.

Therefore,

$$\frac{\rho l\pi d^2}{4} \times \frac{v^2}{2} = \rho V_0 gh$$

where V_0 is the volume of mass m.

The average volume raised in two seconds is

$$V_0 = \frac{l\pi d^2 v^2}{8gh} = 1.7 \times 10^{-3}\ \text{m}^3$$

One hour of ram operation will raise

$$V = 1.7 \times 10^{-3} \times 30 \times 60 \cong 3\ \text{m}^3$$

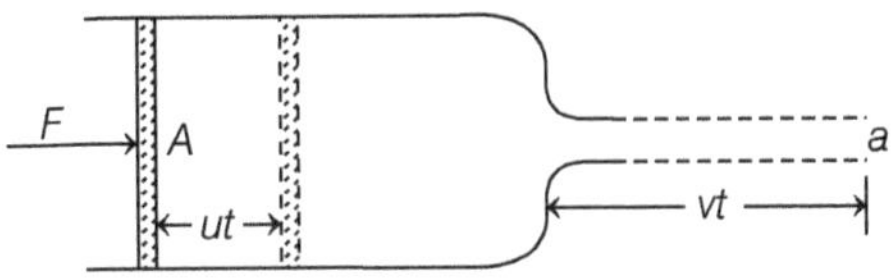

Fig. 389

281. The pressure of the air streaming over the roof is less than of air at rest. It is this surplus pressure of the stationary air under the roof that causes the described phenomena.

282. Since the gas in the stream has a high velocity, the pressure inside the stream is below atmospheric. The ball will be supported from the bottom by the thrust of the stream, and on the sides by the static atmospheric pressure.

283. When air flows between the disks, its velocity diminishes as it approaches the edges of the disks, and is minimum at the edges. The pressure in a jet of air is the lower, the higher its velocity. For this reason the pressure between the disks is lower than atmospheric. The atmospheric pressure presses the lower disk against the upper one, and the flow of the air is stopped. After this the static pressure of the air again moves the disk away, and the process is repeated.

284. The pressure diminishes in a stream of a flowing liquid with an increase in its velocity. The velocity with which the water flows in the vessel is much smaller than in the tube and, therefore, the pressure of the water in the vessel is greater than in the tube. The velocity increases at the boundary between the vessel and tube, and the pressure drops. For this reason the ball is pressed against the screen and does not rise.

285. The piston will cover the distance ut during the time t (Fig. 389). The force F will perform the work $W = Fut$. The mass of the liquid flowing out during the time t is ρAut. The outflow velocity of the liquid v can be found from the equation $Au = av$. The change in the kinetic energy of the liquid during the time t is

$$\rho Aut\left(\frac{v^2}{2} - \frac{u^2}{2}\right)$$

This change should be equal to the work performed by the force F:

$$Fut = \rho Aut\left(\frac{v^2}{2} - \frac{u^2}{2}\right)$$

Upon eliminating u, we find that

$$v^2 = \frac{2F}{A\rho} \times \frac{1}{1 - \frac{a^2}{A^2}}$$

If $a \ll A$, then $v = \sqrt{\frac{2F}{A\rho}}$.

286. It was assumed in solving Problem 285 that the velocity of any element of a liquid in the pump is constant. The velocity changes from u to v when the liquid leaves the pump. This does not occur immediately after the force begins to act on the piston, however. The process requires some time to become stable, i. e., after the particles of the liquid in the cylinder acquire a constant velocity. When $a \to A$ this time tends to infinity and, for this reason, the velocity acquired by the liquid under the action of the constant force becomes infinitely high.

287. Let us introduce the coordinate system depicted in Fig. 390. According to Torricelli's formula, the outflow velocity of a liquid is $V = \sqrt{2gy}$, where y is the thickness of the water layer in the upper vessel. Since water is incompressible, $aV = Av$, where v is the velocity with which the upper layer of the water lowers, A is its area, and a is the area of the orifice.

If we assume that the vessel is axially symmetrical, then $A = \pi x^2$, where x is the horizontal coordinate of the vessel wall.

Therefore,

$$\frac{\pi x^2}{\sqrt{2gy}} = \frac{a}{v} = \text{const}$$

Since in conformity with the initial condition, the water level should lower with a constant velocity. Hence, the shape of the vessel can be determined from the equation

$$y = kx^4$$

where

$$k = \frac{\pi^2 v^2}{2ga^2}$$

288. The pressure changes in a horizontal cross section depending on the distance to the axis r according to the law

$$p = p_0 + \frac{\rho\omega^2}{2} r^2$$

where p_0 is the pressure on the axis of the vessel and ρ the density of the liquid.

The compressive deformation of the liquid will be maximum near the walls of the vessel, while the tensile deformation of the revolving rod (Problem 203) is maximum at the axis.

289. The excess pressure at a distance r from the axis of rotation is

$$p = \frac{\rho\omega^2}{2} r^3$$

(see the solution to Problem 288). On the other hand, this pressure is determined by the elevation of the liquid level in this section above the level on the axis : $p = \rho gh$ (Fig. 391).

Upon equating these expressions, we have

$$h = \frac{\omega^2}{2g} r^2$$

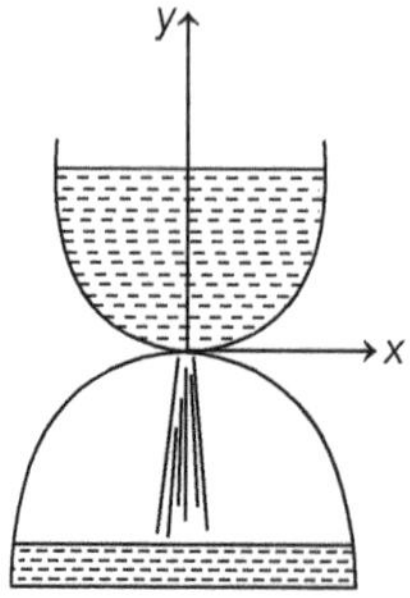

Fig. 390

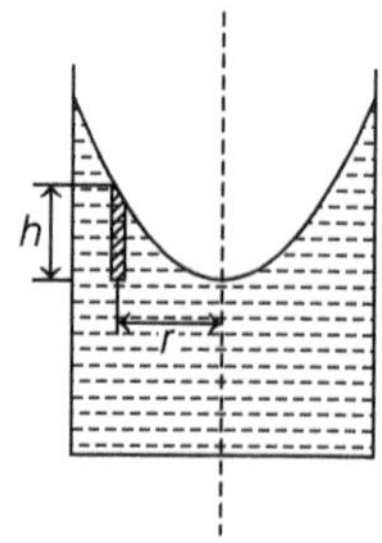

Fig. 391

This is the equation of a parabola, and the surface of the liquid in the rotating vessel takes the form of a paraboloid of revolution.

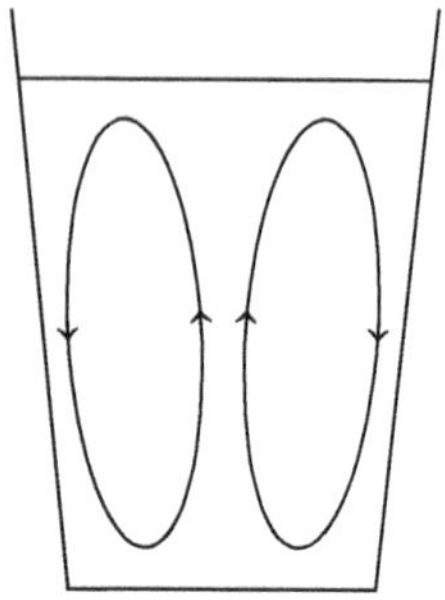

Fig. 392

290. The stirring imparts a certain angular velocity ω to the particles of the water in the glass. The pressures in the liquid will be distributed in about the same way as in the solution to Problem 288. The excess pressure inside the liquid balances the pressure due to a higher level of the liquid at the edges of the glass (see Problem 289).

When stirring is stopped, the velocity of rotation of the liquid near the bottom will begin to decrease owing to friction, the greater the farther the elements of the liquid are from the centre.

Now the excess pressure caused by rotation will no longer balance the weight of the liquid column near the edge of the vessel. This will cause the liquid to circulate as shown in Fig. 392. This is why the tea leaves will gather in the middle of the glass.

2. HEAT: MOLECULAR PHYSICS

2-1. Thermal Expansion of Solids and Liquids

291. $\Delta t \cong 420°\,C$.

292. Reinforced-concrete structures are very strong because the expansion coefficient of concrete is very close to that of iron and steel.

293. The quantity of heat transferred from one body to another in a unit of time is proportional to the difference between the temperatures of these bodies. When the temperatures of the thermometer and the surrounding objects differ appreciably, the volume of the mercury will change at a fast rate. If the temperature of the thermometer is nearly the same as that of the surrounding bodies, the volume of the mercury changes slowly.

For this reason it takes longer for the thermometer to take the temperature of a human body. If the warm thermometer is brought in contact with relatively cool air in the room, the mercury column "drops" so fast due to the great temperature difference that it can be shaken down in a moment.

294. When the scale cools down from t_1 to $t_0 = 0°\,C$ the value of each graduation diminishes. Therefore, the height of the mercury column read off the scale with a temperature of $t_0 = 0°\,C$ will be different and equal $H = H_1 \times (1 + \alpha t_1)$. The heights of the mercury columns at different temperatures and identical pressures are inversely proportional to the densities :

$$\frac{H_0}{H_1} = \frac{\rho}{\rho_0} = \frac{1}{1 + \gamma t_1}$$

Hence,
$$H_0 = \frac{H_1(1 + \alpha t_1)}{1 + \gamma t_1} \cong H_1(1 + \alpha t_1 - \gamma t_1)$$

295. The thermometer can be precooled in a refrigerator and shaken. If no refrigerator is available, put the thermometer into your mouth or in your arm-pit for a time sufficient for the entire thermometer to reach the body temperature, then take it out and shake it immediately. The thermometer will show the temperature of the body.

296. The difference in the lengths of the rulers at a temperature t_1 is

$$l_0'(1 + \alpha_1 t_1) - l_0''(1 + \alpha_2 t_1) = l$$

At a temperature t_2 this difference is equal to

$$l_0'(1 + \alpha_1 t_2) - l_0''(1 + \alpha_2 t_2) = \pm\, l$$

The plus sign corresponds to the case when the difference in lengths is constant (see Fig. 393*a*). The relation between the lengths and the temperature shown in Fig. 393*b* corresponds to the minus sign.

In the first case the system of equations gives

$$l'_{0(1)} = \frac{\alpha_2}{\alpha_2 - \alpha_1} l = 6.8 \text{ cm}; \quad l''_{0(1)} = \frac{\alpha_1}{\alpha_2 - \alpha_1} l = 4.8 \text{ cm}$$

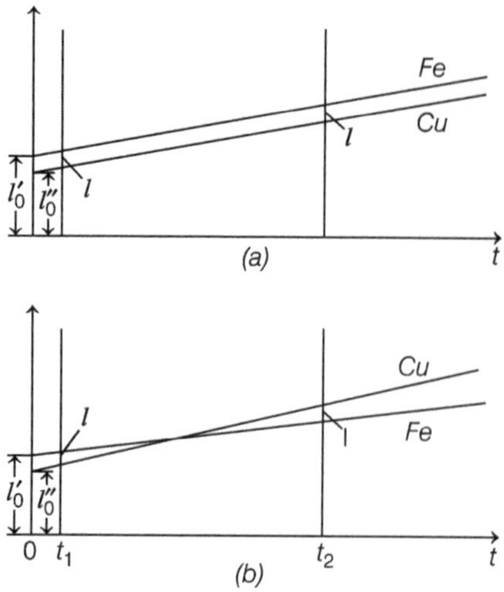

Fig. 393

In the second case

$$l'_{0(2)} = \frac{2 + \alpha_2(t_1 + t_2)}{(t_2 - t_1)(\alpha_2 - \alpha_1)} l = 208.5 \text{ cm};$$

$$l''_{0(2)} = \frac{2 + \alpha_1(t_1 + t_2)}{(t_2 - t_1)(\alpha_2 - \alpha_1)} l = 206 \text{ cm}$$

When $t = 0°$ C, the iron ruler should be longer than the copper one.

297. A possible way of suspension is depicted in Fig. 394, where *1* and *2* are rods with a small coefficient of linear expansion α_1 (e. g., steel rods), and 3 are rods with a high coefficient of expansion α_2, (e.g., zinc or brass rods). The lengths of the rods can be so selected that the length of the pendulum does not change with the temperature. With this aim in view it is essential that $\alpha_1(l_1 + l_2) = \alpha_2 l_3$.

298. When the cylinder is heated, its volume increases according to the same law as that of the glass: $v_1 = v_0(1 + \gamma t_1)$, where γ is the coefficient of volume expansion of glass. If the densities of mercury at the temperatures t_0 and t_1 are denoted by ρ_0 and ρ_1, we can write that $m_0 = v_0\rho_0$ and $m_1 = v_1\rho_1$, where

$$\rho_1 = \frac{\rho_0}{1 + \gamma_1 t_1}$$

This system of equations will give the following expression for γ:

$$\gamma = \frac{m_1(1+\gamma_1 t_1) - m_0}{m_0 t_1} \approx 3\times 10^{-5}\,\text{deg}^{-1}$$

The coefficient of linear expansion

$$\alpha = \frac{\gamma}{3} \cong 10^{-5}\ \text{deg}^{-1}.$$

299. Let the pendulum of an accurate clock perform N oscillations a day. At the temperature t_1 the pendulum of our clock will perform N oscillations in $n-5$ seconds (where $n = 86{,}400$ is the number of seconds in a day) and at the temperature t_2 in $n+10$ seconds. The periods of oscillations will respectively be equal to

$$T_1 = \frac{n-5}{N} \text{ and } T_2 = \frac{n+10}{N}.$$

Hence, $$\frac{T_1}{T_2} = \frac{n-5}{n+10} \cong 1 - \frac{15}{n}.$$

On the other hand, bearing in mind that the period of pendulum oscillations $T = 2\pi\sqrt{\frac{l}{g}}$, we obtain

$$\frac{T_1}{T_2} = \sqrt{\frac{1+\alpha t_1}{1+\alpha t_2}} \cong \sqrt{1+\alpha(t_1-t_2)}$$

$$\cong 1 + \frac{\alpha}{2}(t_1 - t_2)$$

Upon equating the expressions for the ratio of the periods, we find that

$$\alpha \cong \frac{30}{(t_2-t_1)\,n} \cong 2.3\times 10^{-5}\ \text{deg}^{-1}$$

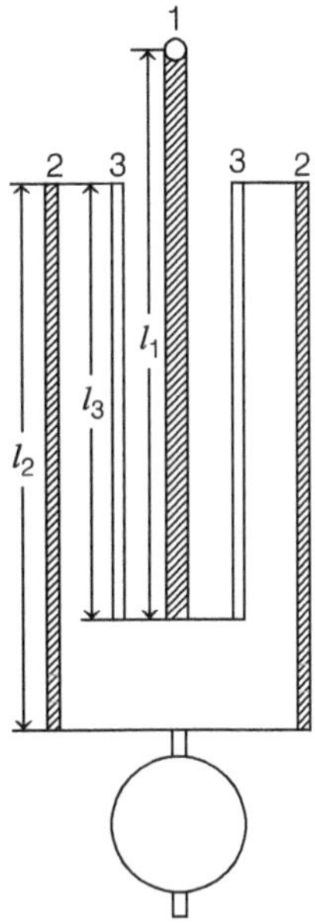

Fig. 394

2-2. The Law of Conservation of Energy. Thermal Conductivity

300. According to the law of conservation of energy, the liberated heat is equal to the loss of kinetic energy

$$Q = \frac{Mv_0^2}{2} - \frac{(M+m)v^2}{2}$$

where v is the velocity of the cart after the brick has been lowered onto it.

This velocity can be found from the law of conservation of momentum :

$$v = \frac{Mv_0}{M + m}.$$

In mechanical units $Q = \frac{Mmv_0^2}{2(M + m)}$ and in thermal units $Q = j\frac{Mmv_0^2}{2(M + m)}$, where j is the thermal equivalent of work.

301. On the basis of the law of conservation of energy,

$$mgl = \frac{mv^2}{2} + \frac{k(l - l_0)^2}{2} + Q$$

where l is the length of the cord at the moment when the washer leaves it. On the other hand, we can write that

$$mgl = \frac{mv^2}{2} + W_1 + W_2$$

where $W_1 = fl_0$ is the work of the force of friction acting on the washer (the washer travels a path of l_0 relative to the cord), and $W_2 = f(l - l_0)$ is the work of the force of friction acting on the cord.

Therefore, $$Q = W_1 + W_2 - \frac{k(l - l_0)^2}{2}$$

Using Hooke's Law, $f = k(l - l_0)$

we find that $$Q = fl_0 + \frac{f^2}{2k}$$

The work W_1 is used entirely to liberate heat. Only half of the work $W_2 = \frac{h_2}{k}$, however, is converted into heat, the other half producing the potential energy $\frac{k(l - l_0)^2}{2}$.

302. The electric current performs the work $W = P\tau$. At the expense of this work the refrigerator will lose the heat $Q_2 = qH + qct$, where c is the heat capacity of water and H is the heat of fusion of ice. According to the law of conservation of energy, the amount of heat liberated in the room will be

$$Q_1 = W + Q_2 = P\tau + qct + qH$$

since in the final run the energy of the electric current is converted into heat.

303. The temperature in the room will rise. The quantity of heat liberated in a unit of time will be equal to the power consumed by the refrigerator, since in the final run the energy of the electric current is converted into heat, and the heat removed from the refrigerator is returned again into the room.

304. It is more advantageous to use a refrigerator that removes heat from the outside air and liberates it in the room. The heat liberated in the room in a unit of time is $P + Q_2$, where P is the power consumed by the refrigerator and Q_2 is the heat removed from the outside air in a unit of time (see Problem 302).

It is only the high cost and complicated equipment that prevent the use of such thermal pumps for heating at present.

305. When salt is dissolved, its crystal lattice is destroyed. The process requires a certain amount of energy that can be obtained from the solvent.

In the second case, part of the intermolecular bonds of the crystal lattice have already been destroyed in crushing the crystal. For this reason, less energy is required to dissolve the powder and the water will be higher in temperature in the second vessel. The effect, however, will be extremely negligible.

306. The quantity of heat removed from the water being cooled is $m_2c(t_2 - \theta)$, where θ is the final temperature. The cold water receives the heat $m_1c(\theta - t_1)$. The heat imparted to the calorimeter is $q(\theta - t_1)$. On the basis of the law of conservation of energy,

$$m_1c(\theta - t_1) + q(\theta - t_1) = m_2c(t_2 - \theta)$$

whence, $$\theta = \frac{(m_1t_1 + m_2t_2)c + qt_1}{(m_1 + m_2)c + q} \cong 4°\text{C}$$

307. The power spent to heat the water in the calorimeter is

$$P_1 = \frac{\rho VctJ}{\tau}$$

where ρ is the density of the water, c is its specific heat, and $J = 4.18$ J/cal is the mechanical equivalent of heat. The sought value is $Q = \frac{P - P_1}{P} = 1 - \frac{\rho VctJ}{P\tau} \cong 5$ per cent

308. $Q = \frac{\lambda}{d}(T_1 - T_0)At \cong 9{,}331$ kcal

309. The quantity of heat Q passing through the first panel a second is $Q = \lambda_1 \frac{T_2 - T_1}{d_1} A$, where A is the area of a panel. Since the process is stationary, the same amount of heat passes through the second panel:

$$Q = \lambda_2 \frac{T_0 - T_2}{d_2} A.$$

We find from the condition

$\lambda_1 \frac{T_2 - T_1}{d_1} A = \lambda_2 \frac{T_0 - T_2}{d_2} A$ that $\Rightarrow T_2 = \frac{\lambda_2 d_1 T_0 + \lambda_1 d_2 T_1}{\lambda_2 d_1 + \lambda_1 d_2}$

310. Upon inserting the temperature T_2 into the expression for Q (see Problem 309) when $d_1 = d_2 = d$, we find that

$$Q = \frac{2\lambda_1\lambda_2}{\lambda_1 + \lambda_2} \frac{T_0 - T_1}{2d} A$$

Therefore, the coefficient of thermal conductivity of the wall is

$$\lambda = \frac{2\lambda_1\lambda_2}{\lambda_1 + \lambda_2}$$

311. The quantity of heat passing a second through the cross-sectional areas of the blocks with coefficients of thermal conductivity λ_1 and λ_2 is equal, respectively, to

$$Q_1 = \frac{\lambda_1}{d}(T_1 - T_0)A \text{ and } Q_2 = \frac{\lambda_2}{d}(T_1 - T_0)A$$

The quantity of heat passing through two blocks whose entire cross-sectional area is $2A$ is

$$Q = Q_1 + Q_2 = \frac{\lambda_1 + \lambda_2}{2} \frac{T_1 - T_0}{d} 2A$$

Hence, the coefficient of thermal conductivity of the wall is equal to

$$\lambda = \frac{\lambda_1 + \lambda_2}{2}$$

312. The coefficients of thermal conductivity of walls I and II are equal to

$$\lambda_I = \frac{\lambda_1 + \lambda_2}{2} \text{ and } \lambda_{II} = \frac{2\lambda_1\lambda_2}{\lambda_1 + \lambda_2}$$

(see the solutions to Problems 310 and 311). It follows from the obvious inequality $(\lambda_1 - \lambda_2)^2 > 0$ that

$$(\lambda_1 + \lambda_2)^2 > 4\lambda_1\lambda_2$$

Hence,

$$\frac{\lambda_1 + \lambda_2}{2} > \frac{2\lambda_1\lambda_2}{\lambda_1 + \lambda_2}, \text{ i.e., } \lambda_1 > \lambda_2$$

313. The quantity of heat supplied by the heater into the water through the pan bottom is

$$Q = \frac{\lambda}{d}(T - T_1)A = mr$$

where T_1 is the boiling point of the water and r is the specific heat of vaporization.

Therefore, $$T = T_1 + \frac{mrd}{\lambda A}$$

2-3. Properties of Gases

314. The removable cap acts as a pump and under it a rarefied space is formed that sucks out the ink. The orifice serves to maintain a constant pressure under the cap.

315. Assuming that the temperature remains constant, let us apply Boyle's law to the volume of air above the mercury:

$$(p_{01} - p_1)(l - 748 \text{ mm}) = (p_{02} - p_2)(l - 736 \text{ mm})$$

whence, $l = 764$ mm.

316. In a position of equilibrium $f - G - F = 0$, where f is the force of expulsion equal to $\gamma h_1 A$ (here γ is the specific weight of the water, and h_1 the height of the air column in the tube after submergence). In our case the force of expulsion is built up by the difference of pressure on the soldered end of the tube from below and from above: $f = p_1 A - (p_0 + \gamma h) A$, where p_1 is the air pressure in the tube after submergence.

According to Boyle's law, $p_0 l A = p_1 h_1 A$.

It follows from this system of equations that

$$E = \frac{A}{2}\left[\sqrt{(p_0 + \gamma h)^2 + 4p_0\gamma l} - (p_0 + \gamma h)\right] - G = 8.65 \text{ gf}$$

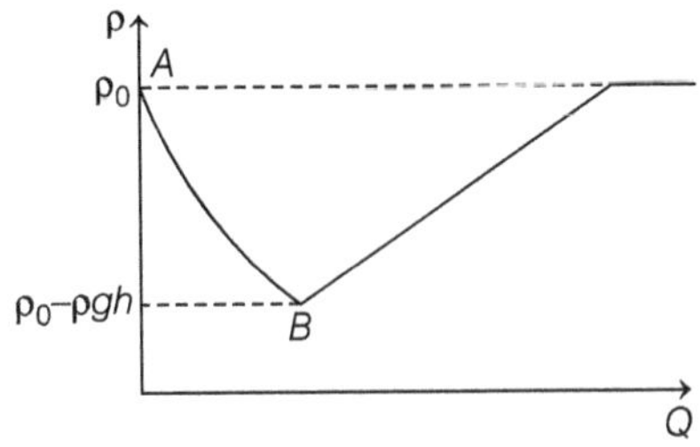

Fig. 395

317. First, the pressure p of the air will decrease approximately isothermally owing to the drop of the level of the water in the vessel. This will continue until the total pressure at the level of the lower end of the tube becomes equal to the atmospheric pressure p_0; i.e., $p + \rho g h = p_0$, where h is the height of the water column in the vessel above the level of the lower end of the tube. From this moment on air bubbles will begin to pass into the vessel. The pressure at the level of the lower end of the tube will remain equal to the atmospheric pressure, while the air pressure $p = p_0 - \rho g h$ will grow linearly with a drop in the water level. The water will flow out from the vessel at a constant velocity.

The relation between p and Q is shown in Fig. 395. The negligible fluctuations of pressure when separate bubbles pass in are not shown in Fig. 395.

318. When the air is being pumped out of the vessel, the pressure in the vessel after one double stroke will become equal to $p_1 = \frac{p_0 V}{V + v_0}$.

After the second double stroke $p_1 V = p_2(V + v_0)$ and, consequently,

$$p_2 = p_0\left(\frac{V}{V + v_0}\right)^2, \text{ etc.}$$

After n double strokes the pressure in the vessel will be

$$p' = p_0\left(\frac{V}{V + v_0}\right)^n$$

When air is being delivered into the vessel, after n double strokes the pressure will be

$$p = p' + \frac{p_0 n v_0}{V} = p_0\left\{\left(\frac{V}{V + v_0}\right)^n + \frac{n v_0}{V}\right\}$$

Here $p > p_0$ at any n, since during delivery the pump during each double stroke sucks in air with a pressure p_0, and during evacuation v_0 of the air is pumped out at pressures below p_0.

319. Applying Boyle's law to the two volumes of gas in the closed tube, we obtain

$$p\frac{L - l}{2}A = p_1\left(\frac{L - l}{2} - \Delta l\right)A$$

$$p\frac{L - l}{2}A = p_2\left(\frac{L - l}{2} + \Delta l\right)A$$

$$p_1 = p_2 + \gamma l$$

Here p is the pressure in the tube placed horizontally, p_1 and p_2 are the pressures in the lower and upper ends of the tube placed vertically when its ends are closed, γ is the specific weight of the mercury, A is the cross-sectional area of the tube.

Hence, the initial pressure in the tube is

$$p = \gamma\frac{l}{2}\left(\frac{l_0}{\Delta l} - \frac{\Delta l}{l_0}\right)$$

Here, for the sake of brevity we have designated $\frac{L - l}{2}$ by l_0.

If one end of the horizontal tube is opened, the pressure of the gas in the tube will become equal to the atmospheric pressure.

According to Boyle's law, $\rho l_0 A = \gamma H l_1 A$ (here H is the atmospheric pressure), and, therefore,

$$l_1 = \frac{l l_0}{2H}\left(\frac{l_0}{\Delta l} - \frac{\Delta l}{l_0}\right)$$

The mercury column will shift through the distance

$$\Delta l_1 = l_0 - l_1 = \frac{ll_0}{2H}\left[\frac{2H}{l} - \left(\frac{l_0}{\Delta l} - \frac{\Delta l}{l_0}\right)\right]$$

For the mercury not to flow out of the tube, the following condition is required

$$\frac{l_0}{\Delta l} \le \sqrt{\left(\frac{H}{l}\right)^2 + 1} + \frac{H}{l}$$

When the upper end of the vertical tube is opened

$$pl_0A = \gamma(H + l)l_2A$$

Hence,

$$\Delta l_2 = l_0 - l_2 = \frac{ll_0}{2(H + l)}\left[\frac{2H}{l} - \left(\frac{l_0}{\Delta l} - \frac{\Delta l}{l_0}\right) + 2\right]$$

The mercury will not flow out of the tube if

$$\frac{l_0}{\Delta l} \le \sqrt{\left\{\frac{2(H + l)}{l}\right\}^2 + 1} + \frac{2(H + l)}{l}$$

When the lower end is opened

$$pl_0A = \gamma(H - l)l_3A$$

whence,

$$\Delta l_3 = l_0 - l_3$$
$$= \frac{ll_0}{2(H - l)}\left[\frac{2H}{l} - \left(\frac{l_0}{\Delta l} - \frac{\Delta l}{l_0}\right) - 2\right]$$

The following condition should be satisfied to prevent the mercury column from being forced out of the tube

$$\frac{l_0}{\Delta l} \le \sqrt{\left\{\frac{4(H - l)}{l^2}\right\}^2 + 1} + \frac{2(H - l)}{l}$$

320. Since for one gramme-molecule of any gas at $p = 1$ atm and $T = 237°$ K we have $V_\mu = 22.4$ litres, then for one mole

$C = \frac{pV_\mu}{T} = 0.082$ lit · atm/mole.deg.

This constant is usually denoted by R and called the universal gas constant. The values of R in the various systems of units are:

$$R = 0.848 \text{ kgf-m/mole·deg} = 8.3 \times 10^7 \text{ erg/mole·deg}$$
$$= 1.986 \text{ cal/mole·deg}$$

321. At a fixed pressure and temperature the volume occupied by the gas is proportional to its mass. A volume of V_μ corresponds to one gramme-molecule and a volume of V to an arbitrary mass m. Obviously, $V_\mu = V\frac{\mu}{m}$, where μ is the molecular weight expressed in grammes.

Upon inserting this expression into the equation of state for one gramme-molecule, we have

$$pV = \frac{m}{\mu} RT$$

322. If the attraction between the molecules suddenly disappeared, the pressure should increase. To prove this, let us mentally single out two layers *I* and *II* inside a fluid (Fig. 396). The molecules penetrating from layer *I* into layer *II* owing to thermal motion collide with the molecules in layer *II*, and as a result this layer is acted upon by the pressure forces p_t that depend on the temperature. The forces of attraction act on layer *II* from the side of the molecules in layer *I* in the opposite direction. The resultant pressure of layer *I* on layer *II* $p = p_t - p_i$, where p_i is the pressure caused by the internal forces of attraction. When p_i disappears, the pressure grows.

323. If the forces of attraction between the molecules disappeared the water would be converted into an ideal gas. The pressure can be found from the equation of state of an ideal gas :

$$p = \frac{m}{\mu}\frac{RT}{V} \cong 1{,}370 \text{ atm}$$

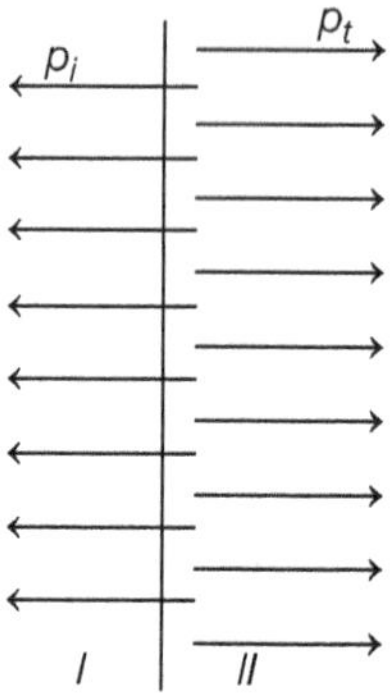

Fig. 396

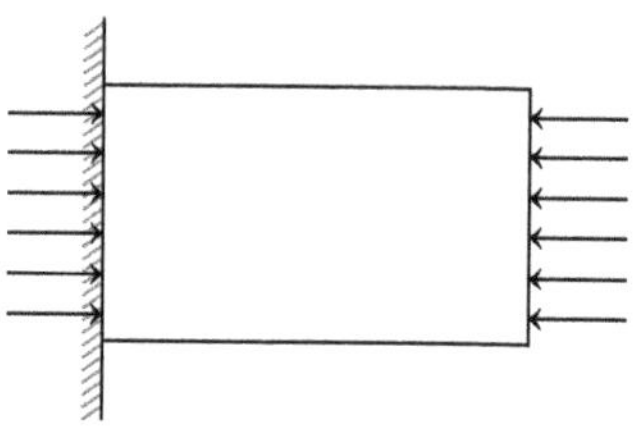

Fig. 397

324. Let us separate a cylindrical volume of the gas in direct contact with the wall (Fig. 397). The forces acting on the side surface of the cylinder are mutually balanced. Since the volume is in equilibrium, the pressure on the gas from the side of the wall should always be equal to the pressure on the other base of the cylinder from the side of the gas. We can conclude, on the basis of Newton's third law, that the pressure of the gas on the wall is equal to the pressure inside the vessel.

325. The pressure in the gas depends on the forces of interaction between the molecules (see Problem 322). The forces of mutual interaction of the molecules and of the molecules with the wall are different, however. Hence the pressures inside the gas and at the walls of the vessel (see Problem 324) can be identical only if the concentrations are different.

326. Since the volume is constant

$$\frac{p_2}{p_1} = \frac{T_2}{T_1} \quad \text{or} \quad \frac{p_2 - p_1}{p_1} = \frac{T_2 - T_1}{T_1} = 0.004$$

Hence, $$T_1 = \frac{T_2 - T_1}{0.004} = 250° \text{ K}$$

327. From Archimedes' law, $mg + G = \gamma V$, where γ is the specific weight of water and V is the volume of the sphere. The equation of state gives

$$(p_0 + \gamma h)V = \frac{m}{\mu} RT$$

Upon deleting V from these equations, we find that

$$m = \frac{G\mu(p_0 + \gamma h)}{\gamma RT - \mu g(p_0 + \gamma h)} \cong 0.666 \text{ g}$$

and equilibrium will be unstable.

328. When the tube is horizontal, the device cannot be used as a thermometer, since the pressures exerted on the drop from the right and from the left will be balanced at any temperature.

If the tube is placed vertically, the pressure of the gas in the lower ball will be higher than in the upper one by a constant magnitude. If the volume is the same, the pressure will grow with a rise in the temperature the faster, the higher is the initial pressure. To maintain a constant difference of the pressures in the balls, the drop will begin to move upward, and in this case the device can be employed as a thermometer.

329. Since the masses of the gas are the same in both ends and the piston is in equilibrium,

$$\frac{V_2}{V_1} = \frac{T_2}{T_1}$$

Hence, $$T_2 = \frac{V_2}{V_1} T_1 = 330° \text{ K}$$

Applying Boyle's law to the volume of the gas whose temperature does not change, we obtain

$$p = \frac{p_0 V_0}{V_1} = 1.05 \text{ atm}$$

330. When the external conditions are the same, equal volumes of various gases contain an equal number of molecules (Avogadro's law). Therefore, $V_1 : V_2 : V_3 : V_4 = N_1 : N_2 : N_3 : N_4$, where V_i is the volume of a gas and N_i the number of molecules of this gas.

The mass of a certain amount of a gas is proportional to the number of its molecules and the molecular weight of the gas:

$$m_1 : m_2 : m_3 : m_4 = N_1\mu_1 : N_2\mu_2 : N_3\mu_3 : N_4\mu_4$$

On the other hand, denoting the relative volume of this gas in per cent by $n_i = \frac{V_i}{V} 100\%$, we have

$$n_1 : n_2 : n_3 : n_4 = \frac{V_1}{V} : \frac{V_2}{V} : \frac{V_3}{V} : \frac{V_4}{V} = \frac{N_1}{N} : \frac{N_2}{N} : \frac{N_3}{N} : \frac{N_4}{N}$$

If the composition of air in per cent is described by $n_i = \frac{m_i}{m} 100\%$ (composition by weight), we can obtain from the previous ratios that

$$n_1' : n_2' : n_3' : n_4' = \frac{m_1}{m} : \frac{m_2}{m} : \frac{m_3}{m} : \frac{m_4}{m}$$

$$= \frac{N_1\mu_1}{N} : \frac{N_2\mu_2}{N} : \frac{N_3\mu_3}{N} : \frac{N_4\mu_4}{N}$$

$$= n_1\mu_1 : n_2\mu_2 : n_3\mu_3 : n_4\mu_4$$

Hence, $$n_i = \frac{n_1' + n_2' + n_3' + n_4'}{n_1\mu_1 + n_2\mu_2 + n_3\mu_3 + n_4\mu_4} n_i\mu_i$$

Remembering that $n_1' + n_2' + n_3' + n_4' = 100$ per cent,

we obtain $$n_i = \frac{n_i\mu_i 100\%}{n_1\mu_1 + n_2\mu_2 + n_3\mu_3 + n_4\mu_4}$$

Therefore, $n_1' = 75.52\%$; $n_2' = 23.15\%$; $n_3' = 1.28\%$; $n_4' = 0.05\%$.

331. For each gas, the equation of state can be written as follows:

$$p_1V = \frac{m_1}{\mu_1}RT; \quad p_2V = \frac{m_2}{\mu_2}RT;$$

$$p_3V = \frac{m_3}{\mu_3}RT; \quad p_4V = \frac{m_4}{\mu_4}RT$$

Hence, $$(p_1 + p_2 + p_3 + p_4)V = \left(\frac{m_1}{\mu_1} + \frac{m_2}{\mu_2} + \frac{m_3}{\mu_3} + \frac{m_4}{\mu_4}\right)RT$$

On the other hand, for a mixture of gases $pV = \frac{m}{\mu}RT$, where $m = m_1 + m_2 + m_3 + m_4$ and μ is the sought molecular weight.

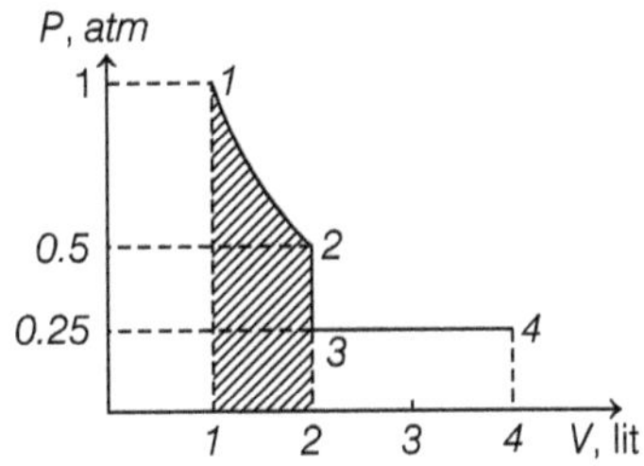

Fig. 398

According to Dalton's law,

$$p = p_1 + p_2 + p_3 + p_4.$$

Therefore, $$\mu = \frac{m_1 + m_2 + m_3 + m_4}{\frac{m_1}{\mu_1} + \frac{m_2}{\mu_2} + \frac{m_3}{\mu_3} + \frac{m_4}{\mu_4}} = \frac{n_1' + n_2' + n_3' + n_4'}{\frac{n_1'}{\mu_1} + \frac{n_2'}{\mu_2} + \frac{n_3'}{\mu_3} + \frac{n_4'}{\mu_4}} = 28.966$$

where $n_i = \frac{m_i}{m} 100\%$ is the composition of the air in per cent by weight. The result obtained in the previous problem allows us to find μ from the known composition of the air by volume

$$\mu = \frac{\mu_1 n_1 + \mu_2 n_2 + \mu_3 n_3 + \mu_4 n_4}{n_1 + n_2 + n_3 + n_4} = 28.966$$

332. On the basis of Clapeyron's equation,

$$\mu = \frac{mRT}{pV} = \frac{\rho RT}{p}$$

$$= 72 \text{ g / mole}$$

The sought formula is C_5H_{12} (one of the pentane isomers).

333. When the gas is compressed in a heat-impermeable envelope, the work performed by the external force is spent to increase the internal energy of the gas, and its temperature increases. The pressure in the gas will increase both owing to a reduction in its volume and an increase in its temperature. In isothermal compression the pressure rises only owing to a reduction in the volume.

Therefore, the pressure will increase more in the first case than in the second.

334. A diagram of p versus V is shown in Fig. 398. The greatest work equal to the hatched area in Fig. 398 is performed during the isothermal process (*1-2*).

The temperature does not change on section *1-2*, and is halved on section *2-3*. After this the temperature rises, and $T_4 = T_1$ when $V_4 - 4$ lit.

335. Line *1-2* is an isobaric line (Fig. 399). The gas is heated at a constant pressure, absorbing heat.

Line *2-3* is an isochoric line. The gas is cooled at a constant volume, the pressure drops and heat is liberated.

Line *3-1* is an isothermal line. The volume of the gas diminishes at a constant temperature. The pressure rises. The gas is not heated, although it is subjected to the work of external forces. Hence, the gas rejects heat on this section.

336. The amount of heat liberated per hour upon combustion of the methane is

$$Q_1 = \frac{q_0 p V_0 \mu}{RT}$$

where $\mu = 16$ g/mole is the mass of one mole of the gas and $T = t + 273° = 284°$ K is its temperature.

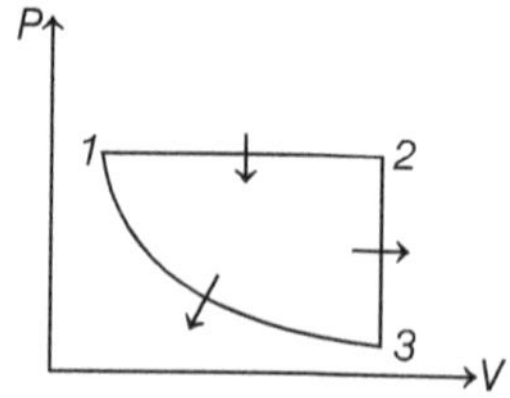

Fig. 399

The amount of heat received by the water in one hour is

$$Q_2 = \frac{\pi D^2}{4} v\rho c(t_2 - t_1) 3{,}600$$

where $\rho = 1\ \text{g/cm}^3$, is the density of the water and $c = 1$ cal/deg·g is the specific heat.

According to the condition, $\frac{Q_2}{Q_1} = \eta = 0.6$

Upon solving these simultaneous equations, we find that

$$t_2 = t_1 + \frac{q_0 p V_0 \mu\eta}{900\pi D^2 v\rho c RT}$$

$$\cong 93° \text{C}$$

337. In the initial state $p_1 V = \frac{m}{\mu_1} RT_1$, where μ_1 is the molecular weight of the ozone. In the final state, $p_2 V = \frac{m}{\mu_2} RT_2$, where μ_2 is the molecular weight of the oxygen. The heat balance equation gives

$$\frac{m}{\mu_1} q = \frac{c_V}{\mu_2} m(T_2 - T_1)$$

Upon solving these simultaneous equations, we find that

$$\frac{p_2}{p_1} = \frac{q}{c_V T_1} + \frac{\mu_1}{\mu_2} = 10$$

338. In view of the linear dependence of pressure on volume we can write :

$$p = aV + b$$

The constants a and b can be found from the condition of the problem:

$$a = \frac{p_1 - p_2}{V_1 - V_2} \cong -0.5 \text{ atm/lit}$$

$$b = \frac{p_2V_1 - p_1V_2}{V_1 - V_2} \cong 20 \text{ atm}$$

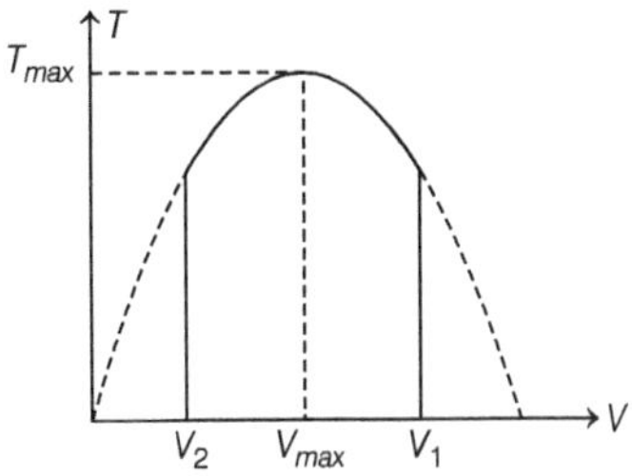

Fig. 400

Upon inserting the expression for p into the equation of state of an ideal gas

$$pV = \frac{m}{\mu} RT = \text{const } T,$$

we find that $$aV^2 + bV = \text{const } T \qquad (1)$$

The relation between T and V (see Fig. 400) is a parabola. The curve reaches its maximum at $V_{max} = -\frac{b}{2a} \cong 20$ lit when the roots of quadratic equation (1) coincide. Here

$$p_{max} = aV_{max} + b = \frac{b}{2} \cong 10 \text{ atm}$$

Therefore, $$T_{max} = \frac{p_{max}V_{max}\mu}{mR} \cong 490° \text{ K}$$

339. The energy of a unit volume of gas $u_1 = CT\rho$, where ρ is the density of the air. According to the equation of state of an ideal gas, $\frac{pV}{T} = mB$ (B is a constant).

Since $\rho = \frac{m}{V}$, then $\rho T = \frac{p}{B}$. Therefore,

$u_1 = \frac{C}{B} p$ is determined only by the pressure. The energy of all the air in the room is also determined only by the pressure. The pressure in the room is equal to the atmospheric pressure and does not change when the air is heated. For this reason the energy of the air in the room also does not change. As the air is heated, some of it flows outside through cracks, and this ensures a constancy of energy despite the heating. The energy would increase with heating only in a hermetically closed room.

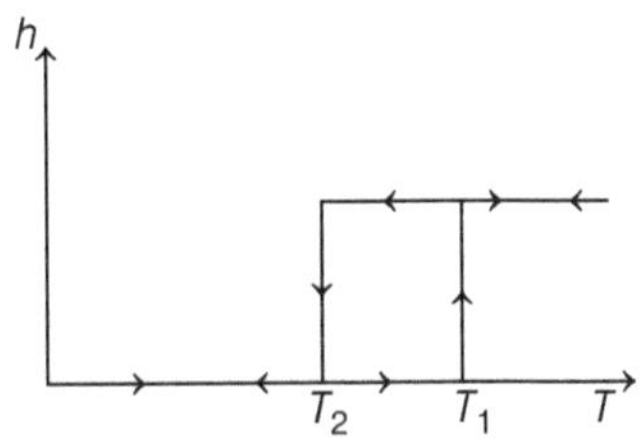

Fig. 401

340. On the basis of the equation of state, the sought mass of the air will be

$$\Delta m = \frac{\mu p V}{R} \frac{T_2 - T_1}{T_1 T_2} \cong 1.3 \text{ kg}$$

341. Let the tube first be near the bottom in a state of stable equilibrium. Upon heating, the air pressure in the tube and, correspondingly, the force of expulsion increase. At a certain temperature T_1 the tube begins to rise to the surface. Since the pressure of the water gradually decreases upwards from the bottom, the volume of the air in the tube and, therefore, the force of expulsion continue to increase. The tube will quickly reach the surface of the water. Upon a further increase in the temperature, the tube will be at the surface. If the temperature lowers, the tube will not sink at T_1, because it has a great reserve of buoyancy caused by an appreciable increase in the force of expulsion as the tube rises. It is only when $T_2 < T_1$ that the tube begins to sink. Here the force of expulsion will drop because, as submergence continues, the air in the tube will occupy a smaller volume. The tube will reach the bottom very quickly.

The relation between the depth of submergence h of the tube and the temperature T is shown in Fig. 401.

The tube will always be at the bottom when $T < T_2$ and at the surface when $T > T_1$. If $T_2 < T < T_1$, the tube will be either at the bottom or at the surface, depending on the previous temperatures.

342. The gas expands at a certain constant pressure p built up by the piston. The work $W = p(V_2 - V_1)$, where V_1 and V_2 are the initial and final volumes of the gas. By using the equation of state, let us express the product pV through the temperature T. Then,

$$W = \frac{m}{u} R(T_2 - T_1) \cong 33.9 \text{ kgf-m}$$

343. The heat imparted to the gas is used to heat it and perform mechanical work. According to the law of conservation of energy,

$$Q = \frac{m}{\mu} C_V (T_2 - T_1) + \frac{m}{\mu} R(T_2 - T_1) = \frac{m}{\mu}$$

$$(T_2 - T_1)(C_V + R) \cong 354.6 \text{ cal}$$

2-4. Properties of Liquids

344. It is more difficult to compress a litre of air, since more work has to be done in this case.

Water has a small compressibility, and a small reduction in volume is required to increase the pressure inside it to three atmospheres.

345. A maximum thermometer can be made as follows. A small unwettable freely moving body is placed inside the tube of a horizontal thermometer (Fig. 402). The position of the body will show the maximum temperature, since the body will move along the tube when the liquid expands and will remain in place when the liquid in the tube is compressed.

To make a minimum thermometer, a wettable body should be placed inside the liquid in the tube.

346. When an elastic rubber film is stretched, the force of tension depends on the amount of deformation of the film. The force of surface tension is determined only by the properties of the liquid and does not change with an increase of its surface.

347. The surface tension of pure petrol is less than that of petrol in which grease is dissolved. For this reason the petrol applied to the edges will contract the spot towards the centre. If the spot itself is wetted, it will spread over the fabric.

348. Capillaries of the type shown in Fig. 403 form in a compact surface layer of soil. They converge towards the top, and the water in them rises to the surface, from which it is intensively evaporated. Harrowing destroys this structure of the capillaries and the moisture is retained in the soil longer.

349. Leather contains a great number of capillaries. A drop of a wetting liquid inside a capillary having a constant cross section will be in equilibrium. When the liquid is heated, the surface tension diminishes and the liquid is drawn towards the cold part of the capillary. The grease will be drawn into the leather if it is heated outside.

350. The grease melts, and capillary forces carry it to the surface of the cold fabric placed under the clothing (see Problem 349).

351. The end of the piece of wood in the shade is colder, and the capillary forces move the water in this direction.

352. The hydrostatic pressure should be balanced by the capillary pressure:

$$\rho g h = \frac{4\alpha}{d}. \text{ Hence, } h = 30 \text{ cm.}$$

353. The following forces act vertically on section *abcd* of the film: weight, surface tension F_{ab} applied to line *ab* and surface tension F_{cd} applied to *cd*.

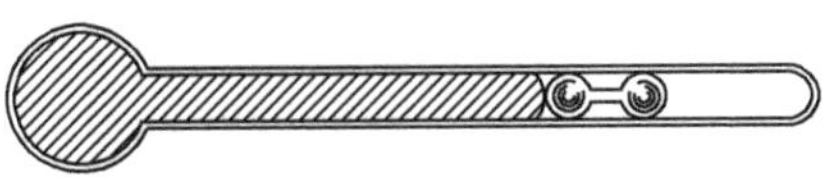

Fig. 402

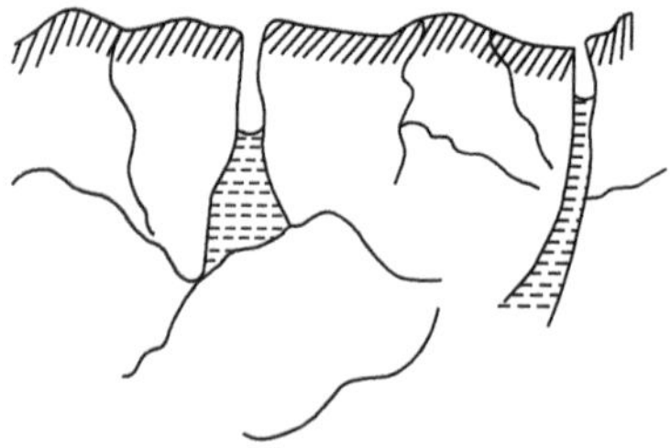

Fig. 403

Equilibrium is possible only if F_{ab} is greater than F_{cd} by an amount equal to the weight of the section of the film being considered.

The difference between the forces of surface tension can be explained by the difference in the concentration of the soap in the surface layers of the film.

354. The force of expulsion balances the weight of the cube mg and the force of surface tension $4a\alpha$, *i.e.*, $a^2x\rho g = mg + 4a\alpha$, where x is the sought distance.

Therefore, $$x = \frac{mg + 4a\alpha}{a^2\rho g} \cong 2.3 \text{ cm}$$

The forces of surface tension introduce a correction of about 0.1 cm.

355. The water rises to a height $h = \dfrac{2\alpha}{\rho g r}$. The potential energy of the water column is

$$E_p = \frac{mgh}{2} = \frac{2\pi\alpha^2}{\rho g}$$

The forces of surface tension perform the work $W = 2\pi r\alpha h = \dfrac{4\pi\alpha^2}{\rho g}$.

One half of this work goes to increase the potential energy, and the other half to evolve heat.

Hence, $$Q = \frac{2\pi\alpha^2}{\rho g}$$

356. The pressure inside the liquid at a point that is at a height h above a certain level is less than the pressure at this level by ρgh. The pressure is zero at the level of the liquid in the vessel. Therefore, the pressure at the height h is negative (the liquid is stretched) and is equal to $p = -\rho gh$.

357. The forces of attraction acting on a molecule in the surface layer from all the other molecules produce a resultant directed downward. The closest neighbours, however, exert a force of repulsion on the molecule which is therefore in equilibrium.

Owing to the forces of attraction and repulsion, the density of the liquid is smaller in the surface layer than inside. Indeed, molecule *1* (Fig. 40 4) is acted upon by the force of repulsion from molecule *2* and the forces of attraction from all the other molecules (3 , 4, . . .). Molecule *2* is acted upon by the forces of repulsion from *3* and *1* and the forces of attraction from the molecules in the deep layers. As a result, distance 1 – 2 should be greater than 2 – 3, etc.

This course of reasoning is quite approximate (thermal motion, etc., is disregarded), but nevertheless it gives a qualitatively correct result.

An increase in the surface of the liquid causes new sections of the rarefied surface layer to appear. Here work should be performed against the forces of attraction between the molecules. It is this work that constitutes the surface energy.

358. The required pressure should exceed the atmospheric pressure by an amount that can balance the hydrostatic pressure of the water column and the capillary pressure in the air bubble with a radius r.

The excess pressure is $p + \rho g h + \frac{2\alpha}{r} = 4{,}840 \text{ dyne}/\text{cm}^2$.

359. Since in this case $\rho g h < \frac{2\alpha}{r}$, the water rises to the top end of the tube.

The meniscus will be a part of a spherical segment (Fig. 405): The radius of curvature of the segment is determined from the condition that the forces of surface tension balance the weight of the water column:

$$2\pi r\alpha \cos\varphi = \pi r^2 h\rho g.$$

Hence, $\cos\varphi = \frac{rh\rho g}{2\alpha}$. It is obvious from Fig. 405 that the radius of curvature of the segment $R = \frac{r}{\cos\varphi} = \frac{2\alpha}{hg\rho} = 0.74$ mm.

360. When the tube is opened, a convex meniscus of the same shape as on the top is formed at its lower end. For this reason the length of the water column remaining in the tube will be $2h$ if $l \geq h$, and $l + h$ if $l \leq h$.

361. (1) The forces of surface tension can retain a water column with a height not over h in this capillary tube. Therefore, the water will flow out of the tube.

(2) The water does not flow out. The meniscus is convex, and will be a hemisphere for an absolutely wetting liquid.

(3) The water does not flow out. The meniscus is convex and is less curved than in the second case.

(4) The water does not flow out. The meniscus is flat.

(5) The water does not flow out. The meniscus is concave.

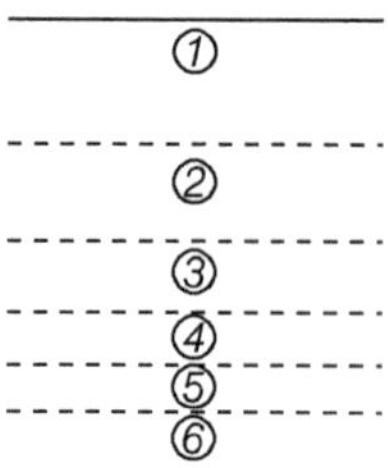

Fig. 404

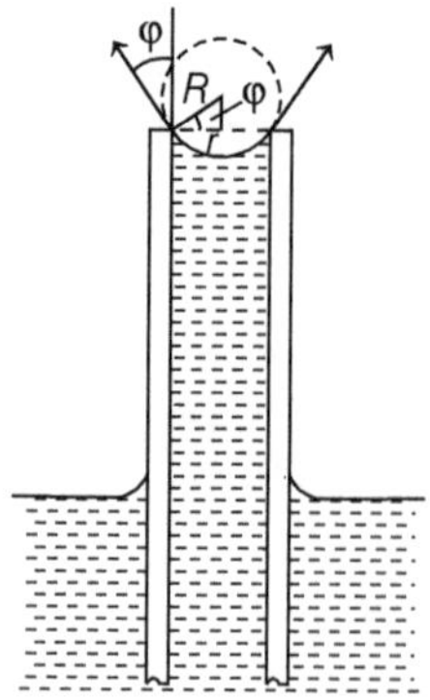

Fig. 405

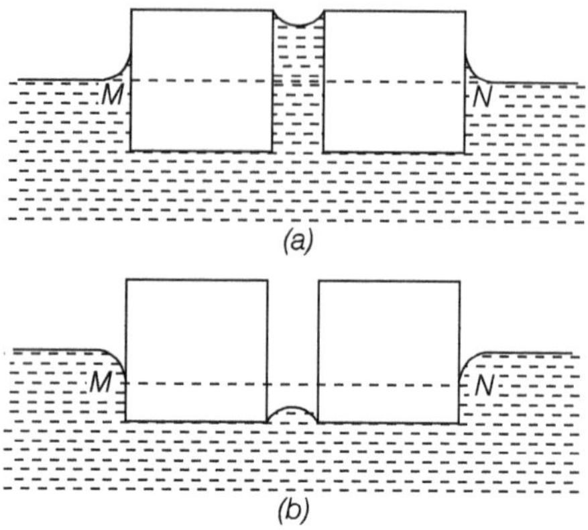

Fig. 406

362. The pressure p inside the soap-bubble with a radius R exceeds the atmospheric pressure by the amount of the double capillary pressure, since the bubble film is double: $p = p_0 + \frac{4\alpha}{R}$.

The pressure inside the bubble with a radius R together with the pressure of the section of the film between the bubbles should balance the pressure inside the smaller bubble.

Therefore, $\frac{4\alpha}{R} + \frac{4\alpha}{R_x} = \frac{4\alpha}{r}$ where R_x, is the radius of curvature of section AB. Hence,

$$R_x = \frac{Rr}{R - r}$$

At any point of contact the forces of surface tension balance each other and are mutually equal. This is possible only when the angles between them are equal to 120°.

363. According to the law of conservation of energy, the cross will not rotate. The components of the forces of surface tension are balanced by the forces of hydrostatic pressure, since the hydrostatic pressure of the water higher than the level in the vessel is negative (see Problem 356).

364. If the bodies are wetted by water, its surface will take the form shown in Fig. 406*a*. Between the matches, above level *MN*, the water is tensioned by the capillary forces, and the pressure inside the water is less than the atmospheric pressure. The matches will be attracted toward each other, since they are subjected to the atmospheric pressure on their sides. For unwetted matches, the form of the surface is shown in Fig. 406*b*. The pressure between the matches is equal to the atmospheric pressure and is greater than the latter on the sides below level *MN*.

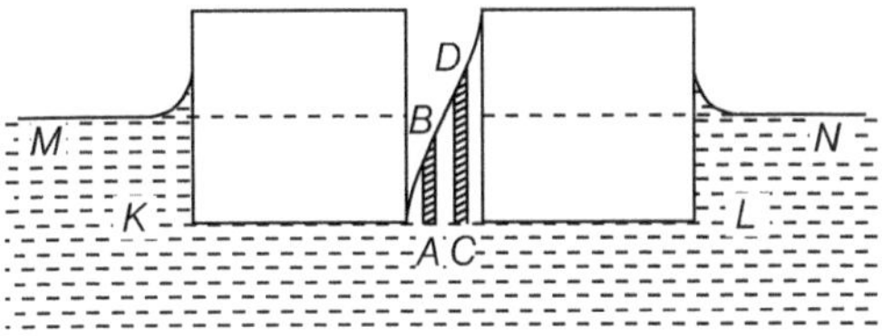

(a)

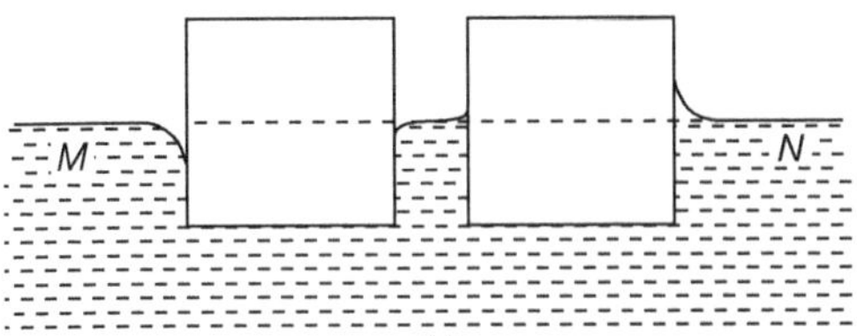

(b)

Fig. 407

In the last case two various forms of the surface correspond to the wetting angles when the matches approach each other (Fig. 407). One of them, however, cannot be obtained in practice (Fig. 407*a*). The pressure at level *KL* should be the same everywhere. In particular, the pressure of columns *AB* and *CD* of different height should be the same. But this is impossible, since the position of the column can be so selected that their surfaces are identical in form. In this case the additional pressure of the surface forces will be the same, and the hydrostatic pressure different. As a result, when the matches approach each other, the surface of the water between them will tend to assume a horizontal form (Fig. 407*b*). In this case, as can be seen from the figure, the pressure between the matches at level *MN* is equal to the atmospheric pressure.

The pressure exerted from the left on the first match is also equal to the atmospheric pressure below level *MN*. The pressure acting on the second match from the right is less than the atmospheric pressure above level *MN*. As a result, the matches will be repulsed.

2-5. Mutual Conversion of Liquids and Solids

365. Water will freeze at zero only in the presence of centres of crystallization. Any insoluble particles can serve as such centres. When the mass of the water is great, it will always contain atleast one such centre. This will be enough for all the water to freeze. If the mass of the water is divided into very fine drops, centres of crystallization will be present only in a comparatively small number of the drops, and only they will freeze.

366. The water and the ice receive about the same amount of heat in a unit of time, since the difference between the temperatures of the water and the air in the room is approximately the same as that for the ice and the air. In 15 minutes the water receives 200 calories. Therefore, the ice receives 8,000 calories in ten hours.

Hence, $H = 80$ cal/g.

367. $v = 2{,}464$ m/s.

368. The heat balance equations have the form

$$Q_1 = m_1 c_1 \Delta t + C\Delta t$$

$$Q_2 = m_1 c_1 \frac{\Delta t}{2} + m_1 H + m_1 c_2 \frac{\Delta t}{2} + C\Delta t$$

where m_1 and c_1 are the mass and heat capacity of the ice, C is the heat capacity of the calorimeter, c_2 is the heat capacity of the water, and $\Delta t = 2°$ C. Hence,

$$C = \frac{Q_1\left(\frac{c_2}{2c_1} + \frac{H}{c_1 \Delta t} + \frac{1}{2}\right) - Q_2}{\frac{c_2}{c_1}\frac{\Delta t}{2} - \frac{\Delta t}{2} + \frac{H}{c_1}}$$

$$= 150 \text{ cal/deg}$$

369. The amount of heat that can be liberated by the water when it is cooled to 0° C is 4,000 cal. Heating of the ice to 0°C requires 12,000 cal. Therefore, the ice can be heated only by the heat liberated when the water freezes. One hundred grammes of water should be frozen to produce the lacking 8,000 calories.

As a result, the calorimeter will contain a mixture of 500 g of water and 500 g of ice at a temperature of 0°C.

370. The final temperature of the contents in the vessel is $\theta = 0°\text{C}$. The heat balance equation has the form

$$m_1 c_1 (t_1 - \theta) = m_2 c_2 (\theta - t_2) + (m_2 - m_3)H$$

where m_1 is the sought mass of the vessel and c_2 is the heat capacity of the ice. Therefore,

$$m_1 = \frac{m_2 c_2 (\theta - t_2) + (m_2 - m_3)H}{c_1 (t_1 - \theta)} = 200 \text{ g}$$

371. (1) The sought mass of the ice m can be found from the equation $mH = Mc(-t)$. Hence, $m = 100$ g.

(2) The heat balance equation can be written in this case as $MH = Mc(-t)$. Hence, $t = -80°\text{C}$.

372. The melting point of the ice compressed to 1,200 atm will drop by $\Delta t = 8.8°\text{C}$. The ice will melt until it cools to $-8.8°\text{C}$. The amount of heat $Q = m_1 H$ if is absorbed, where m_1 is the mass of the melted ice and H is the specific heat of fusion. From the heat balance equation $m_1 H = mc\Delta t$, where c is the heat capacity of the ice.

Hence, $$m_1 = \frac{cm\Delta t}{H} \cong 5.6 \text{ g}$$

2-6. Elasticity and Strength

373. $F = \dfrac{AE(R - r)}{r} = 60$ kgf.

374. When the rod with fastened ends is heated by t degrees, it develops an elastic force F equal, according to Hooke's law, to

$$F = \frac{AE\Delta l}{l} = AE\alpha t$$

where E is the modulus of elasticity of steel and α is its coefficient of thermal expansion.

If one of the rod ends is gradually released, the length of the rod will increase by $\Delta l = l\alpha t$. The force will decrease linearly from F to zero and its average magnitude will be $F/2$.

The sought work $W = \dfrac{F}{2}\Delta l = \dfrac{1}{2} AEl\alpha^2 t^2$.

375. The tension of the wire $T = \dfrac{Mg}{2\sin\alpha}$.

It follows from Hooke's law that $T = \dfrac{\Delta l}{2l} EA$.

Since, $$\Delta l = 2\left(\frac{l}{\cos\alpha} - l\right),$$

then $$T = \frac{1 - \cos\alpha}{\cos\alpha} AE = \frac{Mg}{2\sin\alpha}.$$

At small angles, $\sin\alpha \cong \alpha$, and

$$\cos\alpha = 1 - 2\sin^2\frac{\alpha}{2} \cong 1 - \frac{\alpha^2}{2}.$$

Bearing this in mind, we obtain $\alpha = \sqrt[3]{\frac{Mg}{AE}}$

376. The rod heated by Δt would extend by $\Delta l = l_0\alpha\Delta t$ in a free state, where l_0 is the initial length of the rod. To fit the heated rod between the walls, it should be compressed by Δl. In conformity with Hooke's law, $\Delta l = \frac{lF}{EA}$ Therefore, $F = EA\alpha\Delta t = 110$ kgf.

377. When the rods are heated in a free state, their total length will increase by

$$\Delta l = \Delta l_1 + \Delta l_2 = (\alpha_1 l_1 + \alpha_2 l_2)t.$$

Compression by the same amount Δl will reduce the lengths of the rods by $\Delta l_1'$ and $\Delta l_2'$, where $\Delta l_1' + \Delta l_2' = \Delta l$.

This requires the force $F = \frac{E_1 A}{l_1}\Delta l_1' = \frac{E_2 A \Delta l_2'}{l_2}$

Upon solving this system of equations, we find that

$$F = \frac{\alpha_1 l_1 + \alpha_2 l_2}{\frac{l_1}{E_1} + \frac{l_2}{E_2}} At$$

The rods will act upon each other with this force.

378. It is obvious from considerations of symmetry that the wires will elongate equally. Let us denote this elongation by Δl.

On the basis of Hooke's law, the tension of a steel wire $F_s = \frac{\Delta l}{l} AE_s$

and of a copper one $F_c = \frac{\Delta l}{l} AE_c$.

It follows that the ratio between the tensions is equal to the ratio between he respective Young's moduli

$$\frac{F_c}{F_s} = \frac{E_c}{E_s} = \frac{1}{2}$$

In equilibrium $2F_c + F_s = mg$.

Therefore, $$F_c = \frac{mg}{4} = 25 \text{ kgf}$$

and $$F_s = 2F_c = 50 \text{ kgf}$$

379. On the basis of Hooke's law, $F_c = \frac{\Delta l}{l} A_c E_c$ and $F_i = \frac{\Delta l}{l} A_i E_i$ It follows that $\frac{F_c}{F_i} = 2$. Thus, two-thirds of the load are resisted by the concrete and one-third by the iron.

380. The compressive force F shortens the tube by $\frac{Fl}{A_c E_c}$ and the tensile force F extends the bolt by $\frac{Fl}{A_s E_s}$. The sum $\frac{Fl}{A_s E_s} + \frac{Fl}{A_c E_c}$ is equal to the motion of the nut along the bolt: $\frac{Fl}{A_s E_s} + \frac{Fl}{A_c E_c} = h$

Hence,
$$F = \frac{h}{l} \frac{A_s E_s A_c E_c}{A_s E_s + A_c E_c}$$

381. Since the coefficient of linear thermal expansion of copper α_c is greater than that of steel α_s, the increase in temperature will lead to compression of the copper plate and tension of the steel ones. In view of symmetry, the relative elongations of all the three plates are the same. Denoting the compressive force acting on the copper plate from the sides of the steel plates by F, we shall have for the relative elongation of the copper plate; $\frac{\Delta l}{l} = \alpha_c t - \frac{F}{AE_c}$.

Either steel plate is subjected to the tensile force $F/2$ from the side of the copper one. Upon equating the relative elongation of the plates, we obtain:
$$\alpha_c t - \frac{F}{AE_c} = \alpha_s t + \frac{F}{2AE_s}$$

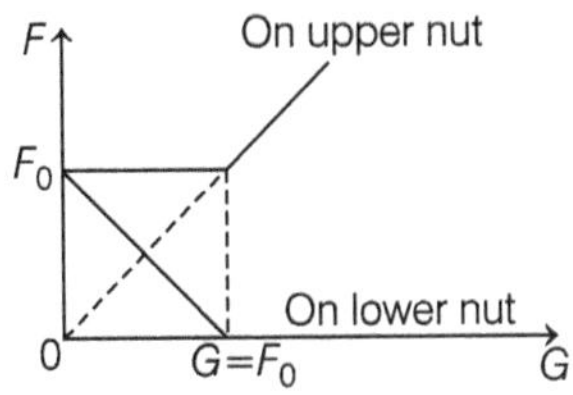

Hence,
$$F = \frac{2AE_c E_s (\alpha_c - \alpha_s)t}{2E_s + E_c}$$

382. When the ring rotates, the tension $T = \frac{mv^2}{2\pi r}$ appears in it (see Problem 201). For a thin ring $m = 2\pi r A\rho$, where A is the cross section of the ring. Therefore, $\frac{T}{A} = \rho v^2$.

Hence, the maximum velocity $v = \sqrt{\frac{\sigma_\alpha}{\rho}} \cong 41$ m/s.

383. Initially, an elastic force F_0 acts on each nut from the side of the extended bolt.

The load $G \le F_0$ cannot increase the length of the part of the bolt between the nuts and change its tension. For this reason the force acting on the upper nut from the side of the block will not change as long as $G \le F_0$.

The lower nut is acted upon by the force F_0 from the side of the top part of the bolt and by the force G from the bottom part. Since the nut is in equilibrium, the force exerted on it from the block is $F = F_0 - G$. Thus the action of the load $G \le F_0$ consists only in reducing the pressure of the lower nut on the block.

When $G > F_0$, the length of the bolt will increase and the force acting on the lower nut from the side of the block will disappear. The upper nut will be acted on by the force G.

The relation between the forces acting on the nuts and the weight of the load G is shown in Fig. 408.

2-7. Properties of Vapours

384. The calorimeter will contain 142 g of water and 108 g of vapour at a temperature 100° C.

385. By itself, water vapour or steam is invisible. We can observe only a small cloud of the finest drops appearing after condensation. When the gas burner is switched off, the streams of heated air that previously enveloped the kettle disappear, and the steam coming out of the kettle is cooled and condenses.

386. On the basis of the equation of state of an ideal gas $\rho = \frac{m}{V} = \frac{\rho\mu}{RT}$. If the pressure is expressed in mm Hg and the volume in m^3, then

$$= \frac{760 \times 0.0224}{273} \frac{\text{mm Hg} \cdot \text{m}^3}{\text{deg} \cdot \text{mole}}$$

Therefore, $\rho = 1.06p\frac{273}{T}$.

At temperatures near room temperature, $\rho \cong p\text{g/m}^3$.

387. It seems at first sight that the equation of state of an ideal gas cannot give values of the density or specific volume of saturated vapours close to the actual ones. But this is not so. If we calculate the density of a vapour by the formula $\rho = \frac{m}{V} = \frac{\mu p}{RT}$ and compare the values obtained with those in Table 2 (p. 69), we shall observe good agreement.

This is explained as follows. The pressure of an ideal gas grows in direct proportion to the temperature at a constant volume of the gas and, therefore, at a constant density. The relation between the pressure of saturated vapours and the temperature depicted in Fig. 146 corresponds to a constant volume of a saturated vapour and the liquid which it is in equilibrium with. As the temperature increases, the density of the vapour grows, since the liquid partially transforms into a vapour. An appreciable increase in the mass of the vapour corresponds to a small change in the volume it occupies. The pressure-density ratio becomes approximately proportional to the temperature, as with an ideal gas.

The Clapeyron-Mendeleyev equation mainly gives a correct relationship between p, V and T for water vapour up to the values of these parameters that correspond to the beginning of condensation. This equation, however, cannot describe the process of transition of a vapour into a liquid and indicate the values of p, V and T at which this transition begins.

388. At 30°C the pressure of saturated vapours $p = 31.82$ mm Hg. According to the equation of state of an ideal gas,

$$V = \frac{m}{\mu}\frac{RT}{p} \cong 296 \text{ lit}$$

389. When the temperature gradually increases, the pressure of the water vapours in the room may be considered constant.

The vapour pressure $p = \frac{w_0 p_0}{100}$ corresponds to a humidity of $w_0 = 10$ per cent, where $p_0 = 12.79$ mm Hg is the pressure of the saturated vapours at 15° C. At a temperature of 25°C the pressure of the saturated vapours is $p_1 = 23.76$ mm Hg.

For this reason the sought relative humidity is

$$w = \frac{p}{p_1}100\% = \frac{w_0 p_0}{p_1} = 5.4\%$$

390. According to the conditions of the problem, the relative humidity outside and in the room is close to 100 per cent. The pressure of saturated water vapours outside, however, is much smaller than in the room, because the temperature of the air in the room is higher and much time is required to equalize the pressures owing to penetration of the vapours outside through slits. Therefore, if the window is opened, the vapours will quickly flow out from the room and the washing will be dried faster.

391. (1) The water levels will become the same as in communicating vessels. The water vapours in the left-hand vessel will partly condense, and some water will evaporate in the right-hand vessel.

(2) The levels will become the same because the vapours will flow from one vessel into the other.

At a given temperature the pressure of saturated vapours is identical in both vessels at the surface of the water and will decrease at the same rate with height. For this reason the pressure of the vapours in the vessels at the same level is different.

This causes the vapour to flow over and condense in the vessel with the lower water level.

392. When $t_2 = 30° C$, the pressure of the vapours is equal to the pressure p_{20} of saturated vapours ($p_{20} = 31.8$ mm Hg) only if the air pressure is 10 atm. Upon isothermal reduction of the air pressure to one-tenth, the volume of the air will increase ten times. Hence, at atmospheric pressure and a temperature of 30°C, the pressure of the water vapour is $p = 3.18$ mm Hg. It follows from the Clapeyron equation that at a temperature of $t_1 = 10° C$, the vapour pressure $p_1 = p\frac{T_1}{T_2}$, where $T_1 = 283° K$ and $T_2 = 303° K$.

The sought relative humidity is $w = \frac{p_1}{p_0} 100\%$

$$= \frac{p}{p_0}\frac{T_1}{T_2} 100\% \cong 32.6\%$$

where, $p_0 = 9.2$ mm Hg is the pressure of saturated vapours at $t_1 = 10° C$.

393. The pressure $p = 6.5$ mm Hg is the pressure of saturated water vapours at $t = 5° C$. A sharp drop of the pressure shows that all the water has been converted into vapour. The volume of the vapour pumped out until the water is evaporated completely is $V = 3{,}600$ litres. On the basis of the Clapeyron-Mendeleyev equation of state, the sought mass of the water is

$$m = \frac{pV\mu}{Rt} \cong 23.4 \text{ g}$$

394. An amount of heat $Q_1 = mc\Delta t = 3{,}000$ cal is required to heat the water to 100° C. Therefore, $Q_2 = Q - Q_1 = 2{,}760$ cal will be spent for vapour formation. The amount of the water converted into vapour is

$$m_1 = \frac{Q_2}{r} = 5.1 \text{ g}$$

In conformity with the equation of state of an ideal gas, this vapour will occupy a volume of $V = \frac{m_1}{\mu}\frac{RT}{p}$. Upon neglecting the reduction of the volume occupied by the water, we can find the height which the piston is raised to: $h = \frac{V}{A} = 17$ cm.

3. ELECTRICITY AND MAGNETISM

3-1. Electrostatics

395. $F = \frac{Q^2}{r^2} = 918$ kgf.

The force is very great. It is impossible to impart a charge of one coulomb to a small body since the electrostatic forces of repulsion are so high that the charge cannot be retained on the body.

396. The balls will be arranged at the corners of an equilateral triangle with a side $\frac{\sqrt{3}}{2}l$. The force acting from any two balls on the third is

$$F = \frac{4Q^2}{l^2\sqrt{3}}.$$

The ball will be in equilibrium if $\tan\alpha = \frac{F}{mg}$ (where $\alpha = 30°$). Hence,

$$Q = \frac{1}{2}\sqrt{mg} \cong 100 \text{ CGS}_Q.$$

397. Since the threads do not deflect from the vertical, the coulombian force of repulsion is balanced by the force of attraction between the balls in conformity with the law of gravitation.

Therefore, in a vacuum

$$\frac{Q^2}{r^2} = \gamma\,\frac{\rho^2 V^2}{r^2}$$

and in kerosene (taking into account the results of Problem 230)

$$\frac{Q^2}{\varepsilon_r r^2} = \gamma\,\frac{(\rho - \rho_0)^2 V^2}{r^2}$$

where V is the volume of the balls.

Hence,

$$\rho = \frac{\rho_0\sqrt{\varepsilon_r}}{\sqrt{\varepsilon_r} - 1} \cong 2.74 \text{ g/cm}^3$$

398. The conditions of equilibrium of the suspended ball give the following equations for the two cases being considered:

$$T_1 \sin\alpha_1 - \frac{QQ_s}{2a^2} \times \frac{\sqrt{2}}{2} = 0$$

$$T_1 \cos\alpha_1 + \frac{QQ_s}{2a^2} \times \frac{\sqrt{2}}{2} - \frac{QQ_s}{a^2} - mg = 0$$

$$T_2 \sin \alpha_2 - \frac{QQ_s}{2a^2} \times \frac{\sqrt{2}}{2} = 0$$

$$T_2 \cos \alpha_2 + \frac{QQ_s}{a^2} - \frac{QQ_s}{2a^2} \times \frac{\sqrt{2}}{2} - mg = 0$$

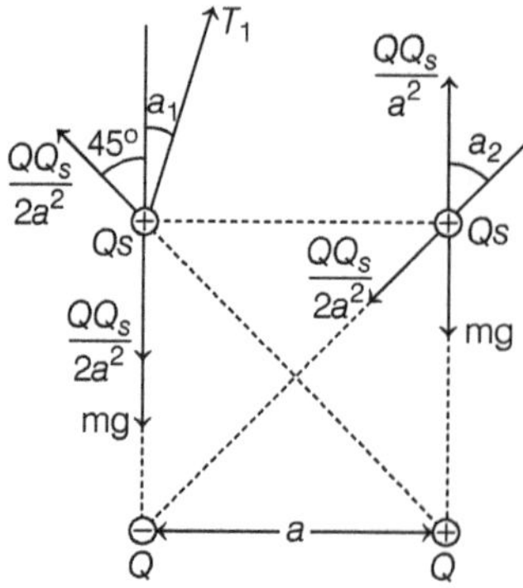

Fig. 409

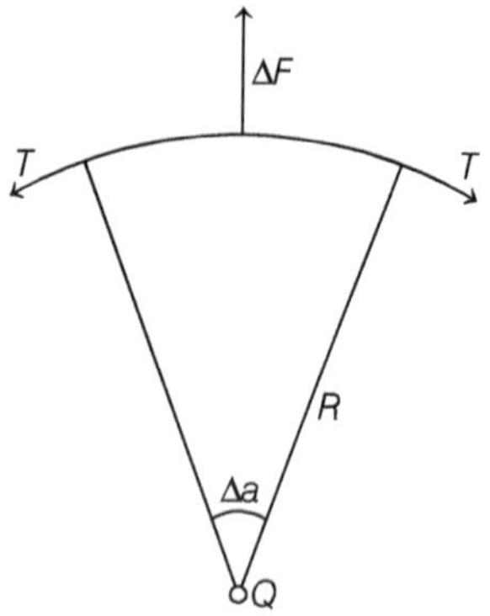

Fig. 410

where T_1 and T_2 are the tensions of the thread, α_1 and α_2 the angles of deflection of the thread, $+Q$ and $-Q$ the charges of the fixed balls, $+Q_s$ the charge of the suspended ball, and mg is the weight of the suspended ball (Fig. 409).

Upon excluding the unknowns from the above simultaneous equations, we get

$$\cot \alpha_1 - \cot \alpha_2 = \cot \alpha_1 - \cot 2\alpha_1 = 2(\sqrt{2} - 1)$$

whence, $$\cot \alpha_1 = 2(2\sqrt{2} - 1) \pm \sqrt{35 - 16\sqrt{2}}$$

Thus, $\alpha_1 = 7°56'$ and $\alpha_2 = 15°52'$ when $mg > \frac{QQ_s}{a^2}\left(1 - \frac{\sqrt{2}}{4}\right)$, and

$\alpha_1 = 82°04'$ and $\alpha_2 = 164°08'$ when $mg < \frac{QQ_s}{a^2}\left(1 - \frac{\sqrt{2}}{4}\right)$.

399. In uniform motion the drop is acted upon by the force of gravity G, the expulsive force of the air (Archimedean force) F, the force of the electrostatic field eE and the force of friction against the air $kv = k\frac{s}{t}$.

All the forces are balanced. Therefore,

$$G - F - eE + k\frac{s}{t_1} = 0$$

$$G - F + eE - k\frac{s}{t_2} = 0$$

$$G - F - k\frac{s}{t} = 0$$

where e is the charge of the drop, E the intensity of the electric field, and s the distance covered by the drop.

Upon solving the equations, we get

$$t = \frac{2\,t_1 t_2}{t_1 - t_2}$$

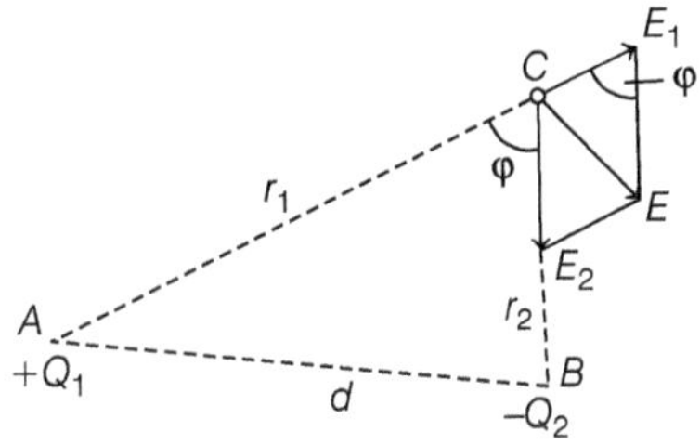

Fig. 411

400. It can, if we use the phenomenon of electrostatic induction. Bring a conductor on an insulated support up to the charged body and connect the conductor to the earth for a short time. The conductor will retain a charge opposite in sign to the given one, while the like charge will pass into the earth.

The charge can be removed from the conductor by introducing the latter into a metallic space. The operation may be repeated many times with a charge of any magnitude.

Electrostatic machines operate on a similar principle.

401. The energy is produced by the mechanical work that has to be performed in moving the conductor from the oppositely charged body to the body that accumulates the charge.

402. They can, if the charge of one ball is much greater than that of the other. The forces of attraction caused by the induced charges may exceed the forces of repulsion.

403. Since $Q \gg q$, the interaction between the separate elements of the ring can be neglected. Let us take a small element of the ring with a length $R\Delta\alpha$ (Fig. 410). From the side of the charge Q it is acted upon by the force $\Delta F = \frac{Q\Delta q}{R^2}$, where $\Delta q = \frac{q\Delta\alpha}{2\pi}$. The tension forces of the ring T balance ΔF.

From the condition of equilibrium, and remembering that $\Delta\alpha$ is small, we have $\Delta F = 2T\sin\left(\frac{\Delta\alpha}{2}\right) \cong T\Delta\alpha$

The sought force is the tension $T = \frac{Q_q}{2\pi R^2}$.

404. Let us consider the case of opposite charges $Q_1 > 0$ and $Q_2 < 0$. The intensities created by the charges Q_1 and Q_2 are equal, respectively, to

$E_1 = \frac{Q_1}{r_1^2}$ and $E_2 = \frac{Q_2}{r_2^2}$. A glance at Fig. 411 shows that

$$E^2 = E_1^2 + E_2^2 - 2E_1E_2 \cos\phi$$

From triangle ABC

$$\cos\phi = \frac{r_1^2 + r_2^2 - d^2}{2r_1r_2}$$

Hence,

$$E = \sqrt{\frac{Q_1^2}{r_1^4} + \frac{Q_2^2}{r_2^4} - \frac{Q_1Q_2}{r_1^3r_2^3}(r_1^2 + r_2^2 - d^2)}$$

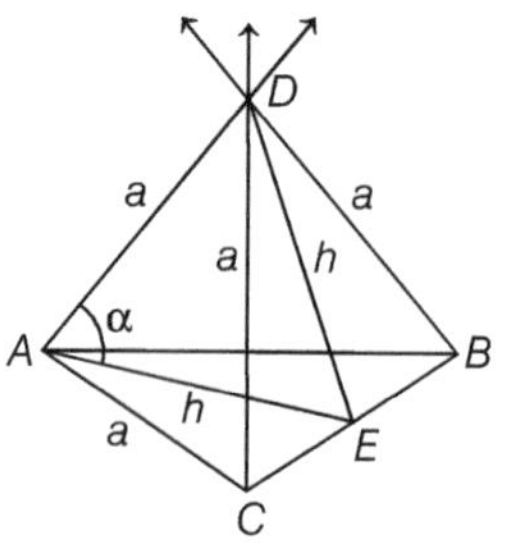

Fig. 412

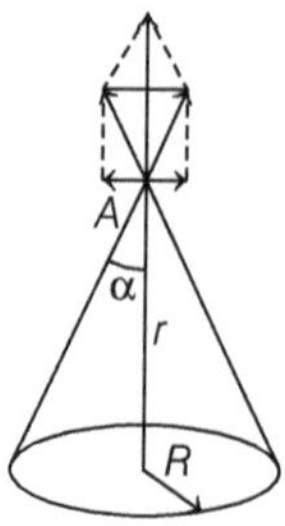

Fig. 413

If the charges are like

$$E = \sqrt{\frac{Q_1^2}{r_1^4} + \frac{Q_2^2}{r_2^4} + \frac{Q_1Q_2}{r_1^3r_2^3}(r_1^2 + r_2^2 - d^2)}$$

405. Each charge creates at point D a field intensity of $E_1 = \frac{Q}{a^2}$. The total intensity will be the sum of three vectors (Fig. 412). The sum of the horizontal components of these vectors will be zero, since they are equal in magnitude and form angles of 120° with each other. The vectors form angles of $90° - \alpha$ with the vertical, where α is the angle between the edge of the tetrahedron and the altitude h of triangle ABC.

The vertical components are identical, each being equal to $\frac{Q}{a^2}\sin\alpha$ It follows from triangle ADE that $\sin\alpha = \sqrt{\frac{2}{3}}$. Therefore, the sought intensity of the field is $E = \sqrt{6}\,\frac{Q}{a^2}$

406. The intensity of the field E at an arbitrary point A on the axis of the ring can be found as the geometrical sum of the intensities produced by separate small elements of the charged ring (Fig. 413).

Upon summation of the vectors of intensity at point A, account should be taken only of the components directed along the axis of the ring. Owing to symmetry, the components of the intensity vectors perpendicular to the axis will be zero.

Therefore, the intensity of the field at point A is

$$E = \frac{Q}{R^2 + r^2} \cdot \cos\alpha = \frac{Qr}{(R^2 + r^2)^{3/2}}$$

407. The intensity of the field at an arbitrary point A on the axis of the ring is

$$E = \frac{Q}{R^2 + r^2} \cos\alpha = \frac{Q}{R^2} \sin^2\alpha \cos\alpha \qquad (1)$$

(see Problem 406).

Obviously, E reaches its maximum at the same values of α as the expression $\frac{2E^2R^4}{Q^2}$. But

$$\frac{2E^2R^4}{Q^2} = 2\sin^4\alpha\cos^2\alpha = 2\cos^2\alpha(1 - \cos^2\alpha)(1 - \cos^2\alpha) \qquad (2)$$

is the product of three positive factors

$$a = 2\cos^2\alpha \qquad (3)$$

$$b = 1 - \cos^2\alpha \qquad (4)$$

$$c = 1 - \cos^2\alpha \qquad (5)$$

whose sum is constant ($a + b + c = 2$), and $b = c$.

The product $abc = ab^2$ will be maximum if the factors are equal, i.e.,

$$a = b = c = \frac{2}{3} \qquad (6)$$

and, therefore,

$$abc = ab^2 = \frac{8}{27} \qquad (7)$$

Let us prove this. If

$$a = \frac{2}{3} + 2d$$

where d is a certain number that, as follows from Eq. (3), can be within

$$-\frac{1}{3} < d < \frac{2}{3} \qquad (8)$$

then, on the basis of Eq. (4)

$$b = \frac{2}{3} - d$$

The product $$ab^2 = \left(\frac{2}{3} + 2d\right)\left(\frac{2}{3} - d\right)^2 = \frac{8}{27} + 2d^2(d-1)$$

is maximum, as follows from Eq. (8), when $d = 0$. Hence,

$$a = \frac{2}{3} \text{ and } \cos\alpha = \frac{\sqrt{3}}{3}$$

The maximum intensity of the field will be observed at points at a distance of $r = \frac{\sqrt{2}}{2}R$ from the centre of the ring. This intensity is equal to

$$E_{max} = \frac{2\sqrt{3}}{9}\frac{Q}{R^2}$$

408. $E_B = \frac{2\pi}{A}(Q_1 - Q_2)$, $E_C = \frac{2\pi}{A}(Q_1 + Q_2)$

$$E_A = -\frac{2\pi}{A}(Q_1 + Q_2)$$

The intensity is positive if it is directed from the left to the right.

409. The charges induced on the surfaces of the other plate are $+\frac{Q}{2}$ and $-\frac{Q}{2}$. Only in this case will the electric field inside the plate be zero, as it should be when the charges are in equilibrium.

410. The molecule will be attracted to the charged cylinder. The force of attraction is

$$F = 2qQ\left(\frac{1}{r} - \frac{1}{r+l}\right) = \frac{2qQl}{r(r+l)}$$

In this expression we can neglect the quantity $l(l \cong 10^{-8}\text{cm})$ as compared with r (r cannot be smaller than the cylinder radius). We finally obtain

$$F = \frac{2qQl}{r^2}$$

411. At the initial moment the forces acting on both molecules are identical. When the molecules approach the cylinder, the force F_1 acting on the molecule with a constant electric moment grows in proportion to $1/r^2$:

$$F_1 = \frac{2qQl}{r^2}$$

(see Problem 410). The force F_2 that acts on the "elastic" molecule grows faster, in proportion to $1/r^3$, owing to the continuous increase in the electric moment of this molecule

$$\left(F_2 = \frac{4q^2 Q^2}{kr^3}\right).$$

The masses of the molecules are the same, and for this reason the acceleration of the second molecule when it approaches the cylinder grows faster than that of the first one and it will reach the surface of the cylinder quicker.

412. Since the thickness of the plate is small it may be assumed that the charge is uniformly distributed on both surfaces, each having an area *ab*.

Thus, the surface density of the charge $\sigma = \frac{Q}{2ab}$. The field inside the metal is zero, and the intensity outside the metal is

$$E = 4\pi \frac{Q}{2ab} = \frac{2\pi Q}{ab}$$

413. The negative charges induced on the surface of a conductor are so distributed that the field inside the conductor resulting from a positive point charge and induced negative charges is zero. (The induced positive charges will move to the remote edges of the plate and their field may be neglected.) This distribution of the induced charges does not depend on the thickness of the plate.

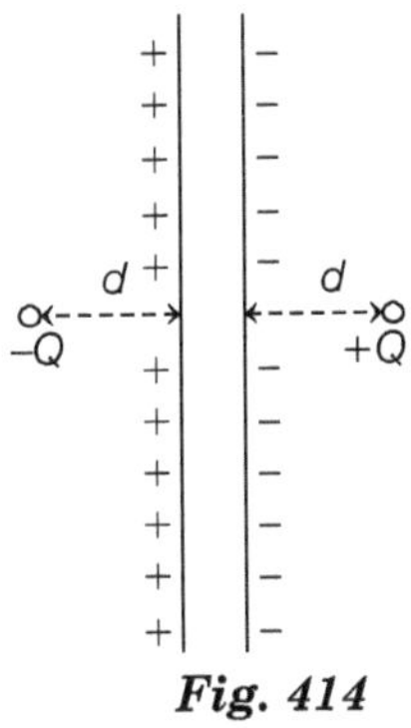

Fig. 414

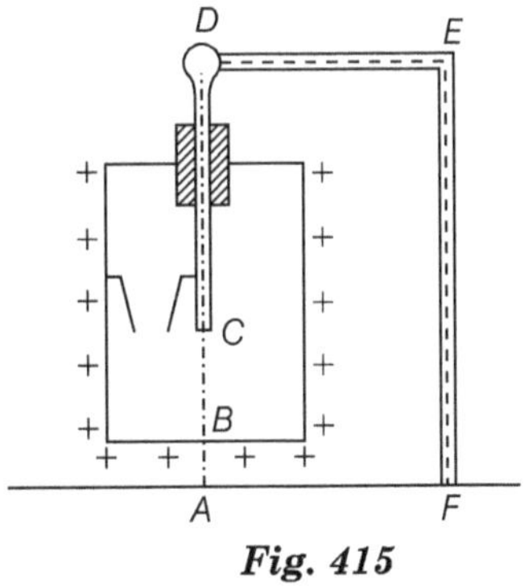

Fig. 415

Let us place a charge $-Q$ on the left of the plate and at the same distance *d*. Obviously, the induced positive charges will be distributed on the left side of the plate in the same manner as the negative charges on the right side. The charge $-Q$ placed on the left of the plate will not cause a change in the electric field at the right of the plate. Thus, at the right of the plate, the electric field induced by the charge $+Q$ and the induced negative charges coincides with the field produced by the charges $+Q$ and $-Q$ and the charges induced on the surfaces of the plate (Fig. 414). If the thickness of the plate is small as compared with d, the plate may be regarded as infinitely thin and hence the field created by the induced charges outside the plate is zero.

It has been shown that the field on the right of the plate produced by the charge $+Q$ and the induced negative charges is equal to the field caused by the point charges $+Q$ and $-Q$.

Since the intensity of the field caused by the induced negative charges at the point where the charge $+Q$ is equals the intensity of the field caused by the point charge $-Q$ at a distance of $2d$ from $+Q$, the sought force of attraction will be $F = \frac{Q^2}{rd^2}$.

414. Since a and b are much greater than c and d, it can be assumed that the plate is infinitely large. Remembering that the intensity of a field caused by several charges is equal to the sum of the intensities produced by each of these charges, and using the results of the solutions of Problems 412 and 413, we can obtain the sought force:

$$F = \frac{2\pi QQ_1}{ab} - \frac{Q_1^2}{4d^2}$$

The plus sign corresponds to the force of repulsion and the minus sign to the force of attraction.

The positively charged plate will attract the positive point charge if

$$\frac{Q_1^2}{4d^2} > \frac{2\pi QQ_1}{ab} \text{ or } \frac{Q_1}{d^2} > \frac{8\pi Q}{ab}$$

415. The maximum charge that can be imparted to the sphere is determined by the equation $E_0 = \frac{Q}{R^2}$

The potential will be $V = \frac{Q}{R} = E_0 R = 30{,}000\ RV$ if the radius of the sphere is in centimetres.

416. If the charged body is placed into the centre of the sphere, an additional charge $-q$ uniformly distributed over the surface will obviously appear on the external surface of the sphere, and a charge $+q$ on its internal surface. The potential V_R at a distance of R from the centre of the sphere will be

$$V_R = \frac{Q-q}{R}$$

When the body moves inside the sphere, the outside field will not change. Therefore, the potential will be V_R at any position of the charged body in the sphere.

417. The housing and the rod connected by the conductor will have equal potentials. For this reason the leaves will not deflect.

After the conductor is removed and the rod is earthed, both leaves will deflect. This is the result of a potential difference appearing between the rod and the housing, since the work of the electrostatic field is zero when the charge moves along a closed path *ABCDEFA* shown by the dotted line in Fig. 415. The work on section *AC* is zero, and the work on section *AB* is equal to that on *BC* taken with the reverse sign. The potential difference between the earth and the housing is equal to that between the housing and the rod.

418. When the housing of the electrometer is given, for instance, a positive charge, electrostatic induction will charge the ball of the electrometer rod positively and the end of the rod negatively. A potential difference will appear between the housing and the rod, and for this reason both leaves will deflect. The potentials of the housing and the rod are positive with respect to the earth, the potential of the housing being higher (that of the earth may be considered as zero). When the rod is connected to the earth, the potential difference between the rod and the housing will increase, as can be proved by the method used in Problem 417. Therefore, the angle of deflection of the leaves will be greater.

419. The electrometer measures the potential difference between the given body and the earth. Since the surface of the wire is equipotential, the leaves will deflect in the first case through the same angle with the ball in any position (if the capacitance of the wire is negligibly small).

In the second case the deflection of the leaves is determined by the potential of the ball with respect to the earth at the moment when the ball is brought in contact with the electrometer. This potential depends on the charge of the ball, its dimensions and the arrangement of the surrounding objects. With a constant arrangement of the objects, the potential changes only when the charge of the ball changes. When the ball touches the bucket, the former acquires the potential of the latter, but the charge of the ball will depend on the section of the surface being touched. If the internal surface of the bucket is touched, the charge of the ball will be zero, and if the external surface is touched, the charge will obviously be other than zero.

As the charge is transferred by the ball, its potential continuously changes, since its position with respect to the surrounding objects also changes. As a result, this method can be used to measure the distribution of the charge on the surface of the metal, but not its potential.

420. The answer follows from the solution of Problem 419.

In the second case the potential of the ball and the readings of the electrometer are determined by the magnitude of the charge carried by the ball from the surface of the wire being investigated. Upon contact with the wire, the potential of the ball is the same irrespective of the point of contact. The capacitance of the ball, however, depends on the shape of the section of the surface (particularly, on its curvature) which the ball is brought in contact with. Correspondingly, the charge transferred to the ball is also determined by the curvature of the section of contact.

421. The potential of all the points of the sphere is the same. To solve the problem it is sufficient to find the potential of only one point. It is easiest to find the potential of the centre of the sphere. It is equal to the potential created at the centre of the sphere by a point charge $U = \frac{Q}{d}$ plus the potential created by the charges appearing at the surface of the sphere owing to electrostatic induction. The latter potential, however, is equal to zero, since the total charge on the sphere is zero and all the elements of the charge are at equal distances from the centre.

Therefore, the potential of the sphere $U = \frac{Q}{d}$.

422. The energy of the charged sphere is equal to the work that can be performed by the charges on the sphere if they leave it and move away over an infinitely great distance.

Let the charges flow off the sphere gradually. Then, as it moves to infinity, the first charge $+q_1 (q_1 \ll Q)$ will perform work equal to $q_1\phi_1$, where $\phi_1 = \frac{Q}{C} = \frac{Q}{R}$ is the initial potential of the sphere. The following charges will perform less work, because the sphere gradually loses its charge and its potential decreases. The work of the last charge $q_n\phi_n = 0$, since the potential of an uncharged sphere is zero. The average potential of the sphere is $\phi = \frac{Q}{2R}$. Upon multiplying it by the initial charge, we obtain the energy of a charged sphere:

$$W_e = Q\phi = \frac{Q^2}{2R}$$

(This energy in known as the intrinsic energy.)

The same result can be obtained if we use a diagram showing the change in the potential of the sphere as the charge decreases. The diagram will have the form of a straight line passing at an angle to the axis of the charges, and the work will be numerically equal to the area limited by the diagram and the axes.

423. The energy of the charged sphere $W_e = \frac{Q^2}{2R} = \frac{RU^2}{2}$, where R is the radius of the sphere and U its potential (see Problem 422).

Upon a discharge this energy will be liberated as heat. Expressing the energy in calories, we obtain $W_e = 0.13$ cal.

424. The potential difference between the balls should be E. Therefore, $\frac{Q_1}{r_1} - \frac{Q_2}{r_2} = E$, where Q_1 and Q_2 are the charges of the balls. According to the law of conservation of charge,

$$Q_1 + Q_2 = 0.$$

Hence,

$$Q_1 = -Q_2 = \frac{Er_1r_2}{r_1 + r_2}$$

According to Coulomb's law,

$$F = \frac{E^2 r_1^2 r_2^2}{R^2(r_1 + r_2)^2} \cong 0.0044 \text{ dyne}$$

425. Let the initial charges of the balls be q_1 and q_2. Then the work $W_1 = \frac{q_1 q_2}{l}$, where l is the distance between the balls. After the balls are connected, their charges become identical: $q = \frac{q_1 + q_2}{2}$ and the work $W_2 = \frac{(q_1 + q_2)^2}{4l}$. It is obvious that $W_2 > W_1$. Besides, the heat Q is liberated in the conductor.

According to the law of conservation of energy, however, the total amount of energy in the balls should be the same in both cases. Since the work W_1 and, correspondingly, W_2, is the potential energy of the second ball in the field of the first one in the first and second cases, then

$$W_1 + W_{e1} = W_2 + Q + W_{e2}$$

where $W_{e1} = \frac{q_1^2}{2r} + \frac{q_2^2}{2r}$ is the intrinsic energy of the balls before connection and $W_{e2} = \frac{q^2}{2r} + \frac{q^2}{2r}$ is the intrinsic energy of the balls after the charges are redistributed (see Problem 422).

The energy liberated as heat is $Q = W_{e1} - W_{e2} + W_1 - W_2$

$$= \frac{(q_1 - q_2)^2}{4}\left(\frac{1}{r} - \frac{1}{l}\right)$$

426. Assume that the radius of the envelope increases by δ, which may be an infinitely small quantity. The expanding force will perform the work $W = 4\pi R^2 f\,\delta$, where f is the force per unit of area. This work is done at the expense of a reduction in the electrostatic energy. First the electrostatic energy is $\frac{Q^2}{2R}$, and after expansion $\frac{Q^2}{2(R+\delta)}$

The change in the energy $\frac{Q^2}{2R} - \frac{Q^2}{2(R+\delta)} = \frac{Q^2}{2}\,\frac{\delta}{R(R+\delta)}$

is equal to the work W, i.e., $4\pi R^2 f\,\delta = \frac{Q^2\delta}{2R(R+\delta)}$

Taking into account the fact that δ can be infinitely small, we obtain the following expression for the force : $f = \frac{Q^2}{8\pi R^4} = 2\pi\sigma^2$

Here, $\sigma = \frac{Q}{4\pi R^2}$ is the density of the electricity, i.e., the charge per unit of area.

The sought force can also be found directly. Let us consider a small area a on the sphere (Fig. 416).

Let us find the intensity E_1 of the electric field created on the area being considered by all the charges except the ones on the area itself. To introduce definiteness, let us consider the case when the sphere carries a positive charge.

Let us denote by E_2 the intensity of the electric field created by the charges on the area itself. Since the resulting intensity is zero inside the sphere, then $E_1 = E_2$.

The resulting intensity on the sphere $E_1 + E_2 = \frac{Q}{R^2}$, and, therefore,

$$2\,E_1 = \frac{Q}{R^2} = 4\pi\sigma.$$

Hence,

$$E_1 = 2\pi\sigma.$$

To find the force that acts from all the charges outside the area on the charges on the area, the intensity E_1 should be multiplied by the magnitude of the electric charge of the area σa:

$$F = E_1\sigma a = 2\pi\sigma^2 a$$

The force per unit of envelope area will be $f = 2\pi\sigma^2$.

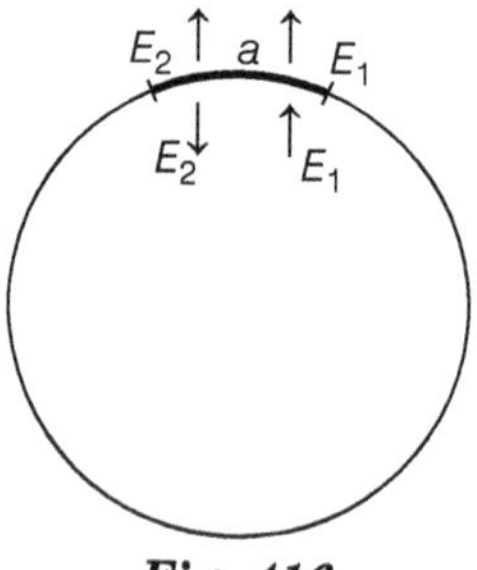

Fig. 416

427. For the charge q to be in equilibrium, the charges $-Q$ should be at equal distances a from it (Fig. 417). The sum of the forces acting on the charge $-Q$ is also zero:

$$\frac{Q^2}{4a^2} - \frac{Qq}{a^2} = 0$$

Hence, $q = \frac{Q}{4}$. The distance a may have any value. Equilibrium is unstable since when the charge $-Q$ is shifted along OO_1 to the left over a distance x, the force of attraction

$$F_q = \frac{Q^2}{4(a+x)^2}$$

acting from the side of the charge q is less than the force of repulsion

$$F_Q = \frac{Q^2}{(2a+x)^2}$$

and the charge $-Q$ moves still further from the position of equilibrium. When the charge $-Q$ is shifted along OO_1 over a distance x towards the charge q, then $F_q > F_Q$ for $x \le a$, and the system does not return to the position of equilibrium.

As can easily be seen, equilibrium is also violated by arbitrary motion of the charge q.

The potential energy of the charge $-Q$ in the field of the other two charges is

$$W_{e1} = -Q\left(\frac{q}{y} - \frac{Q}{a+y}\right) = \frac{Q^2}{4}\frac{3y-a}{y(a+y)}$$

where y is the distance between the charge q and one of the charges $-Q$.

When $0 \le y \le \infty$, the relation between W_{e1} and y is shown by curve ABC for one charge and curve DEF for the other (Fig. 417).

When the charges $-Q$ are stationary, the energy of the charge q is

$$W_{e2} = q\left(\frac{-Q}{a-z} - \frac{Q}{a+z}\right) = -\frac{Q^2}{2}\frac{a}{a^2+z^2}$$

where z is the displacement of the charge q from the position of equilibrium. When z changes from 0 to a, the energy changes according to curve MNP (Fig. 417).

It is interesting to note that the maxima of all the three potential curves correspond to the position of the charges in equilibrium. It is for this reason that equilibrium is not stable.

428. The work performed by the field of induced negative charges when the charge $+q$ moves is equal to the work done by the field of the charge $-q$ (see Problem 413). The work performed during the motion of both $+q$ and $-q$ is $\frac{q^2}{2d}$. Hence, the sought kinetic energy of the charge, equal to the work of motion of only one charge, will be $\frac{q^2}{4d}$.

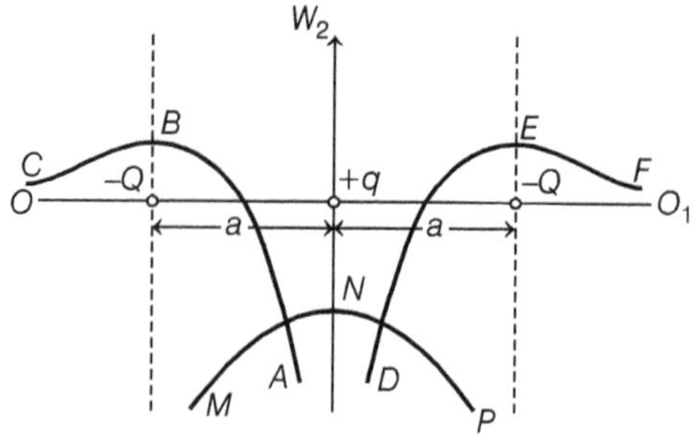

Fig. 417

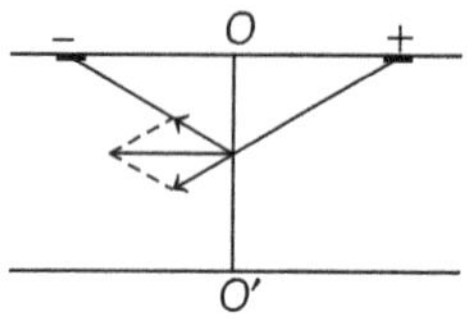

Fig. 418

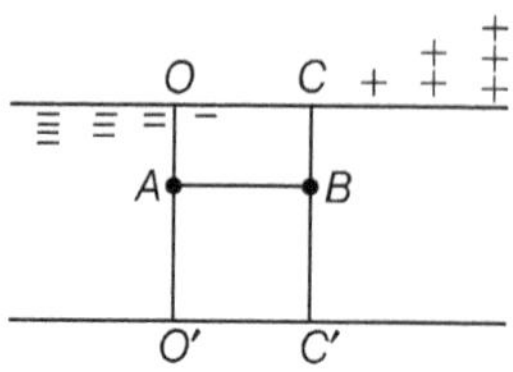

Fig. 419

429. Let us first prove that the intensity of the electric field at all points in the plane of section OO' is directed perpendicular to this plane.

Let us take an arbitrary point in the plane of the section and two small areas arranged arbitrarily but symmetrically on the cylinder with respect to section OO'. It is easy to see that the resulting intensity of the field induced by the charges on these areas will be directed along the axis of the cylinder (Fig. 418). Since another element arranged symmetrically with respect to the plane of the section can be found for each element, it follows that the intensity produced by all the elements will be parallel to the axis of the cylinder.

Let us now prove that the intensity will be the same at all points equidistant from the axis of the cylinder.

Let A and B be two such points (Fig. 419). The intensity of the field inside the cylinder will not change if, apart from the available charge, each square centimetre of the cylinder surface receives the same additional negative charge so that the density of the charges at point C is zero. This is obvious from the fact that the field inside an infinite uniformly charged cylinder is zero.

In this case the densities of the charges will be distributed on the cylinder surface as shown in Fig. 152. Therefore, the intensities at points A and B are the same.

It now remains to prove that the intensities of the fields at points at different distances from the axis of the cylinder are identical.

For this purpose let us consider circuit $BKLD$ (Fig. 420). With an electrostatic field the work in a closed circuit is known to be zero. The work is zero on sections KL and DB because the intensity of the field is perpendicular to the path. The work on section BK is $-E_B l$ and on section LD it is $E_D l$ (as proved above, $E_B = E_K$, and $E_D = E_L$).

Hence, $-E_B l + E_D l = 0$, i.e., $E_B = E_D$.

It has thus been proved that the intensity of the electric field at all points inside the cylinder will be the same everywhere and directed along the axis of the cylinder. It should be noted that such an arrangement of the charge on the surface of a conductor appears when direct current passes through it.

430. The concept of capacitance can be used because the ratio between the charge imparted to a conductor and the increment of the potential induced by this charge does not depend on the magnitude of the charge. In the same way, the ratio between the quantity of liquid poured into a vessel and the increase of the level in it should be a constant quantity. This will be true for any vessel with a constant cross section.

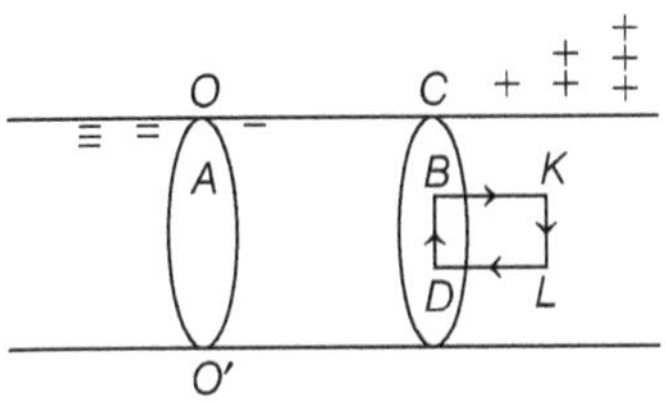

Fig. 420

431. It is impossible to calculate accurately the capacitance of a human body because it is extremely complicated in form. But the capacitance can be estimated in the order of its magnitude.

Let us find the capacitance of the body if it is shaped as a sphere. It should be expected that this assumption will give the approximate value of the capacitance. Since the mean specific weight of a human body $\gamma \cong 1 \text{gf} / \text{cm}^3$, the radius of a sphere whose weight is equal to that of a body can be found from the equation

$$\frac{4}{3}\pi R^3\gamma = 60{,}000 \text{ gf}$$

Hence, bearing in mind that the capacitance of the sphere is equal to its radius, we find that

$$C = \sqrt[3]{\frac{3\times 6\times 10^4}{4\pi}} \cong 25 \text{ cm}$$

Measurements give a close value: $C \cong 30$ cm.

432. The electrometer will show the e.m.f. of the galvanic cell irrespective of the capacitance.

433. (1) $U = 4E_a d = 8.4\times 10^4$ V.

(2) A voltage of $U_1 = E_a d = 2.1\times 10^4$ V can be supplied to each "air" capacitor.

The charge on the capacitor will be $Q = C_1U_1 = \frac{AU_1}{4\pi d}$. Upon series connection, the charge of all the capacitors is the same. Therefore, the voltage on the capacitor with glass will be $U_2 = \frac{Q}{C_2} = \frac{U_1}{E_r}$.

The entire battery may thus receive not more than

$$U = 3U_1 + U_2 = 6.6\times 10^4 \text{ V}.$$

If the voltage exceeds 6.6×10^4 V, all four capacitors will be punctured. The capacitor with the glass dielectric will be punctured last.

434. When a charge moves in a closed circuit, the work of the forces of an electrostatic field is zero. Therefore

$$E_1 - U_1 + E_2 - U_2 = 0$$

The charges on the capacitors are the same, since the sum of the charges present on the upper and the lower conductors is zero. Hence,

$$Q = C_1U_1 = C_2U_2.$$

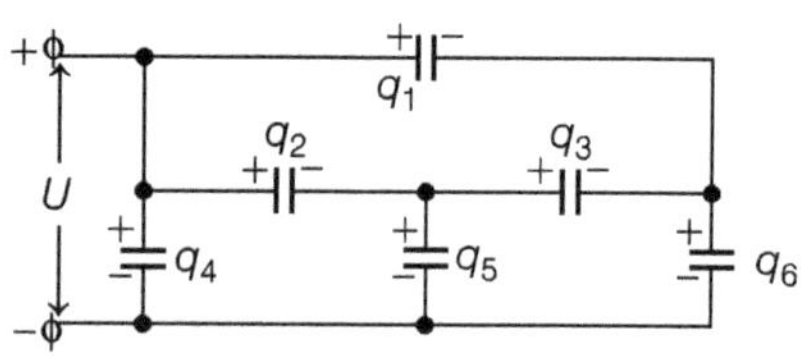

Fig. 421

Therefore,

$$U_1 = \frac{C_2}{C_1 + C_2}(E_1 + E_2) = 17.5 \text{ kV}$$

$$U_2 = \frac{C_1}{C_1 + C_2}(E_1 + E_2) = 7.5 \text{ kV}$$

435. Let the potential difference across the battery terminals be U and the charge of the battery Q. To find the capacitance of the battery means to find the capacitance of a capacitor which would have the same charge Q on its plates as the battery at the voltage U. Hence,

$$C_0 = \frac{Q}{U}$$

where

$$Q = q_1 + q_2 + q_3 = q_4 + q_5 + q_6$$

(Fig. 421) and $U = U_4 = \frac{q_4}{C}$.

In a closed circuit the work of the forces of an electrostatic field is zero. Therefore,

$$\frac{q_1}{C} - \frac{q_2}{C} - \frac{q_3}{C} = 0,$$

$$\frac{q_2}{C} - \frac{q_4}{C} + \frac{q_5}{C} = 0,$$

and

$$\frac{q_3}{C} - \frac{q_5}{C} + \frac{q_6}{C} = 0$$

Besides, the conductor that connects the second, third and fifth capacitors is electrically neutral. Hence,

$$q_3 + q_5 - q_2 = 0$$

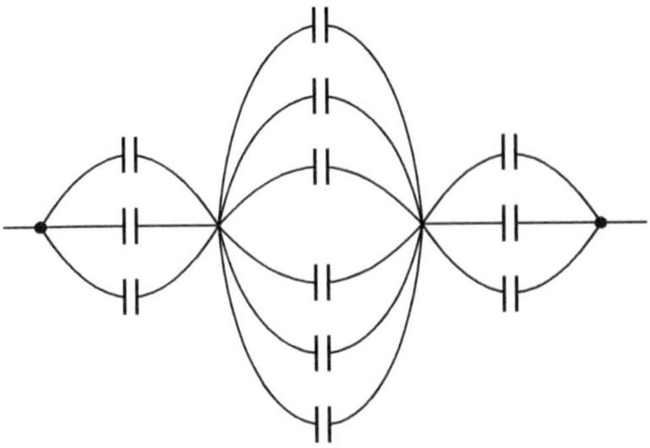

Fig.422

Upon solving these equations, we obtain

$$q_1 = q_2 = q_5 = q_6 = \frac{q_4}{2},$$

and

$$q_3 = 0$$

Therefore,

$$C_0 = 2C.$$

436. Let the battery of capacitors be charged. Points *1*, *2* and *3* will have the same potential and they can be connected to one another. Points *4*, *5* and *6* can also be interconnected (Fig. 155). The result will be the equivalent diagram shown in Fig. 422.

The capacitance of the separate sections is $3C$, $6C$ and $3C$. The total capacitance can be found from the formula

$$\frac{1}{C_0} = \frac{2}{3C} + \frac{1}{6C}$$

Hence, $$C_0 = 1.2C.$$

437. When the spark gaps are punctured, the parallel connection of the capacitors automatically changes to series connection; the voltage between the corresponding plates of the capacitors grows, since the capacitance of the system drops.

Indeed, the high resistance of conductors *AB* and *CD* makes it possible to neglect the currents flowing through them during the time of discharge and consider them as insulators through which the capacitors are not discharged.

An equivalent diagram after the first spark gap is punctured is shown in Fig. 423.

Upon puncturing of the first gap, the potential difference across the second gap will be equal to the sum of the voltages across the first and the second capacitors, i.e., it will double. This will cause puncturing of the second gap.

When the n-th gap is punctured, the voltage in it will reach $V = nV_0$.

The resistances of conductors *AB* and *CD* should be high enough so as not to allow the capacitors to be discharged through them when the gaps are punctured with the plates connected in series.

438. Yes, it will. Each plate has a definite, usually small, capacitance with respect to the earth (the force lines are distorted near the edges of the plates and reach the earth). An equivalent diagram is shown in Fig. 424. The capacitance of the plates with respect to the earth is shown in the form of small capacitances C_1 and C_2.

When the left-hand plate is short-circuited, part of the charge present in it is neutralized. This will also occur if the right-hand plate is short-circuited. The capacitor will continue to be discharged the slower, the higher is the capacitance of the capacitor as compared with that of the plates relative to earth.

439. The initial state of the system is illustrated by an equivalent diagram (Fig. 425*a*). The full charge of the capacitor is $Q + q$. The force lines of the charge $+Q$ terminate on the other plate of the capacitor. The force lines of the charges $+q$ and $-q$ terminate or start on the earth. Since $C \gg C_1 = C_2$, then $Q \ll q$.

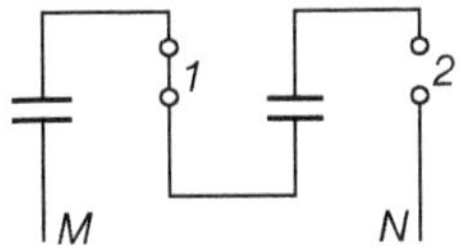

Fig. 423

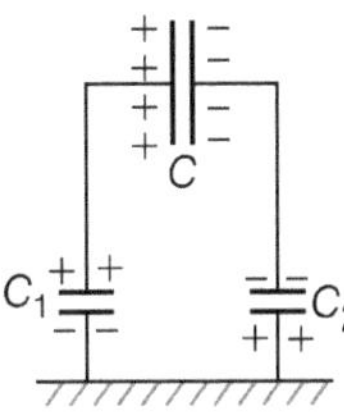

Fig. 424

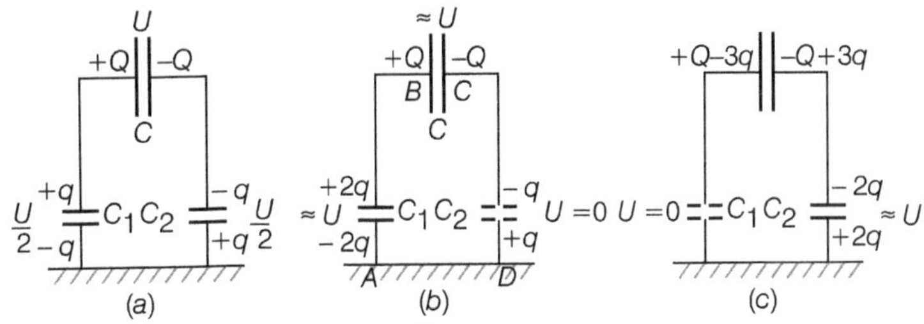

Fig. 425

When the capacitor C_2 is short-circuited, the charge $-q$ is neutralized. The potential difference between the plates of the capacitor C should remain approximately equal to U, since $q \ll Q$. The work required to move the charge along the circuit $ABCDA$ is zero. Therefore, the voltage across the capacitor C_1 should become equal to $\approx U$ and the charges on it to $+2q$ and $-2q$. The charges on the plates of the capacitor C will be $+Q-q$ and $-Q+q$, respectively (Fig. 425*b*).

When the capacitor C is disconnected from the earth, the distribution of the charges and, hence, of the potentials will not change. if the capacitor C_1 is shorted, the charges will be redistributed as shown in Fig. 425*c*. It is only in this case that the required potentials relative to the earth will be obtained: zero on the left plate and $\approx U$ on the right one. When the plates are earthed alternately, the potential difference between the plates will gradually drop because the charge decreases.

440. No, they will not. When the plates are alternately earthed the same processes will take place as in the absence of the battery (see Problem 439). The only difference is that the potential difference between the plates is always kept constant.

441. The total energy of the two capacitors before connection is

$$W_{e0} = \frac{1}{2}(C_1U_1^2 + C_2U_2^2)$$

and after connection

$$W_e = \frac{1}{2}\frac{Q^2}{C_1 + C_2} = \frac{1}{2}\frac{(C_1U_1 + C_2U_2)^2}{C_1 + C_2}$$

It is easy to see that $W_{e0} > W_e$.The difference in the energies is

$$W_{e0} - W_e = \frac{C_1C_2}{C_1 + C_2}(U_1^2 + U_2^2 - 2U_1U_2) > 0$$

When $U_1 = U_2$, we have $W_{e0} - W_e = 0$, and when $C_1 = C_2$ and $U_2 = 0$, then $W_{e0} = 2W_e$.

The electrostatic energy diminishes because when the capacitors are connected by conductors, the charges flow from one capacitor to the other. Heat is liberated in the connecting conductors. The quantity of heat evolved will not depend on the resistance of the conductors. When the resistance is low, the conductors will allow greater currents to flow through them, and vice versa.

442. Since the dielectric is polarized, the intensity will increase at points A and C and decrease at point B.

443. $E = \dfrac{2\pi Q}{\varepsilon_r A} = 50.2\ \text{CGS}_\text{E}$.

444. The capacitances and, therefore, the charges of the balls immersed in kerosene increase ε_r times:

$$q_1' = \varepsilon_r q_1 \text{ and } q_2' = \varepsilon_r q_2$$

The force of interaction of the charges in the dielectric, on the contrary, diminishes ε_r, times.

Hence,

$$F = \frac{q_1'q_2'}{\varepsilon_r R^2} = \frac{\varepsilon_r q_1 q_2}{R^2} = \frac{\varepsilon_r E^2 r_1^2 r_2^2}{R^2(r_1 + r_2)^2} = 0.0088 \text{ dyne}$$

The force of interaction increases $\varepsilon_r$times, while if the balls were disconnected from the battery it would decrease ε_r times.

445. As a result of motion of the plates, the charge on the capacitor will be increased by

$$\Delta Q = Q_2 - Q_1 = \frac{EA}{4\pi}\left(\frac{1}{d_2} - \frac{1}{d_1}\right)$$

The battery will perform the work

$$W = E\Delta Q = \frac{E^2A}{4\pi}\left(\frac{1}{d_2} - \frac{1}{d_1}\right)$$

The electrostatic energy of the capacitor will be increased by

$$\Delta W_e = W_{e2} - W_{e1} = \frac{EQ_2}{2} - \frac{EQ_1}{2} = \frac{E^2 A}{8\pi}\left(\frac{1}{d_2} - \frac{1}{d_1}\right)$$

The mechanical work W_1 was performed when the plates were moved closer to each other. On the basis of the law of conservation of energy,

$$W = W_1 + \Delta W_e.$$

Therefore,

$$W_1 = W - \Delta W_e = \frac{E^2 A}{8\pi}\left(\frac{1}{d_2} - \frac{1}{d_1}\right)$$

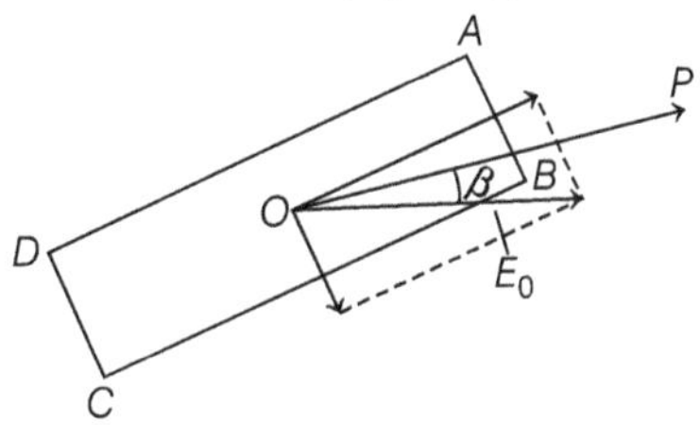

Fig. 426

At the expense of the work of the battery, the electrostatic energy of the capacitor increased and the mechanical work W_1 was done.

446. Let us consider for the sake of simplicity a dielectric in the form of a homogeneous very elongated parallelepiped (Fig. 426).

Let us resolve the field E_0 in which the piece of dielectric (for example, a rod) is placed into components directed along the rod and perpendicular to it. These components will cause bound charges to appear on surfaces *AB*, *CD*, *BC* and *AD*. The field of the bound charges between surfaces *AD*, *BC*, and *AB*, *DC* weakens the components of the field E_0 inside the dielectric, the component perpendicular to the rod being weakened more since the bound charges on surfaces *AD* and *BC* are close to each other and their field is similar to the homogeneous field of a plane capacitor, while the charges on the surfaces of the small area are moved far apart. For this reason the full field inside the dielectric will not coincide in direction with the field E_0. The dipoles appearing will therefore be oriented not along E_0, but along a certain direction *OP* forming the angle β with E_0. (This refers to both ordinary and dipole molecules.) From an electrical standpoint, a polarized dielectric can be regarded as a large dipole forming the angle β with the field E_0.

The dielectric will rotate in this field until it occupies a position along the field. The field of the bound charges is an internal force and cannot cause rotation of the dielectric.

447. (a) The capacitance of the capacitor will be equal to that of capacitors connected in parallel, one of which is filled with a dielectric, and the other not, i. e.,

$$C = \frac{\varepsilon_r A l_1}{4\pi d l} + \frac{A(l - l_1)}{4\pi d l}$$
$$= \frac{A}{4\pi d}\left\{1 + (\varepsilon_r - 1)\frac{l_1}{l}\right\}$$

(b) The electric field between the plates of the capacitor will not change, and, consequently, the capacitance will not change if the upper surface of the dielectric is coated with an infinitely thin layer of a conductor. Therefore, the sought capacitance will be equal to the capacitance of two capacitors connected in series:

$$C = \frac{C_0 C_1}{C_0 + C_1}, \text{ where}$$

$$C_0 = \frac{A}{4\pi(d - d_1)}$$

and
$$C_1 = \frac{\varepsilon_r A}{4\pi d_1}$$

Hence,
$$C = \frac{\varepsilon_r A}{4\pi\{d_1 + \varepsilon_r(d - d_1)\}}$$

448. To simplify reasoning, assume that two parallel metallic plates carrying the charges $+Q$ and $-Q$ are placed into a liquid dielectric. The intensity of the electric field between the plates is $E = \frac{4\pi Q}{\varepsilon_r A}$. The intensity of the field induced by each plate will be

$$E_1 = E_2 = \frac{2\pi Q}{\varepsilon_r A}$$

Let us find the force acting, for example, from the side of the first plate on the second. For this purpose the intensity of the field induced by the first plate should be multiplied by the charge on the second one. Thus,

$$F = \frac{2\pi Q^2}{\varepsilon_r A}$$

Let us assume that the first plate is fixed and the second can slowly move (we disregard the change in the mechanical energy of the dielectric). The work that the electric field can perform in moving the plates up to direct contact is equal to the product of the force F (constant) by the distance d, i.e.,

$$W = fd = \frac{2\pi Q^2 d}{\varepsilon_r A}$$

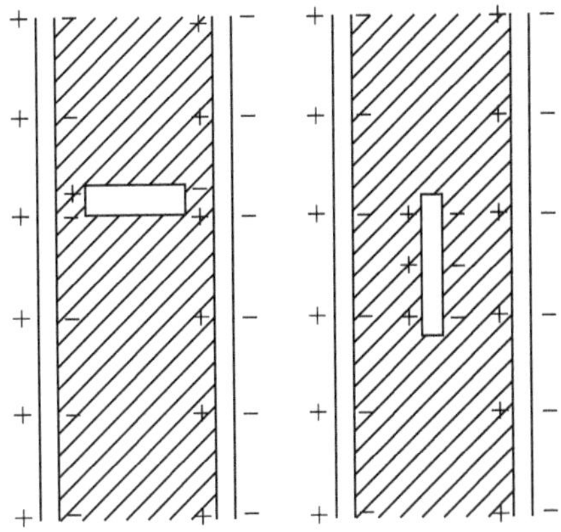

Fig. 427

This work is done at the expense of a reduction in the electric energy of the capacitor. Thus, the electrostatic energy will be equal to

$$W_e = \frac{2\pi Q^2 d}{\varepsilon_r A} = \frac{Q^2}{2C}$$

$$\text{or, } W_e = \frac{QU}{2}$$

where U is the potential difference. The above formula is true for any dielectric.

449. The intensity of the field inside the space consists of E and the intensity induced by the charges that appear on the surfaces of the space owing to polarization of the dielectric (Fig. 427). In the first case the influence of the polarizing charges is negligibly small, and $E_1 = E$. In the second case the action of the charges exerted on the surfaces of the space is fully compensated by the action of the charges on the surfaces of the dielectric adjoining the metal plates of the capacitor. For this reason the sought intensity is equal to that which would be induced if the dielectric were removed altogether, i. e., $E_2 = \varepsilon_r E$

450. If the dielectric is drawn into the capacitor over the distance x, the energy of the capacitor will be

$$W_{e1} = \frac{Q^2}{2C} = \frac{AU^2}{8\pi d} \frac{1}{1 + (\varepsilon_r - 1)\frac{x}{l}}$$

$$\text{since } C = \frac{A}{4\pi d}\left\{1 + (\varepsilon_r - 1)\frac{x}{l}\right\}$$

and $$Q = \frac{A}{4\pi d} U \text{ (see Problem 447).}$$

If x is increased by δ, the energy will be reduced and become equal to

$$W_{e2} = \frac{AU^2}{8\pi d} \frac{1}{1 + (\varepsilon_r - 1)\dfrac{x + \delta}{l}}$$

The difference between the energies $W_{e1} - W_{e2} = \dfrac{AU^2}{8\pi d}$

$$\frac{(\varepsilon_r - 1)\dfrac{\delta}{l}}{\left\{1 + (\varepsilon_r - 1)\dfrac{x + \delta}{l}\right\}\left\{1 + (\varepsilon_r - 1)\dfrac{x}{l}\right\}}$$

will be equal to the work of the sought force F over the path δ. Generally speaking, the magnitude of the force will change over this path, but if a sufficiently small value of δ is taken we can write

$W_{e1} - W_{e2} = F\delta$

It follows therefore that

$$F = \frac{AU^2}{8\pi dl} \frac{\varepsilon_r - 1}{\left\{1 + (\varepsilon_r - 1)\dfrac{x}{l}\right\}^2}$$

If δ is neglected in the denominator.

It should be noted that if when calculating the energy we assumed that the field inside the capacitor was homogeneous and neglected the boundary effects, it is necessary to take into account the heterogeneity of the field near the edges to explain the action of the force acting on the dielectric from the physical viewpoint.

451. When the dielectric is drawn over the distance x into the capacitor, the energy of the latter will be

$$W_{e1} = \frac{1}{2}U^2C = \frac{1}{2}U^2\frac{A}{4\pi d}\left\{1 + (\varepsilon_r - 1)\frac{x}{l}\right\}$$

If x increases by δ, the energy of the capacitor will increase by

$$W_{e2} - W_{e1} = \frac{U^2A}{8\pi d}(\varepsilon_r - 1)\frac{\delta}{l}$$

When the dielectric moves over the distance δ, the charge on the plates of the capacitor will increase by

$$Q_2 - Q_1 = \frac{UA}{4\pi d}(\varepsilon_r - 1)\frac{\delta}{l}$$

The work performed by the battery in moving such a quantity of electricity will be equal to

$$W = (Q_2 - Q_1)U = \frac{U^2A}{4\pi d}(\varepsilon_r - 1)\frac{\delta}{l}$$

Some of this work is spent to increase the electrostatic energy of the capacitor and some to pull in the dielectric. Let us denote, as in the previous problem, the force with which the dielectric is drawn into the capacitor by F. Then, on the basis of the law of conservation of energy, we have: $W = W_{e2} - W_{e1} + F\delta$, i.e.,

$$\frac{U^2 A}{4\pi d}(\varepsilon_r - 1)\frac{\delta}{l} = \frac{U^2 A}{8\pi d}(\varepsilon_r - 1)\frac{\delta}{l} + F\delta$$

Therefore,

$$F = \frac{U^2 A}{8\pi d l}(\varepsilon_r - 1)$$

As can be seen, in this case the force is constant and does not depend on x.

452. Apart from the weight directed downward, the kerosene is acted upon by electrostatic forces directed upward. All the forces are applied to all the elements of the liquid volume. As a result, the hydrostatic pressure in the kerosene will decrease as if its specific weight were reduced. For this reason, the lifting force acting on the balls in the left-hand part will be the same as in the right-hand one despite the fact that the left-hand contains part more balls.

453. The intensity of the field in the dielectric will be

$$E = \frac{U}{d} = \frac{Q}{Cd} = \frac{4\pi Q}{\varepsilon_r A} = \frac{4\pi}{\varepsilon_r}\sigma_0$$

where $\sigma_0 = \dfrac{Q}{A}$.

This intensity is created by the free charges on the plates of the capacitor, and by the bound charges induced owing to polarization of the dielectric. The bound charges are on the surface of the dielectric. Let us denote the density of these charges by σ_1 (Fig. 428). The intensity of the field created by the free charges is $E_0 = 4\pi\sigma_0$, and that of the field created by the bound charges is $E_1 = 4\pi\sigma_1$.

Thus the resulting intensity

$$E = E_0 - E_1 = 4\pi(\sigma_0 - \sigma_1) = \frac{4\pi\sigma_0}{\varepsilon_r}.$$

This equation can be used to find $\sigma_1 = \dfrac{\varepsilon_r - 1}{\varepsilon_r}\sigma_0$.

To determine the force acting, for example, on the upper surface of the dielectric, it is necessary to calculate the intensity of the field on this surface created by all the charges except for the charges on the surface itself. This intensity will be equal to

$$E_0 - \frac{E_1}{2} = 4\pi\sigma_0 - 2\pi\sigma_1 = 2\pi\sigma_0\frac{\varepsilon_r + 1}{\varepsilon_r}$$

The force acting on the upper surface of the dielectric will be directed upwards and be equal to $F = 2\pi\sigma_0 \frac{\varepsilon_r + 1}{\varepsilon_r} \frac{\varepsilon_r - 1}{\varepsilon_r} \sigma_0 A$

An identical force will act on the lower surface of the dielectric. Thus, the dielectric will be stretched and a unit of area of the dielectric will be acted upon by a force equal to $f = 2\pi\sigma_0^2 \frac{\varepsilon_r^2 - 1}{\varepsilon_r^2}$

454. Since the action of the field causes complete orientation of the molecules, the ends of the "dumb-bells" with negative electric charges will be in a layer with a thickness l near the positively charged plate, and the ends of the "dumb-bells" with a positive electric charge near the negatively charged plate (Fig. 429).

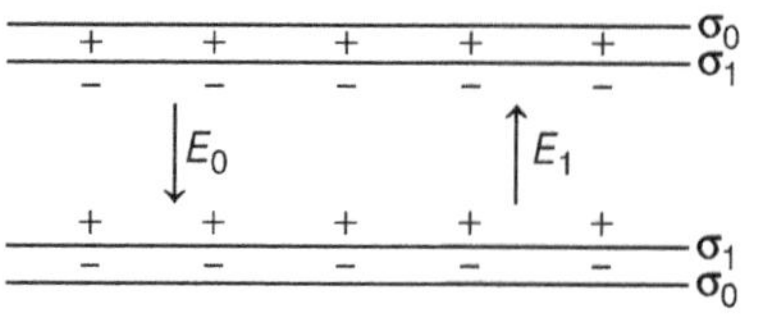

Fig. 428

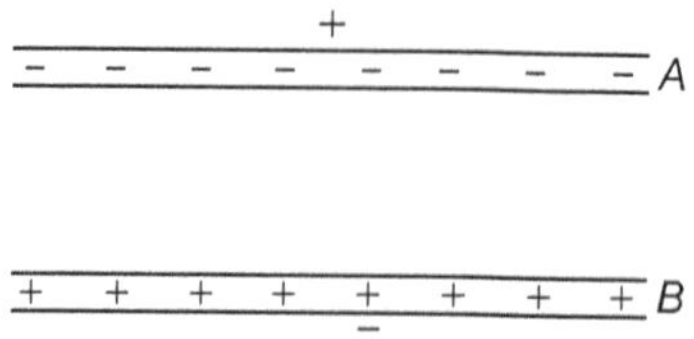

Fig. 429

The quantity of negative and positive charges is the same at a distance greater than l from the plates inside the dielectric.

The intensity of the field inside the dielectric produced by the charges in layers A and B is $E_1 = 4\pi\sigma_1 = 4\pi Qln$ and the total intensity is $E = E_0 - E_1 = E_0 - 4\pi Qln$.

It should be noted, however, that the orienting action of the field will always be hampered by the disorienting thermal motion, which is not taken into account in the calculations.

455. Let us denote the sought full intensity of the field in the dielectric by E. The distance l over which the charges moved apart in each molecule can be found from the expression $kl = qE$. As in solving Problem 454, we obtain

$$E_1 = 4\pi\sigma_1 = 4\pi Qln = 4\pi \frac{Q^2 n}{k} E$$

and $$E = E_0 - E_1 = E_0 - 4\pi \frac{Q^2 n}{k} E$$

The permittivity ε_r can be determined from the ratio

$$E = \frac{E_0}{\varepsilon_r}. \text{ Hence, } \varepsilon_r = 1 + 4\pi \frac{Q^2}{k} n.$$

456. When the charges $+Q$ and $-Q$ move apart in the molecule over the distance l, the work $\frac{kl^2}{2}$ is performed (see Problem 137).

The energy stored in the dielectric is

$$W_{e1} = \frac{kl^2}{2} N$$

where $N = ALn = Vn$ is the number of molecules in the volume V of the dielectric located between the plates of the capacitor. Thus,

$$W_{e1} = n\frac{kl^2}{2} V$$

Since $l = \frac{QE}{k}$, then $W_{e1} = n\frac{Q^2E^2}{2k} V$.

By expressing $\frac{nQ^2}{k}$ through ε_r., i.e.,

$$\frac{\varepsilon_r - 1}{4\pi} = \frac{nQ^2}{k}$$ we obtain for W_{e1}

$$W_{e1} = \frac{\varepsilon_r - 1}{8\pi} E^2V$$

The total energy of the capacitor is

$$W_e = \frac{Q^2}{2C} = \frac{\varepsilon_r}{8\pi} E^2V$$

This energy W_e can be represented as the sum of purely electrostatic energy

$$W_{e0} = \frac{E^2}{8\pi} V$$

and the energy stored in the dielectric

$$W_{e1} = \frac{\varepsilon_r - 1}{8\pi} E^2V$$

3.2. Direct Current

457. When a direct current flows along a conductor, the electric field inside it is constant and directed along it. If the charge moves in a closed circuit *abcd* (Fig. 430) the work of the electric field is zero. Let us assume the sections *ad* and *bc* to be infinitely small so that the work done on them may be neglected. Therefore, the work along *ab* is equal to that along *dc*. For this reason the tangential component of the field near the surface of the conductor should be equal to the field inside it.

458. The arrangement of the force lines is shown in Fig. 431. The increase in the slope of the line near the bend can be explained by the fact that the tangential component of the field at the surface of the conductor with an invariable cross section is constant (see Problem 457), while the normal component decreases towards the curvature since the potential difference between the corresponding sections lying on the opposite sides of the arc diminishes.

459. By applying Ohm's law to section AB of the circuit, we have

$$\frac{1}{V}=\frac{1}{r}+\frac{1}{R}$$

Hence,

$$r=\frac{V}{I}-\frac{1}{1-\frac{V}{IR}}\cong\frac{V}{I}\left(1-\frac{V}{IR}\right)$$

since $\frac{V}{IR}=0.008\ll 1$; finally $r=20.16\,\Omega$.

460. A reduction in the sensitivity n times means that the galvanometer carries a current I_1 which is n times smaller than the current in

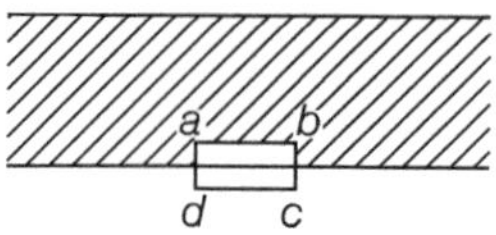

Fig. 430

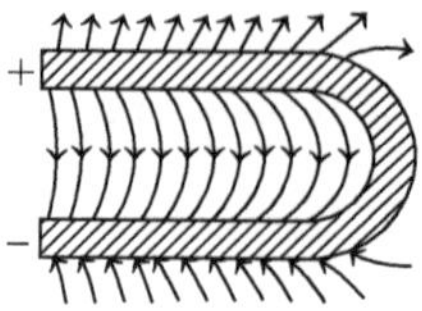

Fig. 431

the rest of the circuit before the branching off. Therefore, the current I_2 through the shunt is $\frac{n-1}{n}$ of the current I in the rest of the circuit.

Hence,

$$\frac{r}{R}=\frac{I_1}{I_2}=\frac{1}{n-1}$$

Therefore,

$$r=\frac{R}{n-1}\cong 204\,\Omega.$$

461. The error sought is $\varepsilon=\frac{V_0-V}{V_0}$, where V_0 is the voltage across the resistance R before the voltmeter is switched on and V the voltage after it is switched on.

According to Ohm's law, $V_0=IR$ and $V=I\frac{RR_0}{R+R_0}$, where R_0 is the resistance of the voltmeter.

Hence,

$$\varepsilon = \frac{R/R_0}{1 + \dfrac{R}{R_0}}$$

is determined only by the ratio between the resistances of the section of the circuit and the voltmeter. When $R_0 \gg R$, the error may be neglected.

462. Before the ammeter is connected, $I_0 = \frac{V}{R}$, and after it is connected $I = \frac{V}{R + R_0}$, where R_0 is the resistance of the ammeter. The error is

$$\varepsilon = \frac{I_0 - I}{I_0} = \frac{1}{1 + \dfrac{R}{R_0}}$$

When $R_0 \ll R$ the error may be neglected.

463. In series connection, the resistance of the circuit is

$$R = R_{01} + R_{02} + \alpha_1 R_{01} t + \alpha_2 R_{02} t$$

On the other hand we can write $R = R_0(1 + \alpha' t)$, where $R_0 = R_{01} + R_{02}$ and α' is the sought temperature coefficient. Therefore,

$$\alpha' = \frac{R_{01}\alpha_1 + R_{02}\alpha_2}{R_{01} + R_{02}}$$

In parallel connection

$$R = \frac{R_{01}R_{02}(1 + \alpha_1 t)(1 + \alpha_2 t)}{R_{01}(1 + \alpha_1 t) + R_{02}(1 + \alpha_2 t)}$$

$$= R_0(1 + \alpha'' t)$$

where

$$R_0 = \frac{R_{01}R_{02}}{R_{01} + R_{02}}$$

Omitting the terms proportional to the products of the temperature coefficients as being small, we obtain $\alpha'' = \frac{R_{02}\alpha_1 + R_{01}\alpha_2}{R_{01} + R_{02}}$

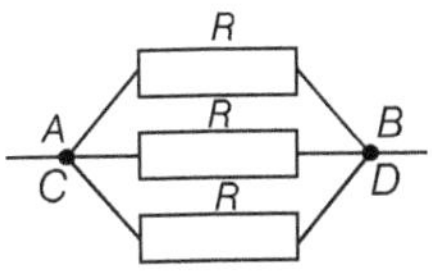

Fig. 432

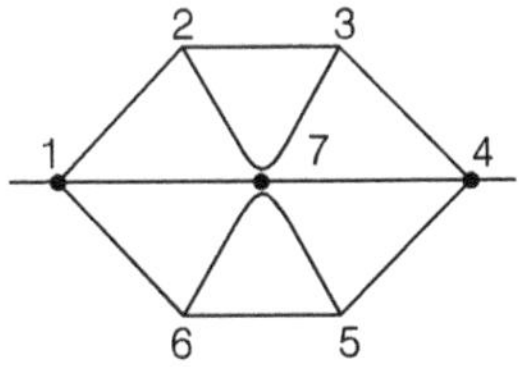

Fig. 433

464. Points A and C have the same potentials because they are connected by a wire whose resistance can be neglected. The potentials of points B and D are also identical. For this reason the ends of resistors A, C and, correspondingly, B, D can be assumed as connected together. Thus, resistors AB, CB and CD are connected in parallel. The equivalent diagram is shown in Fig. 432.

The total resistance is $R/3$.

465. Owing to symmetry, it is obvious that the current in conductor *1-7* is equal to that in *7-4*, the current *2-7* is equal to *7-3*, and the current *6-7* to *7-5* (see Fig. 165). For this reason the distribution of the currents and, hence, the resistance of the hexagon will not change if conductors *2-7*, *7-3*, *6-7* and *5-7* are disconnected from the centre (Fig. 433). It is easy to calculate the resistance of this circuit, which is equivalent to the initial one. The resistance of the upper and lower parts is $8R/3$.

The total resistance R_x can be found from the relation

$$\frac{1}{R_x} = \frac{1}{2R} + \frac{6}{8R}$$

Hence,
$$R_x = \frac{4}{5}R.$$

466. Owing to symmetry, it is obvious that the potentials of the cube vertices *2*, *3* and *6* are the same, as are those of vertices *4*, *5* and *7* (see Fig. 166).

Therefore, vertices *2*, *3*, *6* and *4*, *5*, *7* can be connected by conductors without resistance, i.e., by bus-bars. This will not change the resistance of the cube. The bus-bars will thus be connected by six conductors *2-7*, *2-4*, *3-5*, *3-4*, *6-7* and *6-5*. The resistance of the circuit (Fig. 434) is equal to the sought resistance of the cube:

$$R_x = \frac{R}{3} + \frac{R}{6} + \frac{R}{3} = \frac{5}{6}R$$

467. The resistance of section CD is equal to $R_{CD} = \dfrac{R_2 r}{R_2 + r} = 40\ \Omega$, and that of the entire circuit is $R_{AB} = R_1 + R_{CD} = 100\ \Omega$. The current $I = \dfrac{U}{R_{AB}} = 1.2$ A. The voltage drop across section CD is

$$U_1 = IR_{CD} = 48\text{ V}.$$

468. The resistance between points A and B is

$$R_{AB} = \frac{1}{\dfrac{2}{r_a + r_b} + \dfrac{1}{r_c}} = \frac{(r_a + r_b)r_c}{r_a + r_b + 2r_c}$$

where

$$r_a = \rho \frac{a}{A}, \; r_b = \rho \frac{b}{A}, \; r_c = \rho \frac{\sqrt{a^2 + b^2}}{A}$$

The resistance between points C and D can be found if we consider the currents flowing in the branches of the circuit (see Fig. 435). It is obvious from considerations of symmetry that the currents in conductors DB and AC, and also in AD and BC, are mutually equal, the current in AD being equal to $i_1 + i_2$, since the sum of the currents at junction A is equal to zero.

On section DAC

$$(i_1 + i_2)r_a + i_1 r_b = U_{DC}$$

and on section $DABC$

$$2(i_1 + i_2)r_a + i_2 r_c = U_{DC}$$

Hence,

$$i_1 = \frac{r_a + r_c}{2r_a\, r_b + r_a\, r_c + r_b\, r_c} U_{DC}$$

$$i_2 = \frac{r_b - r_a}{2r_a\, r_b + r_a\, r_c + r_b\, r_c} U_{DC}$$

The sought resistance is

$$R_{CD} = \frac{U_{CD}}{I} = \frac{U_{CD}}{2\,i_1 + i_2} = \frac{2r_a r_b + r_c(r_a + r_b)}{r_a + r_b + 2r_c}.$$

469. If no current flows through the galvanometer, the potentials of points C and D are the same and the current I_1 passing through the resistance R_x is equal to the current flowing through the resistance R_0, while the current I_2 along slide wire AB is the same in all the cross sections.

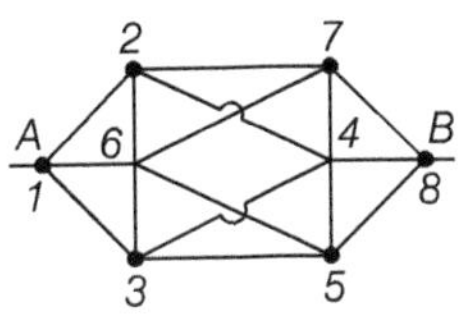

Fig. 434

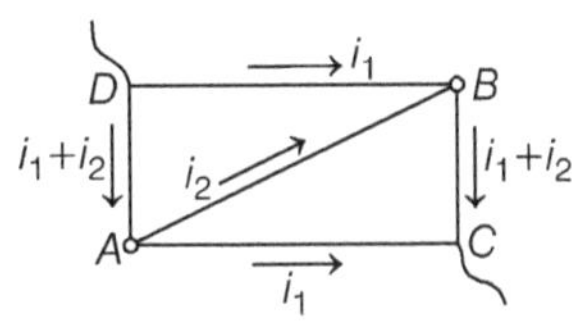

Fig. 435

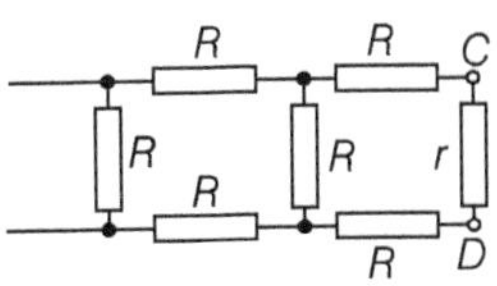

Fig. 436

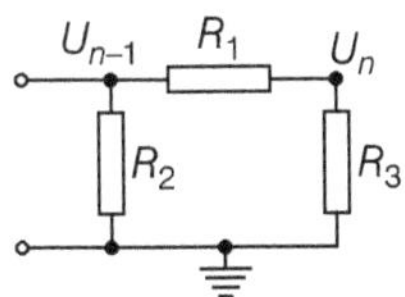

Fig. 437

According to Ohm's law,

$$I_1R_x = I_2l_1\frac{\rho}{A}$$

and
$$I_1R_0 = I_2l_2\frac{\rho}{A}$$

where ρ is the resistivity and A is the cross section of the slide wire.

Therefore,
$$\frac{R_x}{R_0} = \frac{l_1}{l_2}$$

470. Such a resistance r should be connected between points C and D that the resistance of the last cell (Fig. 436) is also r. In this case the last cell can be replaced by the resistance r, then the same can be done with the next to last cell, etc. Now the total resistance of the circuit will not depend on the number of cells and will be equal to r.

The following equation can be written for r:
$$\frac{(2R+r)R}{3R+r} = r$$

Hence,
$$r = R(\sqrt{3}-1) \cong 0.73R.$$

471. The last cell is a voltage divider that reduces the potential of the n-th point k times as compared with the $(n-1)$th point. Hence, $U_n = \dfrac{U_{n-1}}{R_1+R_3}R_3 = \dfrac{U_{n-1}}{k}$ or $\dfrac{R_1}{R_3} = k-1$ (see Fig. 437).

The relation $U_i = \dfrac{U_{i-1}}{k}$ should be true for any cell. For this reason the resistance of the last cell, of the last two, of the last three, etc., cells should also be R_3 (see Problem 470). Therefore,

$$\frac{1}{R_3} = \frac{1}{R_2} + \frac{1}{R_1+R_3}$$

$$R_2 = \frac{R_3(R_1+R_3)}{R_1} = R_3\frac{k}{k-1}$$

and finally $R_1 : R_2 : R_3 = (k-1)^2 : k : (k-1)$

472. Devices whose action is based, for example, on the deflection of a current-carrying conductor in a magnetic field cannot be used. The angle through which the pointer is deflected in such a device is proportional to the current flowing through it. Determination of the potential difference with the aid of this kind of devices is based on Ohm's law: the current flowing through a voltmeter is proportional to the potential difference applied to it. To verify Ohm's law, an electrostatic voltmeter and an ordinary ammeter are required.

473. Let us denote the charges on the first and second capacitors at the moment of time t by q_1 and q_2. The latter are related by the expressions

$$q_1 + q_2 = Q \text{ and } \frac{q_1}{C_1} = \frac{q_2}{C_2}$$

Since $$C_1 = \frac{A}{4\pi(d_0 + vt)} \text{ and } C_2 = \frac{A}{4\pi(d_0 - vt)}$$

then $$\frac{q_1}{q_2} = \frac{d_0 - vt}{d_0 + vt}$$

and it therefore follows that

$$q_1 = Q\frac{d_0 - vt}{2d_0} \text{ and } q_2 = Q\frac{d_0 + vt}{2d_0}$$

The reduction in the charge on the first capacitor is equal to the increase of the charge on the second capacitor. The current intensity is

$$I = -\frac{\Delta q_1}{\Delta t} = \frac{\Delta q_2}{\Delta t} = \frac{Qv}{2d_0}$$

The current will flow from the positively charged plate of the first capacitor to the positively charged plate of the second capacitor.

474. The forces of attraction acting between the plates of the capacitors are equal, respectively, to

$$F_1 = 2\pi\frac{q_1}{A}q_1 = \frac{\pi}{2}\frac{Q^2}{A}\frac{(d_0 - vt)^2}{d_0^2}$$

for the first capacitor and to

$$F_2 = \frac{\pi}{2}\frac{Q^2}{A}\frac{(d_0 + vt)^2}{d_0^2}$$

for the second capacitor (see Problem 473).

Since the plates of the first capacitor move apart, the forces of the electrostatic field perform the negative work W_1. These forces perform the positive work W_2 in the second capacitor. The work ΔW done by the field when each plate moves over a small distance Δx is

$$\Delta W = \Delta W_1 + \Delta W_2 = (F_2 - F_1)\Delta x = \frac{2\pi Q^2}{A}\frac{x}{d_0}\Delta x$$

where $x = vt$.

Thus the work performed on a small section is proportional to the displacement x, as when a spring is stretched. Therefore, the total work can be found with the aid of the method employed in solving Problem 137:

$$W = \frac{\pi Q^2 a^2}{A d_0^2}$$

The work W can also be calculated by another method. Since the resistance of the connecting wires is zero, the quantity of heat liberated is also zero. For this reason the change in the electrostatic energies of the two capacitors will be equal to the work of the electrostatic field.

At the moment of time t, the energies of the first and the second capacitors will be

$$W_{e1} = \frac{q_1^2}{2C_1} = \frac{\pi}{2}\frac{Q^2}{A d_0^2}(d_0 - vt)^2(d_0 + vt)$$

and $$W_{e2} = \frac{q_2^2}{2C_2} = \frac{\pi}{2}\frac{Q^2}{A d_0^2}(d_0 + vt)^2(d_0 - vt)$$

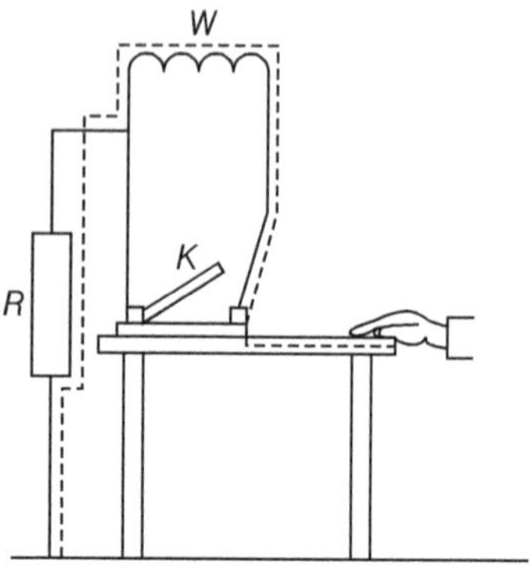

Fig. 438

The total energy

$$W_e = W_{e1} + W_{e2} = \pi\frac{Q^2}{A d_0}(d_0^2 - a^2)$$

Therefore, the energy will drop by $\Delta W_e = \frac{\pi Q^2}{A d_0} a^2$ during the time t.

This change will be equal to the work W of the electrostatic field.

475. Rubbing of the clothes against the chair produces electrification, and the body of the experimenter and the chair form a sort of capacitor. When the experimenter stands up, the capacitance of this capacitor sharply decreases, and therefore the potential difference sharply increases between the chair (i. e., "earth") and the experimenter's body. The body should obviously be well insulated from the earth (rubber soles).

When the experimenter touches the table, the potential difference between his hand and the earth levels out. An electric current is generated, a negligible part of which is branched off into the galvanometer. To deflect the pointer, the resistance between one end of the galvanometer coil and the earth should be lower than that between the other end and the earth.

The path of the current is shown schematically in Fig. 438. Here W is the winding of the galvanometer, K is the key and R shows a very high but finite resistance between one of the winding ends and the earth.

The pointer of the galvanometer deflects despite the tremendous resistance of the circuit owing to the great potential difference appearing when the capacitance is reduced.

476. There is obviously a definite asymmetry between the conductors which the ends of the galvanometer winding are connected to. This occurs when the resistance of the insulation between one end of the coil and the earth is less than between the earth and the other end. It should also be taken into account that the resistance between the conductors coming from the coil is not infinite, despite good insulation.

The path of the current is illustrated in Fig. 439. W is the winding of the galvanometer, C_1 and C_2 are the conductors leading from the ends of the winding, E is the earth and R_1, R_2 and R_3 show schematically very high but finite resistances appearing because the insulation is not ideal; $R_3 \gg R_1 + R_2$.

The dash line shows the path of the current if a negatively charged body is brought up to C_2. If the body is brought up to C_1, the current path is shown by dots. In both cases the current will flow through the winding of the galvanometer in the same direction.

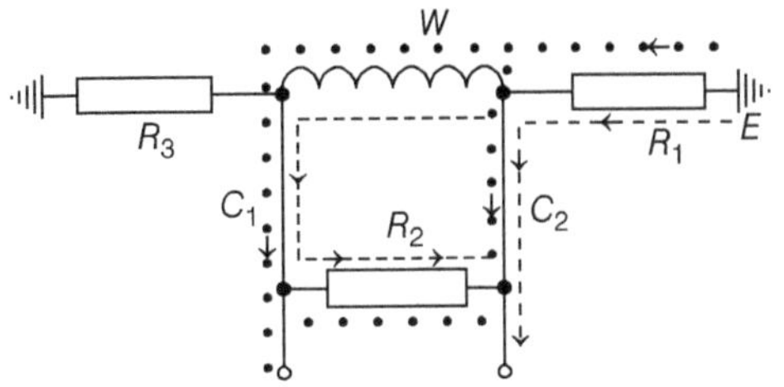

Fig. 439

This problem illustrates the presence of conductivity in all bodies, which is especially important in working with sensitive devices.

477. Point A in Fig. 440 shows the potential of the positive (copper) electrode and point D that of the negative (zinc) electrode. In the $ZnSO_4$ solution, the zinc electrode is charged negatively as a result of evolution of positive ions of Zn, while the copper electrode in the $CuSO_4$ solution is charged positively since it receives positive ions of Cu. The potential of the electrolyte is depicted by the line BC. Lines $AB = E_1$ and $CD = E_2$ show the jumps of the potential on the electrode-electrolyte boundaries. The e.m.f. equal to the potential difference at the ends of the opened cell is

$$E = E_1 + E_2$$

478. When the circuit is closed, a voltage drop* occurs both on the internal resistance of the element r and on the external resistance R of the circuit. The magnitude of the jumps of the potential (and of the e.m.f.) does

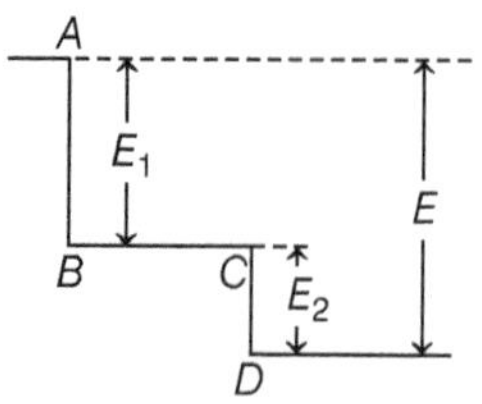

Fig. 440

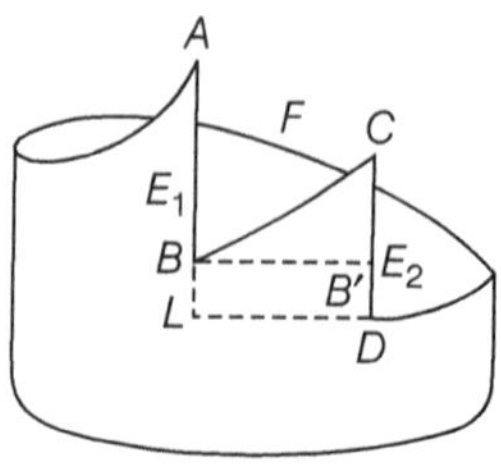

Fig. 441

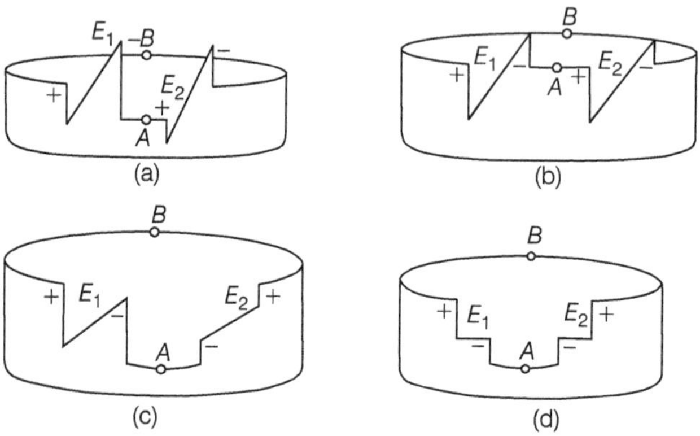

Fig. 442

*Here and below the term "voltage drop" denotes the product ir, while "voltage" is equivalent to "potential difference".

not change. The corresponding distribution of the potential is shown in Fig. 441. The voltage drop on the internal resistance of the element occurs along *CB* and on the resistance *R* along *AFD*. The section *CB′* shows the voltage drop on the internal resistance *r* equal to *Ir* and the section *AL* the voltage drop on the resistance *R*.

Since the line depicting the potential *ABCDFA* is closed, the sum of the voltage drops should be equal to the sum of the potential jumps:

$$E_1 + E_2 = IR + Ir$$

Hence, $$I = \frac{E}{R + r}.$$

479. The potential is distributed as shown in Fig. 442 *a*, *b*, *c* and *d*.

(a) $I = \frac{E_1 + E_2}{r_1 + r_2}$;

$$V_{BA} = V_B - V_A = E_1 - Ir_1 = -(E_2 - Ir_2) = \frac{E_1 r_2 - E_2 r_1}{r_1 + r_2} > 0$$

(b) $I = \frac{E_1}{r_1}; V_{BA} = 0$

The potentials of the conductors connecting the elements are the same, but the current is not zero.

(c) $I = \frac{E_1 - E_2}{r_1 + r_2}; V_{BA} = E_1 - Ir_1 = E_2 + Ir_2 = \frac{E_1 r_2 + E_2 r_1}{r_1 + r_2}$

(d) $I = 0; V_{BA} = E_1 = E_2$

There is a potential difference between the conductors, but no current flows through them.

480. In sections *BA* and *DC* the chemical forces that cause reactions between the electrodes and the electrolyte perform positive work (see Fig. 441). This work is equal to the sum of the potential jumps on these sections, i. e. to the e. m. f., because in a state of equilibrium the chemical forces that act in the layer of the electrolyte adjacent to the electrode are equal to the electrostatic forces.

Since the forces of non-electrostatic origin do not act in the other sections of the circuit, the work performed by these forces is also equal to the e.m.f. of the battery when a single positive charge moves along the closed circuit. (The work of electrostatic forces in a closed circuit is zero.)

481. An energy of $W_e = 106{,}000 - 56{,}000 = 50{,}000$ calories $\cong 2 \times 10^{12}$ ergs is liberated per mole of the substances reacting in the cell. Owing to this energy, the electric current performs the work $W = QE$, where E is the e.m.f. of the cell and Q is the quantity of transferred electricity. Since the copper and the zinc are bivalent, the charges of their ions are equal in magnitude to the doubled charge of an electron. One mole of the substance contains 6.02×10^{23} atoms.

Therefore, $Q = 2 \times 4.8 \times 10^{-10} \times 6.02 \times 10^{23}$ CGS_Q.

Hence, $E = \frac{W_e}{Q} \cong 3.5 \times 10^{-3} CGS_Q = 1.05\ V$.

482. The ratio between the intensities of the currents flowing through the cells is $\frac{I_1}{I_2} = \frac{r_2}{r_1}$, since the e.m.f.s of the cells are the same. According to Faraday's law, the masses of the dissolved zinc are proportional to the currents:

$$\frac{m_1}{m_2} = \frac{I_1}{I_2} = \frac{r_2}{r_1} \cong 1.625$$

483. As it passes into solution in the form of an ion Zn^{++}, each atom of the zinc gives off to the external circuit two electrons carrying a charge of $q = 2e = -3.2 \times 10^{-19}$ C. At the same time the copper ions Cu^{++} are deposited on the copper plate as neutral atoms, owing to which the concentration of the $CuSO_4$ solution decreases. To maintain the concentration constant, it is necessary to continually dissolve crystals of $CuSO_4 \cdot 5H_2O$ in an amount that will compensate for the passing of the ions Cu^{++} and SO_4^{--} out of the solution.

According to the initial conditions, a charge of $Q = 2{,}880$ C passed through the cell. This corresponds to a transfer into the solution of $n = \frac{Q}{q} = 9 \times 10^{21}$ atoms of zinc, i.e., about 0.98 g of the zinc. Correspondingly, the same amount of copper atoms (about 0.95g) will pass out of the solution and 3.73 g of crystals of blue vitriol will have to be dissolved to restore the concentration of the $CuSO_4$ solution.

484. When the zinc is dissolved, the positive ions Zn^{++} pass into solution and the liberated electrons flow along the wire onto the graphite layer and neutralize the positive ions of copper in the $CuSO_4$ solution. Therefore, the graphite will be covered by a layer of copper. This phenomenon can be used in galvanoplasty.

485. The change in the e.m.f. of the battery depends on the ratio between the dimensions of the electrodes and those of the vessel. If the two middle plates are almost equal in size to the section of the vessel, the e.m.f. of the battery will change insignificantly. If they are small, the e.m.f. will be almost halved.

486. The Zinc rod forms a short-circuited galvanic cell with each half of the carbon rod. The resistance of half of the carbon rod, the resistance of the zinc rod and the zinc-carbon contact are the external resistance of the cell (see the equivalent diagram in Fig. 443).

When the zinc rod is vertical, the currents i_1 and i_2 in both halves of the carbon rod are identical and the voltmeter will show zero. If the rod is inclined, the internal resistance of one of the cells will decrease and of the other increase. The currents i_1 and i_2 will be different and a potential difference will arise between the ends of the carbon rod that will be registered by the voltmeter.

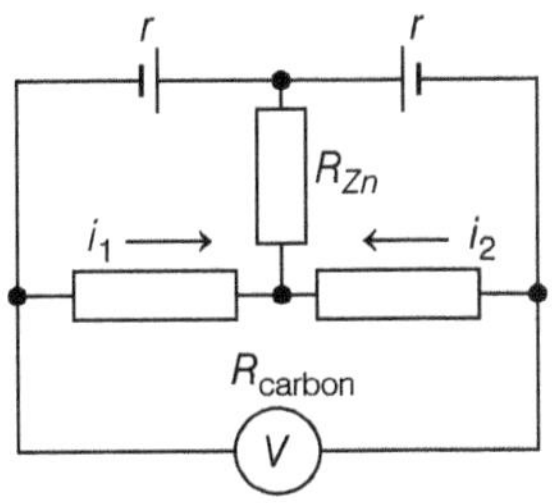

Fig. 443

487. Since $r \ll R$, there will be practically no field inside the sphere and no current on its internal surface. Therefore, the mass of the deposited copper is

$$m = \frac{A}{n}\frac{4\pi R^2 jt}{F} \cong 1.86 \text{ g}$$

where A/n is the electrochemical equivalent of copper (A is the atomic weight and n the valency) and F is Faraday's number.

488. The matter is that the electrodes are polarized during electrolysis and each bath acquires an e.m.f. directed against the current flowing from the capacitor. For this reason the capacitor cannot be discharged completely. The more baths we have, the greater will the total e.m.f. of polarization be, and the greater the charge remaining on the capacitor. The energy of the detonating gas will always be smaller than that of the charged capacitor.

489. In the electrolysis of water, the electrodes are polarized and an e.m.f. of polarization E_p appears that is directed against the e.m.f. of the battery. For this reason electrolysis will occur only if the e.m.f. of the battery is greater than E_p.

When a charge Q flows through the electrolyte, the battery performs work against the e.m.f. of polarization: $W = E_pQ$. This work decomposes the water and detonating gas is formed. On the basis of the law of conservation of energy, the chemical energy of the detonating gas W_e liberated during the flow of the charge Q is equal to E_pQ. In accordance with Faraday's law, the evolution of one gramme of hydrogen on the cathode is accompanied by the flow of electricity $Q = m\frac{n}{A}F = 96{,}500\text{C}$.

Therefore,

$$E_p = \frac{W_e}{Q} \cong 1.5\text{V}.$$

The e.m.f. of the battery should be greater than 1.5 V.

490. A definite concentration of ions is the result of dynamic equilibrium: the number of ions produced by electrolytic dissociation is equal to the reduction in the number of ions due to the reverse process—recombination (when they collide, ions of opposite signs may form a neutral molecule).

Near the electrodes the concentration of the ions drops, and this equilibrium is violated. The number of ions that appear owing to dissociation is greater than the number of recombined ions. It is this process that supplies the electrolyte with ions. The process takes place near the electrodes. Dynamic equilibrium inside the electrolyte is not violated.

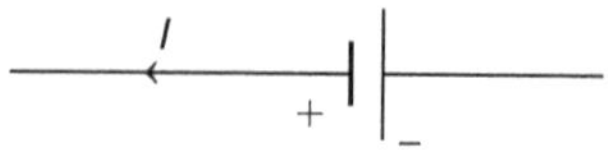

Fig. 444

491. During a second the cathode receives n_+v_+A positive ions (A is the area of the cathode). At the same time $n_-\ v_-A$ negative ions leave it. When the negative ions move away, the dynamic equilibrium between the neutral molecules of the electrolyte and the ions into which they are dissociated is violated (see Problem 490). There again appear n_-v_-A negative ions and the same number of positive ions. The positive ions are also liberated at the cathode and, as a result, the number of positive ions liberated at the cathode per second will be equal to the full current.

492. When the temperature changes by Δt, the current changes by ΔI. On the basis of Ohm's law, $\Delta I = \dfrac{\alpha \Delta t}{R+r}$ ($\alpha = 50 \times 10^{-6}$ V / deg). The minimum change in the current registered by the galvanometer is $\Delta I = 10^{-9}$A. Hence, the minimum change in the temperature that can be registered will be $\Delta t = \dfrac{\Delta I(R+r)}{\alpha} = 5 \times 10^{-4}$deg

493. The maximum theoretically possible efficiency of the thermoelectric battery is $\eta = \dfrac{Eq}{Q} = \dfrac{T_1 - T_2}{T_1}$

where Q is the quantity of heat absorbed in a unit of time by the hot joints, q the charge flowing in the circuit in a unit of time, and T_1 and T_2 the absolute temperatures of the joints. According to Faraday's law, the mass of copper liberated at the cathode per second is $m = \dfrac{A}{n}\dfrac{q}{F}$. Upon inserting the value of q from the first equation, we have $m = \dfrac{AQ}{nFE}\dfrac{T_1 - T_2}{T_1} \cong 1.7 \times 10^{-4}\,g$

494. If the current flows in the direction shown in Fig. 444 (the storage battery is being discharged), then $V = E - IR$. If the current flows in the opposite direction (the battery is being charged), $V = E + IR$ (see the answer to Problem 479*c*).

495. If the voltmeter is connected, then, according to Ohm's law, $V = E - \dfrac{Er}{R_1 + r}$, where r is the internal resistance of the cell.

For the second circuit,

$$I = \frac{E}{R + R_2 + r}$$

and therefore

$$E = \frac{IV(R_1 - R - R_2)}{R_1 I - V} = 2.1\text{V}$$

496. The internal resistance of the galvanic cell is small and that of the electrostatic machine very great. It is the resistance of insulators (tens and hundreds of millions of ohms).

497. At the first moment after the key K is closed, a potential difference will appear between the plates of the capacitors C_1 and C_2. The current will flow in the circuit until the capacitor C_1 is charged. After this the potential difference across the capacitor C_1, becomes equal to the e.m.f. of the battery, and the potential difference between the plates of the second capacitor will be zero.

498. For two cells

$$I = \frac{E_1 + E_2}{r_1 + r_2 + R}$$

where E and r are the e.m.f. and the internal resistances of the cells, and R is the external resistance.

For one cell (the first one, for example)

$$I = \frac{E_1}{r_1 + R}$$

According to the initial condition, $I < I_1$, i.e.,

$$\frac{E_1 + E_2}{r_1 + r_2 + R} < \frac{E_1}{r_1 + R}$$

Therefore, it is necessary that

$$\frac{E_2}{r_2} < \frac{E_1}{R + r_1}$$

499. From Ohm's law

$$I_1\left(2r\frac{l}{L}+\rho+R\right)=E$$

$$I_2\left(2r\frac{l}{L}+\rho+\frac{2r\left(1-\frac{l}{L}\right)R}{2r\left(1-\frac{l}{L}\right)+R}\right)=E$$

$$I_3\rho=E$$

where l is the distance from the battery to the breakdown and ρ is the internal resistance of the battery.

From these simultaneous equations we find that

$$R=\frac{E}{I_1}-\frac{E}{I_2}\pm\sqrt{\left(\frac{E}{I_1}-\frac{E}{I_2}\right)\left(\frac{E}{I_3}-\frac{E}{I_2}+2r\right)}=(4\pm 3)\Omega$$

The value $R=1\ \Omega$ should be discarded, since in this case the point of breakdown will be at a distance of 5.9 km from the battery. Actually, when $R=1$

$$l=L\frac{LE-I_1L_\rho-I_1LR}{2rI_1}=5.9\text{ km}$$

The sought resistance is $R=7\ \Omega$

500. On section AE_2B we have $V_A-V_B=E_2-I_1r_2$, where $I_1=\dfrac{E_1+E_2}{r_1+r_2+R}$ and r_1 and r_2 are the internal resistances of the cells. According to the initial condition, the potential of point A is lower than that of point B. Therefore, $u_1=V_B-V_A>0$. For the other circuit $u_2=V'_B-V'_A=(E_2+I_2r_2)$ where

$$I_2=\frac{E_1-E_2}{r_1+r_2+R}$$

Upon solving this system of equations, we find that

$$u_2=\frac{2E_1E_2+u_1(E_1-E_2)}{E_1+E_2}=+1.86\ V$$

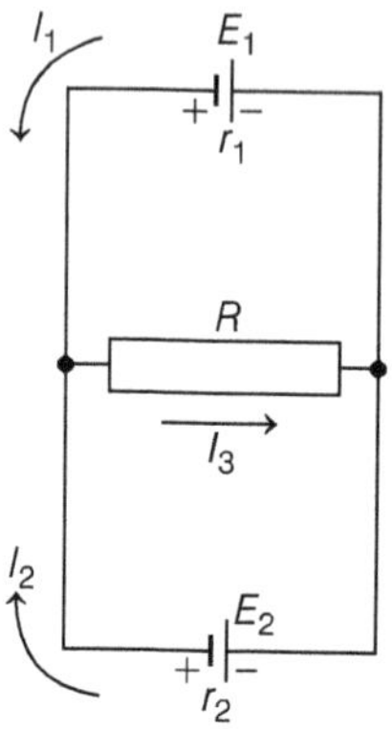

Fig. 445

501. In this case the potential of point A, when the key is closed, is higher than that of point B, since when the key is open $V_B>V_A$. For this reason $E_2-I_1r_2=u_1$. The other equations have the same form as in the solution of Problem 500. Therefore,

$$u_2=\frac{2E_1E_2-u_1(E_1-E_2)}{E_1+E_2}\cong+1.57\ V$$

502. Assuming arbitrarily that the currents are directed as shown in Fig. 445, Ohm's law may be used to write the equalities

$$U_{AB} = E_1 - I_1 r_1$$
$$U_{AB} = E_2 - I_2 r_2$$
$$U_{AB} = I_3 R$$

Since no point in the circuit accumulates a charge

$$I_1 + I_2 = I_3$$

Upon solving these simultaneous equations, we can find the currents I_1, I_2 and I_3:

$$I_1 = \frac{5}{4}\text{ A},\ I_2 = \frac{1}{4}\text{ A and } I_3 = \frac{3}{2}\text{ A}$$

The positive values of the currents obtained show that the directions of the currents selected initially are correct.

503. When $I_2 = 0$, we have $I_1 = I_3$, and $U_{AB} = E_2$. Hence, as before $U_{AB} = E_1 - I_1 r_1$, and

$$R = \frac{E_2 r_1}{E_1 - E_2} = 1\Omega$$

If the current I_2 is directed against E_2, the simultaneous equations will take the form

$$U_{AB} = E_1 - I_1 r_1$$
$$U_{AB} = E_2 + I_2 r_2$$
$$U_{AB} = I_3 R_3$$
$$I_1 = I_2 + I_3$$

Hence,

$$I_2 = \frac{E_1 - E_2 - E_2 \dfrac{r_1}{R}}{r_1 + r_2 + \dfrac{r_1 r_2}{R}}$$

Our condition will be fulfilled if $I_2 > 0$. Therefore the following inequalities should exist

$$E_1 - E_2 - E_2 \frac{r_1}{R} > 0$$

or

$$R > \frac{E_2 r_1}{E_1 - E_2} = 1\Omega$$

504. There are two methods of connecting the storage battery cells. Either the batteries are connected in series in the separate groups, and the groups themselves in parallel, or vice versa.

Denoting the total number of cells by N and the number of cells in a separate group by n, we shall have in the first case:

$$I_1 = \frac{nE_0}{R + \frac{rn^2}{N}} = \frac{E_0}{\frac{R}{n} + \frac{rn}{N}}$$

since the e. m. f. of one group is nE_0, the resistance of the group is rn and the number of groups connected in parallel is N / n. The current I_1 reaches its maximum if $\frac{R}{n} + \frac{r}{N}n$ is minimum.

The minimum of an expression of the type $ax + \frac{b}{x}$ can be found as follows. The relationship

$$y = ax + \frac{b}{x} \tag{1}$$

is shown graphically by the curve in Fig. 446 which has its minimum at point x_0, at which the roots of quadratic equation (1) coincide. For this reason,

$$x_0 = \sqrt{\frac{b}{a}}$$

Therefore,

$$n = \sqrt{\frac{RN}{r}} = 4$$

and

$$I_{1max} = \frac{E_0}{2}\sqrt{\frac{N}{Rr}} = 20\ A$$

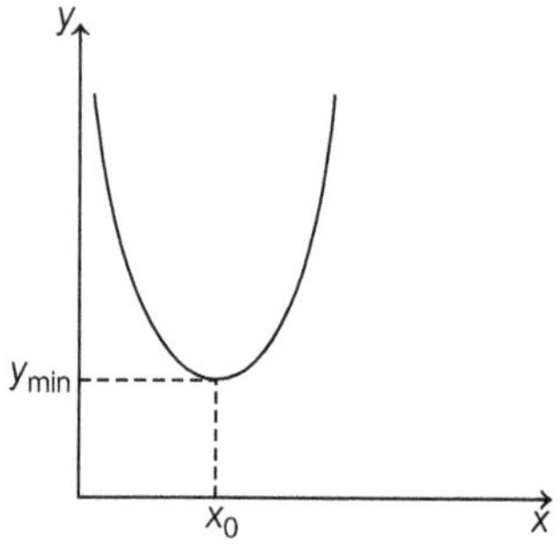

Fig. 446

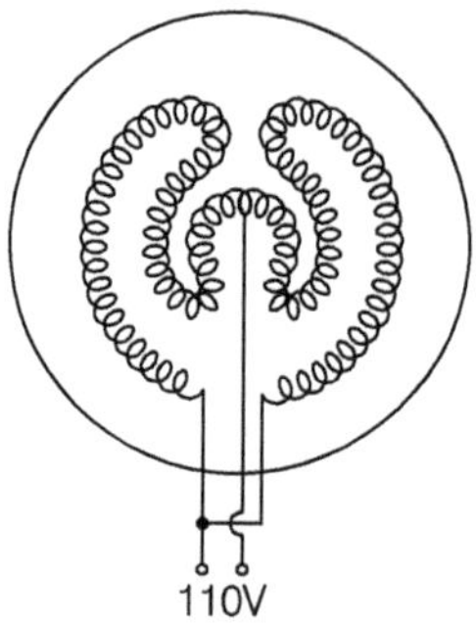

Fig. 447

In the second case

$$I_2 = \frac{\frac{N}{n}E_0}{R + \frac{rN}{n^2}} = \frac{NE_0}{nR + \frac{rN}{n}}$$

The current reaches its maximum when $n = \sqrt{\frac{rN}{R}} = 6.$

Hence,

$$I_{2\,max} = \frac{E_0}{2}\sqrt{\frac{N}{Rr}} = I_{1max}$$

Thus, it is impossible to get a current exceeding 20 A.

505. What is to be done is shown in Fig. 447.

506. The temperature in the calorimeter remains equal to 0°C. Therefore,

$$0.24\,\frac{u^2}{R}t = mH$$

and

$$t = \frac{mHR}{0.24u^2} = 5 \text{ min.}$$

507. At a room temperature of $t_0 \cong 20°$ C (i.e., at the moment the lamp is switched on) a lamp consumes a power of $P_0 = \frac{V^2}{R_0}$ (R_0 is the resistance of the filament at the temperature t_0). When $t = 2{,}500$°C, the consumed power is $P = \frac{V^2}{R}$, where $R = R_0\{1 + \alpha(t - t_0)\}$. Therefore, the sought power is

$$P_0 = P\{1 + \alpha(t - t_0)\} \cong 600 \text{ watts}$$

508. The power consumed by the device at the first moment is very much higher than the rated one (see Problem 507) since the resistance of the cold heating coil is small. Correspondingly, there will be a large drop of voltage in the conductors leading from the mains to the room. As the coil gets heated, the consumed power drops and approaches the nominal rating.

509. The permissible drop of voltage on the feeding wires is $\Delta U = IR_0 =$ 7 V, where I is the maximum current. Hence, the maximum power is

$$P = IU = \frac{\Delta U \times U}{R_0} = 1{,}680 \text{ watts}$$

510. Since in all cases the kettle is connected to the same electric mains, it is more convenient to find the quantity of heat evolved from the formula $Q = 0.24\,\frac{U^2}{R}t$. Hence, $R = 0.24\,\frac{U^2}{Q}t$. Since U and Q are the same in all cases, the latter equation can be rewritten as $R = \alpha t$, where $\alpha = 0.24\,\frac{U^2}{Q}$.

Denoting the resistances of the windings by R_1 and R_2, we have $R_1 = \alpha t_1$ and $R_2 = \alpha t_2$. In parallel connection of the windings

$$R_a = \frac{R_1 R_2}{R_1 + R_2} = \frac{\alpha^2 t_1 t_2}{\alpha(t_1 + t_2)} = \alpha t_a$$

and in series connection

$$R_b = R_1 + R_2 = \alpha(t_1 + t_2) = \alpha t_b$$

Therefore,

$$t_a = \frac{t_1 t_2}{t_1 + t_2}, \text{ and } t_b = t_1 + t_2$$

511. (1) $t_a \cong 57$ minutes, (2) $t_b = 3$ minutes 30 seconds (see the solution to Problem 510).

512. When a direct current flows through a conductor, the potential difference does not change. When a capacitor is discharged, the potential difference changes from U to zero.

513. When the losses of heat in high-voltage wires are calculated by the formula $Q = 0.24\dfrac{U^2}{R}t$, the value of U is the potential difference at the ends of the line (voltage drop in the wires), but not the voltage in the secondary winding of the step-up transformer. This potential difference is small (as distinct from the voltage in the winding of the transformer) and decreases with a reduction of the current flowing in the line.

514. In conformity with the initial conditions, $k = \dfrac{Ir}{E}100$, where E is the e.m.f. of the battery, and $I = \dfrac{P}{U}$ is the current in the circuit.

Remembering that $E = 2U + Ir$, we obtain

$$r = \frac{2k^2}{P(100 - k).}\Omega$$

515. The power liberated in the external resistance R is $P = IU$. In our case $U = E - Ir$ and, therefore, $I = \dfrac{E - U}{r}$

Thus, $$P = \frac{EU - U^2}{r}$$

whence $$U = \frac{E}{2} \pm \sqrt{\frac{E^2}{4} - Pr}$$

$$U_1 = 9V \text{ or } U_2 = 1V.$$

The ambiguity of the result is due to the fact that the same power can be liberated on various external resistances R, each R having its own current:

when $U_1 = 9$ V, we have $I_1 = 1A$ and $R_1 = \dfrac{P}{I_1^2} = 9\,\Omega$

when $U_2 = 1V$, we have $I_2 = 9A$ and $R_2 = \frac{P}{I_2^2} = 1/9\Omega$

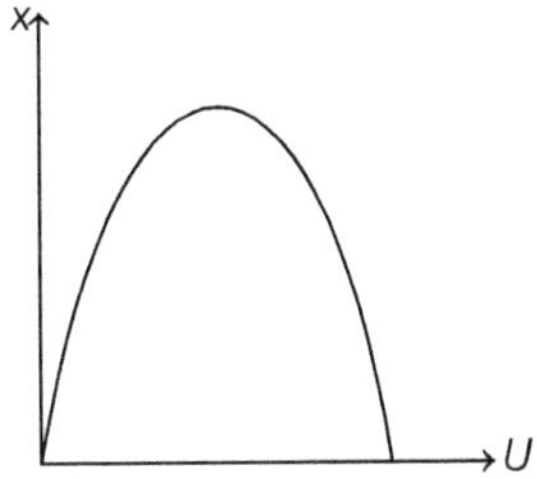

Fig. 448

516. The useful power (see Problem 515) is equal to $P = \frac{EU - U^2}{r}$. For simplicity let us denote $EU - U^2$ by x. It is necessary to determine the value of U at which x will reach its maximum.

Graphically, x versus U is depicted in Fig. 448. The curve has the form of a parabola, two values of U corresponding to each value of x. At the given x we have a quadratic equation with respect to U. The maximum of x is reached when the two roots of the equation coincide. Therefore, when x is maximum the discriminant of the equation should be equal to zero:

$$x_{max} = \frac{E^2}{4}.$$

Hence,

$$U = \frac{E}{2} \text{ and } P_{max} = \frac{E^2}{4r} = 25 \text{ watts}$$

Here

$$I = \frac{E}{2r} \text{ and } R = \frac{P_{max}}{I^2} = \frac{E^2 \times 4r^2}{4rE^2} = r,$$

i.e., the external resistance is the same as the internal one.

517. By definition, the efficiency η is the ratio of the useful power to the entire power produced by a battery:

$$\eta = \frac{IU}{IE} = \frac{U}{E}$$

where $U = \frac{ER}{R + r}$ is the potential difference in the external resistance R.

Therefore,
$$\eta = \frac{R}{r + R}$$

In problem 515, $\eta_1 = 90$ per cent and $\eta_2 = 10$ per cent.

In problem 516, $\eta = 50$ per cent.

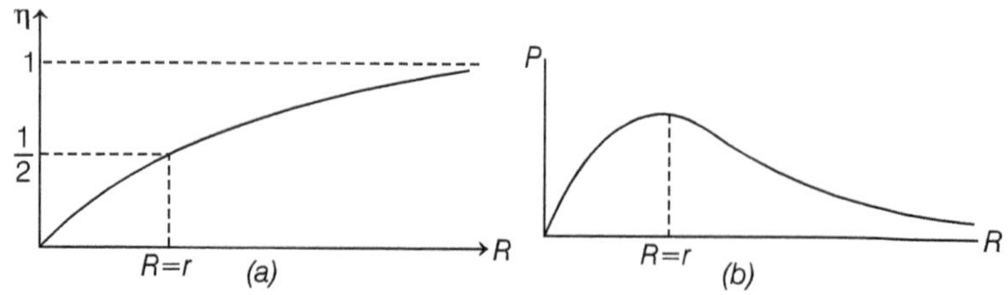

Fig. 449

$\eta \to 1$ when $R \to \infty$, but in this case the useful power $P = \dfrac{E^2R}{(R+r)^2}$ evolved tends (the same as the total power) to zero (Fig. 449).

518. $\eta_1 = \dfrac{R}{R+r} = 0.6$, where r is the internal resistance of the source of current (see Problem 517).

Hence,

$$\eta_2 = \frac{6R}{6R+r} = 0.9$$

In per cent $\eta_2 = 90\%$.

519. According to Ohm's law, $U = E + Ir$.

Hence,

$$I = \frac{U-E}{r}.$$

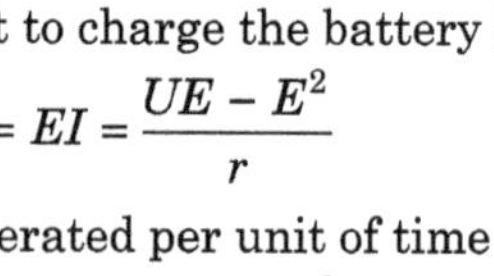

Fig. 450

The useful power spent to charge the battery is

$$P_1 = EI = \frac{UE - E^2}{r}$$

The amount of heat liberated per unit of time is

$$P_2 = I^2r = \frac{(U-E)^2}{r}$$

The total consumption of power is

$$P = IU = P_1 + P_2$$

520. The useful power is

$$P_1 = \frac{E(U-E)}{r}$$

(see Problem 519). The heat evolved per unit of time is

$$P_2 = \frac{(U-E)^2}{r}$$

Ordinarily, $U - E \ll E$ when a storage battery is charged. For this reason $P_1 \gg P_2$. Therefore, only a small part of the power of the charging station is spent to evolve heat.

521. During one second all the electrons contained in the volume $Av1$ will pass through section S of the conductor (Fig. 450).

Therefore, the intensity of the current is $I = Av1ne$ (e is the charge of an electron).

Hence,

$$v = \frac{I}{Ane} \cong 10^{-4}\,\text{cm/s}.$$

522. Electrons in a metal may be considered free. Redistribution of the electrons inside the block will end when the resulting electric field is capable of imparting an acceleration a to the electrons. In this way the sought intensity of the field can be found from the relation $ma = eE$ (m and e are the mass and the charge of an electron).

Therefore,

$$E = \frac{m}{e}a.$$

The front surface of the block perpendicular to the direction of motion will be charged positively, and the rear one negatively. The density of the charges is

$$\sigma = \frac{E}{4\pi} = \frac{1}{4\pi}\frac{m}{e}a$$

523. Free electrons rotate together with a cylinder. Therefore, an electron that is at a distance r from the axis has an acceleration of $a = \omega^2 r$. This acceleration can be produced only by an electric field directed along a radius from the centre of the cylinder and equal to $E = \dfrac{m\omega^2 r}{e}$.

Here e and m are the charge and the mass of an electron.

The potential difference $U = \dfrac{1}{2}\dfrac{m}{e}\omega^2 R^2$, since the average force acting on a single charge when it moves from the axis of the cylinder to its surface is equal to $\dfrac{1}{2}\dfrac{m}{e}\omega^2 R$.

3-3. Electric Current in Gases and a Vacuum

524. During a glow discharge the electrons are knocked out from the cathode by the positive ions. These ions are produced when electrons collide with atoms of the gas. In the region of the cathode dark space there are practically no collisions. For this reason the anode should be placed beyond the region of the potential drop of the cathode. Otherwise, there will be no discharge.

525. For a discharge to occur in the tube, the anode should be placed beyond the region of the cathode drop (see Problem 524). But in this case the electrons approaching the anode upon collision with the molecules of the gas lose energy and will not generate X-rays when they impinge on the anode. Two electrodes are therefore necessary. The anode is located in the region of the glow and serves to maintain the discharge. The anticathode in the region of the cathode drop is bombarded by the electrons that did not lose their energy.

Tubes with a heater cathode have one electrode that acts as an anode and an anticathode.

526. The electrons striking the anticathode flow along the wire onto the anode. If the wire is removed, the anticathode will be gradually charged negatively and retard the electrons. In a certain time after it is switched on the X-ray tube will stop functioning.

527. Before the discharge, the voltage on the counter is equal to the e.m.f. of the source E. At the moment of discharge a current flows through the circuit and the voltage between the housing and the wire becomes equal to $U = E - IR$. The resistance R is very high, and the voltage drop IR is so great that the discharge stops.

528. According to Ohm's law, the sought voltage drop $U = IR$, where I is the current in the circuit.

The current is the same in all the cross sections inside the capacitor. The positive plate owes its current only to the negative ions, and the negative plate to the positive ions. Some positive and negative ions pass through an arbitrary cross section inside the capacitor.

$I = enAd$, where e is the charge of an electron and A is the area of the plates.

For a plane capacitor $Ad = 4\pi Cd^2$.

Therefore,

$$U = en \times 4\pi Cd^2 R \cong 1.4 \times 10^{-11} V.$$

529. If the negative carbon is cooled, the arc will be extinguished, since the arc burns owing to strong thermionic emission from the cathode that ceases upon cooling. Cooling of the positive carbon will not affect functioning of the arc.

530. When the contacts of the controller open, an electric arc may appear, since the current in the iron becomes very high and the distance between the contacts is small. With an alternating current the arc is unstable and extinguishes at once. A direct current produces a stable arc that at least will burn the contacts and put the iron out of commission.

531. 1 eV (one electron-volt) $= 1.6 \times 10^{-12}$ erg.

532. No, it does not. The tangents to the trajectory show the direction of the velocity of the particle and the tangents to the force line show the direction of the force acting on the particle and, therefore, the direction of acceleration.

The trajectory of the particle will coincide with a force line only in a field with straight force lines if the initial velocity of the particle is directed along a force line.

533. As the charge approaches the plate, electrostatic induction causes the charges on the plate of the same sign as the flying one to pass into the earth, while the charges of the opposite sign accumulate on the surface of the plate. A current pulse passes through the galvanometer (Fig. 451). No current flows through the galvanometer when the charge moves above the plate. An opposite current is generated when the charge moves away from the plate.

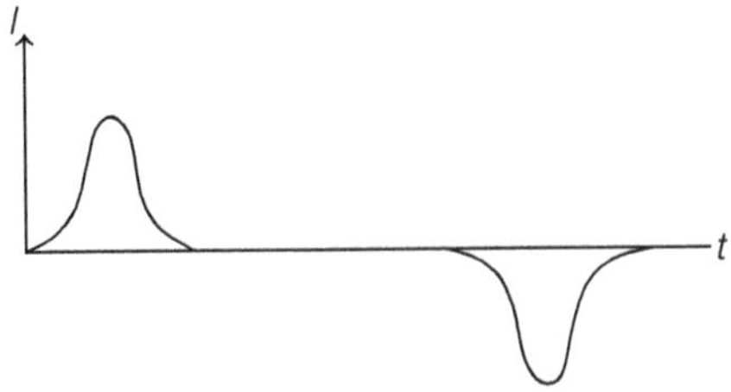

Fig. 451

534. Positive induced charges will appear on the neck of the tube and accelerate the electron. The kinetic energy of the electron will increase owing to the drop in the potential energy of the electron-tube system.

535. The total energy of an electron is equal to the sum of its kinetic and potential energies.

As the electron approaches the ring, its potential energy diminishes in the field of the ring, and as a result the kinetic energy increases. After passing through the ring, the electron moves away from it. The potential energy of the electron increases and the velocity gradually drops to zero.

536. The work done to move the charge $-Q$ is proportional to the potential difference between point O and remote point A on the axis (see Fig. 186). At infinity the potential is taken equal to zero. If the distance $OA \gg R$, the potential of point A can also be assumed to be zero. The potential at point O can be found by summing up the potentials produced by the separate small elements of the ring:

$$U_0 = \Sigma \frac{\Delta Q}{R} = \frac{Q}{R}.$$

On the basis of the law of conservation of energy $\frac{mv^2}{2} = \frac{Q^2}{R}$ we find that

$$v = \sqrt{\frac{2Q^2}{mR}}$$

537. As usual, the potential in infinity is considered to be zero. The potentials of the plates are thus respectively equal to $+\frac{U}{2}$ and $-\frac{U}{2}$, where $U = \frac{Q}{C}$.

The potentials at the points of the initial location of the electron are respectively: 0, $+\frac{U}{4}$, and $-\frac{U}{4}$. The initial full energies of the electron are

(1) $\frac{mv_0^2}{2}$, (2) $\frac{mv_0^2}{2} - \frac{eU}{4}$ and (3) $\frac{mv_0^2}{2} + \frac{eU}{4}$

The final velocities v_1, v_2 and v_3 are determined from the law of conservation of energy:

(1) $\frac{mv_0^2}{2} = \frac{mv_1^2}{2}$; hence $v_1 = v_0$

(2) $\frac{mv_0^2}{2} - \frac{eU}{4} = \frac{mv_2^2}{2}$; $v_2 = \sqrt{\frac{mv_0^2 - \frac{eQ}{2C}}{m}}$

(3) $\frac{mv_0^2}{2} + \frac{eU}{4} = \frac{mv_3^2}{2}$; $v_3 = \sqrt{\frac{mv_0^2 + \frac{eQ}{2C}}{m}}$

In the first case the final velocity is equal to the initial one, in the second case it is lower than the initial velocity and In the third case higher.

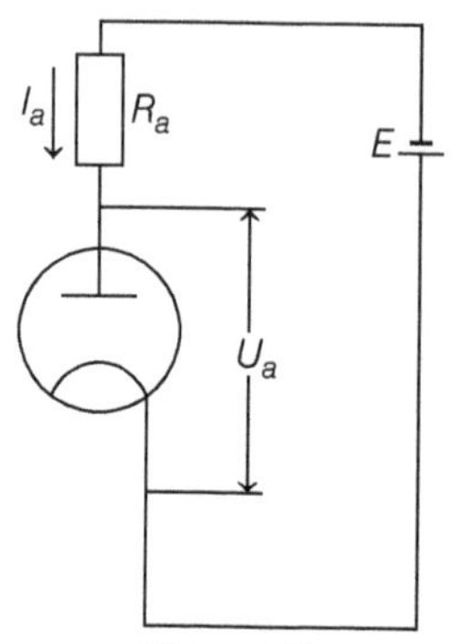

Fig. 452

In all three cases the velocity first grows (during motion in the capacitor) and then decreases.

538. Electrons with energies ranging from 80 eV to 74 eV reach the anode, since a voltage drop of 6 V exists along the filament.

The energy of the electrons at the anode is determined only by the potential difference passed by them and does not depend on the potential of the grid. The latter changes the distribution of the velocities of the electrons at intermediate points of the path and affects the number of electrons reaching the anode.

539. On the basis of Ohm's law, $E = I_a R_a + U_a$ (Fig. 452)

The intensity of the current

$$I_a = AU_a + BU_a^2$$

Hence, $$I_a = \frac{E}{R_a} + \frac{(AR_a + 1) - \sqrt{(AR_a + 1)^2 + 4EBR_a}}{2BR_a^2} = 5 \text{ mA}$$

The second root of the quadratic equation has no physical meaning, since it corresponds to $U_a < 0$.

540. The simultaneous equations determining the currents i_1 and i_2 have the form:

$$i = i_1 + i_2$$
$$i_1 = A_1 U_a + B_1 U_a^2$$
$$i_2 = A_2 U_a + B_2 U_a^2$$
$$U_a = E - iR$$

Hence,

$$U_a = \frac{-(A_1 + A_2)R - 1 + \sqrt{(A_1 R + A_2 R + 1)^2 + 4E(B_1 + B_2)R}}{2(B_1 + B_2)R} = 60 \text{ V}$$

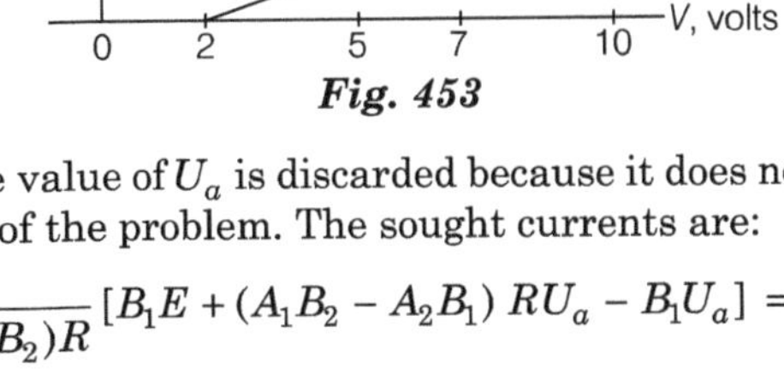

Fig. 453

The negative value of U_a is discarded because it does not correspond to the sense of the problem. The sought currents are:

$$i_1 = \frac{1}{(B_1 + B_2)R}[B_1 E + (A_1 B_2 - A_2 B_1) RU_a - B_1 U_a] = 22.2 \text{ mA}$$

$$i_2 = \frac{1}{(B_1 + B_2)R}[B_2 E + (A_2 B_1 - A_1 B_2) RU_a - B_2 U_a] = 37.8 \text{ mA}$$

541. With the potential of the grid $E_2 = -6\,V$, the current flowing through the valve is $I_2 = \frac{U_2}{R}$ and with $E_1 = -3$ V it is $I_1 = \frac{U_1}{R}$.
Therefore, an increase in the potential of the grid by $E_1 - E_2 = 3$ V raises the anode current of the valve by

$$I_1 - I_2 = \frac{1}{R}(U_1 - U_2) = 3.5 \text{ mA}$$

Since the grid characteristic of the valve in the region being considered is assumed to be linear, the additional increase in the potential of the grid relative to the cathode by 3 V (from – 3 V to zero with the short-circuited grid and cathode) will increase the anode current by another 3.5 mA.

The voltage drop across the resistance R will now increase additionally by $U_1 - U_2 = 35$ V, and become equal to $U_0 = U_1 + (U_1 - U_2) = 130$ V, while the potential difference between the anode and the cathode of the valve will be equal to

$$E - U_0 = 120 \text{ V}.$$

542. The first diode begins to conduct current only when $U_a > 0$, i. e., when $V > E_1$, the second at $V > E_2$ and the third at $V > E_3$. For this reason, the diagram showing the full current versus the voltage will have the form of a broken line (Fig. 453):

$$I = 0 \text{ at } V \le E$$
$$I = k(V - E_1) \text{ at } E_1 \le V \le E_2$$
$$I = k(V - E_1) + k(V - E_2) \text{ at } E_2 \le V \le E_3$$
$$I = k(V - E_1) + k(V - E_2) + k(V - E_3) \text{ at } E_3 \le V$$

Such circuits are sometimes used in radio installations to obtain a functional dependence of the current on the voltage.

543. In Fig. 454, A and B are control grids, MN is the screen and OC the trajectory of an electron. The origin of the coordinate system is at point O.

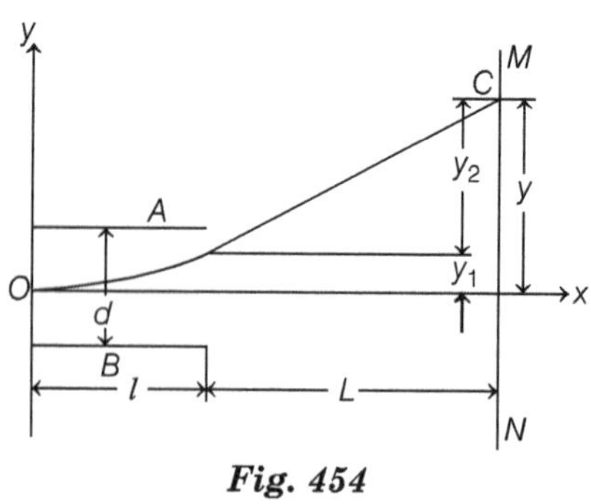

Fig. 454

An electron travelling between the grids moves in the direction of the y-axis with a uniform acceleration of $a = \frac{eU}{md}$, where U is the potential difference between A and B. The electron covers the distance l along the x-axis in the time $t_1 = \frac{l}{v_x}$, when v_x is the horizontal component of the velocity of the electron determined from the condition

$$\frac{mv_x^2}{2} = eU_0$$

During the time t_1 the electron is deflected in the direction of the y-axis by $y_1 = \frac{at_1^2}{2} = \frac{eUl^2}{2dmv_x^2}$. The electron moves outside the grids with a constant velocity during the time $t_2 = \frac{L}{v_x}$. The velocity along the y-axis is $v_y = at_1$.

The deflection outside the grids is $y_2 = v_y t_2 = \frac{eUlL}{dmv_x^2}$

The total deflection is $y = y_1 + y_2 = \frac{eUl}{dmv_x^2}\left(\frac{l}{2} + L\right) \cong \frac{eUlL}{dmv_x^2} = \frac{UlL}{2U_0 d}$

and the sensitivity is $q = \frac{y}{U} = \frac{lL}{2U_0 d}$

3-4. Magnetic Field of a Current. Action of a Magnetic Field on a Current and Moving Charges

544. If the current is expressed in amperes, the coefficient k is numerically equal to 0.1. Since $1A = 3 \times 10^9$ cgs electrostatic units (CGS_I) the sought value of the coefficient will be $k = \frac{1}{3 \times 10^{10}}$ if the current is measured in these units. The dimension of the coefficient can be found directly from the formula for the intensity H:

$$[k] = \frac{[H]\,[l]}{[I]}$$

Bearing in mind that $[H] = [E]$ and $[E] = \frac{[F]}{[Q]} = g^{1/2}\text{cm}^{-1/2} \cdot \text{s}^{-1}$, $[l] = \text{cm}$ and $[I] = g^{-1/2} \cdot \text{cm}^{3/2} \cdot \text{s}^{-2}$, we obtain $[k] = \text{s/cm}$.

If a new constant c is introduced instead of k, so that $k = \frac{1}{c}$, then c will be equal to the velocity of light in a vacuum.

545. The intensity of the field induced by the first winding is

$$H_1 = \frac{0.4\pi I N_1}{2\pi R} = 400 \text{ Oe}$$

The second winding induces the intensity

$$H_2 = \frac{0.4\pi I N_2}{2\pi R} = 200 \text{ Oe}$$

Since the fields H_1 and H_2 are directed oppositely, the sought field will be $H = H_1 - H_2 = 200$ Oe

546. The conductor BC does not induce a field at point M lying on the continuation of BC. According to the rule given in the note, the magnetic field produced by any elements of conductor BC should be perpendicular to line BM .

For this reason the presence of a field other than zero at point M, would disagree with the symmetry of the problem, because all the directions perpendicular to BM have equal rights.

Since the intensity of the field is proportional to the current, then $H_1 = kI$ without conductor BD. The fields from the conductors AB and BD are summated. Therefore after conductor BD is connected

$$H_2 = kI + \frac{kI}{2}$$

whence

$$\frac{H_2}{H_1} = \frac{3}{2}.$$

547. At an arbitrary point on line AB any small element of current of conductor ACB induces a magnetic field perpendicular to the plane of the drawing (see Problem 546).

The element of conductor ADB symmetrical to it induces the same field, but directed oppositely. For this reason the field from any two elements arranged symmetrically will be zero. Hence, the field at the arbitrary point on AB induced by the entire conductor is zero, since the straight sections of the conductor also do not induce a field on AB.

548. In the main, the field of the solenoid will be concentrated inside the toroidal winding and will not act on the magnetic pointer. A single layer winding, however, can simultaneously be regarded as one turn of a large radius that induces a magnetic field perpendicular to the plane of the torus.

The magnetic pointer will be positioned along the axis of the torus. The direction of its poles can be determined by the right-hand screw rule.

549. The current flowing along the pipe can be regarded as the sum of a great number of identical straight currents uniformly distributed over the surface of the pipe. The intensity of the magnetic field at any point of space can be represented as the sum of the intensities of the fields induced by these currents.

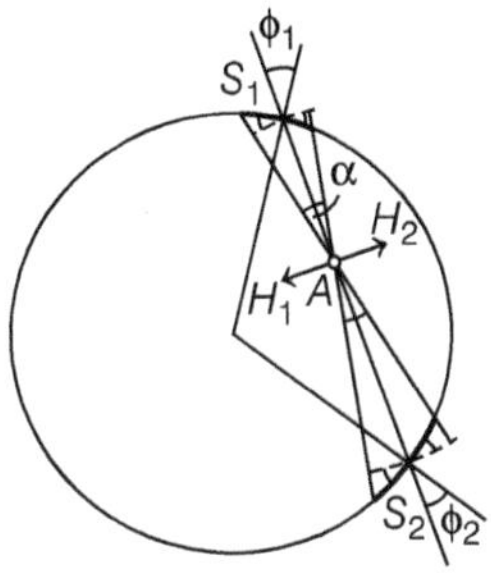

Fig. 455

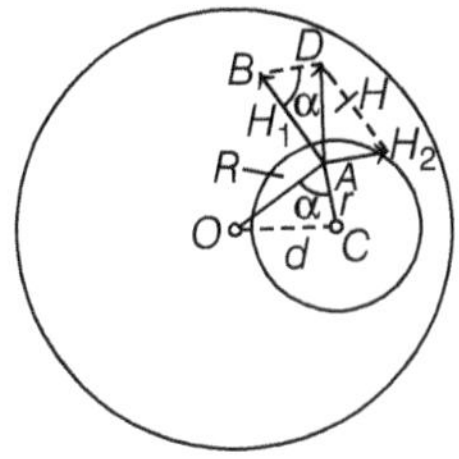

Fig. 456

Figure 455 shows a cross section of the pipe along which the current flows. Let us compare the intensities of the magnetic fields H_1 and H_2 created at point A by the straight currents I_1 and I_2 passing through the small arcs S_1 and S_2. The lengths of the arcs are $S_1 = \frac{\alpha R_1}{\cos\phi_1}$, and $S_2 = \frac{\alpha R_2}{\cos\phi_2}$, where $R_1$and R_2 are the distances to point A. But as can be seen from the drawing, $\phi_1 = \phi_2$. Hence $\frac{S_1}{S_2} = \frac{R_1}{R_2}$. The current in the pipe is distributed uniformly, and therefore $\frac{I_1}{I_2} = \frac{S_1}{S_2}$,

whence $$\frac{I_1}{R_1} = \frac{I_2}{R_2}.$$

The intensities of the magnetic fields created at point A by the currents I_1 and I_2 are equal, namely,

$$H_1 = k\frac{I_1}{R_1} = k\frac{I_2}{R_2} = H_2$$

and directed oppositely.

Since a corresponding element that compensates completely for the magnetic field of the first element at point A can be selected for each element of the pipe cross section, the resulting magnetic field of the current flowing in the pipe will be zero at any point inside the pipe.

550. The conductor with the space is equivalent to a solid conductor which carries a current with a density j, while through the volume that corresponds to the space there also flows a current of the same density in the opposite direction.

The total current in this volume will be equal to zero, and this shows that a space exists in the solid conductor.

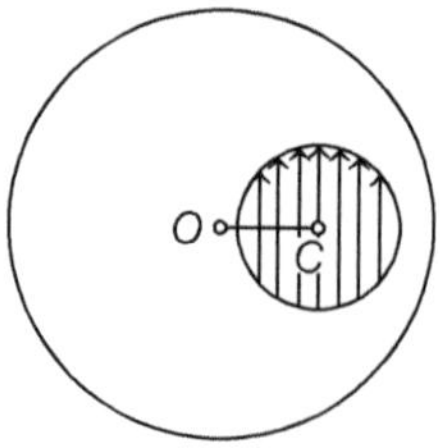

Fig. 457

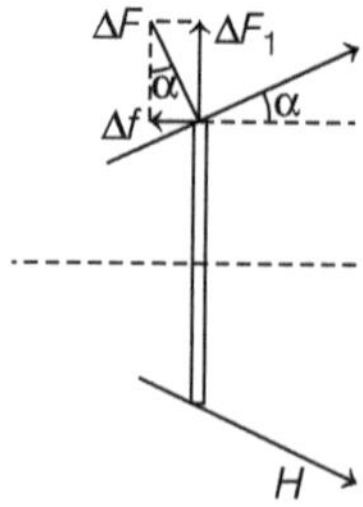

Fig. 458

The field created by the current with a density j at an arbitrary point of space A is equal to $H_1 = k \cdot 2\pi jR$ (Fig. 456). Here R is the distance from the axis of the conductor to point A. (It is assumed that the current flows toward us.)

At the same point, the current flowing through the volume corresponding to the space, but in the reverse direction, induces a field of $H_2 = k \cdot 2\pi jr$. Figure 456 shows that the total intensity of the field is

$$H = \sqrt{H_1^2 + H_2^2 - 2H_1H_2 \cos\alpha}$$

Obviously,

$$\cos\alpha = \frac{R^2 + r^2 - d^2}{2Rr}$$

Therefore, the intensity $H = k \cdot 2\pi jd$ is the same for all the points of the space.

551. Triangle AOC is similar to triangle BAD (Fig. 456) since they have one equal angle, and the sides confining these angles are proportional.

Therefore, $\angle AOC = \angle BAD$. But $R \perp H_1$, and therefore $H \perp d$.

The intensity of the magnetic field at any point of the space is perpendicular to the line that connects the centres of the conductor and the space. The distribution of the lines is shown in Fig. 457.

552. $k = \frac{1}{c}$, where c is the velocity of light in a vacuum.

553. No, it will not. Forces of attraction exist between the separate elements of the current. As a result, the density of the current increases somewhat toward the axis of the conductor. The effect is negligible.

554. When lightning strikes, a very high current flows for an instant through the pipe and the separate elements of the current are mutually attracted with a high force. It is this force that converts the pipe into a rod.

555. The currents in the adjacent turns are parallel and flow in the same direction. For this reason the turns will be mutually attracted. At the same time the currents in opposite sections of the turns flow in different directions. Therefore, opposite sections are repulsed.

The turns of the winding will tend to increase in diameter, and the distance between them along the axis of the solenoid will be reduced.

556. $\alpha = \frac{HnAI}{k}$.

557. The action of the magnetic field will cause the ring to so turn that the force lines of the field are perpendicular to the plane of the ring and form a right-hand screw with the direction of the current. The tension of the ring will be maximum. Upon employing the method used to solve Problem 403, we obtain

$$F = kIRH = 5 \text{ dynes}$$

558. The element of the ring Δl is acted upon by the force $\Delta F = kIH\Delta l$ (Fig. 458). Let us resolve it into the components ΔF_1 and Δf. The component ΔF_1 lies in the plane of the ring and $\Delta f = \Delta F \sin\alpha$ is normal to the plane of the ring. The resultant of the forces ΔF_1 that act on the separate elements of the ring is zero. These forces only stretch the ring. The full force f acting on the ring is equal to the sum of the forces Δf:

$$f = \sum_i kIH \sin\alpha \cdot \Delta l_i = kIH \cdot 2\pi R \sin\alpha \cong 273 \text{ dynes}$$

559. The forces acting on BC and AD are perpendicular to the motion of these sides and, therefore, perform no work.

The forces acting on AB and CD are constant, form a right angle with the direction of the field and are numerically equal to $f = kHIa$ (Fig. 459).

The sought work will be equal to the double product of the force and the motion of side AB or CD in the direction of the force. When the circuit is turned through 180°, this motion is b.

Therefore, $W = 2kHIab$.

560. Assuming that all the electrons move with a velocity v, the intensity of the current can be expressed as follows (see Problem 521):

$$I = nAev$$

Upon inserting the value of I into the formula for F, we get

$$F = kHnAlev \sin\alpha$$

Since a piece of the conductor contains $N = Aln$ electrons, the force acting on one electron is $f = kHev \sin \alpha$.

The force f is known as the Lorentz force. The direction of the Lorentz force is determined by the left-hand rule (the magnetic field intersects the palm, four fingers are directed opposite to the motion of the electrons, or along the motion of a positively charged particle, and the thumb shows the direction of the Lorentz force).

561. The Lorentz force is always perpendicular to the velocity of a particle and therefore performs no work. The kinetic energy and, hence, the absolute velocity of the particle remain constant.

562. The electron is acted upon by the force $f = kevH$. If H is measured in oersteds and the charge in cgs electrostatic units, then $k = \frac{1}{c}$ (see Problem 560). This force is constant in magnitude and perpendicular to the velocity v. For this reason the acceleration of the electron is also constant in magnitude and constantly remains perpendicular to the velocity. The velocity changes only in direction.

With a constant acceleration perpendicular to the velocity, the motion at a velocity constant in magnitude is uniform motion along a circle.

On the basis of Newton's second law, $\frac{mv^2}{R} = \frac{e}{c} \cdot vH$. Therefore, the electron will move along a circle with a radius $R = \frac{mcv}{eH}$.

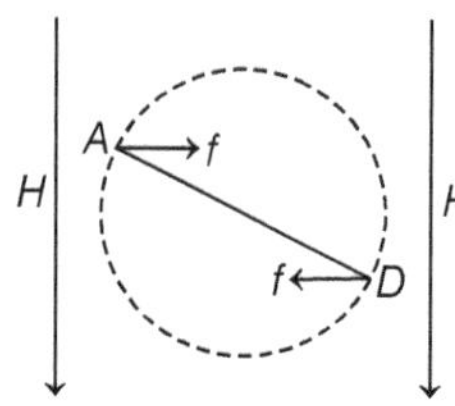

Fig. 459

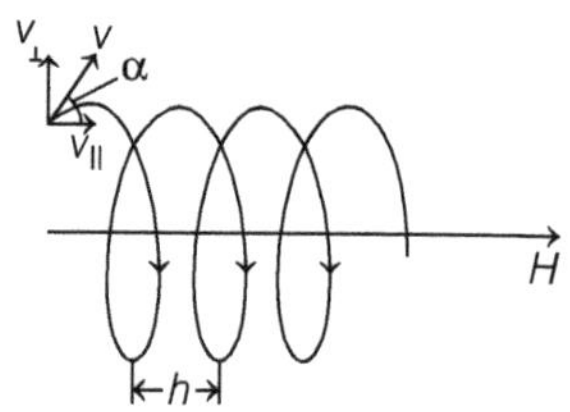

Fig. 460

563. Let us resolve the velocity of the electron into the components $v_{\parallel}$ parallel to H and $v_{\perp}$ perpendicular to H (Fig. 460). The component $v_{\parallel}$ does not change in magnitude or direction since the Lorentz force does not act on a particle whose velocity is directed along the field. The component $v_{\perp}$ changes in direction in the same way as in Problem 562.

Thus, rotation along a circle in a plane perpendicular to H is superposed on the uniform translational motion along H. This produces motion along a helical line with a constant pitch $h = v_{\parallel}\tau$, where τ is the duration of one revolution of the electron along a circle with a radius $R = \frac{mcv \sin\alpha}{eH}$.

Since

$$\tau = \frac{2\pi R}{v_{\perp}} = \frac{2\pi mc}{eH}, \text{ then } h = \frac{2\pi mc}{eH} v \cos\alpha$$

564. The action of the Lorentz force (see Problem 560) will cause the electrons to move towards the edge of the band. For this reason one edge of the band will receive a negative charge and the other a positive one. An additional electric field will be generated inside the band with an intensity E directed perpendicular to the current. The electrons will continue to move until the Lorentz force is equalized by the force acting on the electron from the side of the electric field $E: eE = kevH$. Hence, $E = kvH$.

The potential difference $\phi_A - \phi_B = Ea = kvHa$ or, since $I = nevA$, then

$$\phi_A - \phi_B = kHa\frac{I}{neA}.$$

565. $\phi_A - \phi_B \cong 23\mu V$.

566. The Lorentz force (see Problem 560) acts on both the free electrons and the positive ions at the points of a crystal lattice, since both move in a magnetic field. In accordance with the left-hand rule, the force f that acts on the free electrons will be directed as shown in Fig. 461. The electrons are displaced with respect to the lattice, and one side of the parallelepiped is charged negatively and the other positively. An electric field is produced in the block, and when the intensity of this field satisfies the ratio $eE = kevH$, the electrons will no longer move with respect to the lattice.

The sought intensity $E = kvH$.

The density of the charges σ can be found from the equation $4\pi\sigma = E$.

Therefore,

$$\sigma = \frac{1}{4\pi} kvH.$$

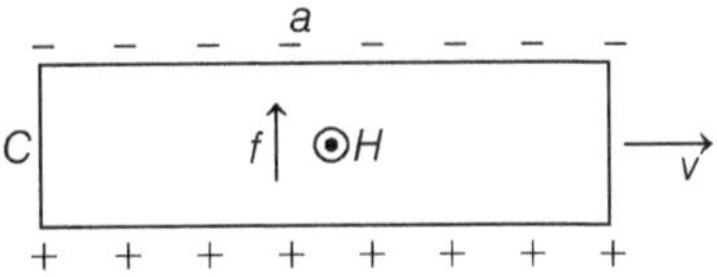

Fig. 461

567. For no electrostatic field to appear, the electrons should not move with respect to the crystal lattice when the cylinder revolves. This motion will be absent if the Lorentz force acting on the electrons is equal to $m\omega^2 r$, i.e., $m\omega^2 r = kevH$.

Since $v = \omega r$, then $H = \frac{m\omega}{ek}$.

The field should be arranged in the direction of the forward motion of a screw rotating in the same direction as the cylinder.

3-5. Electromagnetic Induction. Alternating Current

568. The direction of the intensity of the electric field is shown in Fig. 462.

569. When the circuit moves, the magnetic flux passing through area *ABCD* diminishes. Therefore, in accordance with Lenz's law, the induced current will flow clockwise.

570. As the iron rod flies through the coil, the magnetic flux passing through it changes. This induces an e.m.f. of induction in the circuit. According to Lenz's law, the total current in the coil decreases when the rod enters it, and increases when the rod leaves it.

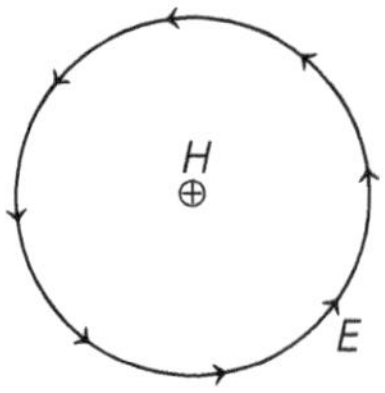

Fig. 462

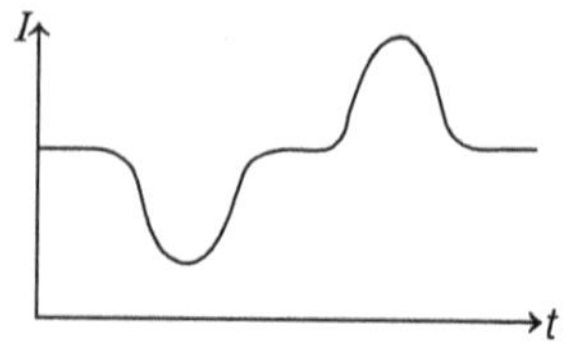

Fig. 463

A diagram of the change in the current is shown in Fig. 463.

571. The magnetic flux changes at a constant rate, and therefore the e. m. f. of induction in the second coil will also be constant. If the coil is connected to a closed circuit, it will carry a direct current, which will set in not at once, but depending on the coefficient of self-induction of the second coil and its resistance.

572. Yes, it will. The e.m.f. of induction is proportional to the rate of change of the magnetic flux, while the magnitude of the magnetic flux in the iron core does not change directly with the current. The relationship will be more complicated.

573. According to Faraday's law,

$$E_i = 10^{-8}\frac{\Delta\phi}{\Delta t} = 10^{-8}kA$$

The e.m.f. of induction is numerically equal to the work performed by the electric field when a single positive charge moves in the turn, i.e.. $E_i = 2\pi rE$.

Hence $$E = \frac{E_i}{2\pi r}.$$

Thus, we finally obtain:

$$E = 10^{-8}\frac{k\pi r^2}{2\pi r} = 10^{-8}\frac{kr}{2}$$

It should be noted that this electric field is induced not by the electric charges, but by a magnetic field varying with time.

Let us recall that when an electric charge moves in a closed circuit in an electrostatic field the work is always equal to zero. By an electrostatic field is meant an electric field induced by electric charges.

574. Let us divide the ring into $n = \dfrac{b-a}{\delta}$ small rings each with a width δ.

Let us consider a ring with a height h whose internal radius is x and external radius is $x+\delta$. If δ is small as compared with x, the resistance of such a ring can be expressed by the formula

$$R = \rho \frac{2\pi x}{\delta h}$$

The e. m. f. of induction acting in this ring (if $\delta \ll x$) is equal to

$$E = 10^{-8} \frac{\Delta\varphi}{\Delta t} = 10^{-8} \pi x^2 k$$

The intensity of the current flowing in such a ring is

$$\Delta I = \frac{E}{R} = 10^{-8} \frac{\pi x^2 k \delta h}{\rho 2\pi x} = 10^{-8} \frac{k\delta h x}{2\rho}$$

The current flowing through the entire ring can be found from the following sum:

$$I = 10^{-8} \frac{kh\delta}{2\rho} \{a + (a+\delta) + (a+2\delta) + \ldots + [a + (n-1)\delta]\}$$

The expression in the braces is an arithmetical progression. Therefore,

$$I = 10^{-8} \frac{kh}{2\rho} (b-a) \frac{2a + b - a - \delta}{2}$$

This result will be the more accurate, the smaller is δ. Assuming δ as tending to zero, we obtain

$$I = 10^{-8} \frac{kh}{4\rho} (b^2 - a^2)$$

575. On the basis of the law of electromagnetic induction and Ohm's law, we have for the quantity of electricity that passed through the galvanometer;

$$\Delta Q = I\Delta t = 10^{-8} \frac{\Delta\varphi}{R}$$

or

$$Q = \frac{10^{-8}}{R} (\varphi - \varphi_0)$$

Since the initial magnetic flux $\varphi_0 = HAn$ and the final flux $\varphi = 0$, the quantity of electricity in coulombs will be $Q = \dfrac{10^{-8}}{R} HAn$ if R is measured in Ohms and H in oersteds.

576. The e. m. f. of induction $E_1 = 10^{-8}ka^2$ acts in circuit $ABCD$, and $E_2 = 10^{-8}k\frac{a^2}{2}$ in circuit $BEFC$.

The simplest equivalent circuit with galvanic cells used as the e. m. f. of induction will for our circuit have the form shown in Fig. 464.

On the basis of Ohm's law,

$$I_3ar = E_1 - I_1\, 3ar = I_2 2ar - E_2$$

Since the charge is retained, $I_1 = I_2 + I_3$. All three currents can easily be found from the given system of equations:

$$I_1 = \frac{6E_1 + 2E_2}{22\, ar},$$

$$I_2 = \frac{2E_1 + 8E_2}{22\, ar}, \text{ and } I_3 = \frac{2E_1 - 3E_2}{11\, ar}$$

Taking into account the expressions for E_1 and E_2, we have:

$$I_1 = \frac{10^{-8}ka}{r} \times \frac{7}{22},$$

$$I_2 = \frac{10^{-8}ka}{r} \times \frac{3}{11}, \text{ and } I_3 = \frac{10^{-8}ka}{r} \times \frac{1}{22}$$

577. The third way is the worst, since eddy currents circulate in the winding turns without meeting an insulated layer.

The first way makes it possible to get rid of most eddy currents, but not all of them, since one layer of the autotransformer winding actually has many turns around the core and one turn along it (see Problem 548). The best is the second way, which is used in practice.

578. The potential difference between any points of the ring should be equal to zero. Otherwise, there will be a contradiction in applying Ohm's law to the short and long sections of the ring. Besides, this is obvious from considerations of symmetry.

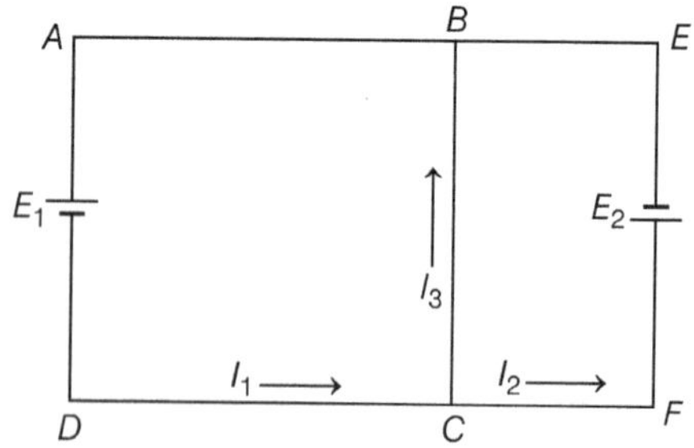

Fig. 464

If there is no potential difference, the electrostatic field inside the ring is zero. The current is produced by the e.m.f. of induction uniformly distributed along the ring:

$$I = \frac{e_i}{r} = \frac{E_i}{R}$$

where e_i and E_i are the e.m.f. s of induction on the short and long sections of the ring, and r and R are the respective resistances of the sections.

Despite the absence of a potential difference between points A and B, the electrometer will register a potential difference between the rod and the housing.

The matter is that the current in conductors AC and BD is zero. Therefore, the applied electric field of inductive origin is equalized at each point of these conductors by the intensity of the electrostatic field generated by the redistribution of the charges in the conductors under the effect of the e.m.f. of induction. The work of the electrostatic forces in moving along closed circuit $ACDBA$ is zero. There is no electrostatic field on section AB. When a charge moves along AC and BD, the work of the electrostatic forces is equal to the e. m. f. of induction in these conductors and has an opposite sign.

Hence, for the work of the electrostatic forces along a closed circuit to be equal to zero, the potential difference between points C and D should be equal to the e. m. f. of induction in conductors AC and DB and should coincide with it in sign. Since the e. m. f. of induction in closed circuit $ACDBA$ is zero (the magnetic field does not pass through this circuit), the e. m. f. of induction on section AB is equal in magnitude and opposite in sign to the e. m. f. in conductors AC and BD, if we neglect the work of the applied forces of induction on the section between the rod and the housing of the electrometer, as compared with the work in conductors AC and BD.

For this reason the electrometer will show a potential difference approximately equal to the e. m. f. on the section AB.

579. As distinct from Problem 578, the potential difference $U_A - U_B$ is not zero.

Let us write Ohm's law for all three sections of the conductor, denoting the currents in ADB, AKB and ACB by I_1, I_2 and I_3, and the respective e. m. f. s of induction by E_1, and E_2 and E_3:

$$I_1 = \frac{\varepsilon_1 + U_A - U_B}{R_1},$$

$$I_2 = \frac{E_2 + U_B - U_A}{R_2}, \text{ and } I_3 = \frac{E_3 + U_A - U_B}{R_3}$$

Since the charge is retained, $I_2 = I_1 + I_3$. Summation of the first two equations gives

$$I_1R_1 + I_2R_2 = E_1 + E_2 = E$$

Upon subtracting the first equation from the third, we get

$$I_3R_3 - I_1R_1 = E_3 - E_1$$

But the e. m. f. of induction in circuit *ACBDA* is zero, since the circuit is not pierced by a magnetic field. Therefore,

$$E_1 = E_3 \text{ and } I_3R_3 - I_1R_1 = 0$$

The system of equations gives the following value for the sought current:

$$I_3 = \frac{ER_1}{R_1R_2 + R_1R_3 + R_2R_3}$$

580. When the resistance R_3 is other than zero, we can find from the equations of Problem 579 that:

$$I_1 = \frac{ER_3}{R_1R_2 + R_2R_3 + R_1R_3}$$

$$I_2 = \frac{E(R_1 + R_3)}{R_1R_2 + R_2R_3 + R_1R_3}$$

$$I_3 = \frac{ER_1}{R_1R_2 + R_2R_3 + R_1R_3}$$

When $R_3 = 0$

$$I_1 = 0,\ I_2 = \frac{E}{R_2} = I_3$$

In the general case

$$U_A - U_B = -\frac{ER_1^2R_2}{(R_1 + R_2)(R_1R_2 + R_2R_3 + R_1R_3)}$$

When $R_3 = 0$

$$U_A - U_B = -\frac{ER_1}{R_1 + R_2}$$

Here $U_A - U_B = -E_1$ (since $I_1 = 0$ on section *ADB*), where E_1, is the e. m. f. of induction on section *ADB*.

581. An alternating current should be passed through the electromagnet. The current should increase comparatively slowly, since at this moment the metal object will be repulsed weakly in view of the current induced in it, and diminish very rapidly, since in accordance with Lenz's law there will appear a high force of attraction proportional to the rate of current change. A possible relation between the current and time is shown in Fig. 465. In principle, a force of attraction also appears when the direct current in the electromagnet is switched off.

582. In both cases equilibrium will set in if the moment of the forces acting from the magnetic field on the current induced in the ring is equal to zero, or if there is no induced current. This will occur only if the plane of the ring is arranged along the force lines of the field (the induced current is zero) or when the plane of the ring is strictly perpendicular to the force lines (the moment of the forces is zero.

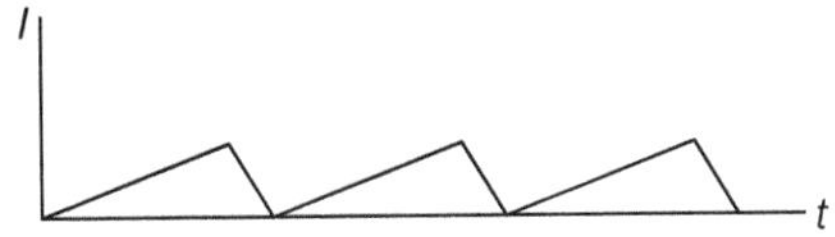

Fig. 465

According to Lenz's law, the first position of the ring will be stable in an increasing magnetic field, and the second unstable.

In a decreasing magnetic field, on the contrary, equilibrium will be stable when the angle between the plane of the ring and the force lines is a right one, and unstable when the plane of the ring is parallel to the force lines.

583. Let the velocity of the conductor be v at a certain moment of time. The e.m.f. (in volts) at the same moment of time will thus be $\varepsilon = 10^{-8}Hlv$, and the current $I = \frac{1}{R} \times 10^{-8}Hlv$. The action of the magnetic field on the conductor carrying a current will induce a force f that prevents free dropping of the conductor:

$$f = 10^{-9}\frac{H^2l^2v}{R}$$

Hence, at the moment of time being considered, the acceleration can be determined from the relationship

$$ma = mg - f = mg - 10^{-9}\frac{H^2l^2v}{R}$$

It is easy to see that as the velocity increases, the acceleration a will diminish and become zero at the moment of equality of the forces $f = mg$. From this moment on, the conductor will move with a constant velocity v_k equal to

$$v_k = \frac{mgR \times 10^9}{H^2l^2}$$

584. The e.m.f. of induction appearing in the conductor (measured in volts) is $E = 10^{-8}Hlv$.

The charge on the plates of the capacitor can be found from the relationship

$$Q = EC = 10^{-8}HlvC$$

The current flowing in the circuit is

$$l = \frac{\Delta Q}{\Delta t} = 10^{-8} HlC \frac{\Delta v}{\Delta t} = 10^{-8} HlCa$$

where a is the sought acceleration.

The interaction of this current with the magnetic field will produce a force F_1 acting on the moving conductor. On the basis of Lenz's law, this force will be directed oppositely to the force F.

The force $F_1 = kIHl = 10^{-9} H^2 I^2 aC$, if C is measured in farads. The sought acceleration can be found from the equation $ma = F - F_1$.

Hence, $a = \dfrac{F}{n + 10^{-9} H^2 l^2 C}$ is a constant quantity.

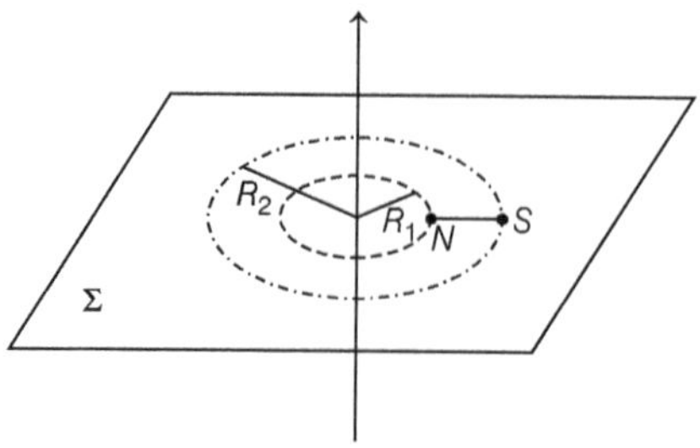

Fig. 466

The work of the force F over the path S is spent to increase the kinetic energy of the conductor and the electrostatic energy of the capacitor.

585. Let the magnet be initially positioned as shown in Fig. 466. Its north end is at a distance R_1 from the current and its south end at a distance R_2, the length of the magnet being $l = R_2 - R_1$. Let us now move the magnet in the plane Σ around the wire, keeping the distances R_1 and R_2 unchanged until after one revolution the magnet occupies its initial position. Since during this motion the total change in the magnetic flux through the area restricted by the straight wire and the conductors that short-circuit the current at a great distance from the magnet is zero, the quantity of induced electricity that has flown through the circuit is also zero. On the basis of the law of conservation of energy, the work of the forces of the magnetic field should also be equal to zero:

$$2\pi R_1 H_1 m - 2\pi R_2 H_2 m = 0$$

where m is the magnetic charge of the pole, and H_1 and H_2 are the intensities of the magnetic field at the distances R_1 and R_2 from the wire. Hence, $\dfrac{H_1}{H_2} = \dfrac{R_2}{R_1}$ which is possible only when H is proportional to $\dfrac{l}{R}$.

586. Since according to the initial condition, the intensity of the magnetic field is directly proportional to time, i. e., $H = 0.4\pi \frac{N}{l} kt$, then the e. m. f. of self-induction is equal to

$$E_i = 10^{-8} \times 0.4\pi \frac{N^2}{l} kA \, (A = \pi r^2)$$

and directed against the current. The voltage across the solenoid terminals should be

$$U = 4\pi 10^{-9} A \frac{N^2}{l} k + kRt$$

In this case $$I = \frac{U - E_i}{R} = kt.$$

587. When $R = 0$, the e.m.f. of self-induction E_i, remains constant, since the voltage across the solenoid terminals is $U = E_i = E$. It follows from the solution of Problem 586 that when E_i is constant, the current changes in proportion to time, i.e., $I = kt$, where $k = \frac{Ul}{4\pi \times 10^{-9} AN^2}$. Therefore, $I = \frac{10^9 lE}{4\pi AN^2} t$. If the resistance is finite and not zero, the current will increase according to the same law until the voltage drop IR across the resistance R becomes negligibly small as compared with E_i.

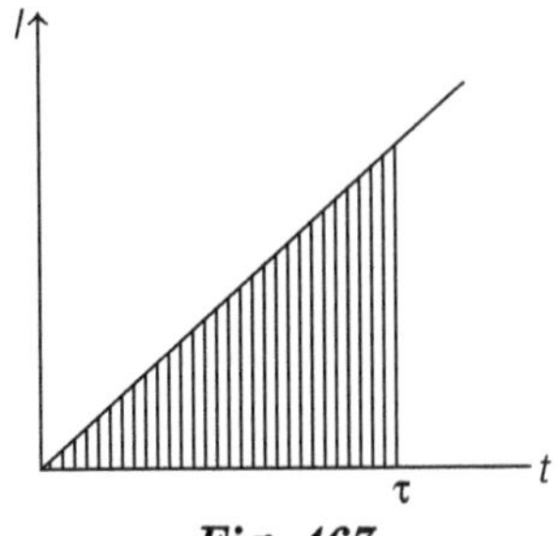

Fig. 467

588. The work of the battery during the time τ is $W = EQ$, where Q is the quantity of electricity that passes through the solenoid during the time τ.

The current in the solenoid grows directly with time: $I = \frac{10^9 l Et}{4\pi \, AN^2}$

(see the solution to Problem 587). Therefore, Q will be equal to the product of the mean intensity of the current

$$\frac{I_{init} + I_{final}}{2} \text{ (here } I_{init} = 0)$$

and the time τ or numerically equal to the area of the hatched triangle (see Fig. 467):

$$Q = \frac{lE\tau^2}{8\pi AN^2} 10^9$$

Hence, the work in ergs will be

$$W = \frac{l\varepsilon^2\tau^2}{8\pi AN^2} 10^{16} \text{erg}$$

This work goes to increase the energy of the magnetic field. Thus, we can write that $W = W_e$, where W_e is the energy of the magnetic field. Bearing in mind that $H = 0.4\pi \frac{N}{l} I$, and inserting the expression for the current, this energy can be written as

$$W_e = \frac{1}{8\pi} H^2 Al$$

(W_e is in ergs, H in oersteds and l in centimetres).

589. Since the resistance of the ring is zero, its total electromotive force should also be zero. This will occur only if the change of the full magnetic flux piercing the ring is zero. For this reason the change in the external magnetic flux φ_0 is equal in magnitude and opposite in sign to the change of the magnetic flux generated by the induced current: $\Delta\varphi_0 = L\Delta I$. Remembering that the flux φ_0 grows from 0 to $\pi r^2 H_0$ and the induced current changes in this case from 0 to I, we obtain $\pi r^2 H_0 = LI$.

Hence $$l = \frac{\pi r^2 H_0}{L}.$$

590. The magnetic flux through the ring cannot change (see Problem 589). Therefore, $\varphi = \pi r^2 H$. First this flux was produced by the external magnetic field, and after it was switched off, by the current induced in the ring.

591. The e. m. f. of induction $E_i = -10^{-8}\frac{\Delta\varphi}{\Delta t}$ if E_i is in volts and φ in oersteds per cm^2. The magnetic flux through N turns of the coil $\varphi = NA\mu_r H$, where $H = 0.4\pi \frac{N}{l} I$ (the current is in amperes).

On the other hand, $E_i = -L\frac{\Delta I}{\Delta t}$ if E_i is in volts, L in henries and I in amperes.

Hence, $$L = 10^{-9}\frac{4\pi N^2 \mu_r A}{l} \text{ henries.}$$

592. $\varphi = N_1 A\mu_r H_2$, where N_1 is the number of turns of the first winding and H_2 is the magnetic field created in the core of the second winding.

Approximately, $H_2 = \frac{4\pi N_2 I}{l}$, where l is the perimeter of the core.

Hence, $$M = \frac{4\pi\mu_r A N_1 N_2}{l}.$$

593. The e. m. f. of induction in the disk is shifted in phase by $\pi / 2$ with respect to the alternating current in the electromagnet.

On the other hand, the phase shift between the e. m. f. and the current in the disk tends to $\pi / 2$ if $L\omega \gg R$. In our case R is small and this inequality is observed. As a result, the eddy current in the disk is shifted in phase by π with respect to the current in the electromagnet. The currents are opposite in direction and will be mutually repulsed. Therefore, the disk will be pushed away and the string on which it is suspended will be deflected from the vertical.

The same result can be obtained by another method. If the resistance of the disk can be neglected as compared with its inductive reactance, the full magnetic flux through the disk undergoes almost no change. (For a super-conductor the change in flux is strictly zero, see Problem 589.) This means that the field of the eddy current in the disk is always directed against the field of the electromagnet.

Hence, the disk will be repulsed.

594. If the self-inductance of the wires can be neglected, the amount of heat $\Delta W_e = u\Delta Q$ will be evolved (u is the potential difference between the plates connected by the wires at a certain moment of time) when the charge $+\Delta Q$ is transferred from one plate to the other, or $-\Delta Q$ for the other pair of the plates.

The transfer of the charge ΔQ leads to a change in the potential difference on both capacitors by $\Delta\varphi = \frac{\Delta Q}{C}$.

Hence, $$\Delta W_e = uC\Delta\varphi$$

A decrease in the potential difference between the plates of one capacitor by ϕ is attended by a similar increase on the other.

Therefore,

$$u = (U - \varphi) - \varphi = U - 2\varphi$$

where U is the initial potential difference between the plates of the charged capacitor and φ is the value by which this potential difference drops at the given moment. Since φ changes from zero to $U/2$, the diagram showing u versus φ has the form shown in Fig. 468 by the dotted line. The relation uC versus φ is shown by the solid line AB. The quantity of heat $\Delta W_e = uC\Delta\varphi$ is shown in Fig. 468 by the area of the trapezium with the middle line ab that corresponds to the average potential in the given range $\Delta\varphi$.

The full amount of liberated heat W_e is depicted by the area of triangle OAB. It is equal to the loss of the electrostatic energy

$$W_e = UC\frac{U}{2}\frac{1}{2} = \frac{U^2C}{4} = \frac{Q^2}{4C} = \frac{W_{e0}}{2}$$

and does not depend on the resistance of the wires.

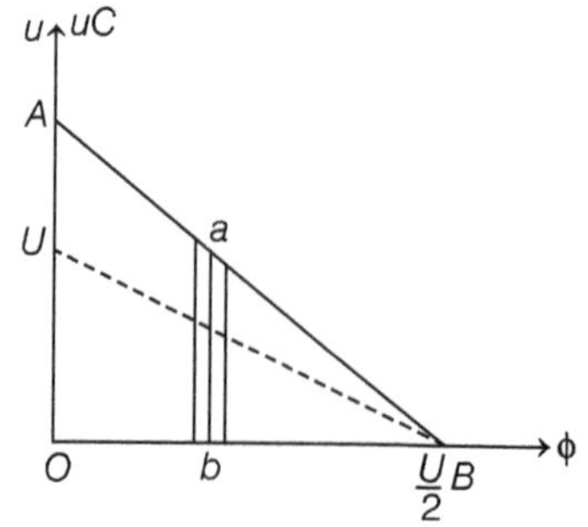

Fig. 468

The resistance must not be regarded as zero or very small, however, since the self-inductance of the wires must be taken into account.

595. The effective value of an alternating current is the value of a direct current that produces the same quantity of heat in a conductor as the alternating current during the same time.

Let us calculate the quantity of heat liberated during a period:

$$Q = 0.24I_0^2R\frac{T}{8} + 0.24I_0^2R\frac{T}{8} = 0.24I_0^2R\frac{T}{4}$$

On the other hand, $Q = 0.24I_{eff}^2RT$

Therefore, $I_{eff} = \frac{I_0}{2}$

596. When a sinusoidal alternating current flows through the circuit, the d-c ammeter will show zero, since its pointer cannot follow the rapid changes in the instantaneous values of the current owing to the inertia of the movable ammeter parts. The thermal ammeter shows the effective value of the alternating current $I_2 = \frac{I_0}{\sqrt{2}}$.

If a direct and an alternating currents flow simultaneously through the circuit, the d-c ammeter will show the mean value of the current equal to the direct current $I_1 = 6\text{A}$.

The current flowing through the thermal ammeter is

$$I = I_1 + I_0 \sin\omega t$$

On an average, it evolves per second the heat

$$Q = kI^2R = kR\,(I_1^2 + 2\,I_1I_0\overline{\sin\omega t} + I_0^2\,\overline{\sin^2\omega t})$$

The line denotes averaging in time.

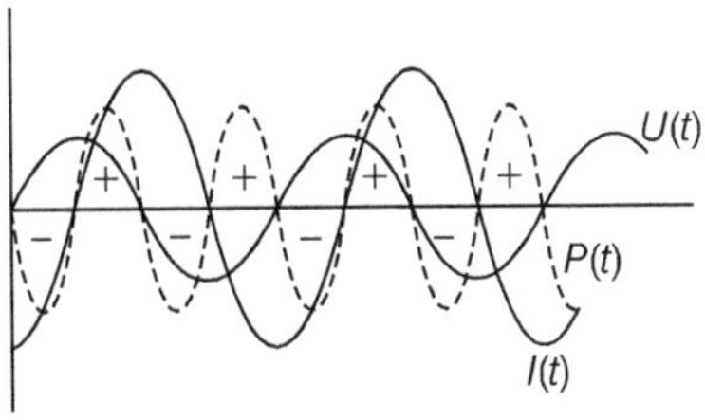

Fig. 469

The average value of $2I_1I_0 \sin \omega t$ a period is equal to zero and the ave rage value of

$$I_0^2\overline{\sin^2 \omega t} = I_0^2 \frac{\overline{1 - \cos 2\omega t}}{2} = \frac{I_0^2}{2} = I_2^2$$

is equal to the square of the effective value of the alternating current. Therefore, $Q = kR(I_1^2 + I_2^2)$

On the other hand, the quantity of heat Q liberated in the ammeter per second is related to the effective intensity of the current I flowing through the ammeter by the formula $Q = kI^2R$.

Hence, the $a - c$ thermal ammeter will show

$$I = \sqrt{I_1^2 + I_2^2} = 10 \text{ A}$$

597. Since $R = 0$, the current I lags in phase behind the voltage U by $\pi / 2$. Diagrams showing $U = U_0 \sin \omega t$, $I = I_0 \sin\left(\omega t - \frac{\pi}{2}\right)$ and the instantaneous power $P = IU$ are shown in Fig. 469. The sign of P changes every quarter of a period. The supply of energy from the source to the coil corresponds to a positive value of P. When P is negative, the energy returns from the coil to the source. On an average, the coil consumes no power during a period. The mean power is equal to zero.

598. The inductive reactance of a choke is much greater than its resistance: $L\omega \gg R$. The advantage of a choke over an ordinary resistor is that no heat is liberated on the inductive reactance of a choke. For this reason a lamp with a choke is much more economical than a lamp with a resistor connected in series.

599. If $L\omega \gg R$, the phase shift between the current and the voltage is great and the power consumed by the mains cannot be high. When capacitors are switched on, the phase shift is reduced, since the current flowing through a capacitor leads the voltage, compensating thereby for the lag of the current in phase in electric devices with a high inductance. As a result, the power consumed by the mains increases.

600. (a) Since ends A and B are open, no current flows in section AC. Therefore, the voltage drop in AC is zero. For this reason $U_2 = U_1$.

(b) When a variable potential difference is applied between points B and C, the current flowing along BC creates a variable magnetic flux that generates an e.m.f. of induction in section AC. Since $L\omega \gg R$, the amplitude of this e.m.f. will also equal U_1. For this reason the amplitude of the voltage U_2 between points A and B will be equal to $2U_1$ (step-up autotransformer).

601. When an alternating current flows in a conductor the amount of heat evolved is $Q = I_{eff}^2 Rt$. The expression for the evolved heat $Q = \frac{U_{eff}^2}{R} t$ is true only when Ohm's law can be applied in the usual form: $I = \frac{U}{R}$.

The winding of a transformer has a very high inductive reactance. For this reason Ohm's law in its usual form and, therefore, the expression $Q = \frac{U_{eff}^2}{R} t$ cannot be applied. The amount of heat evolved is small, since the intensity of the current and the ohmic resistance of the winding are small.

602. If we neglect the ohmic resistance, the voltage across the terminals of the primary winding U_1 can be represented as the algebraic sum of the e.m f. of self-induction of this winding and the e.m.f. of induction generated in it by the current flowing through the secondary winding

$$U_1 = L_1 \frac{\Delta I_1}{\Delta t} - M \frac{\Delta I_2}{\Delta t}$$

The minus sign is due to the fact that the currents I_1 and I_2 have opposite phases.

If the currents change according to the laws $I_1 = I_{01} \sin \omega t$ and $I_2 = I_{02} \sin \omega t$, then

$$\frac{\Delta I_1}{\Delta t} = \omega I_{01} \cos \omega t \text{ and } \frac{\Delta I_2}{\Delta t} = \omega I_{02} \cos \omega t$$

Since the voltage U_1 is shifted in phase relative to the current I_1 by $\pi / 2$, we can write $U_1 = U_{10} \cos \omega t$.

Upon dividing the expression for U_1 by $L_1 \omega \cos \omega t$, we get

$$\frac{U_{01}}{L_1 \omega} = I_{01} - \frac{M}{L_1} I_{02}$$

$\frac{U_{01}}{L_1 \omega}$ is the no-load current if the ohmic resistance of the winding is neglected. Disregarding the no-load current, we find that

$$\frac{I_{01}}{I_{02}} = \frac{M}{L_1}$$

By using the expressions for the coefficient of self-induction and mutual inductance from Problems 591 and 592, we obtain

$$\frac{I_1}{I_2} = \frac{I_{01}}{I_{02}} = \frac{N_2}{N_1}$$

603. The positive half-waves of the current will charge the capacitor to the amplitude voltage of the mains, equal to $127\sqrt{2}\ V = 180$ V. When the diode carries no current, it receives the voltage of the mains (with an amplitude of 180 V) plus the same voltage of the charged capacitor. The change in the potential along the circuit at this moment of time is shown in Fig. 470.

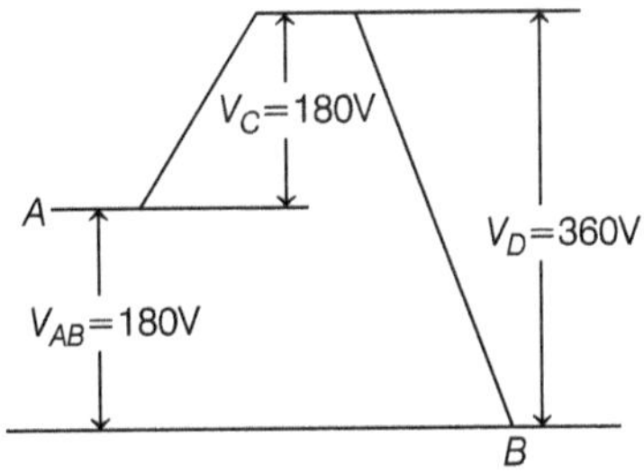

Fig. 470

If the rectifier operates without load, the capacitor should be calculated for a puncturing voltage of at least 180 V, and the diode for a voltage of at least 360 V.

604. The anode voltage of each diode is $U_a = \frac{U}{2}\sin\omega t - IR$. A current flows through the diode when $U_a > 0$ and does not flow through it when $U_a \le 0$. In a quarter of a period the current will not flow during the time interval $0 \le t \le t_1$(Fig. 471), where t_1, is determined by the equation $\frac{U}{2}\sin\omega t_1 - IR = 0$. Hence, $t_1 = \frac{T}{2\pi}\ \text{arc sin}\ \frac{2IR}{U}$.

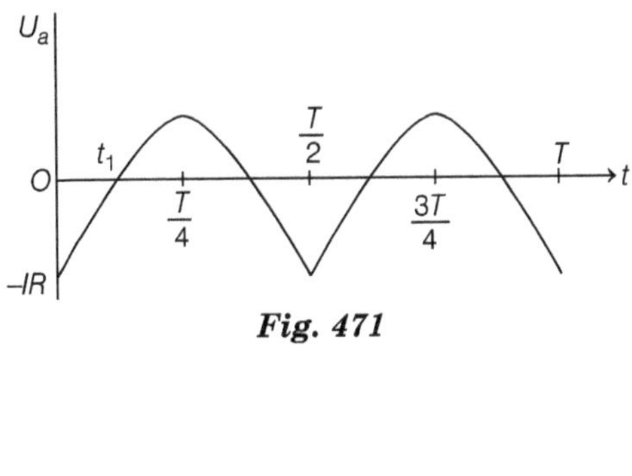

Fig. 471

This is also the time during which the current does not flow in the following quarters of the period. Altogether in a period the current does not flow during

$$\frac{2T}{\pi}\,\text{arc sin}\,\frac{2IR}{U} = 0.465\,T$$

3-6. Electrical Machines

605. If the frequency of the alternating current remains the same, this means that the revolutions of the motor and the generator also remain as before, and the e.m.f. of the generator will not change.

When the external resistance in the circuit is high, the circuit will carry a smaller current and a lower power will be supplied. For this reason the power of the motor that revolves the generator should be reduced.

606. The work performed by a field in moving conductors carrying a current (armature windings) is not equal to the total work of the field. Apart from the work spent to move the conductors, the magnetic field performs work to retard the electrons in the conductor, which produces an e.m.f. of induction in the armature winding. The first part of the work is positive and the second negative. The total work of the magnetic field is zero.

The electromotive force of the source that generates a current in the motor armature performs positive work, and the latter compensates for the negative work of the magnetic field in retarding the electrons.

In essence, the motor does its work at the expense of the energy of the source feeding it.

607. The power consumed by the motor is $P = IU$; here $U = E_i + IR$, where E_i is the e.m.f. of induction appearing in the armature.

Hence,

$$P = IE_i + I^2R$$

Here I^2R is the Joule heat liberated in the windings, and IE_i is the work against the e.m.f. of induction, equal to the mechanical power P_1 developed by the motor.

This power $P_1 = \frac{UE_i - E_i^2}{R}$, since $I = \frac{U - E_i}{R}$. This expression is maximum when $E_i = \frac{U}{2}$ (see the solution to Problem 516). Therefore, the maximum value of $P_1 = \frac{U^2}{4R} = 180\ V$. The motor cannot develop a power of 200 watts.

608. The maximum power developed by a series motor (see Problem 607) is

$$P_{max} = \frac{U^2}{4(R_1 + R_2)}$$

The power consumed by the motor is

$$P = UI = U\frac{U - E_i}{R_1 + R_2} = \frac{U^2}{2(R_1 + R_2)}$$

since $$E_i = \frac{U}{2}.$$

Hence, the efficiency is $\eta = \frac{1}{2}$. For a shunt-wound motor

$$P_{max} = \frac{U^2}{4R_1}$$

The power consumed is

$$P = UI = U\left(\frac{U - E_i}{R_1} + \frac{U}{R_2}\right) = \frac{U^2(2R_1 + R_2)}{2R_1R_2}$$

Therefore,

$$\eta = \frac{1}{2}\frac{1}{1 + 2\frac{R_1}{R_2}}$$

i. e., less than 50 per cent.

609. Let us denote the length of the turn by l and its width by d (Fig. 472). The force F acting on a conductor with a length l is $F = kIHl$. The power is

$$P = 2kIHlv = kIHA\omega$$

The current I can be determined from the formula

$$I = \frac{U - E_i}{R}$$

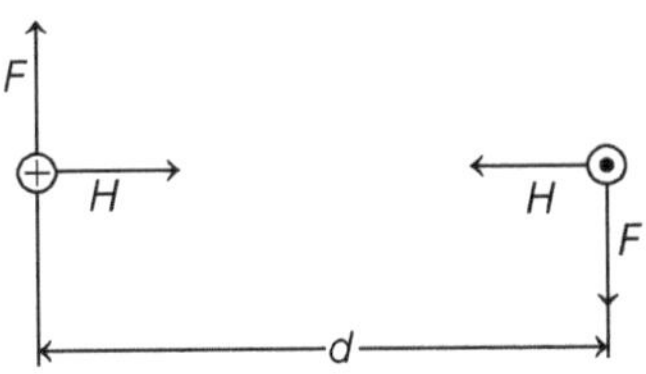

Fig. 472

where $$E_i = 10^{-8}HA\omega$$

Finally P can be written as follows:

$$P = kHA\frac{U}{R}\omega - 10^{-8}k\frac{H^2A^2}{R}\omega^2$$

P reaches its maximum

$$P_{max} = \frac{kU^2}{4R} \times 10^8 = \frac{U^2}{4R} J$$

when $\omega = \frac{U}{2HA} \times 10^8$. Here $E_i = \frac{U}{2}$ and $I = \frac{U}{2R}$. In a unit of time the battery performs the work $\frac{U^2}{2R}$. From this amount, one half is converted into mechanical power and the other half is liberated as heat (see Problem 607).

The relation between P and ω is shown in Fig. 473.

610. $M = \frac{kHAU}{R} - \frac{k \cdot 10^{-8} H^2 A^2}{R} \omega.$

The moment will be equal to zero when $\omega = \frac{U \cdot 10^8}{HA}$ (see Fig. 474).

Here $I = 0$, because $E_i = U$.

611. The nature of the relation between P and H is shown in Fig. 475 (see the solution to Problem 609). The power reaches its maximum when $H = \frac{U}{2A\omega} \times 10^8$. Here $E_i = \frac{U}{2}$ and $P_{max} = \frac{U^2}{4R} J$.

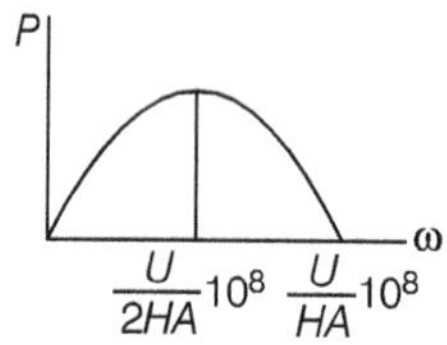

Fig. 473

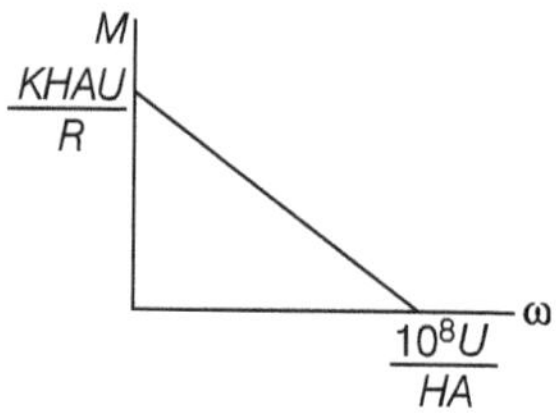

Fig. 474

612. The torque will reach its maximum $M_{max} = 10^7 \times \frac{U^2}{4R\omega}$ dyne · cm

when $$H = 10^8 \times \frac{U}{2A\omega}.$$

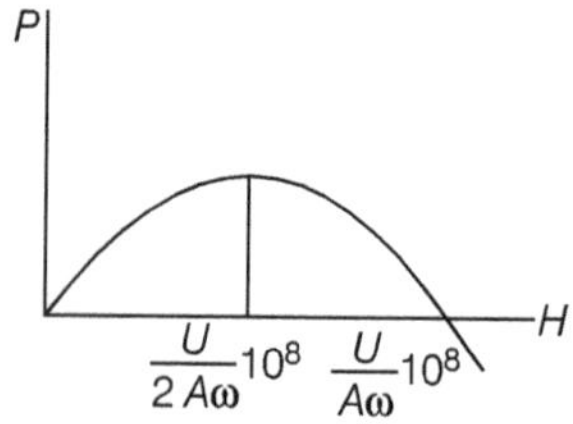

Fig. 475

613. As with a series motor, the power of a shunt-wound motor is

$$P = \frac{UE_i - E_i^2}{R}$$

where R is the resistance of the armature (see Problem 607). Two values of E_i correspond to a power of $P = 160$ watts: $E_1 = 80\ V$ and $E_2 = 40\ V$. Both values depend on the design features of the motor.

According to Faraday's law, E_i is directly proportional to the number of armature revolutions n per second and the intensity of the magnetic field created by the stator. With a shunt-wound motor this intensity does not depend on the load. Therefore, $E_i = an$, where a is a constant quantity determined by the design of the motor and the voltage applied. We obtain from the data in the problem that $a_1 = 8$ and $a_2 = 4$. The value of E_i cannot exceed 120 V.

Therefore, the maximum possible speed is either $n_1 = 15$ rev/s or $n_2 = 30$ rev/s.

614. If the voltage on the stator is kept constant, then at the given speed of the armature, the e.m.f. of induction in it does not depend on whether it is rotated at the expense of the action of the magnetic field of the stator on the current in the armature or with the aid of a mechanical drive.

The power developed by the motor is $P = M\omega$. In our case $P = 160$ watts. The e.m.f. of induction E_i can be determined from the equation

$$P = \frac{UE_i - E_i^2}{R}$$

(see Problem 607). Hence, $E_i = \frac{U}{2} \pm \sqrt{\frac{U^2}{4} - PR}$ has two values: $E_1 = 80\ V$ and $E_2 = 40\ V$. The e.m.f. of the generator will also be equal either to 80 V or 40 V.

The existence of two results is due to the fact that the same power of the motor is obtained with the same product IE_i, while two pairs of possible values I and E_i correspond to this product.

The values of E_i and, hence, the current depend on the design features of the motor, the number of turns, their shape, etc.

615. The mechanical power developed by the motor is

$$M2\pi n = \frac{UE_i - E_i^2}{R}$$

(see the solution to Problem 607).

The e.m.f. of induction in the armature $E_i = kHn$, where k is a proportionality factor determined by the number of turns of the armature winding and their area, and H is the intensity of the magnetic field of the stator, which is directly proportional to the current.

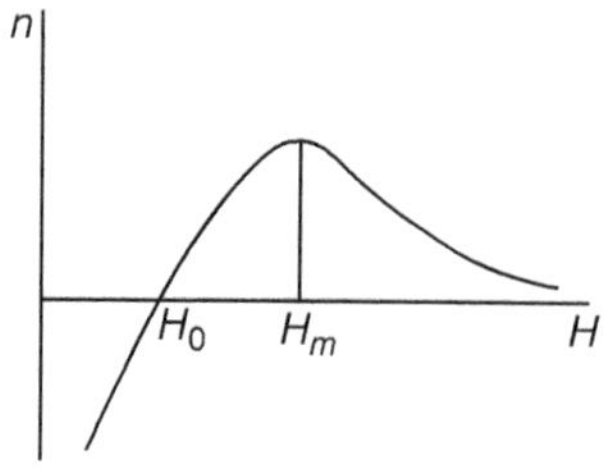

Fig. 476

Upon excluding E_i from these equations, we find that

$$n = \frac{U}{kH} - \frac{2\pi MR}{k^2H^2}$$

The relation between n and H is shown in Fig. 476.

If $H \le H_0 = \frac{2\pi MR}{kU}$, then $n \le 0$. From a physical viewpoint this means that the motor armature will not rotate. When $H = H_m = \frac{4\pi MR}{kU}$, the speed will be maximum. (This value can be found by the method described in the solution of Problem 516.) Therefore, if $H_m > H > H_0$, the speed increases when the current grows in the windings of the stator, and decreases when $H > H_m$.

If the motor operates without load ($M = 0$), the speed will be $n = \frac{U}{kH}$, i.e., it always decreases with an increase of H.

616. The intensity of the current flowing in the winding of the motor will be determined by the e.m.f. of the mains E, their resistance r and the e.m.f. of induction E_i generated in the motor armature: $I = \frac{E - E_i}{r}$.

At any moment, the potential difference U across the terminals of the motor is equal to E_i, since the resistance of the winding is zero. Therefore, the power $P = IU = \frac{EE_i - E_i^2}{r}$ is determined by the e.m.f. of the mains, their resistance and by E_i.

617. Assume that the voltage at the ends of winding W_1 changes in conformity with the law $U_{01} = E_0 \sin \omega t$. Hence $U_{02} = E_0 \sin(\omega t - 120°)$. The potential difference U_{12} is equal to that on the windings W_1 and W_2, i.e., $U_{12} = E_0 \sin \omega t - E_0 \sin(\omega t - 120°) =$

$$= 2E_0 \sin 60° \cos(\omega t - 60°)$$
$$= E_0\sqrt{3} \sin(\omega t + 30°)$$

since $\sin 60° = \frac{\sqrt{3}}{2}$ and $\cos(\omega t - 60°) = \sin(\omega t + 30°)$

Hence, the amplitude of the linear voltage is $\sqrt{3}$ times greater than that of the phase voltage.

618. When the load resistances are the same, the currents I_1, I_2 and I_3 are identical in amplitude and shifted in phase by $\frac{2}{3}\pi$. Therefore,

$$I_1 + I_2 = I_0 \sin \omega t + I_0 \sin\left(\omega t + \frac{2}{3}\pi\right) = I_0 \sin\left(\omega t + \frac{\pi}{3}\right)$$

and $I_1 + I_2 + I_3 = I_0 \sin\left(\omega t + \frac{\pi}{3}\right) + I_0 \sin\left(\omega t + \frac{4}{3}\pi\right)$

$$= 2 I_0 \sin\left(\omega t + \frac{5}{6}\pi\right) \cos\frac{\pi}{2} = 0$$

619. The magnetic fields H_1, H_2 and H_3 can be written as follows:

$$H_1 = H_0 \sin \omega t,$$

$$H_2 = H_0 \sin\left(\omega t + \frac{2}{3}\pi\right), \text{ and } H_3 = H_0 \sin\left(\omega t + \frac{4}{3}\pi\right)$$

Let us select the axes of coordinates x and y as shown in Fig. 220 and find the sum of the projections of the intensities of these fields on these axes:

$$H_x = H_0 \sin \omega t + H_0 \sin\left(\omega t + \frac{2}{3}\pi\right) \cos\frac{2}{3}\pi + H_0 \sin\left(\omega t + \frac{4}{3}\pi\right) \cos\frac{4}{3}\pi$$

$$H_y = H_0 \sin\left(\omega t + \frac{2}{3}\pi\right) \sin\frac{2}{3}\pi + H_0 \sin\left(\omega t + \frac{4}{3}\pi\right) \sin\frac{4}{3}\pi$$

After simple transformations, we have

$$H_x = \frac{3}{2} H_0 \sin \omega t \text{ and } H_y = \frac{3}{2} H_0 \cos \omega t$$

Such projections are possible only if the vector showing the magnetic field rotates clockwise with a constant angular velocity ω.

620. In these conditions the currents in coils *1-2* and *3-4* are shifted in phase by almost $\pi / 2$. Correspondingly, the magnetic fields created by them are shifted by the same magnitude. Thus, the space between the coils contains the fields $H_1 = H_0 \sin \omega t$

directed vertically and $H_2 = H_0 \sin\left(\omega t + \frac{\pi}{2}\right) = H_0 \cos \omega t$

directed horizontally.

This means (see Problem 619) that a revolving magnetic field is produced in the space. As the field rotates, it carries along the cylinder.

This principle underlies the design of single-phase induction motors.

4. OSCILLATIONS AND WAVES

4-1. Mechanical Oscillations

621. The vertical component of the tension force T is equal to $F = t\cos\alpha$ (Fig. 477). For the conical pendulum $F = mg$, since the weight has no acceleration in the vertical plane.

When the mathematical pendulum is deflected the maximum from the position of equilibrium (through the angle α), the resulting force is directed at a tangent to the trajectory of the weight.

Therefore, $T = mg\cos\alpha$.

When the weight is deflected through the angle α, the tension of the thread of the conical pendulum will be greater.

622. The period of oscillations of the pendulum on the surface of the Earth is $T_0 = 2\pi\sqrt{\frac{l}{g_1}}$ and at an altitude of h above the Earth

$T_1 = 2\pi\sqrt{\frac{l}{g_1}}$. The number of oscillations a day is

$$N_1 = 24 \times 60 \times 60 \frac{1}{T_1} = \frac{k}{T_1}.$$

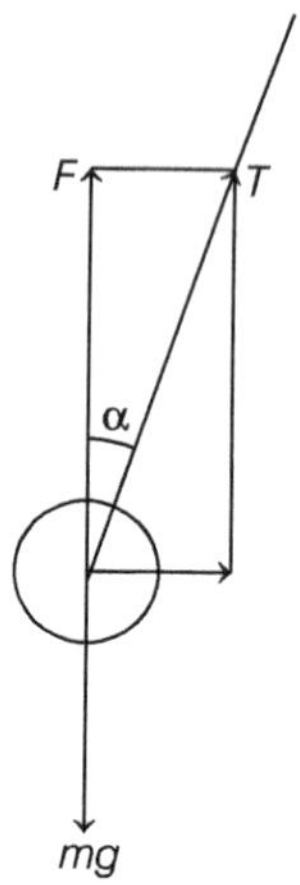

Fig. 477

Therefore, at an altitude of h above the Earth the clock will be slower by the time

$$\Delta t_1 = N_1(T_1 - T_0) = k\left(1 - \frac{T_0}{T_1}\right)$$

The ratio between the periods $\frac{T_0}{T_1} = \sqrt{\frac{g_1}{g}} = \frac{R}{R+h}$ as follows from the law of gravitation. Hence,

$$\Delta t_1 = \frac{kh}{R+h} \cong \frac{kh}{R} \cong 2.7 \text{ seconds}$$

If the clock is lowered into a mine, the acceleration ratio is $\frac{g_2}{g} = \frac{R-h}{h}$, since $g = \gamma\frac{4\pi}{3}R^3\rho\frac{1}{R^2}$ and $g_2 = \gamma\frac{4\pi}{3}(R-h)^3\rho\frac{1}{(R-h)^2}$ (see Problem 234). Hence,

$$\frac{T_0}{T_2} = \sqrt{\frac{g_2}{g}} = \sqrt{\frac{R-h}{R}} \cong 1 - \frac{h}{2R}$$

In this case the clock will be slower by the time

$$\Delta t_2 = k\left(1 - \frac{T_0}{T_2}\right) = \frac{kh}{2R} \cong 1.35 \text{ seconds}$$

623. Each half of the rod with a small sphere on its end is mathematical pendulum with a length $d/2$ that oscillates in the field of gravity of the large sphere. In the field of gravity of the Earth the period of small oscillations of a mathematical pendulum is $T_0 = 2\pi\sqrt{\frac{l}{g}}$.

According to the law of gravitation $mg = \gamma\frac{mM_E}{R^2}$, and therefore

$$T_0 = 2\pi\sqrt{\frac{lR^2}{\gamma M_E}}$$

where $\gamma = 6.67 \times 10^{-8} \text{cm}^2/\text{g}\cdot\text{s}^2$ is the gravity constant, M_E is the mass of the Earth, and R is the distance from the pendulum to the centre of the Earth.

Correspondingly, the period of small oscillations of the mathematical pendulum with a length $l = \frac{d}{2}$ in the field of gravity of the large sphere will be

$$T = 2\pi\sqrt{\frac{dL^2}{2\gamma M}} \cong 5.4 \text{ hours}$$

624. The period of oscillations of a mathematical pendulum is

$$T = 2\pi\sqrt{\frac{l}{g'}}$$

where g' is the gravity acceleration in the corresponding coordinate system. In our case

$$g' = \sqrt{g^2 + a^2}$$

where g is the gravity acceleration with respect to the Earth. Thus,

$$T = 2\pi\sqrt{\frac{l}{\sqrt{g^2 + a^2}}}$$

625. $T = 2\pi\sqrt{\frac{l}{g \pm a}}$. Use the plus sign if the acceleration of the lift is directed upward and the minus sign if it is directed downward.

626. The oscillations of the block in the cup are similar to those of a mathematical pendulum, with the only difference that instead of the tension of the spring the block is acted upon by the reaction of the support. Therefore, the sought oscillation period is

$$T = 2\pi\sqrt{\frac{R}{g}}$$

627. When $M \gg m$, the acceleration of the cup is $a = \frac{F}{M} - g$. Therefore (see Problem 626),

$$T = 2\pi\sqrt{\frac{R}{g+a}} = 2\pi\sqrt{\frac{RM}{F}}$$

If $F = 0$, i.e., with free falling of the cup, $T = \infty$ and there are no oscillations. When $F = Mg$, we have $T = 2\pi\sqrt{\frac{R}{g}}$.

628. The oscillations of the block will periodically displace the cup in a horizontal plane. Hence, the oscillation period of the block will diminish, since an additional variable acceleration directed horizontally will appear in the coordinate system related to the cup (see Problem 624).

629. Let us compare the motion of the centre of the hoop with that of the end of a mathematical pendulum with a length $R - r$. Both points describe the arc of a circle with a radius $R - r$. Let us assume that the hoop and the pendulum are at rest at the angle φ h. On the basis of the law of conservation of energy, we have the following expressions for the velocity v_h of the hoop centre and the velocity v_p of the pendulum end depending on the angle φ

$$v_h = \sqrt{g(R-r)(\cos\varphi_p - \cos\varphi_h)}$$
$$v_p = \sqrt{2g(R-r)(\cos\varphi_p - \cos\varphi_h)}$$

(See Problem 207 for the kinetic energy of a hoop rolling without slipping.)

It follows from these expressions that

$$v_h = \frac{v_p}{\sqrt{2}}$$

Since the centre of the hoop moves $\sqrt{2}$ times slower than the pendulum, the period of motion of the hoop centre will be $\sqrt{2}$ times greater than that of the mathematical pendulum with a length $R - r$. Thus, we have for the sought period:

$$T = 2\pi\sqrt{2\frac{R-r}{g}}$$

Let us note that when $r \to 0$, we have $T = 2\pi\sqrt{\frac{2R}{g}}$, although it may seem at first sight that if $r = 0$ there should exist the equality

$$T = 2\pi\sqrt{\frac{R}{g}}$$

This can be attributed to the fact that the energy of rotational motion of the hoop does not disappear when $r \to 0$.

630. Let the rod be initially deflected from the position of equilibrium through an angle α. At the moment when the rod forms an angle β with the vertical, the angular velocity ω_1 of the rod will, on the basis of the law of conservation of energy, be equal to

$$\omega_1 = \sqrt{\frac{2g(m_1l_1 + m_2l_2)}{m_1l_1^2 + m_2l_2^2}(\cos\beta - \cos\alpha)}$$

Let us now consider a mathematical pendulum with a length l.

In this case at the same angles α and β $\omega_2 = \sqrt{\frac{2g}{l}(\cos\beta - \cos\alpha)}$

Let us so select l that $\omega_1 = \omega_2$. For this it is necessary that

$$l = \frac{m_1l_1^2 + m_2l_2^2}{m_1l_1 + m_2l_2}$$

The angular velocity characterizes the change in the angle β in the course of time. Since $\omega_1 = \omega_2$, the oscillation periods of the two pendulums are the same. For a mathematical pendulum

$$T = 2\pi\sqrt{\frac{l}{g}}$$

Therefore, the sought period is $T = 2\pi\sqrt{\frac{m_1l_1^2 + m_2l_2^2}{m_1l_1 + m_2l_2}\frac{1}{g}}$

631. This problem can be solved by the same method as Problem 630. Let the half-ring be initially deflected from the position of equilibrium through angle α. In motion, all the points on the half-ring have the same linear velocity. The kinetic energy is $\frac{mr^2\omega^2}{2}$.

When the half-ring is turned through the angle $\alpha - \varphi$, the change in the potential energy is $mg\frac{2}{\pi}r(\cos\varphi - \cos\alpha)$ since the centre of gravity is at a distance of $\frac{2}{\pi}r$ from point O (see Problem 115).

Equating the changes in the kinetic and potential energies, we obtain for ω

$$\omega = \sqrt{\frac{2g}{\left(\frac{\pi r}{2}\right)}(\cos\varphi - \cos\alpha)}$$

It follows from the above that a mathematical pendulum with a length $\frac{\pi r}{2}$ will have the same period of oscillations as the half-ring.

Thus, the sought period is equal to $T = 2\pi\sqrt{\frac{\pi r}{2g}}$

632. In the position of equilibrium the spring will be stretched by the amount l that can be determined from the expression $kl = mg$.

Let us assume that the weight is at rest at the initial moment of time and the length of the spring changes by x_0 as compared with the position of equilibrium. If the system is now left alone, the weight will oscillate near the position of equilibrium with an amplitude equal to $|x_0|$. With a weightless pulley ($M = 0$) the period of oscillations is

$$T_0 = 2\pi\sqrt{\frac{m}{k}}$$

Let us denote the displacement of the weight measured from the position of equilibrium by x. The velocity of the weight as a function of x can be found from the law of conservation of energy

$$\frac{k(x_0 + l)^2}{2} - mgx_0 = \frac{mv^2}{2} + \frac{k(x+l)^2}{2} - mgx$$

Bearing in mind that $mg = kl$, we find: $v = \sqrt{\frac{k(x_0^2 - x^2)}{m}}$

If $M \neq 0$, the law of conservation of energy can be written as

$$\frac{k(x_0 + l)^2}{2} - mgx_0 = \frac{(m + M)v^2}{2} + \frac{k(x + l)^2}{2} - mgx$$

It follows that $v = \sqrt{\frac{k(x_0^2 - x^2)}{M + m}}$

Thus, in the second case ($M \neq 0$) the weight moves as if its mass had increased by M as compared with the first case.

Hence, the sought period is $T = 2\pi\sqrt{\frac{m + M}{k}}$

633. When the bottle is displaced from the position of equilibrium by x, the force acting on the bottle will be equal to $F = -\gamma_0 A_x$, where γ_0 is the specific weight of the water. The minus sign means that the force is directed against the displacement x. According to Newton's second law, the oscillations of the bottle are determined by the equation $ma = -\gamma_0 Ax$. This equation is absolutely similar to the equation for the oscillation of a weight on a spring: $ma = -kx$. Since for the weight we have $\omega = \frac{2\pi}{T} = \sqrt{\frac{k}{m}}$, then the oscillation frequency of the bottle will be

$$v = \frac{\omega}{2\pi} = \frac{1}{2\pi}\sqrt{\frac{\gamma_0 A}{m}} \cong 2.5\,\frac{1}{\text{second}}$$

634. The equation of motion of the mercury has the form $ma = -\gamma A2x$ where x is the displacement of the mercury level from the position of equilibrium. The equation of motion has the same form as in the case of oscillations of a weight on a spring. Therefore (see Problem 632),

$$T = 2\pi\sqrt{\frac{m}{2\gamma A}} \cong 1.54 \text{ seconds}$$

635. The force acting on the body is $F = \gamma\frac{4}{3}\pi\rho mr$ where r is the distance from the centre of the Earth and γ is the gravity constant (see Problem 234).

Remembering that $g = \gamma\frac{4\pi}{3}\rho R$, this expression can be written as

$$F = mg\frac{r}{R}$$

Here R is the radius of the Earth. The equation of motion of the body has the form :

$$ma = -\frac{mg}{R}r = -kr$$

The force is proportional to the displacement from the position of equilibrium and directed toward the centre of the Earth. Therefore, the body will perform harmonic oscillations with a frequency

$$\omega = \sqrt{\frac{k}{m}} = \sqrt{\frac{g}{R}}$$

Hence, the period of oscillations is $T = 2\pi\sqrt{\frac{R}{g}}$

The body will reach the centre of the Earth during

$$\tau = \frac{T}{4} = \frac{\pi}{2}\sqrt{\frac{R}{g}} \approx 21 \text{ minutes}$$

It is of interest that the time τ does not depend at all on the distance from the centre of the Earth at which the body begins to move, provided this distance is much greater than the size of the body.

636. The force F acting on the weight deflected from the position of equilibrium is $2f\sin\varphi$ (Fig. 478). Since the angle φ is small, it may be assumed that $F = \frac{4fx}{l}$ or $F = kx$, where $k = \frac{4f}{l}$.

By using the formula $T = 2\pi\sqrt{\frac{m}{k}}$

we get the following expression for the sought quantity:

$$T = 2\pi\sqrt{\frac{ml}{4f}}$$

637. The period of oscillations of the weight on a spring is

$$T = 2\pi\sqrt{\frac{m}{k}}$$

where k is the coefficient of elasticity of the spring equal to the ratio between the force that caused the spring to stretch and its elongation:

$$k = \frac{F}{x}.$$

When two identical springs stretched by the force F are connected in series

$$k_1 = \frac{F}{x_1} = \frac{F}{2x} = \frac{k}{2}$$

since the length of each spring increases by x. When such springs are connected in parallel, the force F_1 required to increase the length of each by x should be two times greater than F.

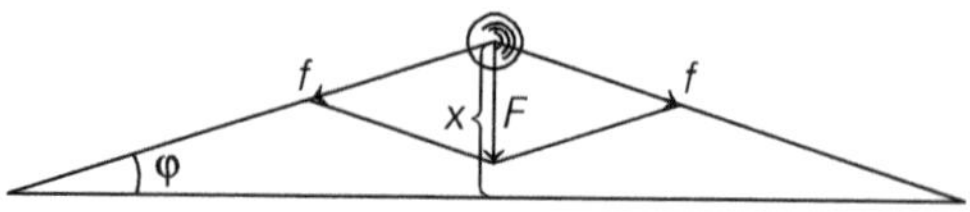

Fig. 478

Hence,

$$K_2 = \frac{F_1}{x} = \frac{2F}{x} = 2k$$

With series connection

$$T_1 = 2\pi\sqrt{\frac{m}{k_1}} = 2\pi\sqrt{\frac{2m}{k}}$$

and with parallel connection

$$T_2 = 2\pi\sqrt{\frac{m}{2k}}$$

Hence, $$\frac{T_1}{T_2} = 2.$$

The period is halved.

638. Let us deflect both pendulums from the vertical in the same direction and through the same angle. In this case the spring will not be deformed. It is easy to see that the pendulums released from this position will oscillate in phase with a frequency of $\omega = \sqrt{\frac{g}{l}}$. If the pendulums are deflected in opposite directions through the same angles they will oscillate in antiphase and the spring will be deformed. To calculate the frequency of these oscillations, let us find the force that returns the pendulums to the position of equilibrium. Upon deflection through the angle φ the force acting on the mass m from the side of the spring is $2kl\sin\varphi$. The sum of the projections of the forces of gravity and elasticity on a tangent to the circumference, the so-called "restoring force" F_r will be equal to

$$F_r = mg\sin\varphi + 2kl\sin\varphi\cos\varphi$$

(Fig 479). Since $\cos\phi \cong 1$ at small angles,

$$F_r = (mg + 2kl)\sin\varphi$$

or
$$F_r = m\left(g + \frac{2kl}{m}\right)\sin\varphi$$

For a mathematical pendulum the restoring force is $mg\sin\varphi$. The frequency of oscillations at small angles φ can be found from the formula $\omega = \sqrt{\frac{g}{l}}$.

In our case the part of g is played by

$$g + \frac{2kl}{m}.$$

Hence,

$$\omega = \sqrt{\frac{g + \frac{2kl}{m}}{l}}$$

The period of oscillations is

$$T = 2\pi\sqrt{\frac{l}{g + \frac{2kl}{m}}}$$

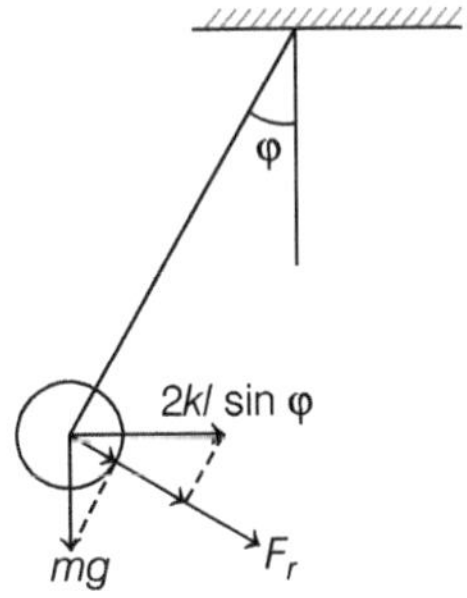

Fig. 479

639. Yes, it can. For this purpose the door should be gradually swung with a frequency equal to the natural frequency of oscillations of the door. Upon resonance, the amplitude of oscillations may be very high.

640. From the law of conservation of energy:

$$\frac{\omega^2}{2}(ml^2 + Mr^2) = Mgr\alpha - mgl(1 - \cos\varphi)$$

where ω is the angular velocity of pulley rotation. Hence,

$$\omega = \sqrt{\frac{2\left(mgr\alpha - 2mgl\sin^2\frac{\alpha}{2}\right)}{ml^2 + Mr^2}}$$

Oscillatory motion will be obtained if the angular velocity is zero at a certain value of the angle α. Here $Mgr\alpha = 2mgl\sin^2\frac{\alpha}{2}$ or, upon introducing $a = \frac{Mr}{ml}$, we obtain $\frac{a\alpha}{2} = \sin^2\frac{\alpha}{2}$. To each value, of a there corresponds a definite maximum deflection from the position of equilibrium α that is determined by the given transcendental equation.

This equation is solved the easiest by the graphical method. For this purpose, plot the curve $y=\sin^2\dfrac{\alpha}{2}$ (Fig. 480). Hence intersection of this curve with the straight line $y=\dfrac{a\alpha}{2}$ will give us point A that determines the value of α at the given a. (The value of α that corresponds to the intersection of this straight line with another branch of the curve $y=\sin^2\dfrac{\alpha}{2}$ cannot be obtained with the initial conditions given in the problem.)

Obviously, our solution of the equation will be other than zero only if a is smaller than a certain limiting value of a_0 determined from the condition that the straight line $y=\dfrac{a_0\alpha}{2}$ is tangent to the curve $y=\sin^2\dfrac{\alpha}{2}$ at point C.

Figure 480 shows that $\alpha_0 \cong 133°$.

Hence, $a_0 = \dfrac{2CD}{\alpha_0} \cong 0.73$. The oscillations are possible when

$$\frac{Mr}{ml} \le a_0 \cong 0.73\,.$$

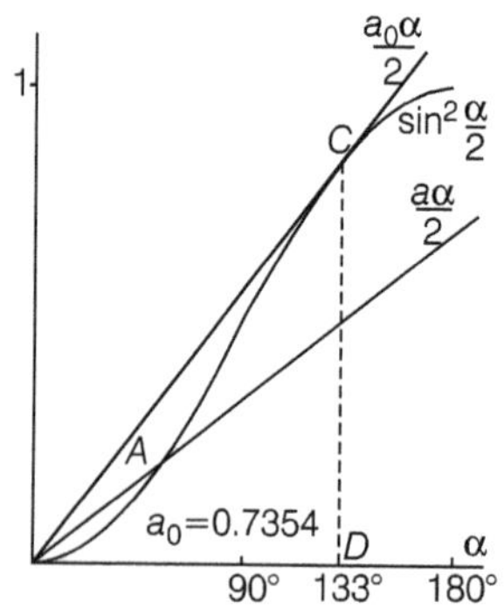

Fig. 480

4-2. Electrical Oscillations

641. Without a permanent magnet the oscillation frequency would be doubled. When a sinusoidal current flowed through the coil of the telephone it would cause the membrane to make two oscillations during one period of current oscillations because the intensity of the magnetic field H created by this current would have the form shown in Fig. 481*a*, and the force of attraction of the membrane does not depend on the sign of H. When there is a permanent magnet that generates an intensity of the magnetic field exceeding the maximum intensity of the current field, the diagram of the resulting intensity has the form shown in Fig. 481*b*.

For this reason one oscillation of the current will correspond to one of the membrane, and the sound will be much less distorted.

642. The frequency of natural oscillations is $\omega = \frac{1}{\sqrt{LC}}$ if L is in henries and C in farads. As has been shown in Problem 591, for a solenoid $L = 10^{-9}\frac{4\pi N^2 A_1}{l}$, henries. The capacitance of the capacitor is $C = \frac{A_2}{4\pi d} \times \frac{1}{9 \times 10^{11}}$ farads.

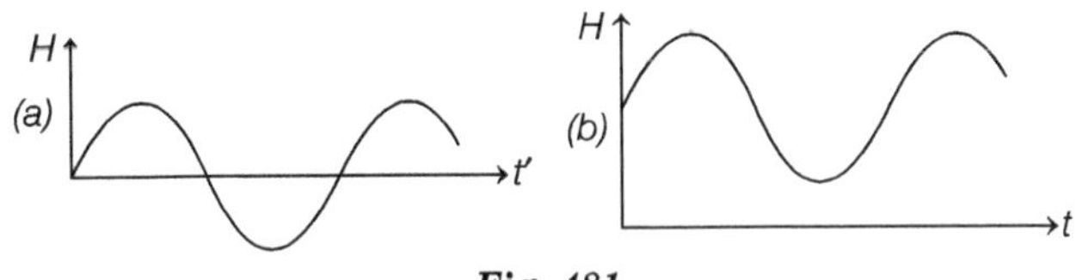

Fig. 481

Hence, $$\omega = 3 \times 10^{10}\sqrt{\frac{ld}{N^2 A_1 A_2}} = 3 \times 10^{6}\frac{1}{\text{second}}$$

643. The frequency of natural oscillations of a circuit is determined from the Thomson formula

$$\omega = \frac{1}{\sqrt{LC}}$$

(a) If the coil contains the copper core, the periodic changes of the magnetic field of the coil will produce in the core eddy currents whose magnetic field will weaken the magnetic field of the coil. This will reduce the inductance of the coil and, consequently, increase the frequency ω.

(b) If the ferrite core is moved into the coil, the magnetic field of the latter will increase. Accordingly, the inductance L of the coil will grow and the frequency ω will decrease.

644. Undamped oscillations will appear in the system (if the small losses of energy for the radiation of electromagnetic waves are neglected.) When the charge is distributed equally between the capacitors, the energy of the electrostatic field is minimum, but the current intensity and the energy of the magnetic field are maximum. The total energy does not change, but one kind of energy is converted into another.

645. The displacement of the electron beam by the voltage supplied along the vertical can be written as

$$x = \frac{lL}{2dV} V_{01} \cos \omega t = a \cos \omega t$$

(see Problem 543). The displacement of the beam along the horizontal (axis y) is

$$y = \frac{lL}{2dV} V_{02} \cos(\omega t - \varphi)$$
$$= b \cos(\omega t - \varphi)$$

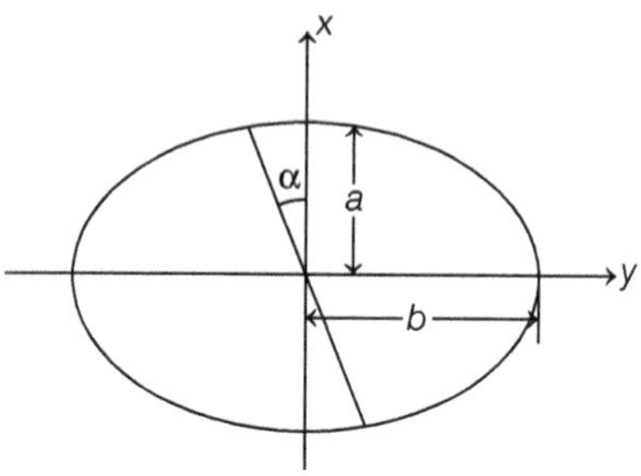

Fig. 482

To find the trajectory, the time should be excluded from the data of the equations. After simple transformations we have

$$\frac{x^2}{a^2} + \frac{y^2}{b^2} - \frac{2xy}{ab} \cos \varphi = \sin^2 \varphi$$

If $\varphi_1 = \frac{\pi}{2}$, then $\frac{x^2}{a^2} + \frac{y^2}{b^2} = 1$. This is the equation of an ellipse.

If $\varphi_2 = \pi$,

then $x = -\frac{a}{b} y$, i. e., the beam oscillates along the straight line forming with the x-axis an angle α determined by the equation $\tan \alpha = \frac{b}{a}$ (see Fig. 482).

646. The relation between the voltage and time is shown in Fig. 483. The voltage across the capacitor (curve *Oa*) increases until it reaches V_S. At this moment the tube ignites and the capacitor is discharged through the tube (curve *ab*) until the voltage drops to V_e. The process is then repeated and relaxation oscillations with a period τ appear.

The charging and discharging current of the capacitor is not constant, because it depends on the voltage across the capacitor, and decreases with a growth of the voltage. For this reason *Oa*, *ab*, *bc*, etc., are not sections of straight lines.

647. When the capacitance grows, the time needed to charge the capacitor to V_s and discharge it to V_e increases. The period will therefore be longer.

An increase in R will reduce the charging current of the capacitor and will increase the period.

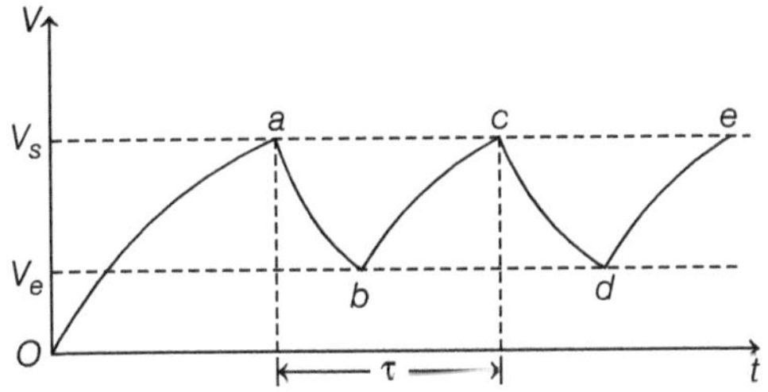

Fig. 483

648. When the charge across the plates of the capacitor reaches its maximum, the plates should be moved apart. Here work must be performed to overcome the forces of attraction between the plates. This work is spent to increase the energy of the circuit. When the charge is zero, the plates should be moved together to their original position. The energy in the circuit will not change.

4-3. Waves

649. The tension of the string should be increased four times.

650. $\upsilon = \upsilon_0 n$, where $n = 1, 2, 3, 4, \ldots,$ and

$$\upsilon_0 = \frac{1}{ld}\sqrt{\frac{T}{\pi\rho}}$$

$$= 2\,\frac{1}{\text{second}}$$

651. The pipe should accommodate a whole number of half-waves:

$$\frac{\lambda}{2}k = l\ (k = 1, 2, 3, \dots)$$

The frequencies of the natural oscillations are

$$\upsilon_k = \frac{c}{\lambda} = \frac{kc}{2l} = k \times 50\frac{1}{\text{second}}$$

($c = 340$ m/s is the velocity of sound in air).

652. The sound of the tuning fork will be intensified when the frequency of the natural oscillations of the air column in the vessel coincides with the frequency of the tuning fork. The natural oscillation frequency of the air column in a tube closed at one end is $\upsilon_k = \frac{2k+1}{4}\cdot\frac{c}{l}$, where l is the length of the tube and $c = 340$ m/s is the velocity of sound. The quantity k takes values of 0, 1, 2, 3, ... Therefore, the possible water levels in the vessel determined by the distance from the surface of the water to the upper edge of the vessel are

$$l_k = \frac{2k+1}{4}\frac{c}{v}\ (k = 0, 1, 2, \dots)$$

When $l = 1$ metre, two positions of the water level are possible: $l_0 = 25$ cm and $l_1 = 75$ cm.

653. Let us consider a number of consecutive positions of the bullet flying along KA, namely, K, F, E, D, B, and A (Fig. 484). At each point the bullet creates before its front a compression that spreads in all directions in the form of a spherical pulse. Since the velocity of the bullet v is greater than that of sound c, these pulses appear only behind the bullet. At the moment when the bullet is at point A, the separate pulses will be as shown in Fig. 484 by circles of various radii. The wave front has the form of a cone that moves forward with the velocity of the bullet. The apex angle of the cone can be determined from the ratio

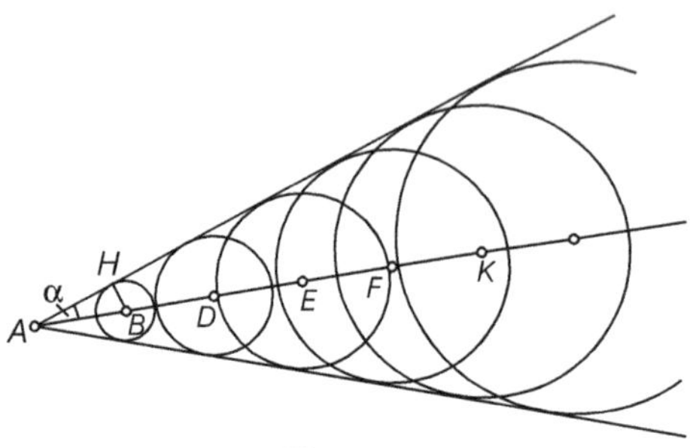

Fig. 484

$$\sin\alpha = \frac{BH}{AB} = \frac{ct}{vt} = \frac{c}{v}$$

654. The sound wave that reaches the man at point B (Fig. 485) is emitted when the plane is at a certain point D (see the solution to Problem 653). The distance $CB = 6$ km.

The sought distance is

$$AB = \frac{BC}{\sin\alpha} = BC\frac{v}{c}$$

where v is the velocity of the plane and c the velocity of sound. Hence, $AB = 9$ km.

655. Ordinarily, the velocity of the wind at a certain altitude above the ground is greater than at its surface. For this reason the wave surfaces which in immobile air have the form of spheres with their centre at the point of the sound source (dotted lines in Fig. 486) change their shape. The velocity of the waves is higher in the direction of the wind than against it, The approximate shapes of the wave surfaces are shown in Fig. 486 by solid lines.

The sound propagates in a direction perpendicular at each point to the wave surfaces. For this reason the sound propagating against the wind is deflected upwards (curve AB) and does not reach the man. If the sound propagates in the direction of the wind the sound is deflected towards the ground (curve AC) and the man hears it.

656. TV stations operate on wavelengths smaller than 10 metres. For such waves the ionosphere is "transparent" and does not reflect the waves. Short waves are practically propagated along a straight line, since they undergo almost no diffraction from such obstacles as houses, etc.

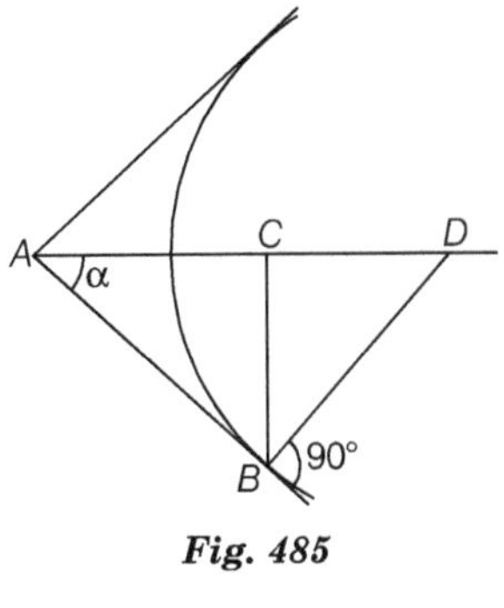

Fig. 485

657. To estimate the distance to an object by the position of the reflected pulse on the screen of a cathode-ray tube, this pulse should arrive not earlier than in the time τ and not later than in the time $T = \frac{1}{f}$ after sending the direct pulse. Hence, the minimum distance to the object is

$$l = \frac{c\tau}{2} = 120 \text{ m}$$

and the maximum distance is

$$L = \frac{cT}{2} \cong 90 \text{ km}$$

658. The wave reflected from the roof will reach the aerial with a lag of $\tau = \frac{AB}{c} = 10^{-5}$ s. The velocity of the cathode beam along the screen is $v = \frac{l}{\Delta t}$, where $\Delta t = \frac{1}{25 \times 625}$ s is the time during which the beam produces one line (neglect the time of reverse travel of the beam). The shift of the images is $\Delta l = v\tau \cong 7.8$ cm.

659. When the vibrator is immersed in kerosene its capacitance C increases E_r times. The frequency of the natural oscillations of the circuit is proportional to $\frac{1}{\sqrt{C}}$. Therefore, the oscillation frequency will decrease $\sqrt{E_r}$ times. The frequency of the natural oscillations of the vibrator in a vacuum is $v_0 = \frac{c}{2l}$, and in a dielectric $v = \frac{c}{2l\sqrt{E_r}}$.

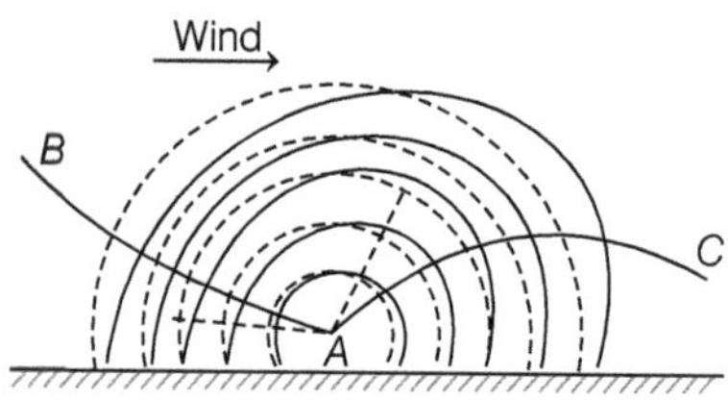

Fig. 486

The wavelength that corresponds to this frequency in a vacuum is

$$\lambda = \frac{c}{v} = 2l\sqrt{E_r} \cong 1.4 \text{ m}$$

In short, this result can be obtained as follows. The wavelength in kerosene is $\lambda = 2l$ and increases $\sqrt{E_r}$ times in a vacuum. Therefore, $\lambda_0 = 2l\sqrt{E_r}$.

660. The horizontal position of the aerial means that the electric vector of the wave oscillates mainly in horizontal planes. Therefore, the magnetic vector oscillates along a vertical.

5. GEOMETRICAL OPTICS

5-1. Photometry

661. The minimum illumination of a wall (Fig. 487) is

$$E_1 = \frac{I\cos\alpha}{r^2}$$

The minimum illumination of the floor is

$$E_2 = \frac{I\cos\beta}{r^2}$$

According to the condition,

$$\frac{E_1}{E_2} = \frac{\cos\alpha}{\cos\beta} = \frac{D}{2h} = 2$$

Hence,

$$h = \frac{D}{4} = 7.5 \text{ m}$$

662. The illumination of the middle of the table is

$$E = \frac{I_1}{H_1^2} = \frac{I_2}{H_2^2}$$

where H_2 is the height of the second lamp above the table.

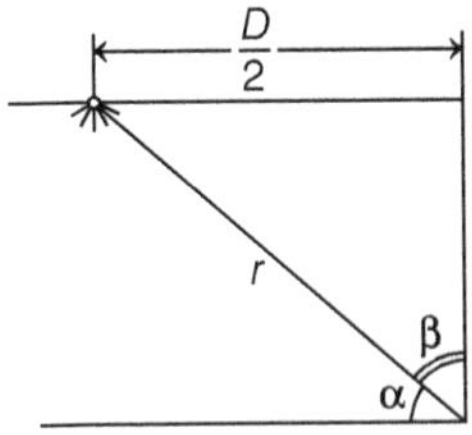

Fig. 487

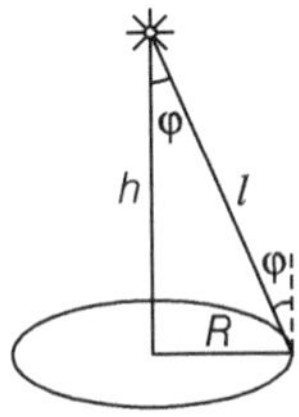

Fig. 488

In both cases, the illumination of the edge of the table is

$$E_1 = -\frac{I_1 H_1}{\left(H_1^2 + \frac{D^2}{4}\right)^{3/2}}, \text{ and } E_2 = \frac{I_2 H_2}{\left(H_2^2 + \frac{D^2}{4}\right)^{3/2}}$$

Therefore, $$\frac{E_1}{E_2} = \frac{\left(H_1^2 + \frac{I_1}{I_2}\frac{D^2}{4}\right)^{3/2}}{\left(H_1^2 + \frac{D^2}{4}\right)^{3/2}} = 3$$

The illumination of the edge of the table will be decreased three times.

663. If a normal to the plate forms an angle α with the direction AS_1, the illumination of the plate will be

$$E = \frac{I}{a^2}[\cos\alpha + \cos(90° - \alpha)] = \frac{I}{a^2} \times 2\cos 45° \cos(\alpha - 45°)$$

For this reason the illumination of the plate will be maximum if it is parallel to side S_1S_2 of the triangle. The illumination is

$$E_{max} = \sqrt{2}\,\frac{I}{a^2}$$

664. When the auxiliary and standard sources are used together the illuminations will be equal if

$$\frac{I_0}{I_1} = \frac{r_1^2}{r_2^2}$$

where I_0 is the luminous intensity of the standard source and I_1 that of the auxiliary source.

In the second case the illuminations are equal when

$$\frac{I_x}{I_1} = \frac{r_3^2}{r_4^2}$$

where I_x is the sought luminous intensity.

Hence $$I_x = \frac{r_2^2 \cdot r_3^2}{r_1^2 \cdot r_4^2} = 400 I_0.$$

665. The illumination will be 25 times smaller.

666. The total luminous flux from the lamp is $\phi_0 = 4\pi I$. If the lamp is fastened to the ceiling, the walls and the floor receive half of this flux. Therefore, the sought flux is $\phi = 2\pi I = 628$ lm.

667. The Earth receives $\frac{1}{2.25 \times 10^9}$ of the total energy of the Sun.

668. The quantity of light energy absorbed by the internal walls of the cylinder in a unit of time (luminous flux) is the same in both cases. The area of the internal surface of the cylinder will change, however, R_1/R_2 times. For this reason the illumination will increase R_1/R_2 times, i. e., $\frac{E_1}{E_2} = \frac{R_2}{R_1}$.

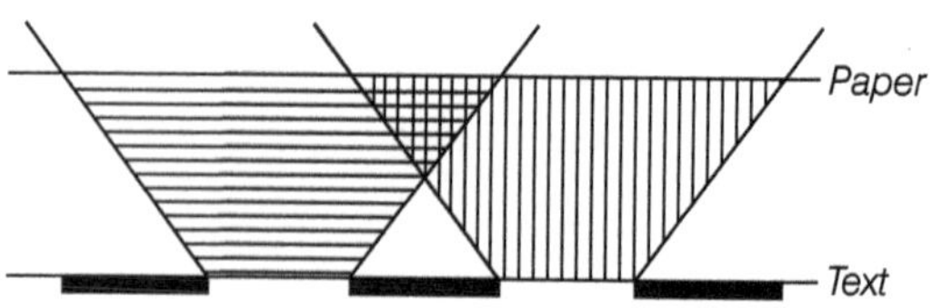

Fig. 489

669. The illumination on the edge of the table is

$$E = \frac{I\cos\varphi}{l^2}$$

$$= \frac{I}{R^2}\cos\varphi\sin^2\varphi$$

where I is the luminous intensity of the lamp, R the radius of the table, and φ the angle of incidence of the rays (Fig. 488).

The maximum value of E is attained if the angle φ satisfies the equation

$$1 - \sin^2\varphi = \frac{1}{2}\sin^2\varphi$$

i. e., when $\varphi = \arcsin\sqrt{\frac{2}{3}}$ (see Problem 407).

The lamp should be hung above the table at a height of

$$h = \frac{\sqrt{2}}{2}R \cong 0.71\ R.$$

670. Tissue paper diffuses the incident light rays in all directions.

If the paper is at some distance from the page, the diverging beams of light reflected from the white portion of the page (between the letters) overlap on the side of the paper facing the text (Fig. 489).

As a result the paper will be illuminated more or less uniformly, and the diffusion of the light will make it impossible to read the text.

If the paper is placed on the text, the illumination of the paper side adjoining the text will not be uniform. Accordingly the intensity of the diffused light will be different at various portions of the paper, and the text can be read.

5-2. Fundamental Laws of Optics

671. The shadow will be equally distinct everywhere only if a point source of light is used. The separate sections of an extended source throw shadows that are superimposed on one another. The boundary of the shadow will be the sharper, the smaller is the distance from the object to the surface on which the shadow is formed, since the distances between the boundaries of the shadows produced by the various sections of the source will be minimum. It is for this reason that a man's legs give a sharper shadow than his head.

672. The pencil should be held parallel to the lamp and as close to the table as possible. As a result the shadows sent by separate portions of the lamp will be almost accurately superimposed.

If the pencil is perpendicular to the lamp, the shadows from it will be so mutually shifted that practically no shadow will appear.

673. This phenomenon can be observed only if the angular distance between the branches is less than the angular diameter of the Sun. Let us assume, to introduce definite conditions, that the lower branch is thicker than the upper one.

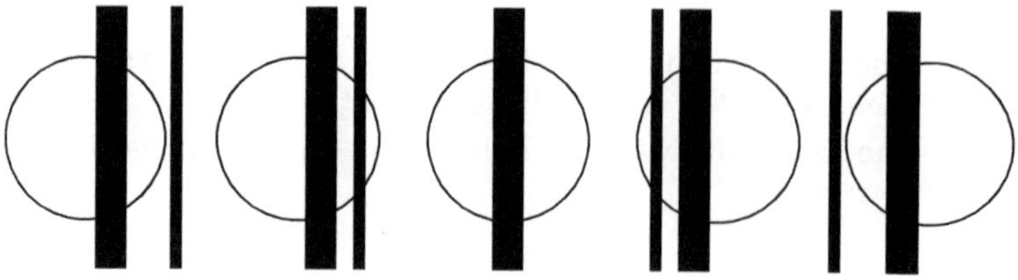

(a) (b) (c) (d) (e)

Fig. 490

To understand why the illumination inside the shadow changes as stated in the problem, assume that we look at the Sun alternately from different sections of the shadow.

The Sun's disk can be seen entirely outside of the shadow. In section A of the shadow (Fig. 235) the eye is in the half shadow cast by the lower branch, and only this branch is visible in front of the Sun's disk (Fig. 490*a*). Since the branch covers a part of the Sun's disk, the illumination of this point will be weaker. Moving the eye farther to position B (Fig. 235), we shall see that the other branch also partly covers the Sun's disk (Fig. 490*b*), and for this reason the illumination will be still less. Moving farther, the eye will occupy position C (Fig. 235) in which both branches will be superimposed (Fig. 490*c*). Now, the part of the Sun's disk covered by the branches is smaller and the illumination greater. The disk as viewed from D and E is shown in Fig. 490*d* and *e*. This explains why the central stripe of the shadow is brighter than the adjacent parts.

674. As can be seen from Fig. 491, we have $H = L\sin\alpha$, while $\sin\alpha = \frac{b}{a}$, since $DE = b$ is the cross-sectional diameter of the light cone on the ground. With the angular dimensions of the Sun's disk β, we obtain $L = \frac{b}{\beta}$.

Therefore, $$H \frac{1}{\beta}\frac{b^2}{a} = 9 \text{ metres.}$$

675. If the rays are turned in the periscope as shown in Fig. 492, the sought ratio of the widths of the prisms a/b can be found from the similarity of the triangles :

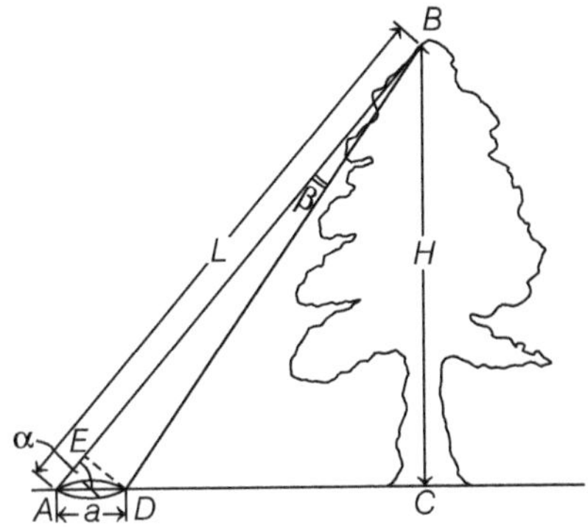

Fig. 491

$$\frac{a}{b} = \frac{L + l}{l}$$

676. The height of the mirror should be equal to half the height of the man. The distance from the lower edge of the mirror to the floor should be equal to half the distance from the man's eyes to his feet (Fig. 493).

677. Let h be the height of the object and α the angle of incidence of the rays on the mirror (Fig. 494).

If the screen is at a distance of $l \geq h \tan \alpha$ from the object, a direct and an inverted shadows with their bases fitted against each other will be seen on the screen. The total length of the shadow is $2h$. The shadow is illuminated by the Sun and contrasts with the other portions of the screen illuminated by both direct and reflected rays.

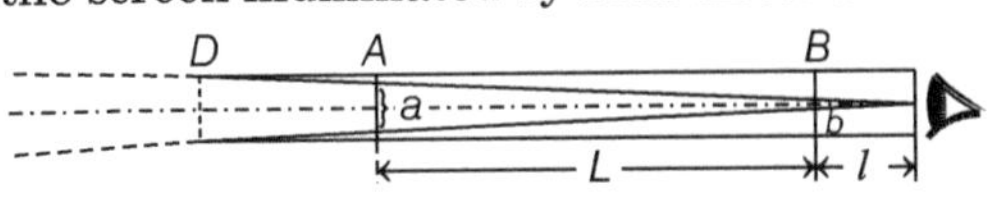

Fig. 492

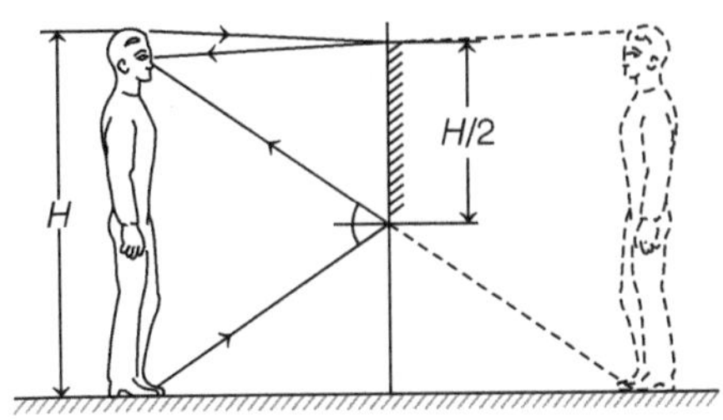

Fig. 493

If the screen is nearer, the length of the shadow is smaller than $2h$ and will have portions that are illuminated neither by direct nor reflected rays.

678. A point source of light always produces a reflection that depends on the shape of the mirror. The dimensions of the Sun are finite. Each small section of the luminescent surface produces a bright spot that gives the shape of the mirror. These spots from various portions of the Sun are superimposed and produce a more or less diffused pattern.

If the surface on which the reflection is observed is far from the mirror, the shape of the bright spot will not depend on the shape of the mirror. It is only at a small distance from the mirror that the spot will reproduce the shape of the mirror, since the angles at which the rays from the various portions of the Sun fall onto the mirror differ very slightly from one another.

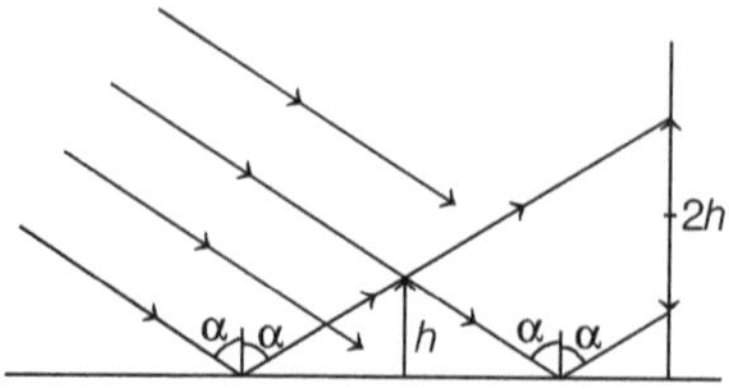

Fig. 494

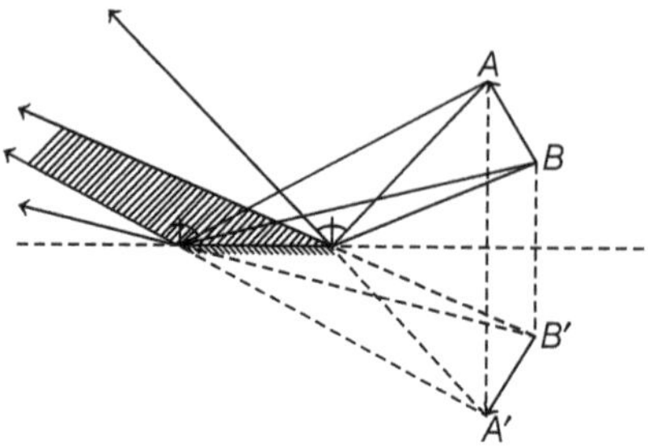

Fig. 495

679. The reflected landscape is seen as if it were viewed from a point below the water surface at a distance equal to that from the camera lens to the water.

680. The whole of the image of straight line AB can be seen only if the eye is inside the hatched area in Fig. 495.

681. When mirror MN moves toward the wall, the position of light spot AB on the wall will be invariable as can be seen in Fig. 496 (S_1 and S_2 are the images of source S with the mirror in two positions: MN and $M'N'$).

The dimensions of the light spot will not change either, constantly remaining equal to the double dimensions of the mirror.

682. If the losses in reflection are neglected, the illumination of the light spot will always be one-fourth of the illumination of the mirror. At the same time the illumination of the mirror changes in view of the change in the distance from the lamp to the mirror and the change in the angle of incidence of the rays. With a small mirror the maximum illumination will be observed when the distance from the mirror to the wall is $l = \frac{\sqrt{2}}{2} d$, where d is the distance from the source of light to the point on the wall which the mirror is brought up to.

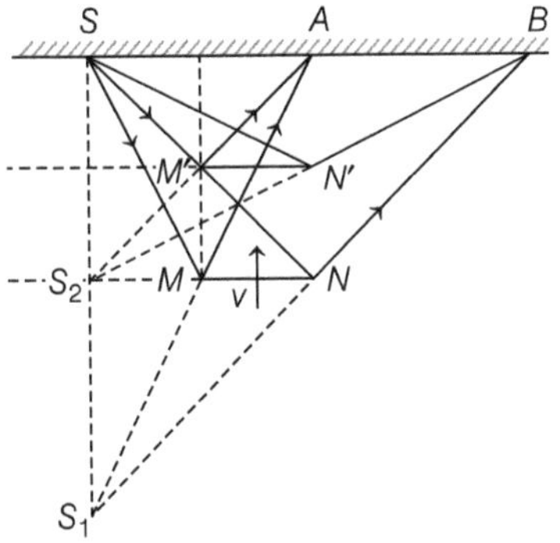

Fig.496

683. When the mirror turns through an angle α the reflected ray will turn through 2α since the angles of incidence and reflection increase by α. Hence, the angular velocity of rotation of the reflected ray is $\omega = 2\pi n \times 2$. The linear velocity with which the light spot moves along the screen is $v = 4\pi nR \cong 62.8$ m/s.

684. (a) The beam reflected from the first mirror forms an angle 2α with the incident beam (α is the angle of incidence). During the time t the mirror will turn through an angle ωt and the new angle of incidence will become equal to $\alpha + \omega t$, as will the angle of reflection. Therefore, the angle between the incident and reflected beams will increase by $2\omega t$, i.e., the reflected beam will turn through an angle $2\omega t$.

In view of this, the angle of incidence on the second mirror, provided it does not rotate, would be $\beta + 2\omega t$, where β is the angle of incidence with immobile disks. But the mirror also revolves through the angle ωt during the time t, and therefore the angle of incidence becomes $\beta + 3\omega t$. The angle of reflection will be the same. Thus, after two reflections the beam will turn through the angle $3\omega t$ from its direction with immobile mirrors. After three reflections the beam will turn through $5\,\omega t$ and after n reflections through $(2n - 1)2\omega t$. In this way its angular velocity will be $\Omega = (2n - 1)2\omega$.

(b) When the mirror moves from the source with a velocity v, the image will move away from the source with a velocity $2v$ and from the second mirror with a velocity $3v$. Therefore, the second image moves with a velocity $3v$ with respect to the second mirror and with a velocity $4v$ with respect to the source. The velocity of the third image with respect to the source will be $6v$ and the velocity of the n-th image $2nv$.

685. (a) When the first mirror turns through an angle ωt the reflected beam will turn through an angle $2\omega t$ (see the solution to Problem 684). Hence, the angle of incidence on the second mirror will also increase by $2\omega t$, and, if the mirror did not revolve, the angle of reflection would also increase by $2\ \omega t$. After two reflections the beam would turn through $2\omega t$ as compared with the case of immobile mirrors.

Since the second mirror does rotate, however, the angle of the beam incident on it decreases by ωt during the time t. The angle of reflection decreases by the same amount and for this reason the reflected beam will travel in the same direction as with immobile disks. Since this line of reasoning may be adopted for any two consecutive reflections, the angular velocity of rotation of the beam subjected to n reflections will be $\Omega = 0$ if n is even, and $\Omega = 2\omega$ if n is odd.

(b) The first image moves away from the source with a velocity $2v$ and from the second mirror with a velocity v. Therefore, the second image moves with respect to the second mirror with a velocity $- v$, i.e., it is immobile with respect to the source.

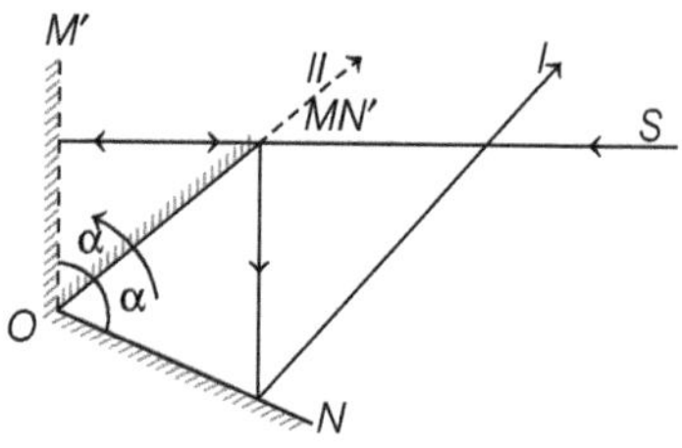

Fig. 497

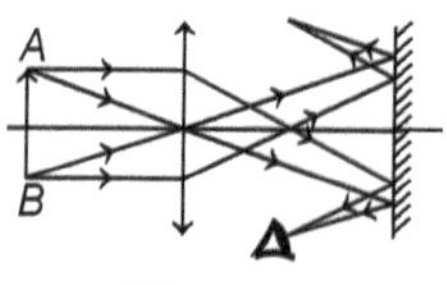

Fig. 498

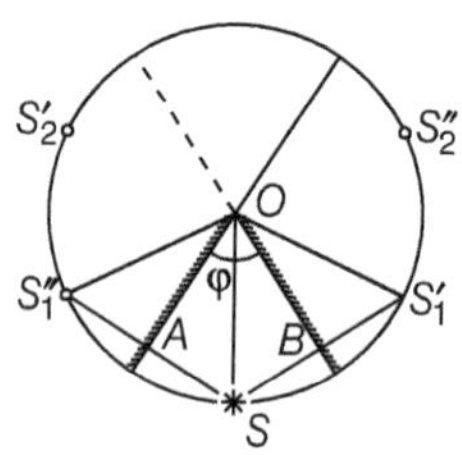

Fig. 499

Reasoning similarly, we find that the sought linear velocity of the n-th image is zero if n is even, and $2v$ if n is odd.

686. The beam reflected from mirror ON forms with the incident beam an angle φ (see Fig. 240) that does not depend on the angle of incidence i. Indeed, as can be seen from triangle ABC, we have $\varphi = 180° - 2(i + r)$. On the other hand, in triangle OAB, we have $\alpha + (90° - i) + (90° - r) = 180°$.

Hence, $\varphi = 180° - 2\alpha = 60°$. When the mirror rotates, the direction of the reflected beam does not change.

Thus, if the beam that fell on mirror OM is reflected from mirror ON, it will always get into the receiver. As can easily be seen in Fig. 497 showing two extreme positions of the mirrors at which the beam gets into the receiver (OM, ON and OM', ON'), this occurs during one-sixth of a revolution. For this reason one-sixth of all the energy of the beam will get into the receiver during one revolution, which is a sufficiently large interval of time.

687. No, it cannot, since rays will reach the eye that produce the image of only small portions of the frame (see Fig. 498 showing the path of the rays from the extreme portions of the frame A and B).

688. $4\ \text{m}^2$.

689. It follows from the similarity of triangles SOA, SOB, $S_1''OA$ and $S_1'\ OB$ (Fig. 499) that the source of light S, its image S_1' in mirror OB and the image S_1'' in mirror OA lie on a circle with its centre at point O. We have $\angle S_1''OS = \angle SOS_1' = \varphi$. The virtual source S_1' is reflected from mirror AO and produces image S_2' lying on the same circle at a distance of 2φ arc degrees from source S. Image S_2'' of virtual source S_1'' is formed in mirror OB in the same way.

Continuing our construction, we obtain the third images S_3' and S_3'' removed from the source by 3φ degrees and the fourth S_4' and S_4'' (removed by 4φ degrees), etc.

If n is even ($n = 2k$), then image S_k' coincides with S_k'' and will be on one diameter with the source. Altogether there will be $2k - 1 = n - 1$ images.

If n is odd ($n = 2i + 1$) it can easily be seen that the i-th images lie on the continuations of the mirrors and thus coincide with the $(i + 1)$ th and all the subsequent images. For this reason there will be $2i$ images, i.e., $n - 1$ as before.

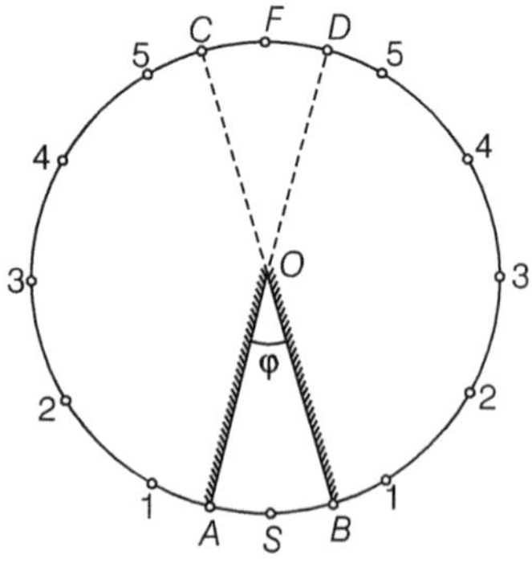

Fig. 500

690. Using the solution of Problem 689, let us plot consecutively the first, second, third, etc., images of source S in the mirrors (Fig. 500). All of them will lie on a circle with a radius OS and the centre at point O.

If a is an integer, the last i-th images will either get onto points C and D at which the circle intersects the continuations of the mirrors, or will coincide with point F diametrally opposite to the source. In both cases the number of images will be $a - 1$.

If a is not an integer, for example $a = 2i \pm \xi$, where $\xi < 1$, and i is an integer, the last i-th images will lie on arc CFD that is behind both the first and the second mirrors and there will be no more reflections. Thus, the total number of images will be $2i$.

691. Let us plot the image of point B in mirror bd (Fig. 501). Let us then construct image B_1 in mirror cd. Also, B_3 is the image of B_2 in mirror ac and B_4 is the image of B_3 in mirror ab.

Let us connect points A and B_4. Point C is the point of intersection of ab with line AB_4. Let us now draw line B_3C from B_3, and connect point D at which this line intersects ac with B_2. E with B_1, and F with B.

It can be stated that broken line $ACDEFB$ is the sought path of the beam. Indeed, since B_3CB_4 is an isosceles triangle, CD is the reflection of beam AC.

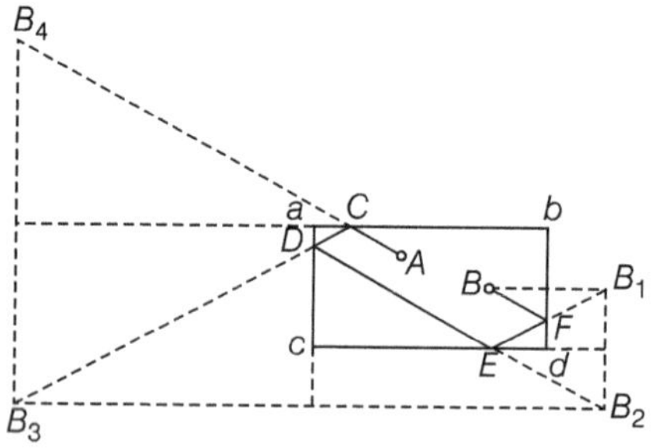

Fig. 501

Similarly, it is easy to show that DE is the reflection of CD, etc.

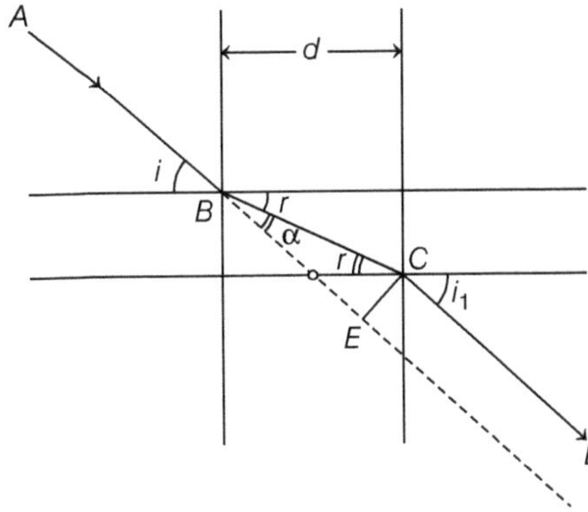

Fig.502

This solution of the problem is not unique, since the beam should not necessarily be sent initially to mirror *ab*.

692. The coefficient of reflection of light from the surface of water diminishes with a reduction in the angle of incidence.

If the observer looks down, rays reflected at small angles reach his eyes. The rays reflected from the sea water at the horizon reach the eyes at greater angles.

693. According to the law of refraction $\frac{\sin i}{\sin r} = n$ (Fig. 502).

Upon exit from the plate $\frac{\sin r}{\sin i_1} = \frac{1}{n}$. Upon multiplying these expressions, we get $i = \sin i_1$, i.e., beam CD leaving the plate is parallel to incident beam AB. A glance at the drawing shows that $\alpha = i - r$. The sought displacement of the beam is $x = EC = BC \sin(i - r)$.

Since $BC = \frac{d}{\cos r}$, then

$$x = \frac{d \sin(i - r)}{\cos r} = d \sin i \left(1 - \frac{\cos i}{\sqrt{n^2 - \sin^2 i}}\right)$$

And a maximum displacement of d can be obtained when $i \to 90°$.

694. The angle of incidence of the ray onto AC and BC is 45°. For total internal reflection it is necessary that $\sin i > \frac{1}{n}$.

Hence, $n > \sqrt{2} \cong 1.4$.

695. The angle of incidence of the ray on face BC is equal to the sought angle α. For the ray to be completely reflected from face BC, the angle α should exceed the limiting one.

Therefore, $\sin\alpha > \frac{n_2}{n_1}$, where n_2 is the refraction index of water.

Hence, $\alpha > 62°30'$.

696. This phenomenon is nothing but a mirage frequently observed in deserts.

The hot layer of air in direct contact with the asphalt has a smaller refraction index than the layers above. Total internal reflection occurs and the asphalt seems to reflect the light just as well as the surface of water.

697. Let us divide the plate into many plates so thin that their refraction index can be assumed constant within the limits of each plate (Fig. 503).

Assume that the beam enters the plate from a medium with a refraction index of n_0 and leaves it for a medium with a refraction index of n_3.

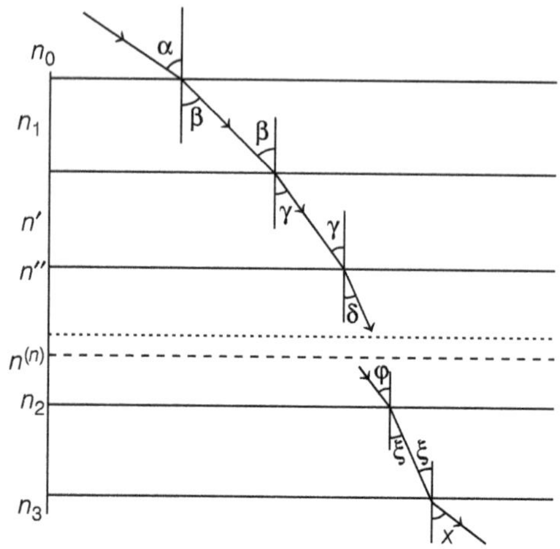

Fig.503

Then, according to the law of refraction,

$$\frac{\sin\alpha}{\sin\beta} = \frac{n_1}{n_0}$$

$$\frac{\sin\beta}{\sin\gamma} = \frac{n'}{n_1}$$

$$\frac{\sin\gamma}{\sin\delta}=\frac{n''}{n'}$$

....

$$\frac{\sin\varphi}{\sin\xi}=\frac{n_2}{n^{(n)}}$$

$$\frac{\sin\xi}{\sin x}=\frac{n_3}{n_2}$$

Upon multiplying these equations we get

$$\frac{\sin\alpha}{\sin x}=\frac{n_3}{n_0}$$

Hence, the angle at which the beam leaves the plate

$$x=\text{arc}\,\sin\left(\frac{n_0}{n_3}\sin\alpha\right)$$

depends only on the angle of incidence of the beam on the plate and on the refraction indices of the media on both sides of the plate. In particular, if $n_3=n_0$, then $x=\alpha$.

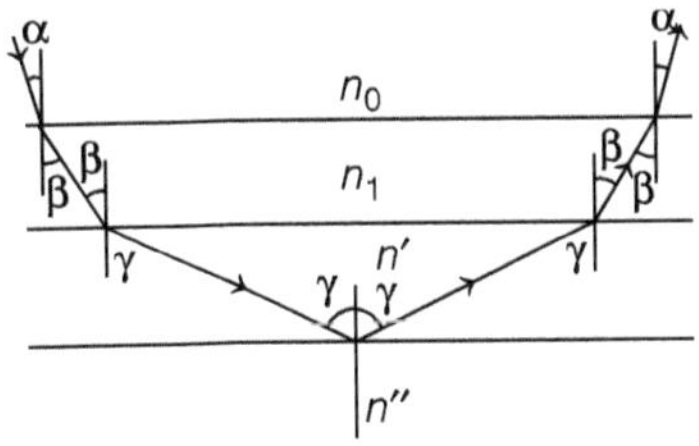

Fig. 504

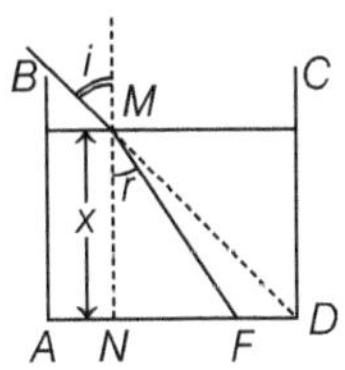

Fig. 505

Generally speaking, the angle θ at which the beam is inclined to the vertical is related to the refraction index n at any point on the plate by the ratio $n\sin\theta=\text{const}=n_0\sin\alpha$. If the refraction index reaches a value of $n=n_0\sin\alpha$ anywhere inside the plate, full internal reflection will take place. In this case the beam will leave the plate for the medium at the same angle α at which it entered the plate (Fig. 504).

698. The minimum amount of water determined by the level x (Fig. 505) can be found from the triangle MNF. We have $NF=x-b=x\tan r$. From the law of refraction

$$\sin r=\frac{\sin i}{n}$$

Therefore,

$$x=\frac{b}{1-\tan r}$$

$$= \frac{b\sqrt{n^2 - \sin^2 i}}{\sqrt{n^2 - \sin^2 i} - \sin i} \cong 27 \text{ cm}$$

Since $i = 45°$ and $n = \frac{4}{3}$.

The amount of water required is $V = xa^2 \cong 43.2$ litres.

699. The man's eyes are reached by rays coming in a narrow beam from an arbitrary point C on the bottom. They seem to the eye as issuing from point C' (Fig. 506). Since Δi and Δr are very small, we can write:

$$AD = AC \cdot \Delta r = \frac{H}{\cos r} \Delta r$$

$$AD' = AC' \cdot \Delta i = \frac{h}{\cos i} \Delta i$$

By equating the values of AB from triangles ABD and ABD', we have

$$\frac{H}{\cos^2 r} \Delta r = \frac{h}{\cos^2 l} \Delta l$$

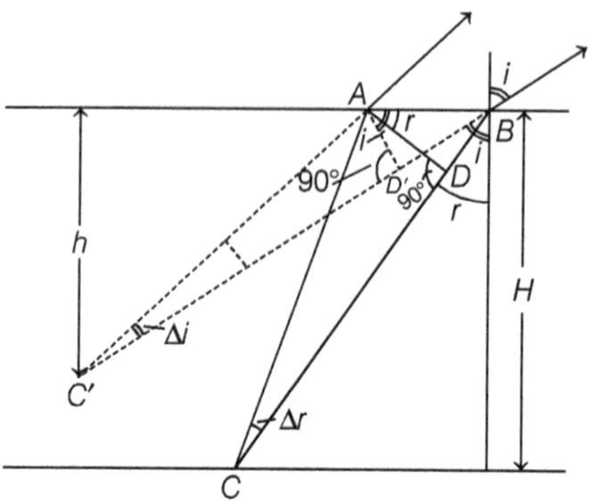

Fig. 506

Using the law of refraction, we can find the ratio $\frac{\Delta i}{\Delta r}$. Indeed,

$$\frac{\sin i}{\sin r} = n \text{ and } \frac{\sin(i + \Delta i)}{\sin(r + \Delta r)} = n$$

Remembering that Δi and Δr are small, we have

$$\sin \Delta i \cong \Delta i, \sin \Delta r \cong \Delta r$$

and $\quad \cos \Delta i \cong \cos \Delta r \cong 1$

Therefore, the last equation can be rewritten as

$$\sin i + \cos i \cdot \Delta i = n \sin r + n \cos r \cdot \Delta r$$

Hence, $$\frac{\Delta i}{\Delta r} = n \frac{\cos r}{\cos i}.$$

Upon inserting this expression into the formula relating

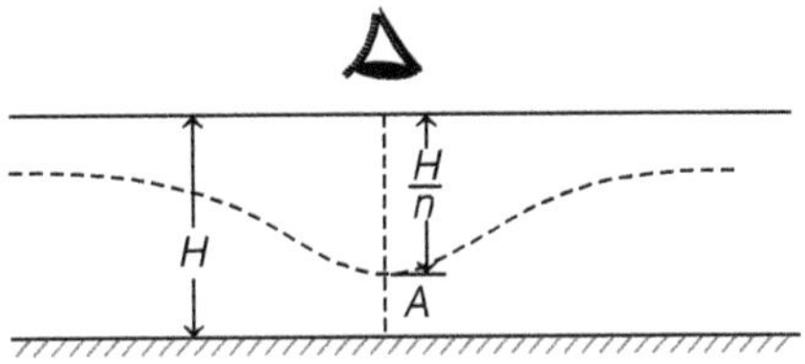

Fig. 507

H and h, we find

$$h = \frac{H \cos^3 i}{n \cos^3 r} = \frac{H}{n} \cdot \frac{\cos^3 i}{\left(1 - \frac{\sin^2 i}{n^2}\right)^{3/2}}$$

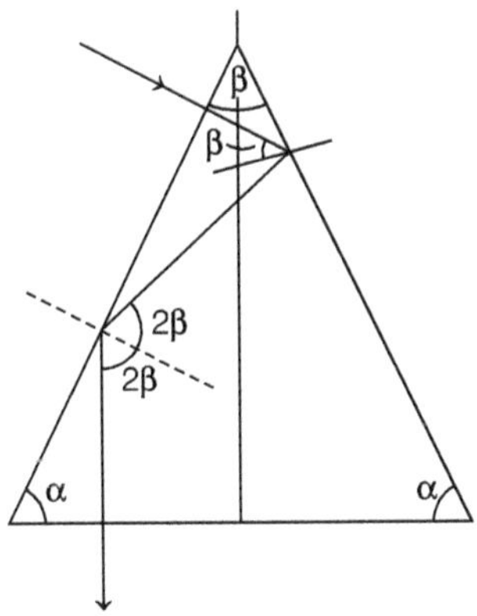

Fig. 508

When $i = 0$, we have $h = \frac{H}{n}$, i.e., the depth seems reduced by n times. As i increases, h diminishes. The approximate dependence of the seeming depth on the angle i is shown in Fig. 507. The man's eye is above point A of the lake bottom.

700. $\varphi = 120°$

701. The path of the ray in the prism is shown in Fig. 508. There is an obvious relationship between the angles α and β, namely, $2\alpha + \beta = 180°$, and $\alpha = 2\beta$.
Hence, $\alpha = 72°$, $\beta = 36°$

702. The path of the ray in the prism is shown in Fig. 509. To avoid full internal reflection on face BN, it is necessary that $\sin\beta \le \frac{1}{n}$. As can be seen from the drawing, $\beta = \alpha - r$. Hence, the greater the value of r, the higher is the permissible value of α. The maximum value of r is determined from the condition: $\sin r = \frac{1}{n}$ (angle of incidence 90°).

Therefore,

$$\alpha_{max} = 2 \text{ arc sin} \frac{2}{3} \cong 83°40'.$$

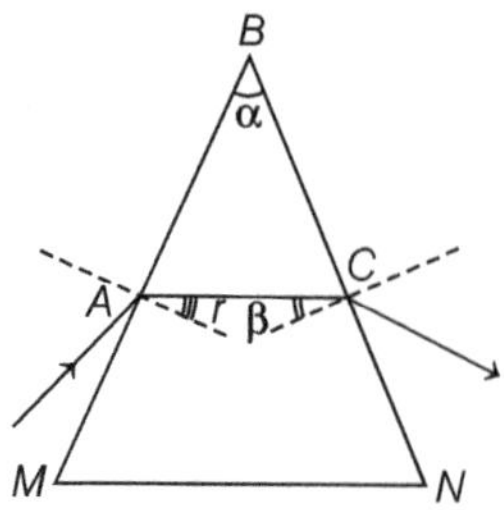

Fig. 509

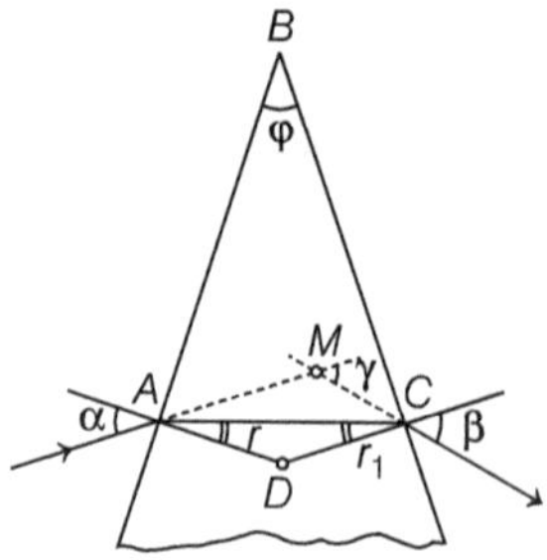

Fig. 510

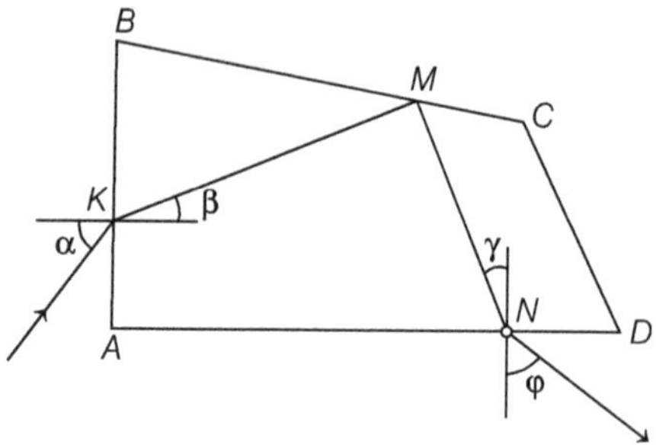

Fig. 511

703. When considering triangles *ABC*, *AMC* and *ADC* (Fig. 510), it is easy to see that $r + r_1 = \varphi$ and $\gamma = \alpha + \beta - \varphi$. According to the law of refraction,

$$\frac{\sin \alpha}{\sin r} = n$$

and

$$\frac{\sin r_1}{\sin \beta} = \frac{1}{n}$$

Upon solving this system of equations, we find that

$$\varphi = \alpha + \beta - \gamma$$

and

$$n = \sin \beta \sqrt{\left\{\frac{\sin \alpha}{\sin \beta \sin(\alpha + \beta - \gamma)} + \frac{1}{\tan(\alpha + \beta - \gamma)}\right\}^2 + 1}$$

704. According to the initial condition, the incident beam and the beam that has passed through the prism are mutually perpendicular. Therefore, $\angle\varphi = \angle\alpha$ and also $\angle\gamma = \angle\beta$ (Fig. 511). The sum of the angles of the quadrangle *AKMN* is 360°. Therefore, $\angle KMN = 90°$ and beam *KM* is incident onto face *BC* at an angle of 45°. If we know the angles of triangle *KBM* it is easy to find that $\beta = 30°$. In conformity with the law of refraction $\dfrac{\sin\alpha}{\sin\beta} = n$.

Hence, $\sin\alpha = 0.5n$ and $\alpha = \arcsin 0.5n$

Since the full internal reflection at an angle of 45° is observed only when $n \geq \sqrt{2}$, the angle α is within $45° \leq \alpha \leq 90°$.

705. Paper partially lets through light. But owing to its fibrous structure and the great number of pores, the dissipation of light is very high in all directions. For this reason it is impossible to read the text.

When they fill the pores, glue or water reduce the dissipation of light, since their refraction index is close to that of the paper. The light begins to pass through the paper without any appreciable deviation, and the text can easily be read.

5-3. Lenses and Spherical Mirrors

706. $n = 1.5$

707. $f = 2R$

708. The convex surface has a radius of curvature of $R_1 = 6$ cm and the concave one $R_2 = 12$ cm.

709. In the first case the focal length is determined from the formula

$$\frac{1}{f_1} = \left(\frac{n}{n_1} - 1\right)\left(\frac{1}{R_1} + \frac{1}{R_2}\right)$$

Since in a vacuum the focal length of the lens is f, then

$$\frac{1}{R_1} + \frac{1}{R_2} = \frac{1}{(n-1)f}$$

Hence,

$$f_1 = \frac{(n-1)f}{\dfrac{n}{n_1} - 1} = 90 \text{ cm}$$

In the second case the sought focal length is

$$f_2 = \frac{(n-1)f}{\dfrac{n}{n_2} - 1} = -102 \text{ cm}$$

The lens will be divergent.

710. As shown in the solution of Problem 709

$$-f = \frac{n_2(n_1 - 1)}{D(n_1 - n_2)}$$

Therefore, $$n_2 = \frac{fDn_1}{fD + 1 - n_1} \cong 1.67$$

711. The image will be $m+1$ times smaller than the object.

712. The lamp should be moved two metres away from the lens.

713. Obviously, one of the image will be virtual. Therefore, denoting the distances from the sources of light to the lens by a_1 and a_2 and from the lens to the images by b_1 and b_2, we obtain:

$$\frac{1}{a_1} - \frac{1}{b_1} = \frac{1}{f}, \text{ and } \frac{1}{a_2} + \frac{1}{b_2} = \frac{1}{f}$$

According to the initial condition, $a_1 + a_2 = l$ and $b_1 = b_2$. Upon solving this system of equations, we get

$$a_1 = \frac{l\left(1 \pm \sqrt{1 - \frac{2F}{l}}\right)}{2}$$

The lens should be placed at a distance of 6 cm from one source and 18 cm from the other.

714. Applying the formula of a lens to both cases, we obtain

$$\frac{1}{a_1} + \frac{1}{b_1} = \frac{1}{f}, \text{ and } \frac{1}{a_2} + \frac{1}{b_2} = \frac{1}{f}$$

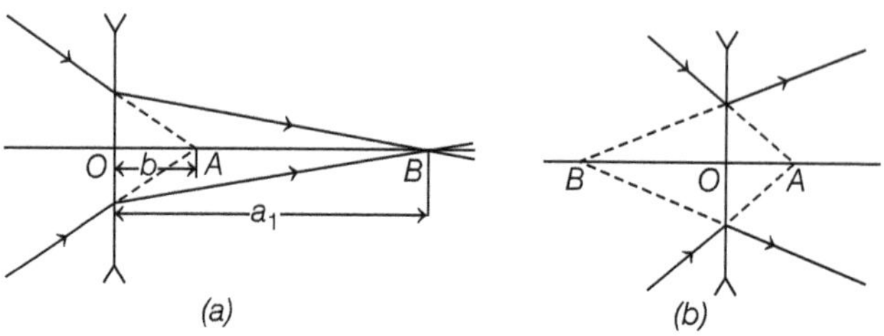

Fig. 512

According to the initial condition,

$$a_2 = a_1 + l \text{ and } \frac{b_1}{a_1} = k_1 = 3$$

(magnification in the first case): $\dfrac{b_2}{a_2} = k_2 = 2$

(magnification in the second case);

Hence, $$f = \frac{k_1 k_2}{k_1 - k_2} l = 9 \text{ cm}$$

715. (1) For this case the path of the rays is shown in Fig. 512*a*. From the standpoint of reversibility of the light beams, point B can be regarded as a source of light and point A as its image. Then, according to the formula of a lens,

$$\frac{1}{a_1} - \frac{1}{b} = -\frac{1}{f}$$

Hence, $$f = \frac{a_1 b}{a_1 - b} = 20 \text{ cm.}$$

(2) The path of the rays is illustrated in Fig. 512*b*. In this case, both the image (point A) and the source (point B) are virtual. According to the formula of a lens

$$-\frac{1}{a_2} - \frac{1}{b} = -\frac{1}{f}$$

Hence, $$f = \frac{a_2 b}{a_2 + b} = 12 \text{ cm.}$$

716. On the basis of the formula of a lens,

$$\frac{1}{a} + \frac{1}{d-a} = \frac{1}{f}$$

where a is the distance between the lens and the lamp. Therefore,

$$a^2 - ad + df = 0$$

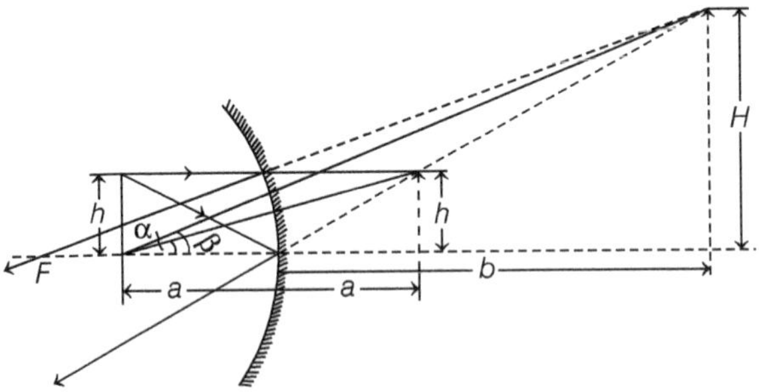

Fig. 513

Upon solving this equation, we obtain

$$a = \frac{d}{2} \pm \sqrt{\frac{d^2}{4} - df}$$

Two positions of the lens are therefore possible: at a distance of $a_1 = 70$ cm and at a distance of $a_2 = 30$ cm from the lamp.

When $f' = 26$ cm, there will be no sharp image on the screen, whatever the position of the lens, since to obtain the image it is necessary that $d \geq 4f$.

717. In the first case $\frac{h_1}{H} = \frac{b_1}{a_1}$, where a_1 and b_1 are the distances from the object and the image to the lens. In the second case $\frac{h_2}{H} = \frac{b_2}{a_2}$.

It follows from the solution of Problem 716 that $a_1 = b_2$ and $b_1 = a_2$. Therefore,

$$H = \sqrt{h_1 h_2}$$

718. On the basis of the formula of a mirror

$$\frac{1}{a} - \frac{1}{b} = \frac{1}{f}$$

The linear magnification of the mirror is

$$\frac{H}{h} = \frac{b}{a}$$

According to the initial condition, the angular dimensions of the image on the concave mirror are 1.5 times greater than those on the flat mirror: $\beta = 1.5\,\alpha$. (Fig. 513)

It is obvious that

$$\tan\alpha = \frac{h}{2a}$$

and

$$\tan\beta = \frac{H}{a+b}$$

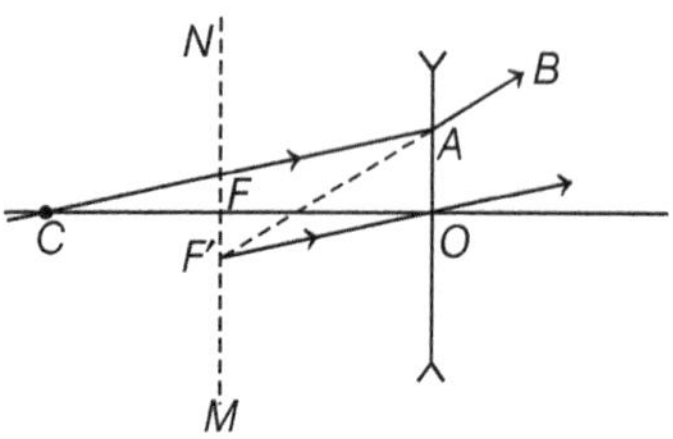

Fig. 514

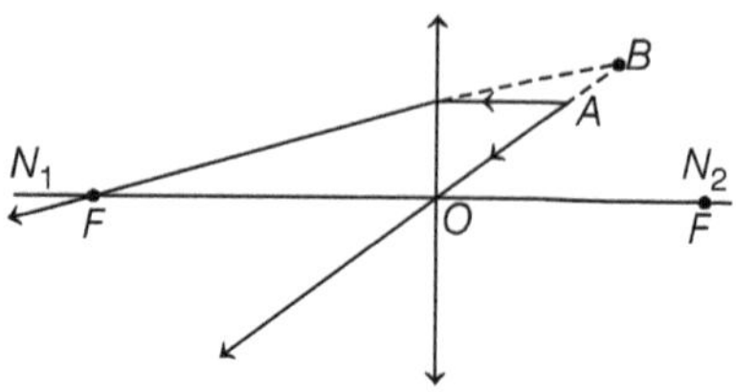

Fig. 515

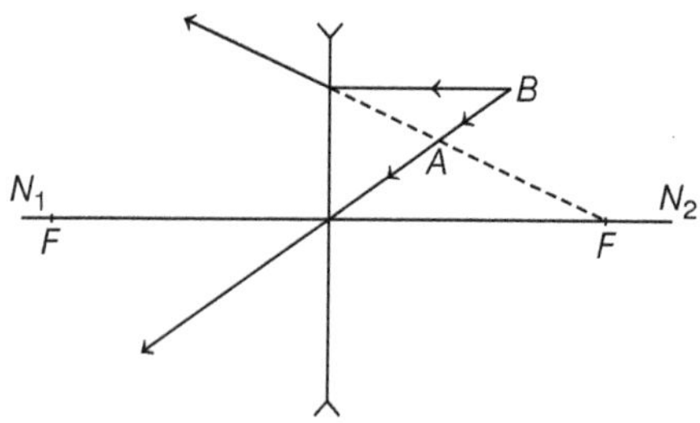

Fig. 516

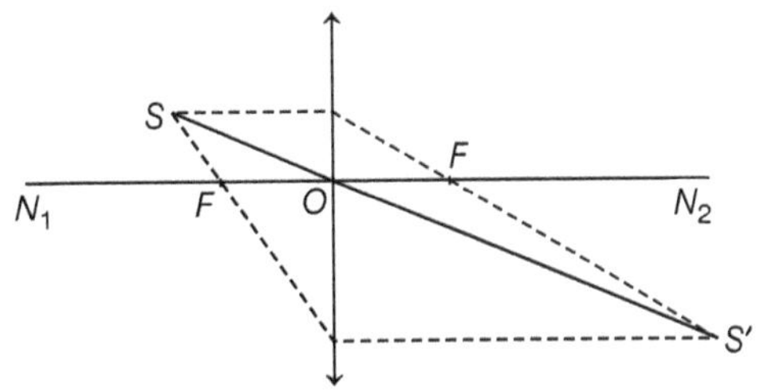

Fig. 517

When $h \ll 2a$, α and β are small. For small angles

$$\frac{H}{a+b} \cong 1.5\frac{h}{2a}$$

Upon excluding the unknown quantities $\frac{H}{h}$ and b from the equations, we find that $f = \frac{3}{2}a$.

Hence, $R = 2f = 3a = 6$ metres.

719. The path of the ray is shown in Fig. 514. Let us continue AB up to its intersection with the focal plane of lens NN. The beam of parallel rays after refraction in the lens so travels that the continuations of the rays should intersect at F'. Ray $F'O$ is not refracted. Thus, ray CA passing to point A is parallel to $F'O$ up to the lens.

720. If A is the source and B is the image, then the lens will be convergent. The position of the optical centre of the lens O and its foci F can be found by construction as shown in Fig. 515.

If B is the source and A is the image, the lens is divergent. The respective construction is illustrated in Fig. 516.

721. The centre of the lens O is the point of intersection of straight lines SS' and N_1N_2. The foci can easily be found by constructing the rays parallel to the major optical axis (Fig. 517).

722. Point O, which is the optical centre of the lens, can be found by dropping perpendicular BO onto straight line N_1N_2 (Fig. 518). Let us draw an auxiliary optical axis DO parallel to ray AB and extend straight line BC until it intersects DO at point E lying in the focal plane. Let us drop a perpendicular from E onto N_1N_2 to find point F, one of the main foci of the lens. By using the property of reversibility of the ray, we can find the other main focus F_1.

723. The image S' may be real or virtual. In both cases let us draw an arbitrary ray ADS' and auxiliary optical axis BOC parallel to it to find the position of the source (Fig. 519). By connecting the points of intersection B and C (of the auxiliary axis with the focal planes) to point D by straight lines, we can find the position of the source S_1 (if the image S' is real) and S_2 (if the image is virtual).

724. Since the ray incident on the mirror at its pole is reflected symmetrically with respect to the major optical axis, let us plot point S_1 symmetrical to S' and draw ray SS_1 until it intersects the axis at point P (Fig. 520). This point will be the pole of the mirror.

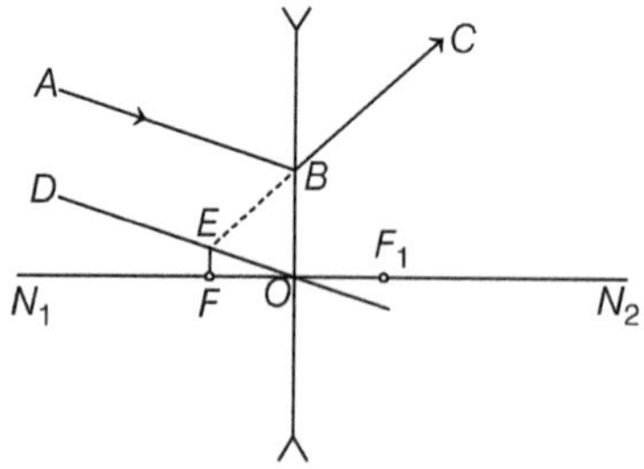

Fig.518

The optical centre C of the mirror can obviously be found as the point of intersection of ray SS' with axis NN'.

The focus can be found by the usual construction of ray SM parallel to the axis. The reflected ray must pass through focus F (lying on the optical axis of the mirror) and through S'.

725. (a) Let us construct, as in the solution of Problem 724, the ray BAC and find point C (optical centre of the mirror) (Fig. 521a).

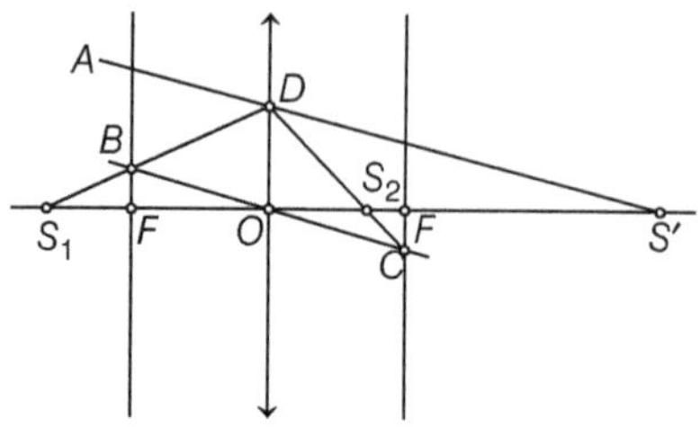

Fig. 519

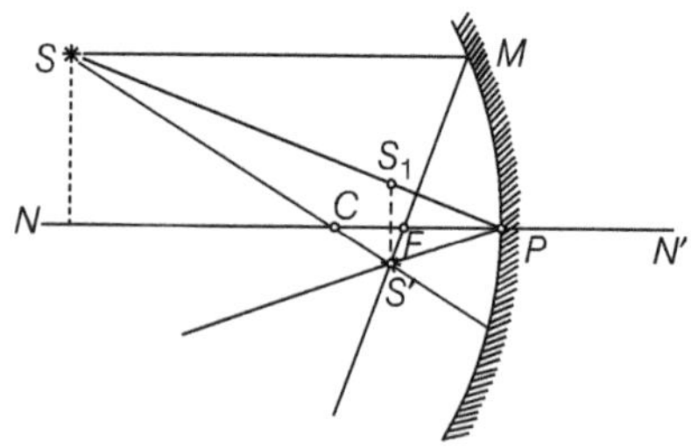

Fig. 520

Pole P can be found by constructing the path of the ray APA' reflected in the pole with the aid of symmetrical point A'. The position of the mirror focus F is determined by means of the usual construction of ray AMF parallel to the axis.

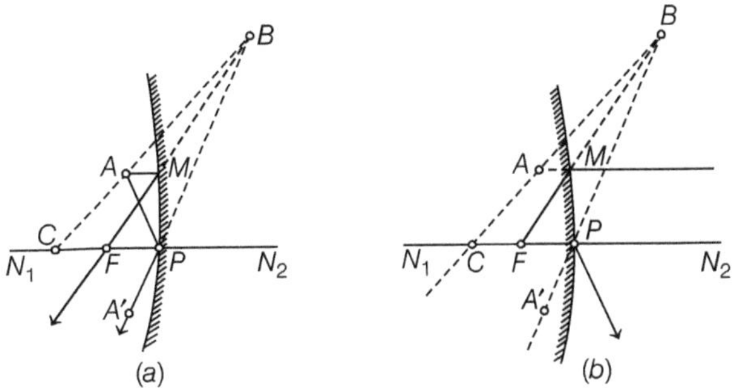

Fig. 521

(b) This construction can also be used to find centre C of the mirror and pole P (Fig. 521*b*). The reflected ray BM will pass parallel to the optical axis of the mirror. For this reason, to find the focus, let us first determine point M at which straight line AM, parallel to the optical axis, intersects the mirror, and then extend BM to the point of intersection with the axis at the focus F.

726. (a) The rays reflected from the flat mirror increase the illumination at the centre of the screen. The presence of the mirror is equivalent to the appearance of a new source (with the same luminous intensity) arranged at a distance from the screen three times greater than that of the first source. For this reason the illumination should increase by one-ninth of the previous illumination, i.e.,

$$E_a = 2.5\ lx$$

(b) The concave mirror is so arranged that the source is in its focus. The rays reflected from the mirror travel in a parallel beam. The illumination along the axis of the beam of parallel rays is everywhere the same and equal to the illumination created by the point source at the point of the mirror closest to it. The total illumination at the centre of the screen is equal to the sum of the illuminations produced by the source at the centre of the screen and reflected by the rays:

$$E_b = 2 \times 2.25\ lx = 4.5\ lx$$

(c) The virtual image of the point source in the convex mirror is at a distance of $2.5r$ from the screen (r is the distance from the screen to the source). The luminous flux Φ emitted by the virtual source is equal to that of the real source incident on the mirror:

$$I_1\omega_1 = I_2\omega_2$$

Since the solid angle ω_1 of the flux incident on the mirror from source S (Fig. 522) is one-fourth of the solid angle ω_2 inside which the rays from virtual source S_1 propagate, the luminous intensity I_2 of the virtual source is one-fourth of the intensity of source S. For this reason the virtual source creates at the centre of the screen an illumination of $4 \times (2.5)^2 = 25$ times smaller that the real source.

Hence, $$E_c = 2.34\ lx$$

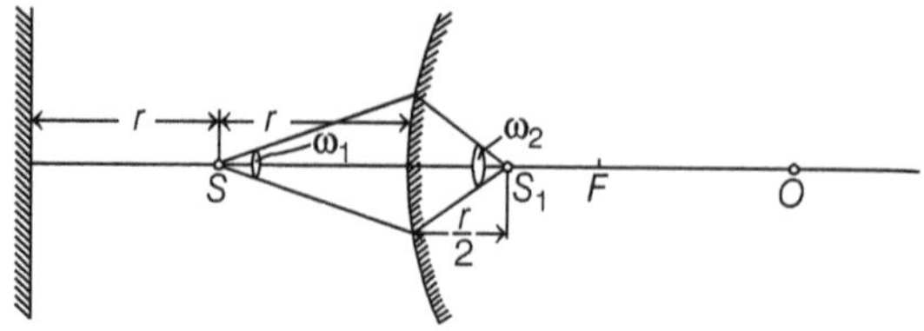

Fig. 522

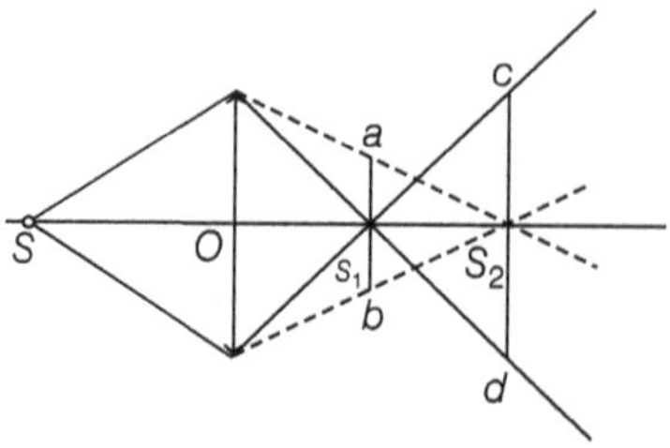

Fig. 523

727. Each section of the lens produces a full image irrespective of the other sections. Therefore, there will be no strips on the photograph, and the image will simply be less bright.

728. Any section of the lens gives an image identical in shape to that produced by the entire lens. The layered lens can be therefore regarded as two lenses with different focal lengths, but with a common optical centre. Accordingly, this lens will produce two images: at point S_1 and at point S_2 (Fig. 523). The image of the source will be surrounded by a bright halo with a diameter of ab or, respectively, cd on a screen arranged perpendicular to the optical axis at point S_1 or S_2.

729. To prove that the visible dimensions of the Sun's disk near the horizon and high above it are the same, let us project the disk in both cases onto a sheet of paper with the aid of a long-focus lens. Both the lens and the sheet should be perpendicular to the sunrays. A long-focus lens is required because the dimensions of the image are proportional to the focal length. If we measure the dimensions of the images, we see that they are equal.

5-4. Optical Systems and Devices

730. The mirror should be placed between the focus and a point lying on the double focal length. The path of the rays is shown in Fig. 524.

731. The divergent lens should be placed at a distance of 25 cm from the convergent one and the foci of both lenses will coincide. The path of the rays is shown in Fig. 525.

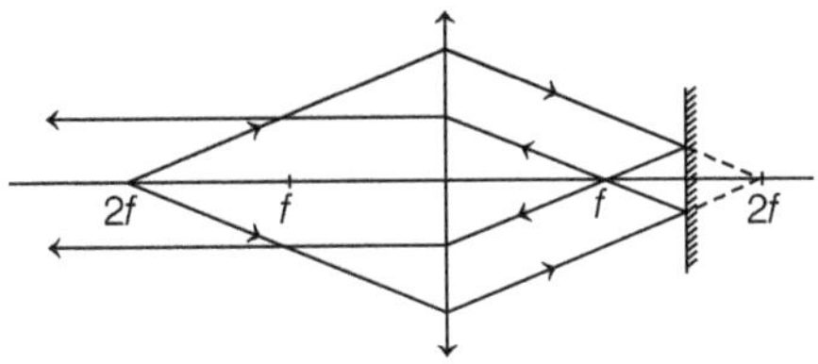

Fig. 524

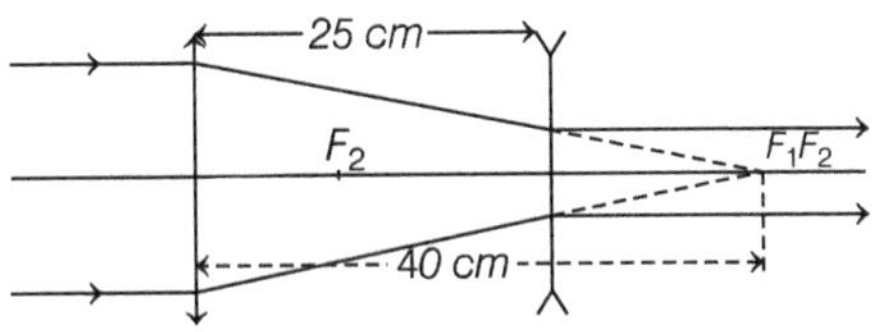

Fig. 525

732. Image $A'B'$ of the object in the spherical mirror will be at a distance b_1 (Fig. 526) from the mirror determined by the formula of the mirror

$$\frac{1}{a_1} - \frac{1}{b_1} = \frac{2}{R}$$

Hence, $b_1 = 8$ cm. Distance AA' is 48 cm. Therefore, the plate should be placed at a distance of 24 cm from object AB.

733. Two cases are possible:

(a) The mirror is at a distance of $d = f + R$ from the lens. The path of the beam parallel to the optical axis of the system and the image of object AB are illustrated in Fig. 527. Image $A'B'$ (direct and real) is obtained to full scale with the object in any position.

(b) The mirror is at a distance of $d = f = R$ from the lens (Fig. 528).

The image of object $A'B'$, also full-scale, will be inversed and virtual with the object in any position.

734. The path of the rays in this optical system is shown in Fig. 529. When the second lens is absent, the first one produces image $A'B'$ that is at a distance of $b_1 = 60$ cm from the lens. This distance can be found from the formula of the lens

$$\frac{1}{a_1} + \frac{1}{b_1} = \frac{1}{f_1}$$

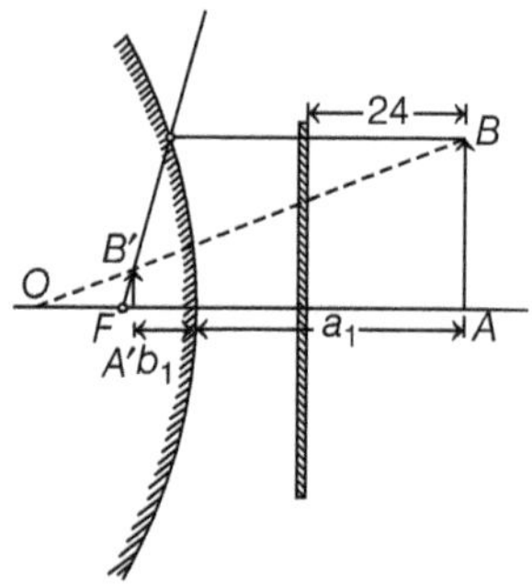

Fig. 526

Image $A' B'$ is virtual with respect to the second lens. Therefore,

$$-\frac{1}{a_2}+\frac{1}{b_2}=\frac{1}{f_2}$$

where $a_2 = b_1 - d = 30$ cm.

Hence, $b_2 = 7.5$ cm.

735. It follows from the solution of the previous problem that in the case of two convergent lenses at a distance d from each other the following equation is true

$$\frac{1}{a_1}+\frac{1}{b_2}=\frac{1}{f_1}+\frac{1}{f_2}+d\,\frac{(a_1-f_1)(f_2-b_2)}{a_1 b_2 f_1 f_2}$$

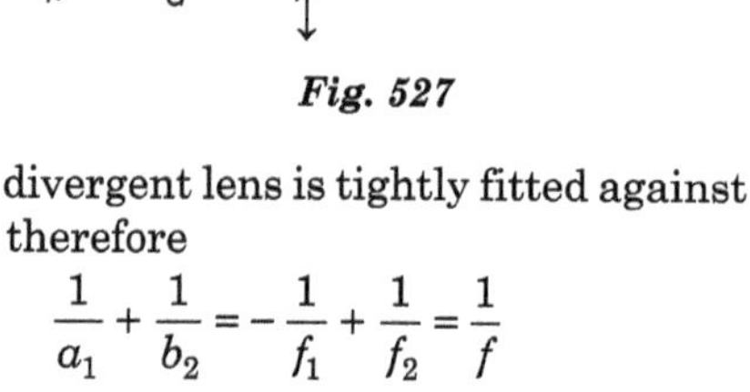

Fig. 527

In our case the divergent lens is tightly fitted against the convergent one ($d = 0$) and therefore

$$\frac{1}{a_1}+\frac{1}{b_2}=-\frac{1}{f_1}+\frac{1}{f_2}=\frac{1}{f}$$

where f is the sought focal length of the system.

Hence, $$f=\frac{f_1 f_2}{f_1-f_2}.$$

736. The light beam issuing from a point at a distance of $a_2 = 5$ cm from the second lens impinges on it. As follows from the formula of a lens, the continuations of the rays refracted by this lens intersect at a distance of $b_2 = 4$ cm from it (Fig. 530). This point coincides with the focus of the third lens. For this reason the rays leaving the system will travel in a parallel beam. The given system is telescopic.

737. Figure 531 illustrates the path of the ray through the plate from point S of the object. As a result of refraction of light by the plate ray BE seems to be issuing from point S' that is the virtual image of S in the plate. Thus, the distance between the image of the object in the plate and the lens is $a' = a - SS'$.

The displacement $SS' = AD = d - DC$.

Assuming the angles of incidence on the plate to be small, we have

$$DC = \frac{BC}{i} = \frac{dr}{r} = \frac{d}{n}$$

since $$\frac{i}{r} \cong n.$$

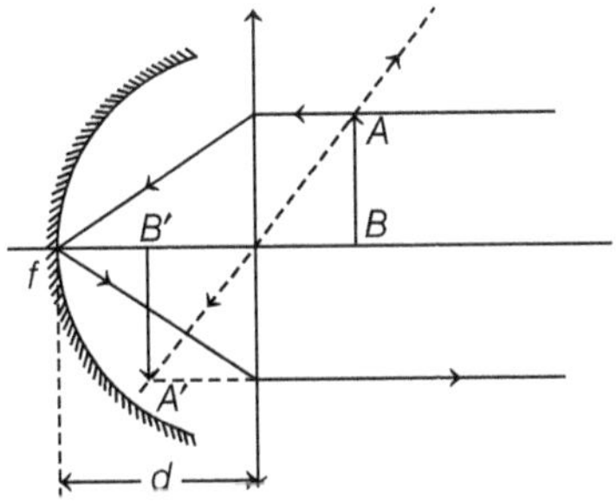

Fig. 528

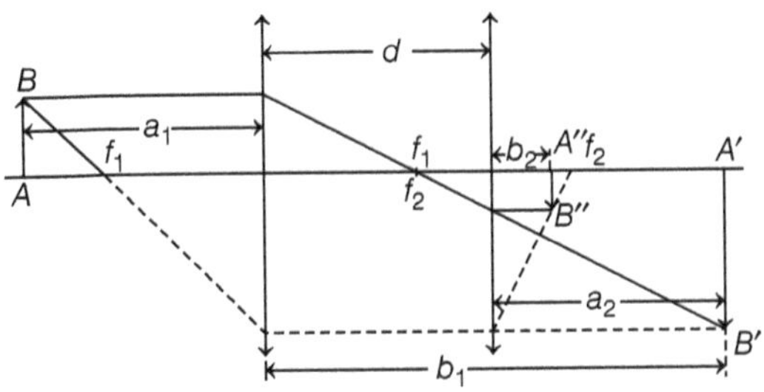

Fig. 529

Therefore, $SS' = d\left(1 - \frac{1}{n}\right) = 4$ cm. Before the plate was inserted, the screen had been at a distance of $b = \frac{af}{a - f} = 120$ cm, and after it was inserted at a distance of $b' = \frac{a' f}{a' - f} = 180$ cm. The screen should therefore be shifted by 60 cm.

738. In the absence of the mirror the image $A'B'$ of object AB (Fig. 532) will be obtained at a distance of $b = \frac{af}{a-f} = 180$ cm from the lens.

After being reflected from the mirror, the image will occupy position $A''B''$ and will be at a distance of $H' = b - l = 80$ cm from the optical axis.

The layer of water with a thickness of d will shift the image over a distance of $H - H' = d\left(1 - \frac{1}{n}\right)$, where n is the refraction index of water. This follows directly from the solution of Problem 737.

Hence,

$$H = H' + d\left(1 - \frac{1}{n}\right) = 85 \text{ cm}.$$

739. Two cases are possible:

(1) The optical axis of the lens is perpendicular to the front face of the wedge. The rays reflected from the front face pass through the lens and produce an image of the point source that coincides with the source itself. The rays reflected from the rear face will be deflected by an angle φ (Fig. 533) determined by the equality $\frac{\sin\varphi}{\sin 2\alpha} = n$. Since the angles are small $\varphi \cong 2\alpha n$.

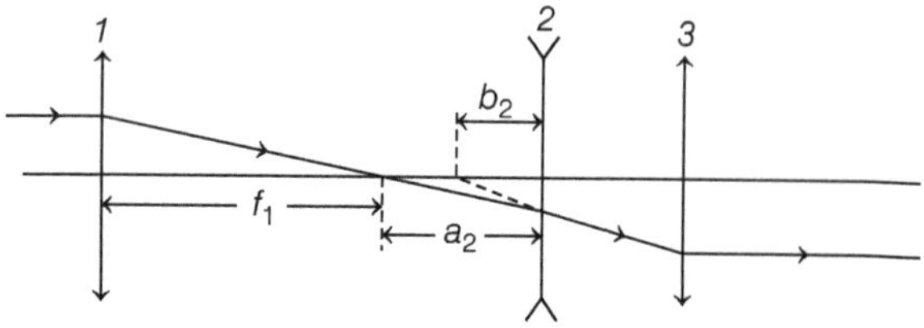

Fig. 530

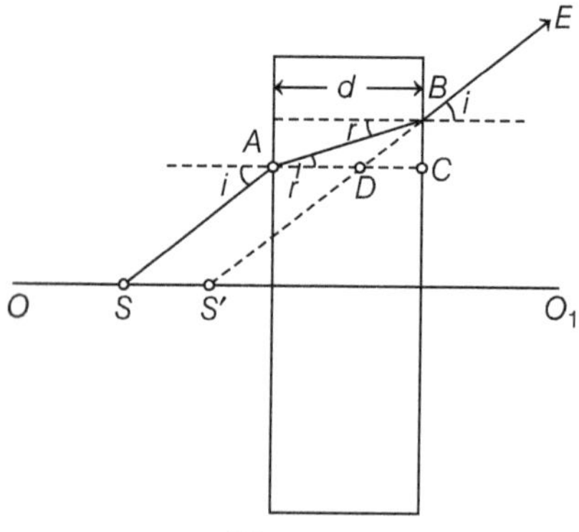

Fig. 531

The second image of the source will be obtained at a distance d from the first image, namely $d = f\varphi = f2\alpha n$.

Hence, $$n = \frac{d}{2\alpha f}.$$

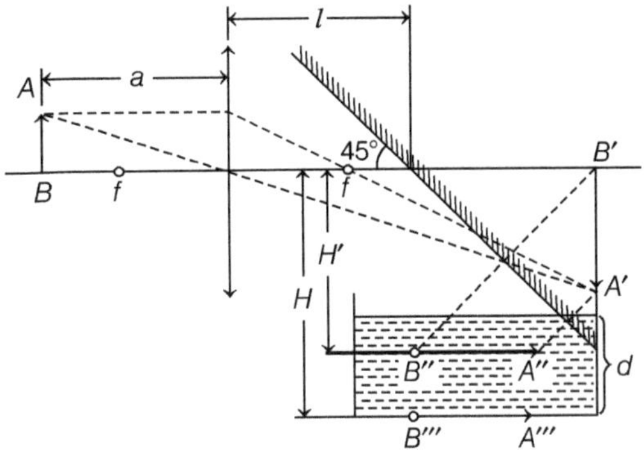

Fig. 532

(2) The optical axis of the lens is perpendicular to the rear face of the wedge. The rays reflected from the front face will be deflected by an angle 2α (Fig. 534) and produce an image at a distance of $d_1 = 2\alpha f$ from the source.

The rays reflected from the rear face will be deflected by an angle θ determined from the equations:

$$\frac{\sin\alpha}{\sin\beta} = n, \text{ and } \frac{\sin(\alpha+\theta)}{\sin(2\alpha-\beta)} = n$$

When the angles are small, $\theta = 2\alpha(n-1)$. For this reason the second image will be at a distance $d_2 = 2\alpha(n-1)f$ from the source. The total distance between the images is $d = d_1 + d_2 = 2\alpha nf$.

Hence, $n = \dfrac{d}{2\alpha f}$ as in the first case.

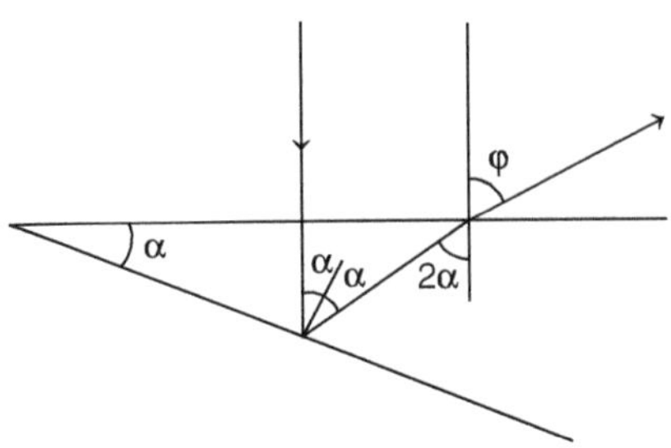

Fig. 533

740. Since the image that coincides with the source is formed owing to reflection from the part of the mirror not covered by the liquid, the source is obviously arranged at centre O of the hemisphere. Let us find the position of the other image (point A in Fig. 535).

According to the law of refraction

$$\frac{\sin\alpha}{\sin\beta} = n \cong \frac{\alpha}{\beta}, \text{ and } \frac{\sin\varphi}{\sin\theta} = n \cong \frac{\varphi}{\theta}$$

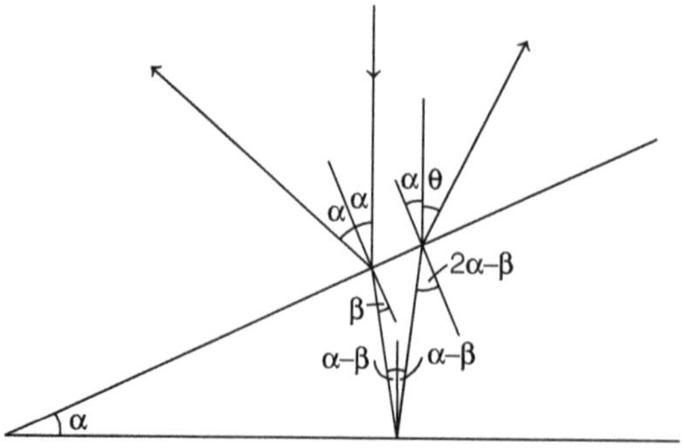

Fig. 534

As can be seen from the drawing, $\theta = \beta + 2\gamma$, where $\gamma = \alpha - \beta$ is the angle of incidence of the refracted ray on the mirror and $(R - l - h)\tan\varphi \cong (R - h)\tan\alpha$.

Neglecting h as compared with R, we can find from this system of equations that

$$n = \frac{2R - l}{2(R - l)} = 1.6$$

741. The image $A'''B'''$ obtained in the system is shown in Fig. 536. F_1 and F_2 are the foci of the lens and the mirror, and $A'B'$ is the image produced by the lens if its surface is not coated with silver.

The image $A''B''$ produced by a concave mirror can be plotted, taking into account that ray BO, after passing through the lens and being reflected from the mirror surface, will travel along OB''; here $\angle BOA = \angle B''OA$. Ray BC emerges from the lens parallel to the optical axis of the system and after reflection passes through F_2.

The rays reflected from the mirror are refracted in the lens once more and produce image $A'''B'''$. Point B''' lies at the intersection of rays OB'' and CD. Ray OB'' passes through the optical centre of the lens after reflection and is thus not refracted. Ray CD can be plotted as follows. After the first refraction in the lens and reflection, ray BC will travel in the direction of F_2 and will be refracted in the lens once more. Its direction after the second refraction can be found by the method described in Problem 723: ray OD parallel to CF_2 is drawn through optical centre O until it intersects the focal plane of the lens. The sought ray is obtained by connecting C and D.

Since the rays are refracted in the lens twice, focal length f of the system can be found from the following ratio (see Problem 735):

$$\frac{1}{f} = \frac{1}{f_1} + \frac{1}{f_2} + \frac{1}{f_1}$$

where $f_2 = \frac{R}{2}$ is the focal length of the mirror

$$f = \frac{f_1 f_2}{f_1 + 2f_2} = 2.5 \text{ cm}$$

Therefore, the distance b to image $A''' \ B'''$ can be found from the formula

$$\frac{1}{a} + \frac{1}{b} = \frac{1}{f}$$

Hence, $$b = \frac{af}{a - f} = 3 \text{ cm}.$$

742. The focal length of the thin lens is $f = \frac{r}{n-1}$, where r is the radius of the spherical surface.

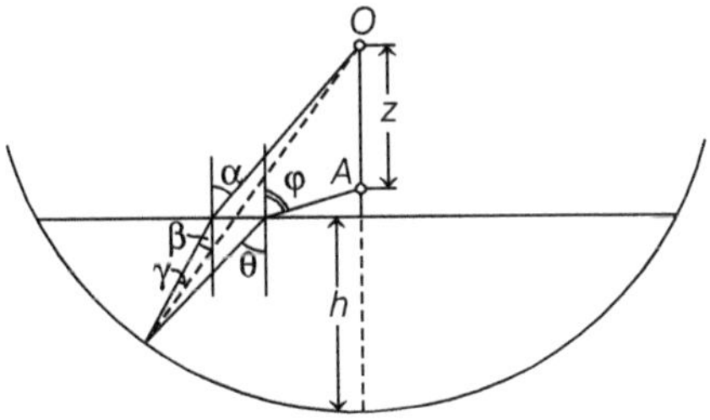

Fig. 535

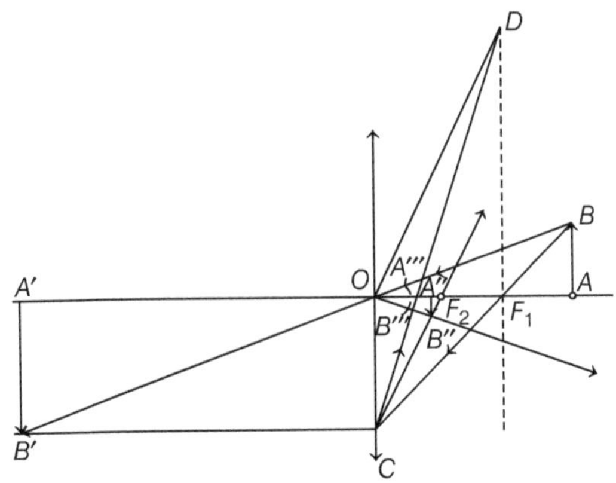

Fig. 536

Let the rays parallel to the optical axis of the spherical surface fall on it from the air (Fig. 537). After being refracted on the surface, ray NK is deflected by an angle $\alpha - \beta$ from the optical axis.

As can be seen in Fig. 537*a*, we have

$$OP \tan\alpha = F_1P \tan(\alpha - \beta).$$

According to the law of refraction,

$$\frac{\sin\alpha}{\sin\beta} = n.$$

Since these angles are small,

$$r\alpha \cong f_1(\alpha - \beta) \text{ and } \alpha = \beta n.$$

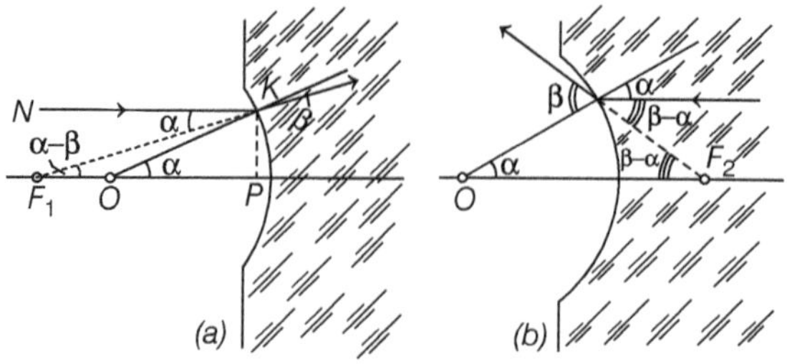

Fig. 537

Therefore,

$$f_1 = \frac{n}{n-1} r = nf$$

If the parallel rays are incident from the glass (Fig. 537b), then similar reasoning will give the equations:

$$\frac{\sin\alpha}{\sin\beta} = \frac{1}{n}, \text{ and } r\tan\alpha = f_2\tan(\beta - \alpha)$$

Since the angles are small, $\beta = n\alpha$, and $r\alpha = f_2(\beta - \alpha)$.

Hence,

$$f_2 = \frac{r}{n-1} = f.$$

743. Two cases are possible: the focus is outside the sphere, and inside it. Let us consider the first case. The path of the ray incident on the sphere at an angle i is shown in Fig. 538.

Bearing in mind that the angles i and r are small, we have, in accordance with the condition of the problem,

$$BC = R\sin\alpha = R\sin(2r - i) \cong R(2r - i) \cong \frac{Ri}{n}(2 - n)$$

The focus obviously lies outside the sphere when $n < 2$ and on the surface of the sphere if $n = 2$.

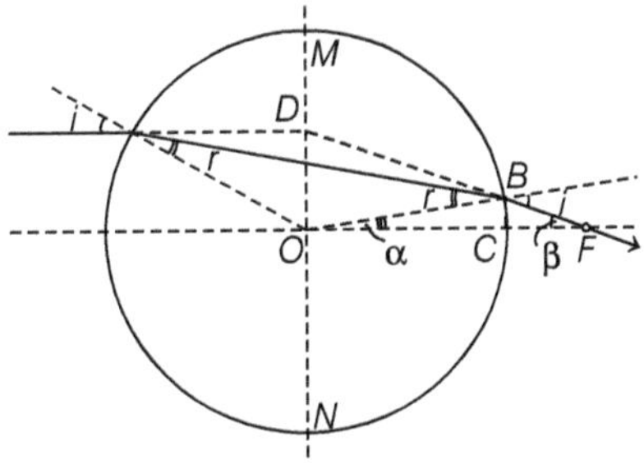

Fig. 538

The distance $CF = BC \cot\beta \cong \frac{BC}{\beta}$, and $\beta \cong 2(i - r) \approx \frac{2i\,(n-1)}{n}$

as can easily be found from Fig. 538.

The sought distance is $f = R + CF = \frac{Rn}{2(n-1)}$

When $n > 2$, the path of the ray is as shown in Fig. 539. The sought distance is $f' = OF \cong CF - R$

As can be seen from Fig. 539,

$$CF = AC \cot(i - r) \cong \frac{AC}{i - r} \cong \frac{Ri}{i - r}$$

Hence, $$f' = \frac{R}{n-1}$$

744. Let us extend ray BF until it intersects the continuation of the ray incident on the sphere parallel to the optical axis (Fig. 538). It can easily be seen that section DO that connects the point of intersection with the centre of the sphere forms a right angle with the direction of the incident ray. Triangle ODF is a right one, since

$$OF\,\beta \cong \frac{R}{2}\frac{n}{n-1}\frac{2i\,(n-1)}{n} = Ri \text{ (see Problem 743)}$$

For this reason the main planes of sphere MN coincide and pass through its centre.

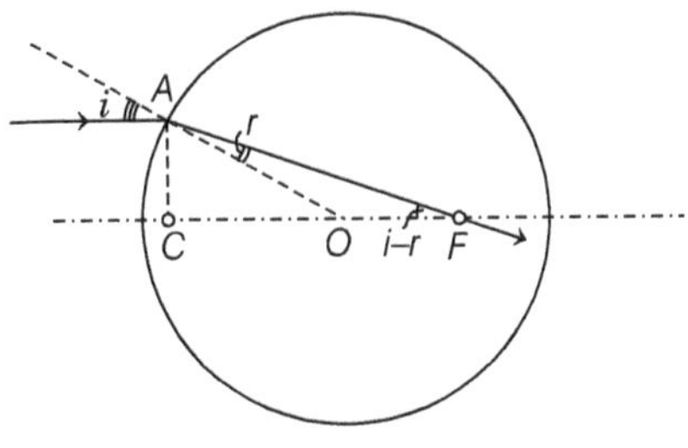

Fig. 539

745. The focal length of the sphere is

$$f = \frac{R}{2}\frac{n}{n-1} = 15 \text{ cm}$$

(see Problems 743 and 744). By using the formula of a lens, and this may be done because the main planes coincide, let us find the distance from the centre of the lens to the image

$$b = \frac{af}{a-f} = -15 \text{ cm}$$

The image is virtual and is in front of the sphere.

746. The thin wall of the spherical flask can be regarded as a divergent lens with a focal length of

$$f_1 = \frac{1}{(n-1)\left(\frac{1}{R_1} - \frac{1}{R_2}\right)} \cong \frac{R^2}{(n-1)\Delta R}$$

After passing through two such lenses at a distance of $2R$ from each other (Fig. 540), the rays parallel to the major optical axis (the flask diameter) will be so refracted that their continuations will intersect at the focus F of the system at a distance b from the second lens. According to the formula of a lens,

$$\frac{1}{f_1 + 2R} - \frac{1}{b} = -\frac{1}{f_1}$$

Hence,

$$b = \frac{f_1(f_1 + 2R)}{2(f_1 + R)}$$

Point D of the intersection of AB (continuation of the incident ray) and CF (continuation of the refracted ray) lies on the main plane of the system at a distance x from the second lens.

It follows from the similarity of triangles ACB and F_1CO, and also of triangles DCB and FCO that

$$\frac{x}{b} = \frac{2R}{2R + f_1}$$

The main plane lies at the following distance from the second lens

$$x = \frac{2Rb}{2R + f_1} = \frac{f_1 R}{f_1 + R}$$

Therefore, the focal length of the system is

$$f = b - x = \frac{f_1^2}{2(f_1 + R)} \cong \frac{f_1}{2} = \frac{R^2}{2(n-1)\Delta R}$$

In view of symmetry of this optical system, the positions of the second focus and of the other main plane are obvious.

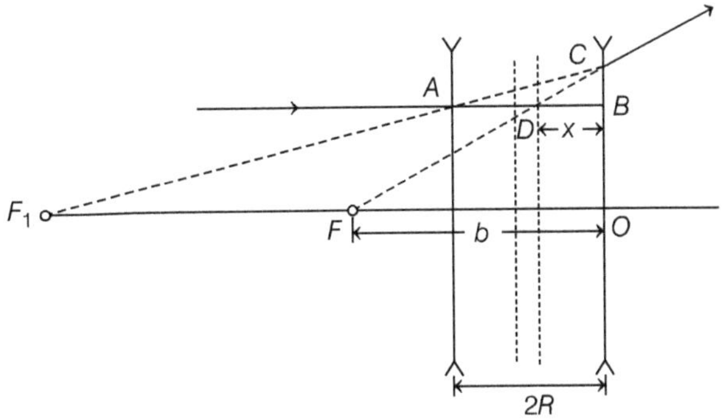

Fig. 540

747. A glance at Fig. 541 shows that the angle of refraction is

$$r = \angle OAB = \angle ABO = \angle OBC = \angle OCB$$

and

$$\angle BAD = \angle BCD = i - r$$

At point A the beam turns through an angle $i - r$, at point B through an angle $\pi - 2r$, and at point C through $i - r$. Therefore, the total angle through which the beam is deflected from the initial direction is

$$\theta = i - r + \pi - 2r + i - r$$
$$= \pi + 2i - 4r$$

The angle r can be found from the ratio $\dfrac{\sin i}{\sin r} = n$.

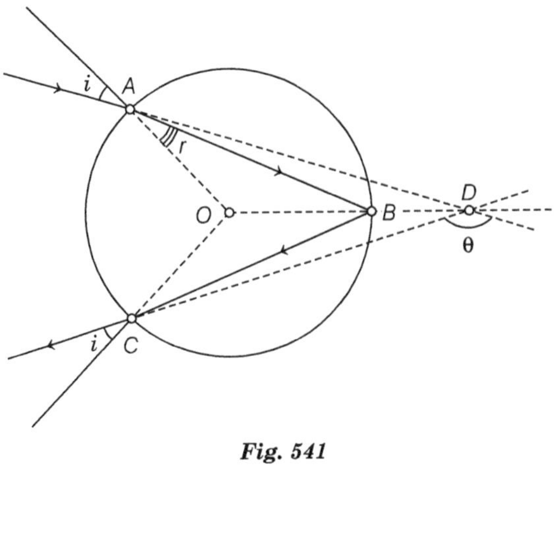

Fig. 541

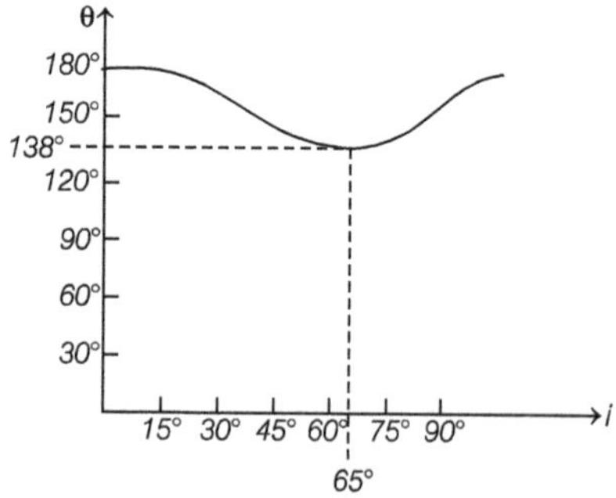

Fig. 542

748. When a parallel beam of rays is incident on the drop, the ray passing along the diameter has an angle of incidence of $i = 0°$, while the angles of incidence of the rays above and below it may range from 0 to 90°.

(1) Using the results of the previous problem and the law of refraction, we can find the values of θ for various values of i:

Table 3

i	θ	i	θ
0°	180°	55°	138°20′
20°	160°24′	60°	137°56′
40°	144°40′	65°	138°40′
50°	139°40′	70°	140°44′

(2) A diagram of θ versus i is shown in Fig. 542.

(3) The minimum value of the angle of deflection is approximately equal to $\theta_{min} = 138°$. The rays leaving the drop are nearly parallel when $\theta = \theta_{min}$, since in this case, as can be seen from Table 3 and the diagram, θ changes the slowest when i changes. An approximate path of the rays in the drop is illustrated in Fig 543.

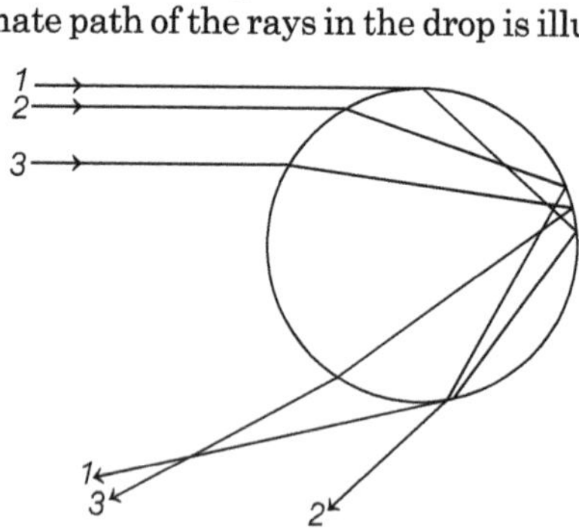

Fig. 453

749. In accordance with the formula of a lens

$$\frac{1}{a}+\frac{1}{b}=\frac{1}{f}$$

The magnification is

$$k=\frac{b}{a}+\frac{b-f}{f}=24 \text{ times}$$

750. The condenser should produce a real image of the source on the lens having the size of the lens. Therefore, using the formula of a lens and the expression that determines its magnification, we can write two equations:

$$\frac{1}{x}+\frac{1}{y}=\frac{1}{f}, \text{ and } \frac{d}{D_0}=\frac{x}{y}$$

Here x is the distance from the source of light to the condenser and y is the distance from the condenser to the lens. According to the initial condition, $x+y=l$.

By cancelling x and y from the expressions obtained, we can find the focal length:

$$f=\frac{dlD_0}{(d+D_0)^2}=7.1 \text{ cm}$$

The diameter of the condenser will be minimum if the slide is behind it. The minimum permissible diameter is $D \cong 11$ cm.

751. Ground glass is needed to fix the plane in which the image is obtained, and to increase the angle of vision.

Transparent glass is used to examine the image produced by a lens in a microscope. For this purpose a line is drawn on a transparent glass to fix the focussing plane, and this line and the adjacent portion of the image produced by the lens are focussed in the microscope. In this case ground glass cannot be used, since the microscope will show all the distortions due to the structure of the ground surface.

752. (1) The lanterns will appear equally bright, since the illumination of the retina of the eye $E=\frac{LA}{b^2}$ is the same for both lanterns. (Here L is the luminance of a lantern, A the area of the pupil, and b the distance from the crystalline lens to the retina.)

(2) The image of a far object is closer to the lens than that of a near object. For this reason a remote lantern produces a higher illumination of the film and its image will be brighter on the photograph.

753. The illumination of the photographic film is

$$E \sim \frac{A}{b^2} \sim S_l\,\frac{(a-f)^2}{a^2}$$

where $S_l = \frac{d^2}{f^2}$ is the lens speed, f is the focal length and a is the distance from the lens to the object being photographed (see Problem 752). It is obvious therefore that the exposure in a camera with a short focal length should be smaller.

754. The distances between the Sun and the Earth and between the Sun and the Moon are practically the same. For this reason if the Moon and the wall had equal coefficients of reflection, their luminance would seem identical. It can be assumed, therefore, that the surface of the Moon consists of dark rock.

755. In air, the external convex cornea of the eye collects the rays and produces an image on the retina. The crystalline lens only helps in this.

The refraction index of the liquid inside the eye is very close to that of water. For this reason the cornea refracts almost no light and the eye becomes very far-sighted.

The refractive properties of the cornea are completely retained when the swimmer is wearing a mask.

756. When the man looks at remote objects through his spectacles, he sees them as he would see objects at a distance of $a_2 = 60$ cm without any spectacles.

Therefore, when the man is wearing spectacles (see the solution to problem 735)

$$\frac{1}{a} + \frac{1}{b} = \frac{1}{f} + \frac{1}{f_0}$$

where $$a = \infty.$$

When the man is without spectacles

$$\frac{1}{a_2} + \frac{1}{b} = \frac{1}{f}$$

Here b is the depth of the eye, $\frac{1}{f}$ the minimum optical power of the eye and $\frac{1}{f_0}$ the optical power of the spectacles. It is assumed that the spectacles are fitted tightly against the eye.

Hence, $$f_0 = -a_2.$$

Let us now determine the position of the nearest point of accommodation of the eye with spectacles

$$\frac{1}{a_1} + \frac{1}{b_1} = \frac{1}{f_1}$$

and $$\frac{1}{a_3} + \frac{1}{b_1} = \frac{1}{f_1} + \frac{1}{f_0}$$

Therefore,

$$\frac{1}{a_3} = \frac{1}{a_1} + \frac{1}{f_0} = \frac{1}{a_1} - \frac{1}{a_2}$$

and $$a_3 = 15 \text{ cm}$$

757. The long-sighted man when wearing his friend's spectacles can see only very far objects. Therefore, the distance a_2 of best vision of the eye of the long-sighted man can be determined from the equation

$$\frac{1}{a_1} - \frac{1}{a_2} = D_1$$

where a_1 is a very great distance ($a_1 \to \infty$) and D_1 is the optical power of the spectacles of the short-sighted man.

The optical power D_2 of the spectacles that correct the defect of vision of the long-sighted man can be found from the formula

$$\frac{1}{a_0} - \frac{1}{a_2} = D_2$$

where $a_0 = 0.25$ metre is the distance of best vision of the normal eye.

The distance a_3 of best vision of a short-sighted eye can be found from the equation

$$\frac{1}{a_0} - \frac{1}{a_3} = D_1$$

If the short-sighted man wears the spectacles of his long-sighted friend, the distance of best vision, i.e., the minimum distance a at which the short-sighted man can easily read a small type, can be determined from the formula

$$\frac{1}{a} - \frac{1}{a_3} = D_2$$

Upon solving these four equations, we get $a = 12.5$ cm.

758. When an object with a height of l is examined from a distance of D, the angle of vision φ_1 is determined by the formula

$$\varphi_1 = \frac{l}{D}$$

If the same object is viewed through a magnifying glass,

$$\varphi_2 = \frac{l'}{b + r} = \frac{l'}{L}$$

where l' is the height of the image (Fig. 544).

The angular magnification is

$$N = \frac{\varphi_2}{\varphi_1} = \frac{l' D}{l L} = k \frac{D}{L}$$

where $k = \frac{l'}{l} = \frac{b}{D} = \frac{f+b}{f}$ is the linear magnification determined from the formula of a lens (f is the focal length).

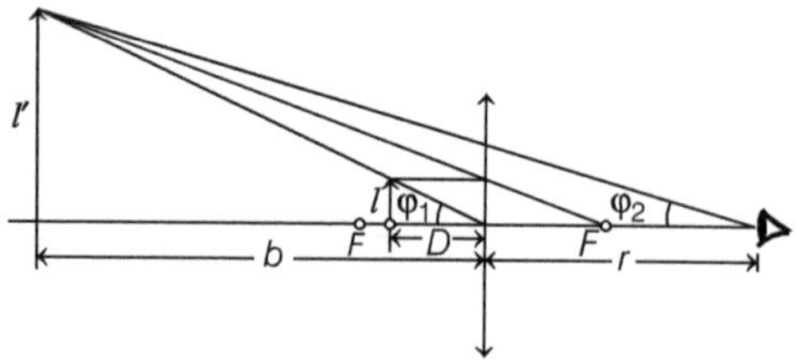

Fig. 544

Hence,

$$N = \frac{D}{f}\frac{b+f}{L} = \frac{D}{f}\frac{L-r+f}{L}$$

(1) When $L = \infty$, we have $N = \frac{D}{f}$.

(2) When $L = D$, we have $N = \frac{D}{f} + 1 - \frac{r}{f}$.

759. The magnification of a telescope is $N = \frac{f_1}{f_2}$, where f_1 is the focal length of the objective and f_2 that of the eyepiece.

Since in a telescope adjusted to infinity the distance between the objective and the eyepiece is $f_1 + f_2$,

$$\frac{D}{d} = \frac{f_1 + f_2}{b}$$

Here b is the distance from the eyepiece to the image of the diaphragm.

According to the formula of a lens,

$$\frac{1}{f_1 + f_2} + \frac{1}{b} = \frac{1}{f_2}$$

Upon cancelling b from these equations, we find that

$$\frac{D}{d} = \frac{f_1}{f_2} = N$$

760. Sharp images of remote objects will be obtained with the convergent lens in three different positions. The lens can be placed in front of the divergent lens or behind it.

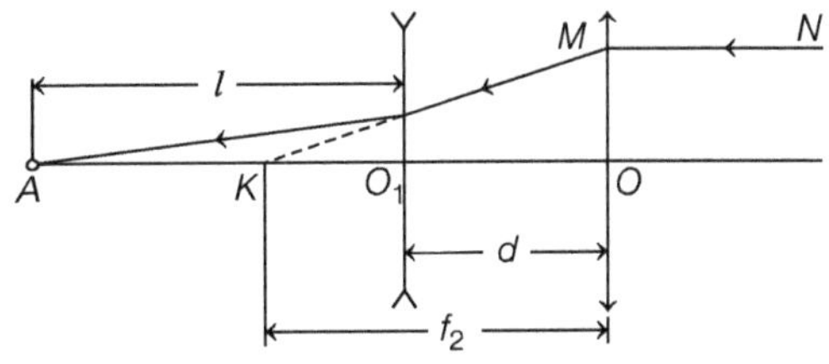

Fig. 545

For the first position the distance d between the lenses can be found considering point K as the virtual image of point A in the divergent lens (Fig. 545).

$$-\frac{1}{f_2 - d} + \frac{1}{l} = -\frac{1}{f_1}$$

Ray MN is parallel to the optical axis of the system.

Hence,

$$d = f_2 - \frac{f_1 l}{f_1 + l} = 3.5 \text{ cm}$$

For the second position (the convergent lens is behind the divergent one) the path of the rays is shown in Fig. 546. Regarding point A as the image of K in the convergent lens, let us use the formula of a lens

$$\frac{1}{f_1 + d} + \frac{1}{l - d} = \frac{1}{f_2}$$

Hence,

$$d = \frac{l - f_1}{2} \pm \frac{l + f_1}{2}\sqrt{1 - \frac{4f_2}{l + f_1}}$$

The distance between the lenses may be $d_2 = 35$ cm or $d_3 = 5$ cm.

761. Let the rays coming from one end of the diameter of the visible disk of the Moon be directed along the optical axis of the system. The rays will produce an image on the optical axis at point A (Fig. 547) removed by a distance of $l = 45$ cm from the divergent lens.

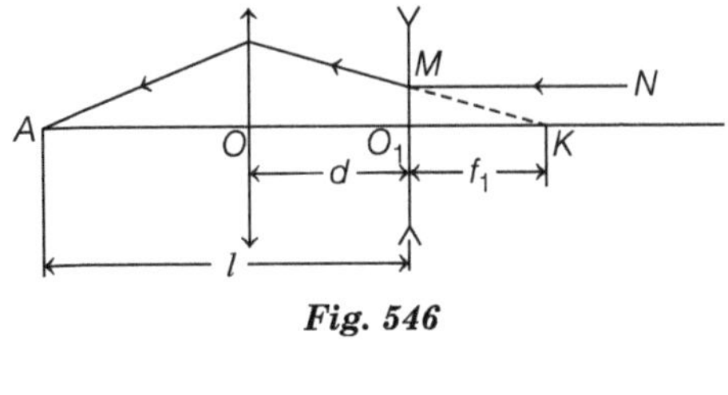

Fig. 546

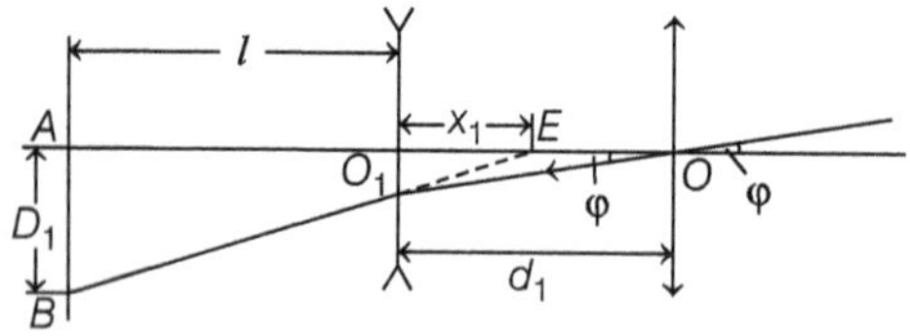

Fig. 547

The rays coming from the other end of the diameter form, according to the condition, an angle φ with the first rays. After passing through the system, these rays will produce an image (point B) lying in a plane perpendicular to the optical axis and removed by the same distance l from the divergent lens.

To find the diameter of the image $D_1 = AB$, let us consider the path of the ray passing through the optical centre of the first lens.

In the first arrangement, the convergent lens is placed in front of the divergent one at a distance of $d_1 = 3.5$ cm. If we consider point E as the virtual image of point O we can write

$$\frac{1}{d_1} - \frac{1}{x_1} = -\frac{1}{f_1}$$

Using the similarity of triangles ABE and O_1PE and remembering that $O_1P = d_1 \tan\varphi$, we obtain

$$\frac{D_1}{l + x_1} = \frac{d_1 \tan\varphi}{x_1} \cong \frac{d_1\varphi}{x_1}$$

Upon cancelling x_1 from these equations, we find that $D_1 = 0.72$ cm.

In the second arrangement of the lenses ($d_2 = 35$ cm), the path of the rays is shown in Fig. 548. The diameter of the Moon's image D_2 can be found from the equations:

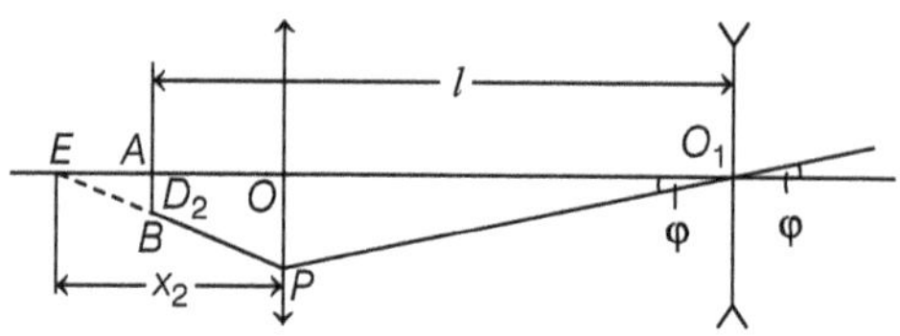

Fig. 548

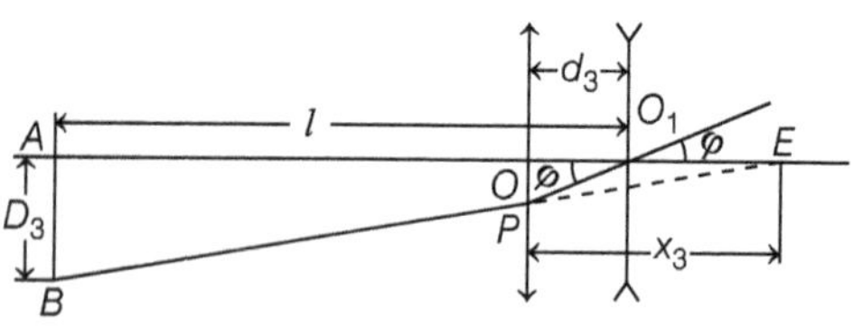

Fig. 549

$$\frac{D_2}{(x_2 + d_2) - l} = \frac{d_2 \tan\varphi}{x_2} \cong \frac{d_2\varphi}{x_2}$$

(considering triangles EOP, EAB and OPO_1) and

$$\frac{1}{d_2} + \frac{1}{x_2} = \frac{1}{f_2}$$

(considering E as the image of O_1). Hence $D_2 \cong 0.011$ cm.

In the third arrangement ($d_3 = 5$ cm) the path of the rays will be some what different (Fig. 549) than that shown in Fig. 548.

The equations for D_3 can be written similar to the preceding cases as follows :

$$\frac{D_3}{(l - d_3) + x_3} = \frac{d_3 \tan\varphi}{x_3} \cong \frac{d_3\varphi}{x_3}, \text{ and } \frac{1}{d_3} - \frac{1}{x_3} = \frac{1}{f_2}$$

Hence, $D_3 = 0.18$ cm.

762. It follows from the formula of a lens

$$\frac{1}{a} + \frac{1}{b} = \frac{1}{F_{ob}}$$

that the magnification of the objective is

$$k_1 = \frac{b}{a} = \frac{F_{ob}}{a - F_{ob}}$$

where b is the distance from the image to the objective.

The real inverse magnification of the object produced by the objective can be viewed through the eyepiece as through a magnifying glass, the virtual image produced by this magnifying glass being arranged from the eye at the distance of best vision $D = 25$ cm. According to the formula of a magnifying glass

$$\frac{1}{a_1} - \frac{1}{D} = \frac{1}{F_{eye}}$$

where a_1 is the distance from the image produced by the objective to the eyepiece.

The magnification of the eyepiece is

$$k_2 = \frac{D}{a_1} = \frac{D + F_{eye}}{F_{eye}}$$

The total magnification of the microscope is

$$k = k_1 k_2 = \frac{F_{ob}(D + F_{eye})}{(a - F_{ob})F_{eye}} = 180 \text{ times}$$

6. PHYSICAL OPTICS

6-1. Interference of Light

763. No, it does not. The presence of illumination minima on the interference pattern means that no quantity of light enters the given section of space.

764. The maximum illumination will be observed at an arbitrary point of the screen C (Fig. 550) if the difference of the paths $d_2 - d_1 = k\lambda$, where $k = 0, 1, 2, \ldots$ is an integer.

According to the Pythagorean theorem

$$d_2^2 = D^2 + \left(h_k + \frac{l}{2}\right)^2$$

$$d_1^2 = D^2 + \left(h_k - \frac{l}{2}\right)^2$$

whence

$$d_2^2 - d_1^2 = (d_2 + d_1)(d_2 - d_1) = 2h_k l$$

In accordance with the initial condition $d_2 + d_1 \cong 2D$. Therefore, $d_2 - d_1 = k\lambda \cong \frac{2h_k l}{2D}$.

The distance to the k-th light band from the centre of the screen is $h_k = \frac{k\lambda D}{l}$. The distance between the bands is $\Delta h = h_{k+1} - h_k = \frac{\lambda D}{l}$.

765. The distance between interference bands is $\Delta h = \frac{\lambda D}{l}$

(see Problem 764).

In our case $D = AB \cong a + b$, and $l = S_1S_2$ is the distance between the images S_1 and S_2 of the source in the flat mirrors (Fig. 551). The value of l can be found from triangle S_1SB:

$$\frac{l}{2} = 2b\frac{\alpha}{2} \text{ or } l = 2b\alpha$$

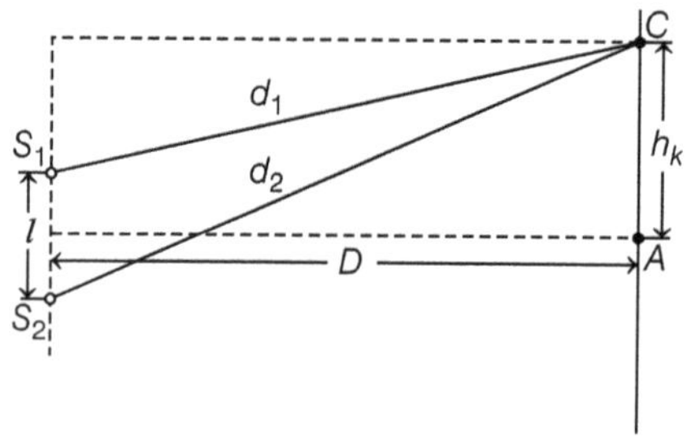

Fig. 550

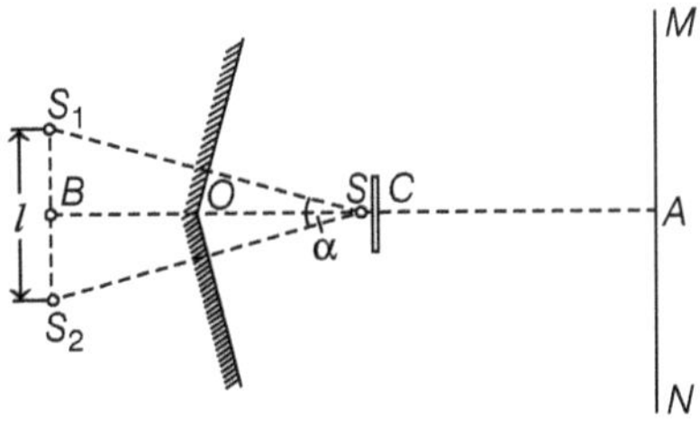

Fig. 551

Hence,

$$\Delta h = \frac{\lambda(a+b)}{2b\alpha}$$

766. The second coherent source is obtained in Lioyd's experiment by reflection of the rays from mirror AO. In reflection the phase changes by π (loss of a half-wave) and the oscillations will be damped at point O where a bright band should be observed (a minimum of illumination). As compared with Problem 764, the entire pattern will be shifted by the width of a light (or dark) band.

767. The illumination on the screen will be intensified when $d_2 - d_1 = k\lambda$. The locus of points on the screen reached by rays from both sources with this difference is a circle with its centre at point A (Fig. 552). For this reason the interference bands will have the form of concentric circles.

When $l = n\lambda$ the illumination will be intensified at point A (an interference maximum of the n-th order). The nearest bright interference band (circle) of the $(n-1)$ th order is at a distance from point A determined from the equation

$$d_2 - d_1 = \sqrt{(n\lambda + D)^2 + h_{n-1}^2} - \sqrt{D^2 + h_{n-1}^2}$$
$$= (n-1)\lambda$$

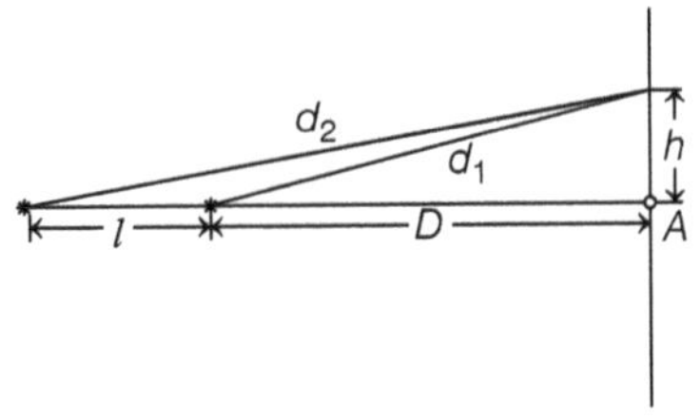

Fig. 552

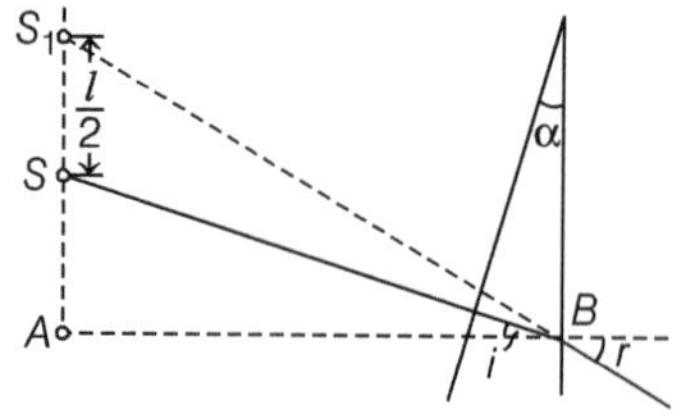

Fig. 553

Bearing in mind the conditions of the problem that $\lambda \ll D$ and $\lambda \ll l$, we get

$$h_{n-1} \cong \sqrt{\frac{2D(D+n\lambda)}{n}} = \sqrt{2D\lambda\left(\frac{D}{l}+1\right)}$$

768. The difference between the path of the rays for the k-th bright ring is

$$d_2 - d_1 = \sqrt{(2n\lambda)^2 + r_k^2} - \sqrt{(n\lambda)^2 + r_k^2} = k\lambda$$

Therefore,

$$r_k = \frac{\lambda}{2k}\sqrt{(9n^2 - k^2)(n^2 - k^2)}$$

769. A semitransparent plate with an aperture can be used to create the second coherent source closer to the screen than the first one. On the basis of Huyghens' principle, the aperture can be regarded as the secondary source and an interference pattern will appear on the screen.

If the distance between the sources is great, an interference pattern can be obtained with the aid of a source producing waves very close to monochromatic ones.

770. To find the distance Δh it is first necessary to calculate the distance l between the virtual sources S_1 and S_2 located at the point of intersection of the continuations of the rays refracted by the prism faces. For this purpose it will be the simplest to consider the path of the ray normally incident on the face of the prism (Fig. 553).

No such ray actually exists, but it can be plotted by mentally extending the upper prism downward. All the beams from a point source refracted by a prism can be assumed to converge at a point, and this method is quite permissible. Since the refracting angle of the prism is small (the prism is thin), the virtual images S_1 and S_2 of the source can be regarded as lying at the same distance from the prism as source S. It can be seen from Fig. 553 that $i = \alpha$ and $SA = a\alpha$. According to the law of refraction, $r \cong n\alpha$. Considering triangle AS_1B, we can write

$$\frac{l}{2} + a\alpha \cong a\alpha n$$

Hence,

$$l = 2a\alpha(n-1)$$

By using the solution of Problem 764, we find:

$$\Delta h = \frac{\lambda D}{l} = \frac{\lambda(a+b)}{2a\alpha(n-1)} = 0.15 \text{ cm}$$

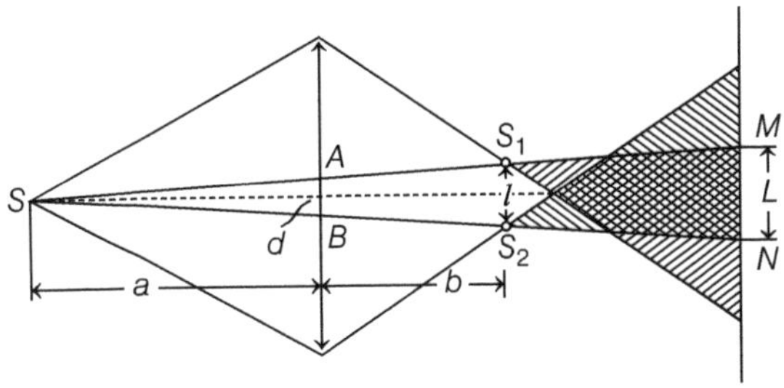

Fig. 554

771. $N = \dfrac{L}{\Delta h}$, where L is the width of the interference pattern.

It can be seen from Fig. 256 that $L = \dfrac{b}{a} l$, where l is the distance between the virtual images S_1 and S_2 of the source.

Using the results of the previous problem, we get

$$N = \frac{4ab\alpha^2(n-1)^2}{(a+b)\lambda} \cong 5$$

772. A biprism made of a substance with a refraction index n_2 deflects rays by an angle

$$\varphi_1 = \left(90° - \frac{\beta}{2}\right)(n_2 - n_1)$$

where n_1 is the refraction index of the medium from which the rays fall. For a biprism in air

$$\varphi_2 = \left(90° - \frac{\delta}{2}\right)(n_2 - 1)$$

If the biprisms are equivalent, $\varphi_1 = \varphi_2$.

Hence,

$$\delta = \beta \frac{n_2 - n_1}{n_2 - 1} + 180° \frac{n_1 - 1}{n_2 - 1}$$

For the values given in the initial conditions, $\delta \cong 179°37'$.

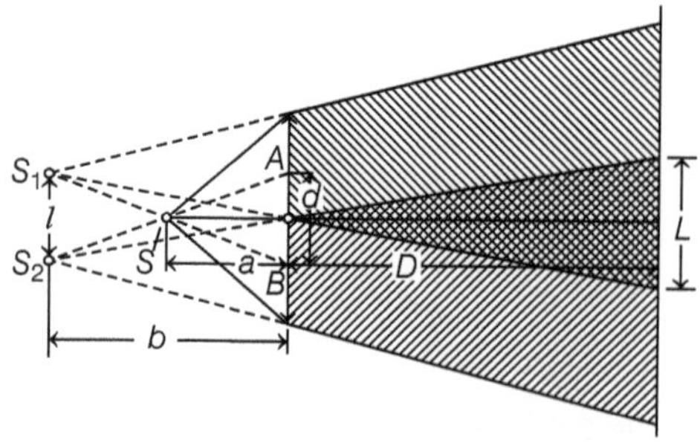

Fig. 555

773. The path of the rays in the system is illustrated in Fig 554. Here S_1 and S_2 are the images of source S in the halves of the lens. Obviously,

$$b = \frac{fa}{a - f}$$

The distance l between S_1 and S_2 can be found from the similarity of triangles SAB and SS_1S_2:

$$l = \frac{ad}{a - f}$$

The distance between adjacent interference bands on the screen is

$$\Delta h = \frac{\lambda(D - b)}{l} = \frac{\lambda}{ad}(Da - Df - af)$$
$$= 10^{-2}\text{cm}$$

(see Problem 764).

The sought number of interference bands is

$$N = \frac{L}{\Delta h} = \frac{d(D + a)}{a\Delta h} = 25$$

774. The distance between the virtual sources S_1 and S_2 can be found by the method described in the solution of Problem 773 (Fig. 555).

The distance between interference bands is $\Delta h = \dfrac{\lambda(Df - Da + af)}{da}$

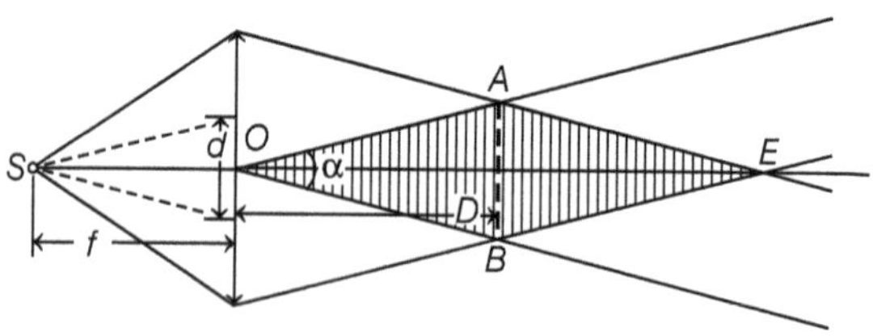

Fig. 556

The number of bands on the screen is

$$N = \frac{L}{\Delta h}$$

where $L = \frac{Dl}{b}$ is the dimension of the portion of the screen on which interference bands are observed.

Therefore,

$$D = \frac{Nab\,f\,\lambda}{adl + abN\lambda - bfN\lambda} = 15 \text{ cm}$$

The maximum possible number of bands can be determined from the condition

$$adi + Nab\lambda - bfN\lambda = 0$$

(here $D \to \infty$).

Hence,

$$N_{max} = \frac{adl}{bf\lambda - ab\lambda} = 5$$

The number of bands is finite, since the distance between them increases as the screen is moved away, and the dimensions of the portion of the screen on which the interference pattern appears grow.

775. The distance between the interference bands will not depend on the position of the screen only if the source is arranged in the focal plane of the lens. This directly follows from the expression

$$\Delta h = \frac{\lambda}{ad}(Df - Da + af)$$

obtained in solving Problem 774. When $a = f$,

$$\Delta h = \frac{\lambda f}{d} = 10^{-2}\text{cm}$$

at any value of D.

For this case the path of the rays is shown in Fig. 556. A glance at this illustration will show that the number of interference bands will be maximum when the screen is in position AB. The distance from the screen to the lens can be found from triangle OAB, remembering that the angle $\alpha \cong \frac{d}{f}$ and $AB = R$:

$$D = \frac{Rf}{d} = 2 \text{ metres}$$

776. The length of a light wave diminishes n times in glass, since the frequency does not change and the velocity decreases n times. This produces an additional difference in the path between the coherent waves in the beams.

At a distance of d_1 the upper beam will accommodate $k_1 = \frac{d_1 n}{\lambda}$ wavelengths, and the lower beam $k_2 = \frac{d_2 n}{\lambda} + \frac{d_1 - d_2}{\lambda}$ wavelengths at the same distance.

At any point on the screen the light waves will additionally be shifted with respect to each other by $k_1 - k_2$ wavelengths. As a result, the entire interference pattern will be displaced by $k_1 - k_2 = \frac{(d_1 - d_2)}{\lambda}(n-1) = 100$ bands

The process of displacement can be observed at the moment when the plates are introduced. After this has been done, the interference pattern on the screen will be as before.

777. The lens is too thick. Interference occurs only with thin films. The air layer between the lens and the glass is thin.

778. No, it will not. The difference of the path between the waves meeting on the screen and emitted by the sources S and S_1 or S and S_2 is great. In these conditions the spectra of various orders that correspond to the spectral interval of the source are superposed in the same way as when the waves are reflected from the boundaries of a thick film.

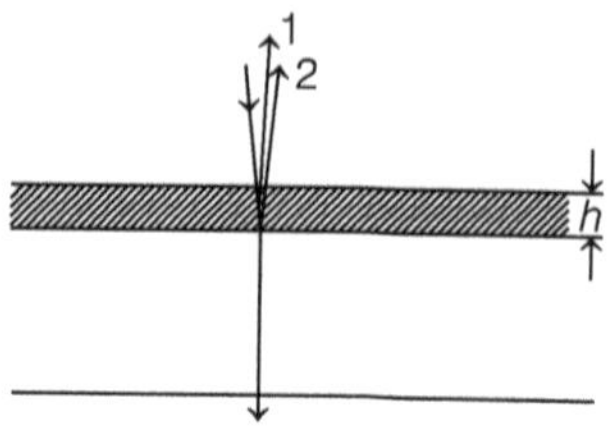

Fig. 557

If the shield is removed, the only result will be superposition of monotonously changing illumination on the interference pattern from sources S_1 and S_2.

779. When the rings are observed in reflected light, the intensity of the interference beams is about the same.

In transmitted light the intensity of one beam that was not reflected is much higher than that of another beam that was reflected twice. As a result, the maxima and minima will appear against the background of uniform illumination, the light will not be extinguished completely and the entire pattern will be less distinct than in reflected light.

780. In the absence of contact, the radius of the fifth ring is determined by the equation $\frac{r_1^2}{R} + 2d = 5\lambda$. If the dust is removed, the radius of this ring can be found from the equation $\frac{r_2^2}{R} = 5\lambda$.

Hence, $$d = \frac{r_2^2 - r_1^2}{2R} = 1.8 \times 10^{-4}\,\text{cm}.$$

781. $$r_k = \sqrt{\frac{k\lambda}{\left(\frac{1}{R_2} - \frac{1}{R_1}\right)}}$$

782. To reduce the reflection factor it is necessary that rays *1* and 2 (Fig 557) reflected from the external and internal surfaces of the coat applied to the optical glass damp each other.

This will occur if
$$2hn = (2k+1)\frac{\lambda}{2} \qquad (1)$$
where $k = 0, 1, 2, \ldots$. Hence, the minimum thickness of the coat is
$$h_{min} = \frac{\lambda}{4n}$$

Condition (1) cannot be satisfied for all wavelengths. For this reason, h is usually so selected as to damp the middle part of the spectrum. The thickness of the coat exceeds h_{min} by an odd number of times, since it is easier to make thick coats than thin ones (a quarter of the wavelength thick).

783. To observe an interference pattern, the maximum of the k-th order corresponding to the wavelength λ should not overlap with the maximum of the $(k+1)$ th order corresponding to the wavelength $\lambda + \Delta\lambda$, where $\Delta\lambda = 100$ Å.

This will occur if $(\lambda + \Delta\lambda)k \le \lambda(k+1)$.

Hence, $$k \le \frac{\lambda}{\Delta\lambda}.$$

The maximum permissible thickness of the layer h_{max} satisfies the equation $2h_{max} = (\lambda + \Delta\lambda)k_{max}$

where $k_{max} = \frac{\lambda}{\Delta\lambda}$. If λ is the wavelength corresponding to the middle of the visible section of the spectrum ($\lambda = 5{,}000$Å), then
$$h_{max} \cong 1.3 \times 10^{-3}\,\text{cm}$$

If a thin film with a refraction index n is used instead of the air layer, the maximum thickness should be n times smaller than in the air layer.

784. Upon the interference of rays *1* and *2* (Fig. 558) reflected from different faces of the wedge, the condition of the minimum can be written as follows:

$$2hn = k\lambda (k = 0, 1, 2).$$

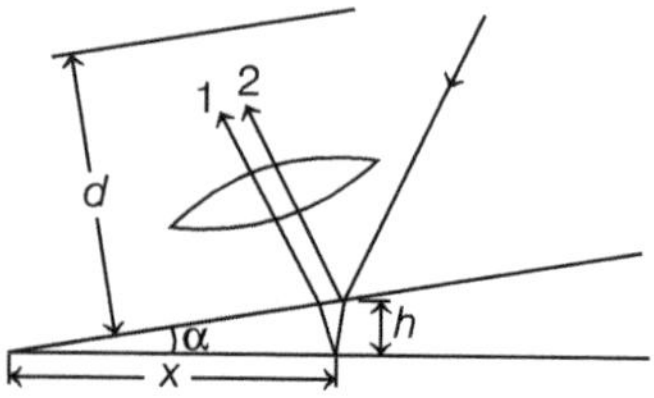

Fig. 558

Since the angle α is small, $h \cong x\alpha$.
Therefore, the distance between the interference bands on the wedge is

$$\Delta x = \frac{\lambda}{2\alpha n}.$$

According to the formula for the magnification of a lens, $\frac{\Delta x}{\Delta l} = \frac{a}{b}$, where a is the distance from the screen to the lens and b from the lens to the wedge. Since $b = d - a$, then according to the formula of a lens, $\frac{1}{a} + \frac{1}{d-a} = \frac{1}{f}$. Upon cancelling a and b from these expressions, we can find the sought value of the angle α

$$\alpha = \frac{\lambda}{2n\Delta l} \frac{d \mp \sqrt{d^2 - 4fd}}{d \mp \sqrt{d^2 - 4fd}}$$

The solution of this problem is not a single one, because a sharp image can be obtained on the screen with fixed d and f when the lens is in one of two positions.

6-2. Diffraction of Light

785. The radius of the first Fresnel zone can be found from triangles *ADE* and *DEB* (Fig. 559):

$$r_1^2 = a^2 - (a - x)^2 = \left(b + \frac{\lambda}{2}\right)^2 - (b + x)^2$$

Since the wavelength is small,

$$x = \frac{b\lambda}{2(a + b)}.$$

Therefore, $$r_1^2 = 2ax - x^2.$$

Neglecting the small value of x^2, we finally obtain

$$r_1 = \sqrt{\frac{ab\lambda}{a+b}}$$

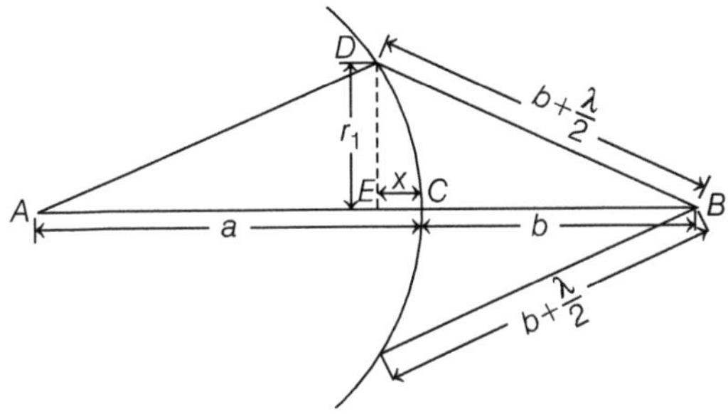

Fig. 559

The same method can be used to obtain the radii of the following Fresnel zones. For zone k

$$r_k = \sqrt{\frac{abk\lambda}{a+b}}$$

786. A distance from a point source to the wave front of $a \to \infty$ corresponds to a plane wave.

The sought radii of the zones are

$$r_k = \lim_{a \to \infty} \sqrt{\frac{abk\lambda}{a+b}} = \sqrt{kb\lambda}$$

(see the solution to Problem 785).

787. To solve the problem it is necessary to calculate the number k of Fresnel zones that can be accommodated in apertures with diameters of D and D_1.

Using the results of Problem 785, we have

$$\sqrt{k\frac{ab\lambda}{a+b}} = \frac{D}{2}$$

Now it is easy to find that $k = 3$ (an odd number). When the aperture is 5.2 mm in diameter, it can contain approximately four zones (an even number). Therefore, a greater aperture will reduce the illumination at point B.

788. When the four Fresnel zones are open the dark spot on the axis of the beam is surrounded by bright and dark rings. When the aperture is increased, the total illumination of the screen grows in magnitude, but the distribution of the luminous energy on the screen so changes that the minimum will be in the centre.

789. The sought illumination will be maximum when one Fresnel zone is accommodated in the diaphragm. With a view to the solution of Problem 786, we have

$$D = 2\sqrt{b\lambda} = 0.2 \text{ cm}$$

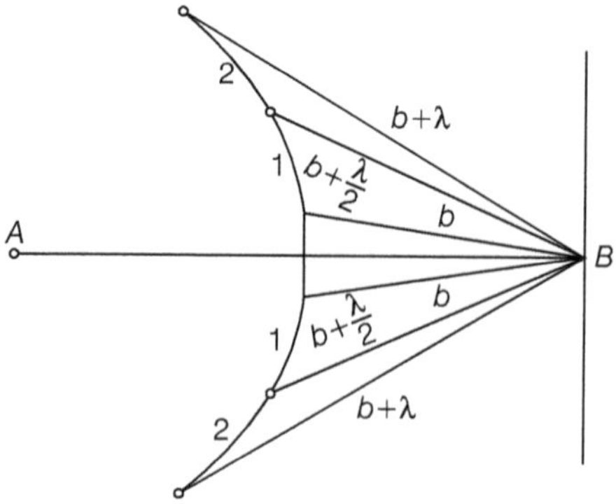

Fig. 560

790. The diffraction will be noticeable if the aperture accommodates a small number of Fresnel zones, i.e., the radius of the aperture will be of the same order as (or smaller than) the radius of the first Fresnel zone:

$$\sqrt{\frac{ab}{a+b}\lambda} \geq R$$

where R is the radius of the aperture.

When $a = b$, we have $a\lambda \geq 2R^2$.

791. The Fresnel zones plotted in Fig. 560 make it possible to determine the luminous intensity at point B. The illumination at this point is created by the first and subsequent Fresnel zones. If the dimensions of the screen do not exceed considerably the radius of the first central zone, found from the formula in Problem 785, a bright spot is sure to appear at point B with an illumination only slightly differing from that which would have appeared in the absence of the screen.

792. Approximately three metres.

793. Here it is more convenient to select the Fresnel zones in the form of bands parallel to the edges of the slit. The minimum will be observed in the direction φ if an even number of zones can be accommodated in slit AB (Fig. 561). This figure shows four Fresnel zones. We have $b = 2kx$, where x is the width of the Fresnel zone, $k = 1, 2, 3, \ldots$. Here AK is the difference in the path between the extreme rays sent by one zone:

$$AK = x \sin \varphi = \frac{\lambda}{2}$$

Hence,

$$x = \frac{\lambda}{2 \sin \varphi}$$

Therefore, the minimum will be observed in the direction φ if $b \sin \varphi = k\lambda$.

794. The rays incident on the pinhole of the camera from a remote point source are nearly parallel. If there were no diffraction the dimensions of the bright spot would be equal to $AB = 2r$ (Fig. 562). In view of diffraction, the dimensions of the spot will increase to DC. The distance OC is determined by the angle φ that gives the direction to the first minimum (dark ring). According to the note, $2r \sin \varphi \cong \lambda$. Therefore, the radius of the spot is

$$OC = r + AC = r + d \sin \varphi \cong r + \frac{d\lambda}{2r}$$

This value reaches the minimum (see the solution to Problem 504) when $r = \frac{\lambda d}{2r}$. The optimum dimensions of the hole $r = \sqrt{\frac{\lambda d}{2}}$.

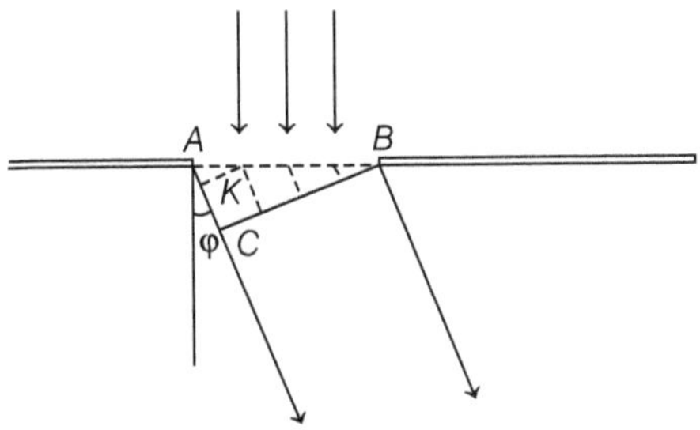

Fig. 561

795. The angles that determine the directions to the maxima of the second and the third orders satisfy the equations:

$$d \sin \varphi_2 = 2\lambda \text{ and } d \sin \varphi_3 = 3\lambda$$

Hence,

$$\begin{aligned}\lambda &= d(\sin \varphi_3 - \sin \varphi_2) \\ &= 2d \cos \frac{(\varphi_2 + \varphi_3)}{2} \sin \frac{(\varphi_3 - \varphi_2)}{2} \\ &\cong d(\varphi_3 - \varphi_2) = d\alpha \cong 1.7 \times 10^{-5} \text{ cm}\end{aligned}$$

796. $\sin \varphi = 1$ corresponds to the maximum value of k.

Hence, $$k = \frac{d}{\lambda} = 4$$

797. For a spectrum of the first order to appear, d should be greater than or equal to λ. Therefore, the sought period of the grating must not be less than 0.02 cm.

798. The direction to the first maximum is determined by the expression $d \sin \varphi = \lambda$. The screen is arranged in the focal plane of the lens. Assuming the angle φ to be small, we have $l = f\varphi$. Hence,

$$\lambda = \frac{dl}{f} = 5 \times 10^{-5} \text{ cm}$$

799. The length of all waves diminishes in water n times (n is the refraction index of water). Hence, the angles φ that determine the directions to the maxima, and the distances from the centre of the diffraction pattern to the maxima corresponding to various wavelengths, also decrease n times, since according to the initial condition the angles φ are small and $\sin \varphi \cong \varphi$.

800. The spectra of different orders will be in contact if $k\lambda_2 = (k + 1)\lambda_1$. Therefore,

$$k = \frac{\lambda_1}{\lambda_2 - \lambda_1} = 5$$

For this reason only the spectra of the sixth and seventh orders can be partially superposed. But this grating (see Problem 796) can give only the spectrum of the fourth order for this range of wavelengths. Therefore the spectra will not be superposed in our case.

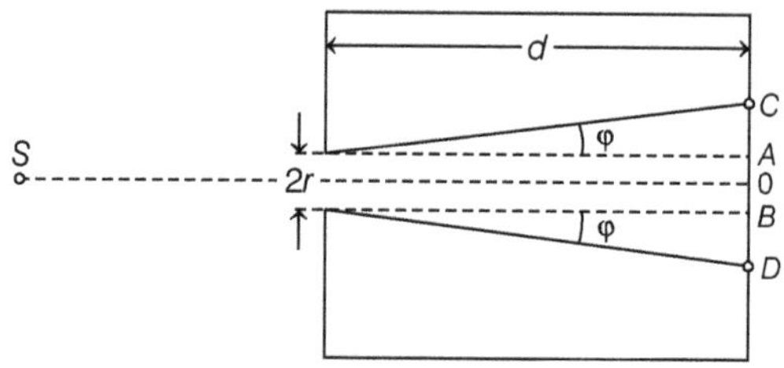

Fig. 562

801. When the rays are incident on the grating at an angle θ (Fig. 563), the difference in the path between the waves issuing from the edges of adjacent slits is

$$\delta = BD - AC = d \sin \phi - d \sin \theta$$

These waves add up and intensify each other when

$$d(\sin \phi - \sin \theta) = k\lambda$$

where $k = 1, 2, 3, \ldots$, for the maxima lying at the right of the central one ($k = 0$) and $k = -1, -2, -3, \ldots$, for those lying at its left.

The maximum order of the spectrum will be observed when $\phi = -90°$.

Thus $d\left(-1-\frac{1}{2}\right)=k\lambda$. Hence, $k=-6$. The spectrum of the sixth order may be observed. The minus sign shows that the spectrum lies to the left of the central one.

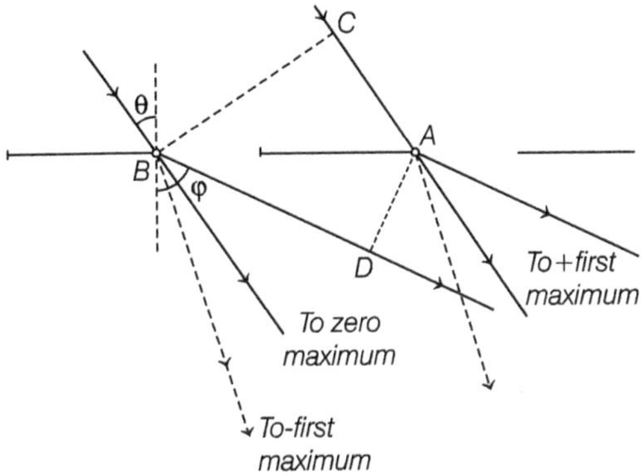

Fig. 563

802. As follows from the formula $d\,(\sin\varphi-\sin\theta)=k\lambda$ (see the solution to Problem 801), the period of the grating will be minimum with tangential incidence of the rays: $\theta=90°$. In this case $d\cong\frac{\lambda}{2}$. Therefore, the period of the grating should satisfy the inequality $d\geq\frac{\lambda}{2}$.

803. In the general case, as shown in the solution of Problem 801, the sought condition will be

$$d(\sin\varphi-\sin\theta)=k\lambda$$

It may be rewritten as

$$2d\cos\frac{\varphi+\theta}{2}\sin\frac{\varphi-\theta}{2}=k\lambda$$

If $d\gg k\lambda$, then $\varphi\cong\theta$. Here $\cos\frac{\varphi+\theta}{2}\cong\cos\theta$ and $\sin\frac{\varphi-\theta}{2}\cong\frac{\varphi-\theta}{2}$.

Therefore. the condition that determines the directions to the principal maxima will take the form

$$d\cos\theta(\varphi-\theta)\cong k\lambda$$

The grating constant diminishes, as it were, and becomes equal to $d\cos\theta$ instead of d. The angles $\varphi-\theta$ are counted off from the direction of the incident light.

6-3. Dispersion of Light and Colours of Bodies

804. As shown in Problem 703, the angle of incidence α, the refraction angle of the prism φ and the refraction index n are related to the angle β at which the beam emerges from the prism by the expression

$$n = \sin\beta\sqrt{\left(\frac{\sin\alpha}{\sin\beta\sin\varphi} + \cot\varphi\right)^2 + 1}$$

Therefore, we obtain the following equation for $\sin\beta$

$$\sin^2\beta(1 + \cot^2\varphi) + 2\sin\beta\frac{\sin\alpha}{\sin\varphi\tan\varphi} + \frac{\sin^2\alpha}{\sin^2\varphi} - n^2 = 0$$

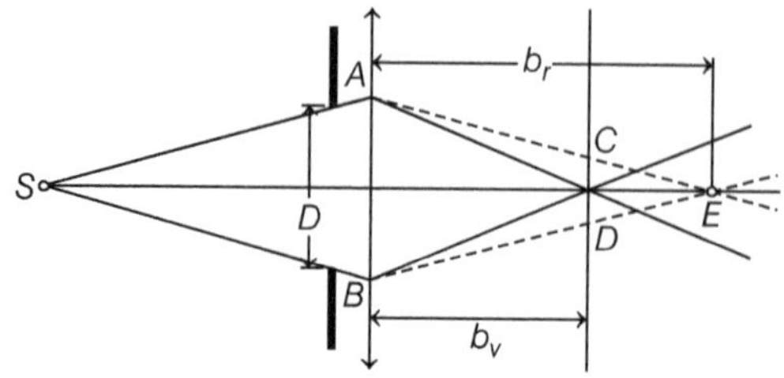

Fig. 564

or
$$2\sin^2\beta + \sqrt{2}\sin\beta + \frac{1}{2} - n^2 = 0$$

Upon solving this equation, we find that

$$\sin\beta = \frac{-1 \pm \sqrt{4n^2 - 1}}{2\sqrt{2}}$$

The solution with the plus sign has a physical meaning. For red rays $\sin\beta_r \cong 0.26$. Therefore, $\beta_r \cong 15°6'$. For violet rays $\sin\beta_v \cong 0.31$ and $\beta_v \cong 18°6'$.

The sought angle is $\theta = \beta_v - \beta_r \cong 3°$.

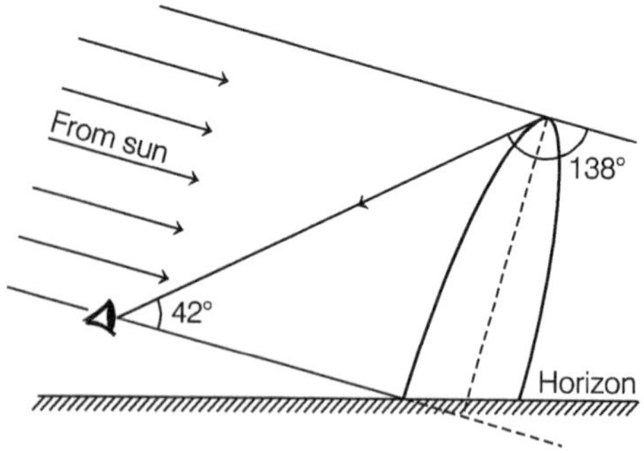

Fig. 565

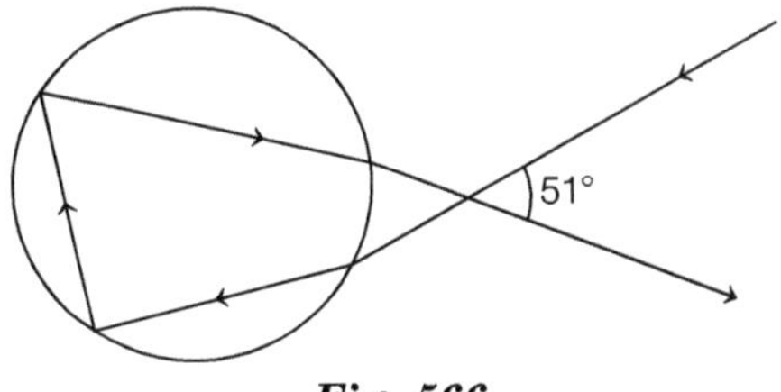

Fig. 566

805. For the red rays the focal length of the lens is

$$f_r = \frac{R}{2(n_r - 1)} \cong 27\text{cm}$$

and for the violet rays $f_v = 25$ cm.

According to the formula of a lens, the image produced by the red rays will be at a distance of $b_r = \frac{af_r}{a - f_r} = 58.7$ cm and that from the violet rays at $b_v = 50$ cm.

The image of the source on the screen (Fig. 564) will have the form of a spot whose edges are coloured red.

The diameter of the spot d can be found from the similarity of triangles ABE and CDE:

$$d = D\frac{b_r - b_v}{b_r} \cong 0.15\text{cm}$$

806. The sunrays falling on rain drops may be assumed to be parallel. As they emerge from a drop after being reflected once on its internal surface, the rays diverge in all directions. Only the rays subjected to minimum deflection are about parallel. When these rays get into the eye, they produce the maximum visual impression. These rays travel, as can be said, with the maximum density. The other rays are diffused in all directions. As shown in Problem 748, the angle of deflection for parallel rays is 138°. Therefore, the angle between the sunrays and the direction to the rainbow is 42° (for red light) (Fig. 565).

The eye will receive light from the drops that are in the direction forming an angle of 42° with the line passing through the eye and the Sun. For violet rays this angle is about 40°.

807. The first (primary) rainbow is observed owing to the rays that were reflected once inside the water drops. Upon refraction, the violet rays under go the greatest deflection from the initial direction (see Problem 747) ($\angle\theta$ grows with n, since r decreases). For this reason the external arc will be red and the internal one violet.

The reflection rainbow is caused by rays that were reflected twice inside the drops. The approximate path of the ray is shown in Fig. 566. As can be shown, the direction to the rainbow is 51° with the line that connects the eye and the Sun. With two refractions and two reflections the colours alternate in the reverse order: the external arc will be violet and the internal one red.

The luminous intensity is much weaker after two reflections, for which reason a reflection rainbow is much less intensive than the primary one.

808. The geographical latitude of Moscow, i.e., the angle between the plane of the equator and a normal to the surface of the Earth, is $\varphi = 56°$. At this moment the Sun is in the zenith above the northern tropic (latitude $\alpha = 23.5°$). Hence, the angle between the direction to the Sun and the horizon (Fig. 567) is

$$\beta = 90° - \varphi + \alpha = 57°30'$$

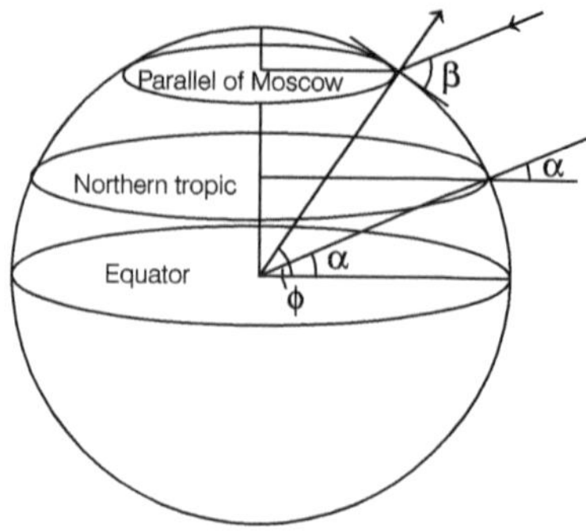

Fig. 567

A rainbow can be observed only when the altitude of the Sun above the horizon does not exceed 42° (see Fig. 565). Therefore, no rainbow can be observed.

809. Our eye perceives a colour when its sensitive elements are irritated by a light wave of a definite frequency. The frequency of light waves, however, does not change during transition from one medium into another.

810. The green glass should be used. In this case the word will appear black against the green background of the paper, since the red colour of the word "excellent" does not pass through green glass.

If the red glass is used, the word written by the red pencil will not be seen against the red background of the paper.

811. Camera lenses predominantly reflect the extreme parts of the visible spectrum: red and violet (see Problem 782). A mixture of these colours produces a lilac tint.

812. The colours of a rainbow are pure spectral colours (see Problem 806) since only a ray of a definite wavelength is seen in a given direction. Conversely, the colours of thin films are produced by extinguishing (totally or partially) of the rays of a certain spectral interval due to interference. The colour of the film will supplement the colour of this spectral interval.

813. Under the force of gravity, the soapy water drains down onto the lower portion of the film, which is always thicker than the upper one. Hence, the bands that show the locus of points of equal thickness should be arranged horizontally. The light-blue (blue-green) tint is obtained when the long-wave (red-orange) part is excluded from the full spectrum (see Problem 812). When the middle (green) part of the spectrum is extinguished, the remaining rays impart to the film a purple (crimson) hue, and when the short-wave (blue-violet) part is excluded from the solid spectrum, the film will appear as yellow. If the difference in the path of the mutually extinguishing rays forms the same number of half-waves in all three cases, the yellow band should be on top, followed by the purple band and by the light-blue band at the bottom.

814. In the daytime the light-blue light diffused by the sky is added to the yellowish light of the Moon itself. This mixture of colours is perceived by the eye as a white colour. After sunset the light-blue colour of the sky is attenuated and the Moon acquires a yellowish hue.

815. The smoke is seen against a dark background because it diffuses the sunrays incident from above. The particles of the smoke diffuse blue light much more intensively than red or yellow light. Therefore, the smoke seems blue in colour.

The smoke is seen in transmitted light against the background of a bright sky. The smoke seems yellowish since the blue light is diffused in all directions and only the long-wave part of spectrum of white light reaches the eye.

816. A thin film of water covering a moist object reflects the incident white light in one definite direction. The surface of the object no longer diffuses white light in all directions, and its own colour becomes predominant. The diffused light is not superposed on the light reflected from the object, and for this reason the colour seems richer.

www.ingramcontent.com/pod-product-compliance
Ingram Content Group UK Ltd.
Pitfield, Milton Keynes, MK11 3LW, UK
UKHW021337190726
13845UKWH00027B/606

9 789324 191779